The Last Seat in the House

The Last Seat in the House

THE STORY OF HANLEY SOUND

JOHN KANE

Edited with Allan DiBiase
Foreword by Ken Lopez

University Press of Mississippi • Jackson

The University Press of Mississippi is the scholarly publishing agency of
the Mississippi Institutions of Higher Learning: Alcorn State University,
Delta State University, Jackson State University, Mississippi State University,
Mississippi University for Women, Mississippi Valley State University,
University of Mississippi, and University of Southern Mississippi.

www.upress.state.ms.us

Designed by Peter D. Halverson

The University Press of Mississippi is a member of
the Association of University Presses.

First printing 2020
∞

Library of Congress Cataloging-in-Publication Data available
LCCN 2019032108

Hardback ISBN	978-1-4968-2680-0
Trade paperback ISBN	978-1-4968-2681-7
Epub single ISBN	9781-4968-2682-4
Epub institutional ISBN	978-1-4968-2683-1
PDF single ISBN	978-1-4968-2684-8
PDF institutional ISBN	978-1-4968-2685-5

British Library Cataloging-in-Publication Data available

supported by grant
Figure Foundation
streaming the arena

For Alice

CONTENTS

FOREWORD

This book is the story of Bill Hanley, his sound company, and other unsung heroes that had an impact on the modern musical experience. The tale is told through the voices of those who were closely involved in this work. These are voices often not heard or recorded. *The Last Seat in the House* gives them a unique opportunity to speak to readers, loud and clear. For almost three decades Hanley and his crews worked tirelessly to enhance audience experiences by bringing performers on stage to "the last seat in the house" vividly.

The scope of *Last Seat* spans the 1950s through the legendary festivals and massive concert gatherings of the 1960s and 1970s. Many of the most important musical events of the times are explored in intimate detail not told before. Readers are given an insider view of Dylan's first electric set at Newport, the Beatles at Shea Stadium, and Woodstock. These were events that changed the course of popular culture and reverberated worldwide. Bill Hanley was a pivotal figure in all of them.

I was a part of this story as a young musician in the early 1960s, sound engineer and musical entrepreneur in the roaring seventies, executive with JBL Professional in the 1980s and 1990s, and then on to higher education for the remainder of my career. Most of the pioneers interviewed for this book are personal and professional friends and acquaintances, including Bill Hanley. I've spent many memorable hours over the years in discussions with him. It has been my great pleasure to host Bill as a guest speaker in classes of the Music Industry Program at the University of Southern California Thornton School of Music, as well as Audio Engineering Society technical and historic presentations. Bill is a pragmatist and dreamer, experimenter and tinkerer, visionary and explorer. His company Hanley Sound was one of the very first companies capable of amplifying large-scale events. Many of his groundbreaking inventions are still used today in thousands of applications.

Author John Kane has painstakingly sought out those who are part of this story. These are the men and women who were intimately familiar with the core technologies involved, the business practices, the logistics, and the need for technical innovation and adaption that were required by the times. Within this larger story, the Woodstock chapters are worth the price of admission alone!

John has created a compelling story that combines historical fact with tales of innovation, trial and error, disruption, leadership, and entrepreneurial brilliance. He allows us to experience, along with those involved, these turbulent times by sharing their visions, challenges, triumphs and failures. The reader will gain insight into a unique man and the many who followed in his steps, eventually making them their own. I thoroughly enjoyed every word. It's a wonderful read.

—KEN LOPEZ

PREFACE

The *Last Seat in the House* is a multilayered journey traveled by one man—Bill Hanley. Since the 1950s his passion for audio has changed the way audiences listen, and how technicians approach, live concert sound.

From his earliest days growing up in the Boston suburb of Medford, to learning his craft and applying it at the Newport Folk and Jazz Festivals, Hanley's quest was to make live sound better. As the music business developed, a new and emerging industry did so as well, with Hanley riding its wave. By the end of the 1960s and early 1970s, musicians, managers, and promoters began to expect more than a typical, primitive PA system. They wanted better sound and Hanley was there to help.

Some of the questions I answer in this book are: What was the state of live sound before Hanley began his career? How did sound reinforcement come to serve the needs of a developing music business? Who were some of the earliest roadies, crew people, and sound pioneers that shaped the billion-dollar music industry as we know it? How were sound systems deployed and what equipment was used at various venues, concerts, and festivals, including Hanley's very important sound application at Woodstock?

Beyond my general interest in popular music history, I suppose I've always had an infatuation with the Woodstock Festival. Ever since I first popped Michael Wadleigh's documentary film into my VHS player as a teenager, I was hooked. Aside from its obviously stellar soundtrack, I thought, "How and why could such an event happen?" From then on, Woodstock was permanently lodged in the back of my mind.

I was a child of the 1970s, and like most latchkey kids I had an intimate and involved relationship with television. I still have vivid memories about the social impact of Vietnam and Nixon-era politics. I remember hitchhikers thumbing rides on the highways. I remember the drive-in. The music of the day and its accompanying technology also affected me. I was intrigued by radio as a child not just for listening pleasure, but even taking them apart to discover how they worked. Each night I would take my pocket radio to bed and tune in faraway stations. This was enthralling to me. A real mystery. Distant signals.

In 2008 I finished grad school and decided to organize a folk festival. An amateur, I reached out to experienced musician friends for advice. One of them was Jeryl Abramson, the current owner of Max Yasgur's homestead in Bethel, New York. Yes! THE Max Yasgur! I was looking for a sound engineer for my event and mentioned this in passing. Casually, she asked, "Why don't you call Bill Hanley?" Ignorant, I needed to be reminded . . . "Who was he?" She said, "You know, the Woodstock sound engineer?" I responded, "Why would the sound engineer of Woodstock, one of the most significant concert events of our lifetime, want to provide sound for my little festival?" She explained to me that he was older now and his business had slowed down. She mentioned that he was very approachable and lived in my area.

Jeryl forwarded me his number and I sat on it for a while. Eventually I called Bill. When he picked up he seemed the perfect gentleman, informative and helpful. However, for whatever reason, we never collaborated.

I finally met Bill in Brookline, Massachusetts, in July 2009 when Michael Lang was signing his book *The Road to Woodstock*. Still having no idea of Hanley's impact beyond Woodstock, I formally introduced myself to Bill and moved on. That August I saw him again at a Woodstock 40th anniversary celebration at the Regent Theater in Arlington, Massachusetts. After hearing him talk on a discussion panel, it finally hit me about how important a role he played in the history of live concert sound. Something clicked.

Later that year I was let go from my full-time job during the nationwide economic downturn. Everyone around me seemed to be getting laid off. Not knowing what direction to take, I decided to get myself into some form of research and study. By 2011, I was at Franklin Pierce University in New Hampshire and in need of a research topic. Enter Bill Hanley.

Much like my curiosity about Woodstock, Hanley's story stayed with me. I called Bill and asked if I could interview him. Remembering who I was, he agreed. I had no idea of how complex his story was until I was finally able to sit with him for several hours. After learning about his role in the evolution of sound reinforcement technology and his influence on popular music, I was committed. Later, I discovered there was no comprehensive book on the subject. As far I could tell there was nothing like it, and this excited me.

In this book, readers will find a wealth of never-before-researched information. There is nothing as collective and comprehensive. Woodstock fans, audiophiles, researchers, sound engineers, music production students, and popular music culture enthusiasts will be entertained and enriched by what *The Last Seat in the House* has to offer—a road untraveled. Realizing there was a lack of reliable resources in this field of study, I had these audiences in mind while composing the text.

Although I am not an audio engineer, I am a passionate learner of technology, media, music, and visual art. After discovering Hanley's innovations, I became even more involved in the content of my research. Music has always been close to my heart. How we listen to it and who was behind its development became even more important to me as time went on.

After eight years, *The Last Seat in the House* is comprised of over a hundred carefully selected interviews. Many of these exchanges occurred during my doctoral research for this book. Its pages are filled with reflections personally gathered from music icons, production, and performance people, friends, and family. Additional material was culled from newspapers, articles, books, and websites.

Writing *The Last Seat in the House* has been humbling. Many of those I have interviewed and who offered their valuable time have passed on. The writing has also been a major sacrifice, mostly for my family. Late hours, early mornings, coffeeshops, libraries until closing, and in between my daughter's nap schedule, have allowed me the time to finish the book. I've met some amazing people on my journey, many who are now close friends, including Bill. After thousands of hours poured into this work, I feel proud of what I have written. I also feel it is important.

While readers are engaged in the text, they will find that Hanley is always in the backdrop. Even when he is not the focus of a particular topic, he most certainly is connected to it, and I discuss how. I have structured *The Last Seat in the House* so audiophiles will find important technical information concentrated in an easy-to-find format, while the more casual reader can skip over and move on without missing important features of the narrative. The landscape of Hanley's story, and that of the developing music and festival business, parallel each other. In this way, the reader will hopefully find the intended chronological structure in the writing helpful. Additionally, discussion about other pioneers who made important contributions to the field of sound can be found throughout the book.

Of the many experiences on this journey, none stands out for me as much as attending the memorial service for legendary folk musician Richie Havens at the Bethel Woods Center for the Arts on August 18, 2013. There I was among Woodstock legends: Hanley, Lang, Rosenman, and others were all reflecting on the singer's impact. As Richie's ashes were scattered over Yasgur's field by a circling small airplane, a recording of his song "Freedom" from the *Woodstock* soundtrack echoed. Chilling. At Richie's heart was Woodstock and on that day he made it back to the Garden. You felt it. Electric.

Just before the event, in my hotel room I realized I had stupidly forgotten an adapter for my microphone. I was to record the event for a film I was working on. I was freaking out. No sooner, Hanley pulled out a medium sized hard shell Samsonite suitcase bursting at the seams with soldering irons, wires, and other various adapters. It came out of nowhere! I don't even think I noticed him putting it in my car before we left Massachusetts. Conveniently, he just happened to have it on him. Looking back it makes perfect sense—it's Bill Hanley we're talking about here.

Amid the typical accommodations any normal hotel room would have, Hanley laid his makeshift electronics station on the coffee table and soldered away. By this time I was in the throes of research about him. Stories of him soldering the electronics of his equipment before many festivals and concerts came up a lot. As I was making these connections, he handed me a modified adapter and confidently said, "Here, that should work." With no time to spare, we hit the road.

At the museum, I set up my video equipment before Richie's service. However, I was still having difficulty with my camera microphone feed. Static. Within minutes of the start of the event, Hanley examined, isolated, and fixed the issue just in the nick of time, like he always had. Wearing his original dark blue Woodstock windbreaker, the eighty-year-old "Father of Festival Sound" shuffled away with his traveling electronics first aid kit. And here I was up and running at the original Woodstock site. Richie was ready and so was Hanley, just as they were almost fifty years ago.

ACKNOWLEDGMENTS

I would like to express my gratitude by thanking the following people who helped, encouraged, and inspired me while writing this book: Jennifer Kane, Bill Hanley, Terry Hanley, Rhoda Rosenberg, Joe Hanley, the Hanley family, Dr. Allan DiBiase, Dr. Rich Abel, Dr. Jim Lacey, Dr. Maggie Moore-West, Dr. Carol Nepton, Dr. Donna Halper, Dr. Doreen Rogan, Jeryl Abramson, David Marks, Bradford Lyttle, Joshua White, Chip Monck, John Morris, Chris Langhart, John Chester, Dr. David Johnson, Elijah Wald, Barry Tashian, Henry Diltz, Ken Davidoff, Stanley Miller, Bob Kiernan, Ken Lopez, Karen Rockow, Geddy, Alex, Neil, Cameron Crowe. To all Hanley Sound, Woodstock, and Fillmore East crews—thank you! I would especially like to thank all the musicians and concert production people who contributed. And of course, the University Press of Mississippi and Will Rigby who did copy editing. Gone but not forgotten: Sam Andrew, Dr. Leo Beranek, Sid Bernstein, Lee Mackler-Blumer, Judi Bernstein-Cohen, Alex Cooley, Major Thomas Convery, Stan Goldstein, Campbell Hair, Tom Hawko, Mel Lawrence, Mike Neal, George Pyche, Pete Seeger, Karl Meagher, Thumper, Fred Taylor, Herb Gart, Julius Chambers, and George Chambers.

PROLOGUE: A SOUNDMAN AND A BOMB

By 1970 Bill Hanley was competing with an industry he helped build. In 1960 the market was small and he was in his stride, yet ten years later the sound engineer faced competition from emerging sound companies that for all intents and purposes morphed his innovations into better designs. Eventually Hanley felt the pressure of maintaining the Medford, Massachusetts, sound company he had founded. By the late 1970s it became a struggle to keep his company, Hanley Sound, afloat and pay his employees.

In 1969 Hanley's name gained prominence because of his work at the Woodstock Music and Art Fair in Bethel, New York, and also at other major festivals. However, after Woodstock, across the country an incessant fear of hippies with insatiable appetites for music and drugs overcame many communities. In New York State, Governor Nelson Rockefeller responded with a Mass Gathering Act, disallowing and discouraging large gatherings from happening. By enacting Part 18 of the New York State Health Sanitary Code, a special permit was now mandatory for any public function with attendance of over 5,000 people. For Hanley a requirement like this meant an inevitable loss of revenue. It meant that he needed to respond with smart, business-savvy decision making in order to adapt to the changing live-music event marketplace.

The early 1970s were a time of instability for the nation as many of the 1960s counterculture ideals faded. Gays, lesbians, women, Native Americans, and African Americans were still powering forward, fighting for equality. With little faith left, the general public was beaten down and insulted by the deceit of the Richard Nixon presidency. Nineteen seventy saw US troops invading Cambodia, the Senate repealing the Gulf of Tonkin Resolution furthering our involvement in the Vietnam War, followed by four students at Kent State University being killed by National Guardsmen. By now Hanley was preoccupied with the massive anti–Vietnam War protests in Washington, DC. Hundreds of thousands of citizens attended.

On September 19, 1970, the sound engineer was on his way back to Boston from a show in Pittsburgh, Pennsylvania, seated on a departing Allegheny

Airline flight that was about to be hijacked. Plane hijackings, even in 1970, were still a rare occurrence, yet statistically if you were traveling as much as Hanley, the chances were greater. This experience—like many in Hanley's story-ridden career in sound—is devoid of any musical escapades. It is, however, a good example of the heated political and cultural atmosphere the country was going through during this pivotal time in history. On the plane, the sound engineer found himself in a uniquely compromising situation like no other he would ever encounter.

Carrying discharge papers, a clean-cut, former Marine in his mid-twenties named Richard Witt, decided to hijack flight 330, the Boeing 727 that Hanley and others were on. The Marine was furious about the treatment he received during his basic training and was described as a "desperate young man" who claimed to "hate Jews" and wanted to fight for the Palestinians. Armed with a pistol and a bomb rigged with nitroglycerin, a stopwatch, and gunpowder he had hidden inside a leather duffel bag, Witt's intentions didn't seem very good. Around 12:45 a.m. as the plane took off, Witt grabbed one of the stewardesses by the neck attempting to pull her into the cockpit. When the stewardess resisted, Witt pulled out a .22 caliber pistol and pressed it against her neck announcing to her that he had a bomb. The plane was due to refuel in fifteen minutes in Philadelphia and was scheduled to let all passengers off.

At that very moment the exhausted thirty-three-year-old sound engineer was comfortably seated in the front cabin. Despite the typical sounds of the plane's engine, the chiming of the intercom, and passenger chatter, something seemed odd to the sound engineer. Sitting close to the cockpit of the plane, Hanley slowly began to catch on as he heard someone softly yell, "There's a man going up the aisle with a gun!" Simultaneously the individual behind Hanley tapped his shoulder and said the same thing. Hanley surveyed the area. He noticed some of the flight attendants looking frightened, chattering away on the plane's telephone that was connected to the cockpit. It then became apparent to him that the hijacker's body was pressed against Evelyn Thornton, one of three flight attendants on board.

And so Hanley had a front-row seat for a potential airline disaster. In a *Philadelphia Daily News* article "The First 15 Minutes," journalist Tom Fox explains: "Hanley was sitting in front of the hijacker in the front cabin. When the hijacking unfolded, Hanley made a move to prevent it, but it didn't work."[1] According to a 1970 *Boston Globe* article "Hijacker Terrorized Boston Stewardess" Hanley said at the time: "I sort of laughed it off, because you know how people are these days. But then I saw a man who kept looking around the corner from the front of the plane. I caught his eye and kept staring at him to get some feedback."[2] By this time Hanley could see the gun up against the frightened Thornton's neck; he heard the hijacker tell the other stewardess to come to his seat.

The stewardess approached Hanley's seat in order to retrieve something beneath it. Quickly making connections, he knew this was something ominous. "When she came over I told her don't give that to him!"[3] At this point neither Hanley nor the stewardess knew the exact contents of the brown paper bag, nevertheless his intuition was kicking in. "The first stewardess went back empty handed and in a sassy manner told him she doesn't 'do that type of work,' that was remarkable to me!"[4] recalls Hanley. Making some sense of the harrowing situation the sound engineer realized the desperate hijacker had planted a bomb under his seat. On the threshold of a potential catastrophe, this was unlike anything he had ever dealt with before.

With no time to spare Hanley slowly crouched over and began to cautiously probe through the contents of the bag. As someone who had been around electronics all of his life, Hanley concluded that the homemade device might be disarm-able. According to Hanley: "I figured if the bomb was electrically wired I could figure a way to disarm it. So I took it out of the bag and looked at it."[5] What Hanley viewed was a square bottle filled with a clear liquid taped to a propane tank around 14-inches long and 5-inches wide. "It had wires wrapped around it, with a stopwatch, and gunpowder. I heard later there were pieces of a chopped-up coat hanger in it which could have acted as shrapnel upon explosion."[6]

The sound engineer had every intention to deactivate the bomb, but there wasn't enough time for Hanley to figure it out. "The hijacker sent over another stewardess to my seat to grab the bag with the bomb in it. It was too heavy for her to lift at first but she got it the second time, then delivered it to the hijacker."[7] Hanley explains what occurred in a 1970 article: "When the stewardess saw what I was doing, she was beautiful. She told the hijacker that the bag was too heavy."[8] Hanley claims he was the only one on the plane aside from the two flight attendants and the hijacker who knew there was a bomb before they landed in Philadelphia. From Hanley's aisle seat he could still see the hijacker's pistol jammed into the flight attendant's throat.

Finally the plane landed at Philadelphia International Airport for refueling. Here the pilot convinced Witt to release all ninety passengers. Among them was the six-foot, 280-pound Hawaiian pro wrestler Toru Tanake. Hanley recalls talking with the famous athlete suggesting that they might be able to tackle the hijacker if they tried. Scared and furious, Tanake had every intention to use his wrestling and jiujitsu skills on the perpetrator, but the wrestler quickly relented when the hijacker aimed the gun at him. Tanake pulled back, claiming that he did not want to "jeopardize the passengers."[9]

Again pressing his pistol against the flight attendant's neck, Witt demanded pilot Capt. John Harkin fly to Cairo, Egypt, insisting that he needed to help Palestinian guerrillas fight Jews. The pilot responded by informing the flight attendant that in order to grant Witt his wishes, they would have to make several stops to refuel. The quick-thinking captain convinced Witt they should go to Havana, Cuba, instead.

Clad in riot helmets and bulletproof vests, hundreds of FBI and local police in unmarked cars occupied Runway 9, quietly waiting for Witt's next move. Around 1:46 a.m., after airport crews facilitated the plane's refueling, it took off for Cuba with only the crew aboard. According to the *Philadelphia Daily News*, after idling for a while its motors began to wind and the jet "roared forward and lurched up into the night. It was headed for Havana."[10]

Once safely off of the plane, Hanley and the other passengers were delayed in the Allegheny Airport terminal lounge for several hours. They were faced with a barrage of questions from custom officials and FBI agents. After that, the sound engineer recalls being inundated with additional questions from the press. Fortunately, Hanley (and others) walked away without a scratch from the incident: "After this all happened I was in a cab in Philadelphia and someone recognized me from my picture in the newspaper. I was on the front page of the paper and I didn't even know it!"[11]

Once Witt and the plane's crew landed in Havana around 4:31 a.m., Cuban soldiers boarded the airliner, capturing the disgruntled Marine. As he was hauled off, Witt professed to the shaken crew: "I'm sorry this had to happen. Thank you for the ride."[12] Imprisoned in Cuba for many years, Witt was released and returned to the United States in 1978. Hanley received a letter from the government asking if he would testify against him in court. The sound engineer decided against it.

Stories like this are not uncommon for Bill Hanley. In fact, to accurately describe the far-reaching scope of his life, the unassuming and prodigiously adventurous character Forrest Gump comes to mind. The comparison of Gump to Hanley is as simple as it is relevant. However, Gump was present only by accident; Hanley was not. Yet an incredible chain of events in the music business pushed Hanley into the spotlight. Like Gump, Hanley witnessed some fairly impressive things in his life. Both are protagonists in a series of connected and iconic adventures. Journalist Adam Zoll reflects on this comparison: "The only difference is Bill Hanley is a real person. Hanley has been the common denominator time and time again in events of rock and roll cataclysm."[13]

How is it that a sound engineer like Hanley experienced so many pivotal and transformative musical and political events in our nation's history? Were these mere coincidences? Was it simply being there at the right time? Or was it just dumb luck? We may never know answers to such questions, but one thing is for certain: Hanley's career is woven like a thread through the fabric of the American popular music landscape.

Throughout history we have learned about innovators who inspired future generations. Some emerged as the vanguard of major industries; some did not. Many, like Hanley, through trial and error paved the way for future specialized fields to flourish. Most are forgotten, buried within our history books.

Bill Hanley was a trailblazer in sound reinforcement. His story is a proto-typical American one in which his local efforts grew in ways far beyond the average person's realm of possibilities. Whether audiophiles or not, readers will find this story a recognizable one. For certain, at one point or another at least one of Hanley's innovations will have reached anyone with a functioning set of ears. His contributions to what we now recognize as a billion-dollar sound reinforcement industry are incomparable, yet he is mostly unknown to people outside his trade.

Prior to the rush of live concert sound during the 1950s, Hanley recognized a host of inadequacies with public address systems widely in use. Even as a child, this oddly sharp-minded audiocentric individual seemed to have an interest in the field. Out of a nondescript Medford location, what engrossed Bill Hanley altered the way in which live music is listened to around the world. Eventually his ideas led to the construction of "systems of sound" that have forever changed what audiences and musicians hear at indoor and outdoor events.

Hanley did not invent any one thing—he custom-built systems of things. Yet his ideas that led him to these outstanding developments were most often responses to what he felt was necessary in the reproduction and amplification of sound. At a time when concert promoters, musicians, and even audiophiles were not focused on high fidelity sound for outdoor events, Hanley revolution-ized "live sound" with extremely limited means.

This book contains insights into how Bill Hanley (and those who worked for him) had a direct, innovative influence on specific sound applications that are widely used and often taken for granted. Many in his field agree that Han-ley's innovations pushed the then nonexistent sound reinforcement industry into a new area of technology, rich with clarity and intelligibility. Hanley's relationship with sound was so highly focused that it often overshadowed any economic, safety, or even political concerns he had. Throughout his career the sound engineer acted selflessly and with extreme passion; but eventually his convictions about sound dismantled his business.

Hanley provided sound reinforcement for over three quarters of the larg-est pop and rock festivals during the golden age of festivals. Being known as "the Father of Festival Sound," a moniker anointed by his peers, makes perfect sense. Most notable were his essential services at the 1969 Woodstock Music & Art Fair. Occasionally drawing negative attention, festivals were sometimes dangerous places. It was these events, and other mass gatherings where Hanley and his crews worked at the height of extreme civil unrest, where he earned his reputation. Hanley provided sound reinforcement for some of the largest anti–Vietnam War demonstrations in US history, a nontypical task for the average soundman of the day.

Even though the music of this period was an important conveyor of political messages, it was a time when the "business" of popular music was evolving. This book examines how Hanley's career paralleled the developing music industry and its growing need for quality sound reinforcement. The musicians' desire for additional amplification came at the perfect time for both parties. As Hanley's reputation grew, so did the need for his unique services in live sound and production. The demands from audience, promoter, and performer for clear and intelligible concert sound became a top priority by the early seventies.

By the mid- to late 1970s, Hanley's efforts were being mimicked nationally by new and smaller sound companies. Unable to successfully adapt to a changing marketplace, Hanley saw the industry he helped create slipping through his grasp. Those who borrowed his ideas moved forward and didn't look back. Not only did his pioneering efforts lay the foundation for this new industry, it gave an identifiable definition to the term "sound engineer." The totality of this story reveals important problems like the lack of financial reward that his company, Hanley Sound, faced.

I do not claim to be an expert in the field of audio engineering. Nonetheless, the years of research that went into producing these findings will serve as a useful guide to the interested reader or emerging scholar. Through a focused collection of reflection-based interviews and articles, Hanley's story will educate enthusiasts and help them understand how he shaped the industry that exists today.

Those involved in the business are mostly in it for the love of music, more than the details of how sound is engineered. This book is not a manual on sound reinforcement. It is an historical recollection on how the industry came to be, with important technical details thrown in. To this day, Hanley's legacy remains an influence on sound reinforcement technology. Although he has been out of the business for over three decades, the "Hanley Sound" name is still renowned—so much so that, on October 20, 2006, Hanley achieved one of the highest honors for live event professionals, the Parnelli Award. Acknowledged by his peers, the sound engineer was recognized for his major contributions to the sound reinforcement industry.

Many professionals have emphasized that if no one complains about the sound, you have done your job effectively. It's only when the sound is bad that the engineer is recognized. This I suppose leaves individuals like Hanley and others with a monumental task, as they are the gatekeepers of quality sound at live events. A distinct and intimate responsibility like this makes them a liaison between artist and audience. It is the job of the sound engineer and his or her equipment to deliver clarity, so that the people sitting in the very last seat can hear just as well as the people sitting in the front—regardless of economics, race, or political belief. This simple yet humble notion happens to be Bill Hanley's *modus operandi* and the title of this book. Therefore I welcome you to *The Last Seat in the House*.

The Last Seat in the House

Chapter 1

FIRST GENERATION, SECOND GENERATION

Prior to Hanley's emergence as a predominant force in his field, the area of sound reinforcement technology was evolving. These innovations etched an inevitable timeline that was foundational to Hanley from the outset of his career. A partial list of these developments follows.

DEVELOPMENTS IN SOUND TECHNOLOGY 1830–1947			
Year	Technology	Manufacturer/Inventor	Significance
1830–40	Telegraph	Sir William Cooke, Sir Charles Wheatstone, Samuel Morse	Long-distance communication
1876	Telephone	Alexander Graham Bell	An outgrowth of harmonic telegraphy, the first production of intelligible speech was made in Bell's lab
1877	Phonograph	Thomas Edison	Recorded sound on a metal cylinder wrapped with tinfoil
1882	DC Power	Thomas Edison	Lower Manhattan is supplied with DC power from Pearl St. Station
1888	AC Polyphase Induction Motor	Nikola Tesla and George Westinghouse	The implementation of AC distribution
1890	Kinetograph (Kinetoscope), Cinématographe	Thomas Edison, Louis and Auguste Lumière	Image projection, image movement, and artistic innovations leading to the silent film era
1900	Chronomégaphone	Léon Gaumont and the Gaumont Film Company	The first public exhibition of projected sound occurs in Paris using compressed air to amplify recorded sound.
1906	Audion Tube	Lee DeForest	The first widely used electrical device that paved the way for long-distance wireless radio
1915	Public Address	Magnavox founders Edwin Jensen and Peter Pridham	The first use of a PA system for 50,000 at a Christmas concert San Francisco City Hall

Year	Technology	Manufacturer/Inventor	Significance
1916	Public Address	Western Electric	PA capable of addressing 12,000 through eighteen loudspeakers
1919	Public Address	Unknown	Distributed sound system. Horns strung along New York City's Park Avenue "Victory Way" as part of a Victory Bond sale.
1920	Public Address	Unknown	Central cluster array, Republican Convention, Chicago, Illinois
1920	Electrical microphones, electronic signal amplifiers, and electromechanical recorders	Western Electric	Electrical recording, which led to sound on film and the beginning of radio broadcasting
1920–21	Public Address	Western Electric	Use of a PA at 1920 Republican presidential convention and at Warren Harding's 1921 inauguration
1923–26	Movietone, Vitaphone	Theodore Case and William Fox. Warner Brothers and Western Electric. Danish engineers Axel Petersen and Arnold Poulsen.	The development of cinema sound systems for short films including *What Price Glory?* and *Don Juan* in 1926. Synchronizing dialogue to film as opposed to disc.
1925	Public Address	Chesapeake and Potomac Telephone Company	Calvin Coolidge's inauguration was heard through an installation of loudspeakers and microphones on the inaugural platform, operated from a room below the Capitol's steps.
1927	Cinema Sound	Bell Telephone Laboratories Manufacturing division of Western Electric	The first "talkie" film, *The Jazz Singer*, debuts.
1933–34	Public Address	Chicago Century of Progress/ Chicago World's Fair	Speakers hung from poles throughout the 427-acre fairground along the shore of Lake Michigan
1947	Point Contact Transistor	American physicists John Bardeen, Walter Brattain, and William Shockley at Bell Laboratories	Allowed for the first wave of electronic miniaturization that evolved into the building block of modern electronics.

A highlight from these developments is sound for cinema. Although others had been experimenting with "talking pictures" since the early 1900s—usually unsuccessfully—most history books credit the 1927 film *The Jazz Singer* as the first successful talking picture. It was around this time, even slightly earlier in 1923, that the foundation of cinema sound loudspeaker technology was introduced. The ability to synchronize pictures with sound raised the cinema industry to new heights. Quality audio became necessary to support the influx of paying audiences, as in Radio City Music Hall, which opened in 1932 and seated around 6,000 people. Various manufacturers surfaced to serve this technological need.

One company that set out to develop products for motion pictures was AT&T's Western Electric; around 1926 it established Electrical Research Products Incorporated (ERPI). In 1936 the ERPI division detached and renamed itself the All Technical Service Company.

In 1941 the Altec Lansing Corporation was created when it joined Lansing Manufacturing. By 1946 it was known as James B. Lansing Sound. When Lansing left, it was renamed JBL. Bell Laboratories, Western Electric, RCA, Altec Lansing, Shure Microphones, Electro-Voice, and other companies, had accommodated the need for public address. Eventually Hanley realized he could do better.[1][2][3][4]

The combination of cinema-quality speakers, home high fidelity, military amplification, and studio-quality microphones were all that early sound reinforcement engineers had to work with. Sound engineers before, contemporaneous with, and after Hanley had to formulate unique ideas on how sound reinforcement should be executed. Some of these groundbreakers include Richard Alderson, Stan Miller, Harry McCune, Art Swanson, and Jim Meagher.

When these various audio elements were intentionally and conspicuously pieced together, the result was a system of sound that could be applied to live music. Acquiring additional power by way of the deliberate placement and organization of such existing technologies was a disruptive tipping point for those embarking on this new technological field.

BEFORE HANLEY, AFTER HANLEY

The term "first generation" sound engineer/company refers to those who were first involved with live concert and festival sound using a specially designed sound system other than a standard public address system. These creative minds carved out their own path toward jazz, big band, folk, and rock and roll sound innovation. In fact, many of these individuals had similar, if not nearly identical, beginnings as Hanley and his "focus" on sound. But during Hanley's golden years (1957–69) the landscape of sound companies for live music remained small. Even though Hanley Sound was the primary East Coast sound company at the time, others were dispersed throughout the country, in New York City, Nebraska, and California. Nevertheless, there was plenty of room for success within this new and emerging industry.

From the late 1960s into the early 1970s, new sound companies surfaced in Florida, Maryland, Pennsylvania, Texas, and Hermosa Beach, California. This is when Hanley began to see his real competition. The developments and innovations of this "first generation" sound engineer became refined by 1965, giving birth to a whole new "second-generation" sound engineer and industry. A partial list of these firms and engineers is shown below.

SOUND COMPANIES 1926–79			
First Generation			
Founded	Company	Owner	Location
1926	Swanson Sound	Art Swanson (founder) and Don Nielson	Oakland, California
1932	McCune Sound Services	Harry McCune Sr.	San Francisco, California
1936	Masque Sound (Broadway)	Mac Landsman, Sam Saltzman, and John Shearing	New York, New York
1946	Sound Associates (Broadway)	Peter Lodge	New York, New York
1949	Meagher Electronics	Jim Meagher	Monterey, California
1951	Watkins Electric Music	Charlie Watkins	United Kingdom
1957	Hanley Sound Inc.	Bill Hanley	Medford, Massachusetts
1961	Independent sound engineer for Harry Belafonte, Albert Grossman, Bob Dylan.	Richard Alderson	New York, New York
1962	Stanal Sound Ltd.	Stan Miller	Kearney, Nebraska
1962	Worcester Sound	Joseph Torterelli	Worcester, Massachusetts
1963	Yorkville Sound	Peter Traynor and Jack Long	Toronto, Ontario
1963	The Festival Group	David Hadler	Philadelphia, Pennsylvania
Second Generation			
1965	Kelly Deyong Sound	Paul Deyong	Vancouver, British Columbia
1966	Clair Brothers	Roy and Gene Clair	Lititz, Pennsylvania
1966	Heil Sound	Bob Heil	Marissa, Illinois
1966	Sound engineers for the Grateful Dead	Owsley Stanley and Dan Healy	San Francisco, California
1966	Maryland Sound International	Bob Goldstein	Baltimore, Maryland

Founded	Company	Owner	Location
1967	Portable Audio Systems	Stanley Andrews	Sturgis, Michigan
1968	Weisberg Sound	Jack Weisberg	New York, New York
1968	Community Light and Sound	Bruce Howze	Philadelphia, Pennsylvania
1968	All Sound Audio	Al Dotoli	Boston, Massachusetts
1969	Carlo Sound	Richard Carpenter John Logan	Nashville, Tennessee
1969	Tycobrahe Sound Company	Jim Gamble and Bob Bogdanovich	Hermosa Beach, California
1969	Sound engineer for The Who	Bob Pridden	United Kingdom
1969	Terry Hanley Audio	Terry Hanley	Cambridge, Massachusetts
1970	Showco	Jack Calmes, Jack Maxson, and Rusty Brutsché	Dallas, Texas
1970	Cameron Sound	Jerry Cameron and Mary Fleming	Miami, Florida
1970	Red Wheeler Sound and Lighting Company	Red Wheeler	Atlanta, Georgia
1971	Silverfish Audio Associates Ltd.	Ross Ritto and Joel Silverman	Rochester, New York
1972	Dawson Sound Company	Dinky Dawson	United Kingdom
1973	Sound on Stage	Jerry Pfeffer	Hayward, California
1976	Pace Sound and Lighting	Peter Schulman	New Orleans, Louisiana
1978	Ultra Sound	Don Pearson and Howard Danchik	San Rafael, California
1979	Meyer Sound	John and Helen Meyer	Berkeley, California

The old cliché "necessity is the mother of invention" rings true when referring to the early days of live concert sound. Just as in the great race for electricity that kept inventors and businessmen like Edison, Tesla, and Westinghouse on their competitive toes, the 1950s and 1960s were an exciting time for this group of sound engineers. They hustled to find inventive solutions, often under extremely adverse conditions. If you couldn't buy it, you built it—or you simply adapted to whatever it was you were doing "until you damn well figured it out." This was their credo. Even before this elite group surfaced, others had been toying with developments related to live sound and high fidelity. Our story begins with Bill Hanley and his company Hanley Sound Inc. of Medford, Massachusetts.

Chapter 2

THE NOT SO GLAMOROUS LIFE
OF A SOUND ENGINEER

In an effort to remain historically accurate, the innovators mentioned above (see Table 1) will be called "soundmen" or "sound engineers" throughout this book. Acclaimed author and scholar on women in communications, Dr. Donna Halper, observes:

> I understand, and fully agree, that in our modern day, we should not use sexist language. But the truth is that in the early days, it was generally a man, and the term was in fact "soundman." In the 1920s, engineers who set up the live remote broadcasts for radio were called "pick-up men" because they were assigned to the site where the broadcast would be picked up and sent by wire to the radio studio.[1]

In the process of writing and researching, many questions surfaced for me: What does a soundman actually do? Who are these people working behind the scenes of our live recordings and sound reinforcement applications? What do they look like? What would the world of media, entertainment, and communications sound like without the mad scientists at the front of the house?

The answer to such questions can be found at the next festival, venue, or concert you attend. While you are nicely settled into your overpriced concert seat or lying on a plush green lawn at a local summer festival, look closely for the individual who walks (or crawls) casually across the stage and scaffolding. Perhaps this person is nimbly scaling a two-hundred-foot speaker array while random tools dangle from his or her tool belt. This individual might be laying hundreds of feet of cable or tweaking knobs. For the most part, their nerves are probably running high come showtime, but in reality it's just another day at work. Their role is to not only interact with the artist but to get down and dirty with the weathered concertgoer. A journal excerpt from one of Hanley

Sound's former crew members, David Marks, insightfully reflects on his early days with the company:

> It is the soundperson who has access to almost every confidential facet of production politics. Good or bad, for better or worse, the sound engineer, mixer, recordist, is the one person married into the production from the start and remains there until the last wire and words are wrapped. The first person to arrive at the gig and the last to leave; with a confidential all-access backstage pass to the dressing rooms, management and performers; as well as all manner of odds and ends, technical and political, that need to be ego and audibly reinforced or inflated. And then he/she goes out into the audience/crowds and feels their way around the rumor and the vibe. The world's most celebrated artists often stoop to whisper in the lowly unkempt soundperson's ear—Hey man, how was that? How am I doing out there? OK? What should I tell the people? A little more voice in the monitor.[2]

Some readers might be left with questions concerning hearing loss. Since these individuals are in front of loudspeakers for hours on end, one could assume they must be deaf. I hold no conclusive findings supporting such supposition, but Hanley's hearing loss seems to be minimal considering the long list of shows he has attended. In general he refused to wear headphones or earplugs while he worked many of these gigs: "If you need earplugs then the music is too loud; loudness should never be confused with quality."[3] Hanley claims a simple yet logical belief that music should be heard in its most pure form: not loud, just clear. Many close to Hanley recall that he often disappeared into the crowd during a performance. At Woodstock, for example, the sound engineer walked the natural bowl listening for balance, clarity, and audibility without earplugs or headphones—for Hanley this was the only way.

Sound engineers and production crews are often the forgotten people of the music arena, although they are usually the first and last to leave the venue, breaking their backs setting up and striking heavy equipment. As facilitators of sound and production, they are there to make sure the show runs efficiently, so that the artists can play to their fullest potential. They are the gear-heads, the technologically savvy scientists within a laboratory of music production. It takes a certain type of person to execute such feats. Unable to fully describe the wizardry carried out by these individuals, this book only scratches the surface of how live concerts are produced. This unique group of people makes up the behind-the-scenes nuts and bolts of the concert business. It is my hope that what is written provides some acknowledgment of their arduous and sometimes thankless work.

Most if not all of the trailblazers mentioned in this book performed under varying degrees of adversity, resulting in some sort of innovation. Many of them share similar stories regarding the early landscape of the emerging sound industry. These include tall tales of "fly by the seat of your pants" innovation and chance encounters. The second-generation music industry people gave us mythical tales of rock and roll debauchery as well as backbreaking work, sweat, and hustle of roadies; riggers, stagehands, truck drivers, tour managers; lighting technicians, and even sound engineers. Forever integral to the classic American songbook is the tune "The Load-Out" by Jackson Browne.

> Let the roadies take the stage
> Pack it up and tear it down
> They're the first to come and the last to leave[4]

A poetically penned anthem commemorating the movers, shakers, and roadies of the concert industry, the artist sings as if the unkempt sweaty stagehand was his muse. Roadies yes, without a doubt. Still, soundmen should not be excluded.

Although we have come to celebrate the roadie in popular culture as the unsung hero of rock and roll myth and lore, it was not easier for people like Hanley decades earlier. In fact Hanley Sound might have even invented the modern roadie, as crew members spent continuous days on the road with bands during the early 1960s. All of it was arduous work, occasionally in disadvantageous and unsafe conditions. Night after night, Hanley and his crews spent endless hours on the road, setting up and breaking down equipment, hoisting heavy speakers onto towering scaffolding; it goes on and on.

Hanley supported numerous tours like the Beatles, the Beach Boys, and Jefferson Airplane, and these gigs were challenging as he and his crew were still finding their footing as a young touring sound company. According to Rusty Brutsché of the Dallas, Texas, second-generation sound company Showco: "I never thought of Hanley as a touring company. The rigors and challenges of touring were a systemization and packaging challenge. I saw Hanley Sound as a festival company."[5] Brutsché is correct, since the majority of what Hanley Sound is known for was a wave of pop festivals occurring in the country in the late 1960s. However, even before that, Hanley Sound supported many short tours via tractor-trailer for big-name performers, a first for the time, for sure.

By the early to mid-1970s, it became easier for those involved in the next phase of the concert production and sound industry. Smaller and more compact equipment in nicely built custom speaker boxes, road cases on rolling casters, better quality staging, portable lighting, and stadiums with an established infrastructure to work from made the job a hell of a lot easier. The aforementioned second-generation firms had begun to deploy concert sound on a more efficient scale. It was the beginning of the organization, systemization, and packaging of touring concert production.

By the mid-1970s to early 1980s, touring musical acts were performing at much larger venues. The result was an upsurge of sound companies that offered a more advanced approach and technical concert innovation designed to service arena rock.

Chapter 3

A BOY WHO LOVED SOUND

MEDFORD

Settled in 1630, Medford—or as Hanley and other blue-collar locals refer to it, "Meffa!"—is around five miles northwest of Boston. It has a population of around 60,000, and occupies about eight square miles. William Francis Hanley Jr. came into this world born on March 4, 1937, at the Lawrence Memorial Hospital, in Medford, Massachusetts. This small city was where Hanley, his family, and the Hanley Sound business lived for decades. It's a diverse, strong, and proud community. As one historian claims: "It is several neighborhoods of variable terrain connected by the Mystic River. In fact Medford is a great deal like the river that flows through its heart. People of different social and economic levels created a community."[1] Hanley often reflects on the many Irish and Italian families that lived near his childhood home at 56 Farragut Avenue.

The city of Medford has boasted many innovators, entrepreneurs, and creative minds. For example, James W. Tufts invented the concept of the drugstore soda parlor here during the mid- to late 1800s.[2] Medford also is home to Tufts University, where Boston's first radio station AMRAD was located. Here the first female broadcast announcer, Eunice Randall, worked during the 1920s. Inspired by the winter sleigh races from Medford to the adjoining town of Malden, local resident James Pierpont penned the holiday song "One Horse Open Sleigh," more commonly known as "Jingle Bells." Other Medford idiosyncrasies include C. E. Twombley, owner of the magazine the *Pigeon News*, est. 1895, who "introduced Medford to the specialized world of pigeon fanciers, enthusiasts dedicated to the breeding, exhibiting and racing of fine homers, tumblers and other varieties."[3] Amelia Earhart lived in Medford with her sister Muriel Earhart Morrissey for a short time during the early 1920s. It is said that she "indulged in flying and wrote aviation articles in Boston magazines."[4] Fannie Farmer, who authored the first modern cookbook, and New York City mayor Michael Bloomberg were residents of Medford as well.

Haines Square, named after the city's longest-serving mayor, Benjamin Franklin Haines (1915–22), is less than a mile from Hanley's childhood home. The Haines Square of his youth was a bustling downtown shopping location for residents who lived close by. Hardware and department stores, insurance agencies, pharmacies, grocery and butcher shops, and a local theater were all active during Hanley's youth. A family-style restaurant called Mueller's, which fed the locals of this bustling Medford village, served as a meeting place for the diverse community. According to many, this was a regular hangout for cops, criminals, families, and much later an overworked and hungry Hanley Sound crew. Hanley's grandmother lived nearby at 409 Salem Street. Often venturing into the square for errands, Hanley remembers these youthful encounters with delight: "Street cars, street vendors and Kennedy's butter and egg; It was a great place to grow up."[5]

HANLEY'S GRANDFATHER

The death of Hanley's grandfather, Patrick J. Hanley, in a tragic trolley accident on Sunday, March 30, 1924, had a great impact on the family. In that same accident Hanley's uncle John lost his leg. The causality occurred just minutes from Haines Square on Salem Street when a trolley struck both individuals as they were stepping onto the sidewalk. The two were on their way to Catholic Mass. Upon hearing of the accident, Bill's father, William F. Hanley, rushed to the scene, recognized his brother's shoes, and then ran home to tell his mother about the horrific incident. Hanley's grandfather was only forty-four years old at the time.

As a result, the family won a $40,000 lawsuit from the Boston Elevated Street Railway Company, an enormous amount of money for the time that equates to around $600,000 today. Hanley's grandfather was an instructor for the Boston Elevated Street Railway Company for over eighteen years and was highly respected in his community. Many who attended his funeral were shaken by the loss. A 1924 *Boston Globe* article states, "The presence of a large concourse of sorrowing men" was evident during his funeral. Mr. Hanley's obituary illustrates the impact it had on his peers: "A delegation of about 100 Elevated car men, representing the Salem St. car barns and other barns of Division 4 and the instructor's department of the Elevated at Bartlett St. station, Roxbury, marched from the house at 409 Salem Street ahead of the funeral cortege as an escort to the church. The flag on the Salem Street car barn flagpole was placed at half-staff."[6]

Hanley recalls the impact this accident had on his father: "Overnight he went from a carefree kid to the head of the household. My father was the oldest of five, four brothers and a sister. When he lost his father in that trolley accident it put a lot of pressure on him to take the reins of the family. He then became the father figure of his brothers and sisters."[7] According to Hanley his dad was often stern with him and his sisters, attributing the accident as the cause of his at times extreme parental behavior. Eventually some of the settlement money allowed the family to acquire a vacation home in the rural northern town of Tewksbury, Massachusetts. This is where Hanley's father and mother met and began their long marriage.

Farragut Avenue is a tree-lined street with reasonably sized two-family dwellings just off of Haines Square. The Hanley residence was a modest two-story house, stained dark green. Space was tight but livable for the large family. Bill shared a room with younger brother Terry, while across the hall his three sisters, Patricia (eldest), Barbara (middle), and Susan (youngest) were settled in their respective rooms. According to Hanley's three sisters, the house was always a busy place, with lots of people coming and going, a bit like Grand Central Station as they remember it. According to Hanley's sister Susan: "Bill would be in and out, and would often bring people home. Other people brought stray dogs home—Billy brought everybody home for dinner."[8] Although Hanley's father could be strict, Bill's upbringing in this blue-collar, Irish Catholic family overall was a pleasant one.

HANLEY'S BASEMENT AND HIS EMERGING INTEREST IN SOUND

As far back as Hanley can remember, he has been interested in sound. Many of his early experiences were fixing radios and televisions between the ages of seven and ten in the basement of his Medford home. In a 1964 *Boston Globe* article "Audio Enthusiast, Lives for Sounds," Hanley emphatically professes, "I've had a mad desire to study electronics since I was in the third grade."[9] The basement Hanley often took over was his father's workshop, affixed with peanut butter jars screwed to the ceiling as an efficient (and frugal) way to store his spare nuts and bolts. Although frustrated, Mr. Hanley tolerated his son invading his work area. It was in this basement that Hanley first learned how to build speakers, radios, and televisions. It was the laboratory of his youth.

His sister Patricia often listened through the heating grates to the goings-on in the basement while her brother was hanging out with friends Tom Hawko and Phil Evans. "My father didn't want the girls and the boys together, so we had to stay upstairs. If we were quiet enough, they couldn't hear or see us. We thought it was really cool that we could listen to everything they were saying."[10]

If Hanley was not at home in Medford, he was at his other grandmother's house, located at 840 Broadway in Chelsea, Massachusetts. According to Hanley, he often drove his sweet grandmother crazy with his addiction to Joe Hayden's 1913 comedy monologue, "Cohen on the Telephone." According to Hanley, "I used to play it over and over on her wind-up Victrola."[11]

Patricia recalls that her brother's interests were growing at a rapid rate even at a young age: "I remember being quite impressed by my brother's first crystal radio set he built out of a Quaker Oats box at around age six or seven."[12] Upon later reflection, she realized how truly special a child he really was. By the mid-1940s Hanley often lay in his bed in the dark of his childhood room listening to WEEI, a local AM radio station, through his self-built radio, attentively immersed in mystery shows like *The Shadow*, *The Green Hornet*, and the *Lux Radio Theater*. "The set had one earphone, which was tucked under my pillowcase. So when my mother and father came in and checked on me they couldn't tell I was listening to the radio. They always thought I was asleep,"[13] adds Hanley.

Continuing with this tradition of building electronics from scratch, the crystal set transformed the young boy at a time when televisions were appearing in homes across America. Projects like this helped channel Bill and Terry's interest in electronics. As Hanley got older, he and Terry innocently deployed a makeshift antenna out of their second-floor shared bedroom window. As the device hung into the cool night air the young men were ecstatic when they finally picked up that mysteriously distant radio signal. Terry learned a lot as he closely watched his older brother build the crystal radio sets. According to Terry, ten years younger than Bill: "We shared a room, he had the top bunk and I had the bottom. We had a crystal radio set up in our room that Bill made from scratch with a cigar box. It had a crystal and some components. We ran an antenna out the window and we were able to pick up radio stations."[14]

It was common for radio- and electronic-minded kids during the mid- to late 1940s to engage in similar activities. According to radio historians: "The early broadcasting stations were weak and the crystal set receivers were weaker. It was not unusual for a radio station to get an excited telephone call from a listener six blocks away, informing the station that they had just been picked up."[15] During this same time Hanley constructed an electric motor, the process by which he learned the principles of magnetism. According to Hanley, the desire to learn about electronics was just in him: "I suppose it was something that I felt I could control. I looked at all of this stuff and learned about it. I loved taking things apart and discovering how things worked. I had an old Atwater Kent radio I used to tinker with."[16]

When Hanley was seven or eight years old, his father took him to the Medford police station to view the intricate radio equipment. Bud Caffarella, who was in charge of the police radio division at the Medford Police Department back then, had a big influence on Hanley. Realizing that he had an emerging interest in radio, Caffarella built the young boy one- and two-tube radio kits, even setting him up with a simple motor to explore. Hanley would go on to build a six-tube radio from scratch himself. Often, Hanley and his dad visited Caffarella at his home to see his amateur ham radio transmitter. Hanley recalls: "My father wasn't too much into electronics so he introduced me to Bud. We drove to Bud's house a different way each time so I couldn't show up on my own. He did this to keep me from knowing the exact location. Eventually, I figured it out, and found my way back the next day on my bicycle!"[17] By eleven years old, Hanley learned amateur radio and Morse code at the Charlestown Boys and Girls Club, claiming that he had "difficulty with the speed of Morse code."

The Christmas season was an important time for Hanley and his family. Bill and Terry affixed a 3x1 ft. high-fidelity speaker box into a small window located in their attic. Powered by one of the early amplifiers they built, the brothers precisely aimed the speaker outward so the community could enjoy the popular holiday music. It was a wholesome activity that gave the brothers' and neighborhood families' great enjoyment. Terry remembers, "We also put sound in the neighborhood square, then the church for meetings, then we were doing little dances."[18]

The cold weather months gave Hanley the opportunity to entertain his friends and family while ice skating. Hanley was famously known for supplying music to ice-skating teens at the Malden Fellsmere Pond Reservoir. This was a local park just a few miles away from his home and a perfect community gathering place during the frigid months. It has been said that if the "safe to skate" sign was pointing in the right direction you knew whether the ice was thick enough to enjoy. Always eager to please, in January 1955 Hanley drove his father's 1937 Buick Roadmaster to the top of the hill overlooking the natural rink. Here he prepared his modest speaker and turntable setup out of the back of the sedan, tapping juice from the nearby Malden Hospital. Before long the entire area was echoing with the sweet sounds of Wurlitzer style organ music by musician Ken Griffin, whom Hanley admired. Not that uncommon, this type of organ music could be heard throughout the roller-skating rinks Hanley and his friends frequented.

In her diary, Hanley's sister Patricia describes how proud she was of her brother that frigid cold January day: "Billy went back to our house and returned with four huge loud speakers, a microphone, several records, and a turntable. Skaters were bewildered at the sound of the music. They couldn't imagine where it could be coming from. It was wonderful fun for everyone to be outside skating to music that soon engulfed the 'Rez' in stereo. This was a big deal."[19] The local police eventually kicked Hanley out after neighbors complained about the loud noise. "I was only playing organ music! Thankfully they never caught me stealing electricity!"[20] laughs Hanley.

HANLEY'S PARENTS

The officer who scolded Hanley fortunately wasn't from the Medford Police Department, where his dad was head of the criminal investigations vice squad. An incident like this would have angered the old man. Remembered as a tightly wound individual, William F. Hanley eventually retired in the 1970s. He was known as a "stand-up guy" that took his job as lieutenant "quite seriously."

According to Hanley, his dad never truly recovered after he saw his father and brother caught under that trolley car so many years prior. Mr. Hanley could never fully grasp the new field of sound that his son Bill was venturing into, often implying that it was an illogical career pursuit. According to Hanley's sister Susan: "My father wanted to be a pilot in the worst way, but he couldn't; he had to take over responsibility of his family. He couldn't relate to Bill's ideas. This was all brand new."[21] Deeply unhappy with Bill's rather unorthodox career choice, Mr. Hanley yelled at the young sound engineer in a vociferating manner: "You might as well be in the horse shoeing business! Can't you see that there is no future in sound!"

Ultimately, Mr. Hanley wanted what was best for his child, even if being a sound engineer didn't make much sense to him. In an effort to deter his son, he bought him various pieces of sports equipment, but Bill didn't want any part of it. Terry reflects: "Sometimes he got a little anxious with both of his sons taking on such a weird business trip. He had hoped we would be lawyers or doctors or something like that."[22] Very rarely did Bill get support from his father.

Hanley's mother, Mrs. Mary McQueeney Hanley, was a librarian and home-maker. A kind and generous woman, she was referred to as "mama bear" by many friends and co-workers at Hanley Sound. As any caring parent would, when Hanley was in town she called him at the shop every night for dinner. Conveniently, a hot meal always seemed to be waiting for him right around the corner—Bill was forever his mother's son.

Other than Hanley's aunt playing trumpet in an all-girl band, the Hanley family was not musical. Music did not come naturally to Hanley like electron-ics did. Susan adds: "Bill had a gift no one in the family had. No one was like Bill. None of my father's or mother's siblings had any technical abilities that resembled Bill's."[23]

UNCLE JOHN'S TV REPAIR SHOP

Hanley's uncle John eventually recovered from losing his leg in the trolley accident in Medford so many years earlier and according to Hanley, "He was able to get around just fine on his prosthetic."[24] Throughout the 1950s Hanley spent his summers at his uncle's radio repair shop, McClellan's Rural Appliance. Here the teenager installed television antennas on rooftops. McClellan's Rural Appliance, known as Town and Country Utilities, in Tewksbury, Massachusetts, sold household items including washing machines and television sets. Hanley's interest in electronics served him well since television was gaining in popularity. At the time, if one could afford it, a seventeen-inch console including an antenna with installation was a whopping $290.96.

Like clockwork every Saturday morning, Bill's mother packed him a half dozen peanut butter sandwiches for a long trek north with his uncle. Early in the morning Hanley patiently waited for the familiar horn as his uncle John picked him up at his Farragut Avenue home. When Terry was old enough, he joined his brother and helped out, and together they went out on calls to the rural communities of Lowell, Tewksbury, Billerica, Burlington, Reading, and North Reading.

The boys additionally worked with a man by the name of Eddie Deltotto, who drove the McClellan Chevy pickup truck delivering gas and bottled water. According to Terry: "I can remember when I was eight years old and Bill was eighteen. At the time this area was still rural and the reception wasn't that good. The broadcasting didn't have as much power as they do now."[25] Longtime television repair man Jack Moore ran the shop for John Hanley as well as providing the boys with additional guidance. According to Moore: "I worked on the television sets that were brought in. If I weren't busy in the afternoon I went out on calls to people's houses. I remember the boys put clamps on the chimneys and mounted a mast with an antenna on top of that. Then they would run the wire down to the television and drill a hole in the wall to get the wire into the TV of people's houses."[26]

Back in Medford, Hanley offered his television repair services to the locals in the neighborhood; most often his clients were the parents of his friends. Owning a television was an expensive luxury then. Hanley would test tubes from a varied assortment in his caddy, feverishly swapping out a blown one for a newer one to fix the problem. Usually Hanley fixed wary neighbors' televisions successfully, but the process made Hanley's parents nervous. According to Patricia: "Dad came home one day and came up from the cellar, shaking his head, sputtering that he was going to have to buy our neighbor Mrs. Westgate a new television! She had a great big console, and Bill had it apart in my father's workshop in the cellar!"[27]

THE MYSTERIOUS JACK BOSTON

Another lasting impression on Hanley was a local sound guy named Jack Boston. Jack owned a small radio and electronics repair shop called Boston Radio located at 7 Cross Street in Malden, Massachusetts. Jack was gruff, big in stature, and always wore a steel grey work suit. With a cigar clenched out of the side of his mouth, Jack drove around town in his 1937 Pontiac affixed with sixteen to eighteen trumpet-style reentrant horns on top of his vehicle. To support this there was a twenty-foot-long flat roof-rack type platform that went from the back bumper to the front bumper of the vehicle. Boston secured the platform by using thin braces at each corner extending down attached to the bumpers. Hanley recalls this time vividly: "In the front seat he had a Masco 75-watt amplifier with a special record player for mobile operations sitting on top of that. Off of the back of the car there was Fairbanks-Morse 2000-watt generator secured to a 4x4-foot piece of plywood off the bumper. It was all tied in to his loudspeakers and sounded great!"[28] This rolling spectacle of sound could often be found at parades, commencements, and local events.

Almost every day Hanley eagerly left the Immaculate Conception School and visited Jack Boston's shop situated right around the corner. The power of Boston's mobile sound system fascinated the young and impressionable boy enough to attract him to the shop every day. Jack saw a spark in Hanley's eye and after school the old soundman played Ken Griffin organ records for him on the 75-watt Masco amplifier. John Charles Boston was born on July 3, 1896, in Northwood, New Hampshire. A World War II veteran, he died in October 1969.[29] Not much else is known about the man.

By the age of fifteen, Hanley was troubled by the poor sound quality of smaller community halls, VFWs, local armories, and churches. He remembers the less-than-audible sound during the Knights of Columbus meetings at the Light Guard Armory in Medford, Massachusetts where his father often took him. Eventually Hanley provided sound for their community breakfasts in an effort to try and rectify the sound situation. Distressed, he recalls being at a function in the basement of a local church and not being able to "understand a word" coming out of the speaker. Instances like this bothered the young sound enthusiast, who later professed his desire to make events like this "sound better." In a 1970 *Boston Globe* article Hanley reflects, "All of the sudden I couldn't hear what the speaker was saying, I figured this was a way I could be of some use."[30]

Not long after, one of the first, and largest, local outdoor jobs the Hanley brothers tackled was an air show at the Hanscom Air Force Base just west of Boston. Jokingly, Terry adds, "We had to buy a lot of equipment for this event and Bill was able to cut labor costs, as I was the unpaid younger brother!"[31] Terry Hanley was in sixth grade at the time of this event.

Chapter 4

POOR SOUND ALL AROUND

PUBLIC ADDRESS

By the time Hanley was in his mid-teens, he was avidly learning about new developments within the emerging field of sound. At any given moment you could find him devouring articles in *Popular Science* and *Popular Mechanics* magazines. Always in tune with his surroundings, Hanley couldn't help but notice the less-than-perfect audibility the public address systems of the day offered. According to Patricia, who watched him absorb all of this information: "We had to go to church every Sunday, and they didn't have a good speaker system there. No matter where you went, or where there was a public function, you couldn't hear it. This really troubled Bill and he felt as though it was his duty to rectify it."[1]

Even at a young age Hanley felt strongly that communications containing sensitive content were being garbled by this poor technology. Terry remembers: "My brother and I liked the challenge of making it possible for people to hear. At the time good sound for events was pretty nonexistent."[2]

When Hanley was a young man during the 1950s, sound installations largely supported spoken word and not music. If you were to perform music out of a typical public address system, it simply lacked quality and intelligibility. As the live concert and festival movement of the late 1950s was expanding, the public address technology clearly could not accommodate it. Barely sufficient for a department store announcement, boxing match, or high school basketball game, these systems were designed so the public could be informed primarily via spoken word.

Between the 1930s and 1950s, in most towns with populations of 10,000 or more, it was common to see a repair shop or retail storefront that sold radios, televisions, and public address systems. Outfits like these rented or installed public address (PA) systems for schools, political rallies, sporting events, and other community functions. PA systems of the time were fairly

simple. According to former JBL vice president, and chair of the Music Industry Program at USC, Ken Lopez: "Typically, these would be public address
horns affixed to a pole or the same would be arrayed on a trailer, a truck, or on
a van. Leo Fender started out much like this."[3] Bogen amplifiers and Rauland
speakers/horns were some of the common brands used.

Like Hanley, West Coast inventor Leo Fender (1909–1991) began experimenting with electronics at a very young age. Also similar to Hanley, Fender's
uncle, John West, was instrumental in supplying Fender used radio parts and
other accessories. During the 1920s, he visited his uncle's shop, where he was
inspired by his collection of radios and loud sounds that were coming out of
the speakers. According to Richard Smith in *Fender: The Sound Heard 'Round
the World*: "At thirteen, Leo took up electronics as a hobby. Leo visited Santa
Monica in 1922 and saw a homemade radio his uncle John West had put on
display in front of the shop. By decree of fate, this loud music made a lasting
impression on the kid from Fullerton."[4]

Fender started a radio service in Fullerton, California, during the 1940s,
where he rented, sold, and serviced radios, appliances, instruments, record
players, and public address systems. He also drove a mobile sound truck that
he rented to politicians so they could project their political campaign messages.
Eventually musicians and bandleaders came from all over to rent his equipment. The California music scene was developing. Later on, Fender became
best known for his craft and development in building the solid body electric
guitar. Although he didn't invent the guitar, like Hanley in sound, he made it
a whole lot better.

Around the age of sixteen Hanley developed a love for big band sound. He
often volunteered his time at local venues like the Hampton Beach Casino
Ballroom in New Hampshire, as well as the famous Boston Garden, where
he recalls the sound system was "terrible!" Usually ballrooms like the Casino,
Nutting's, and Moseley's on the Charles had unrestricted age limits. This was
great for Hanley.

During a Hanley family vacation on Cape Cod, he ventured out and found
himself at the Pine Room at the Cape Codder Hotel, where he saw the Tony
Bruno Band, further advancing his interest in big band sound. Even though
he would have rather been building speakers in his basement back home, this
was a memorable experience for him. Hanley recalls: "Usually they had one
microphone and most had ten-inch loudspeakers. All of the sound was usually
controlled from the side or behind the band. You couldn't see the sound system
at a place like Moseley's. It was pretty poor."[5]

By his late teens, Hanley was immersed in youth based musical gatherings. You could often find him amplifying sound at local record hops, sock
hops, and banquets with Boston radio personality Dave Maynard. Hanley

made the sound system for most of these events, accompanied by Terry and friend Howard Hughes. Working in his basement, Hanley essentially copied and built the speaker cabinets from his old mentor Jack Boston's designs. In the background, Hanley's father stood reluctantly amazed at how well his son constructed his first Rebel speaker cabinets and amplifiers. Hanley recalls: "I found a chassis with some stuff on it that I rebuilt. It was a Masco seventeen-watt power amplifier, that had five or six tubes."[6] In the early days Hanley used Electro-Voice microphones that he conveniently borrowed from his uncle John's store.

By age nineteen Hanley had bought his first microphone. When his uncle John heard about this, he let his nephew borrow an Ampro tape recorder so he could record some of the big bands rolling into town, like the Tommy Dorsey Orchestra. On one occasion, Hanley brought the recording gear to the Hampton Beach Casino Ballroom located on the New Hampshire seacoast. The turn-of-the-century Atlantic resort had boasted vaudeville shows, penny arcades, an opera house, and a shooting range for sizable summer crowds. Like Moseley's on the Charles, the Casino is still one of the oldest running music/ballroom venues in the Northeast.[7]

At some point during the early to mid-1950s, Hanley got permission from the club's owner, John Dineen Jr., and big band leader Tommy Dorsey to allow him to record the orchestra's performance. Dineen was known to be strict about maintaining his club's standards. "The Casino Ballroom was Dineen's first and lasting Casino love. Well into the 1950s, he kept the bands coming. One hot night, Tommy Dorsey dared to remove his suit coat. Dineen upbraided the bandleader and Dorsey promptly put his jacket back on."[8] This experience was great training for Hanley.

After listening to the performance he had recorded, Hanley realized that in order to get a quality mix from Dorsey's orchestra, he needed to use multiple microphones. As Hanley recalls, "They were only using one microphone for the whole band at the time, and that was a real problem; the one microphone on stage was used for the vocalist."[9] Later in his career the young sound engineer tried to sell Dineen on the idea of good sound, but according to Hanley the club owner wouldn't spend a "dime" on a new system.

Even in the early 1960s, PA equipment was still being used for popular music performance. Ken Lopez recalls a James Brown and Ike and Tina Turner performance that he attended: "This was the early 1960s and I remember these performers screeching through the horns of a simple public address system!"[10] If a local sound-savvy person was wise to movie theater technology and had access to Altec Lansing gear, they would have figured out that they could apply it to live concert sound. This was something Hanley, and others, eventually caught on to. At that time the Altec Lansing line was the best in the business for sound installations. The Voice of the Theater speaker, designed for cinema sound, was a logical solution for the few innovators that existed in the country then. The speakers were attractive because Altec offered a variety of models and sizes.

Reflections from a longtime Hanley Sound employee, engineer Harold Cohen, give us an idea of how sound system components were varied in the early days. He remembers when he joined Hanley in 1966: "The consoles for broadcast stations and recording studios were custom, and in many cases were made up of components from multiple companies. It was a slide-pot or rotary-pot from one company, a pre-amplifier and amplifier from another. Mixers were put together unless they were just basic mixers used in broadcast."[11] This arrangement often included the Voice of the Theater speaker by Altec.

The various acts that Hanley worked with during the mid- to late 1960s usually did not travel with their own sound system. Often, they hired a sound company like Hanley's or used whatever primitive PA system was available at the venue. Author Herbert Wise in his 1967 book, *The Complete Guide to the Electric Band: Professional Rock and Roll*, explains what a typical sound system looked like then: "It is an electronic system (sometimes called a PA or Public Address System) designed specifically for the amplification of voices. The basic components usually are microphones, a microphone mixer (when two or more mikes are used) a power amplifier and speakers. There are infinite variations and designs."[12] As rock and roll became more popular, and powerful amplifiers surfaced, the sound levels of instruments in a live setting began to compete with each other. The result often drowned out the vocals, compelling groups to use a more advanced sound system similar to what Hanley was designing.

Chapter 5

ORGAN MUSIC AND TRAINING

THE BAL-A-ROUE ROLLAWAY

Bill Hanley developed an affinity for organ music through his childhood role model Jack Boston and American organist Ken Griffin. That love was later reinforced by his time spent roller skating at the Bal-A-Roue Rollaway, located at 376 Mystic Avenue in Medford. Also referred to as the "Rollerway," the rink was just a few miles west of Haines Square toward the muddy banks of the Mystic River. French for "dance on the round" or "dance on wheels," the "Bal-AH-Roo," as the colorful locals referred to it, was a haven for many. It was a popular weekend hangout for the community, and many skaters forged long lasting relationships here. It was at this rink where the young sound engineer formed a special relationship with high fidelity sound.

A former early nineteenth-century boat manufacturing building, the huge structure, measuring 300x100 feet, sat close to the river edge before the land was filled. It offered skaters great acoustics. During the 1930s, businessman and roller skating entrepreneur Fred H. Freeman transformed the building into a rink that was then named the Mystic Arena. Freeman became known as a popularizer of international-style roller skating.

According to a later owner, George Pyche, Freeman knew what he was doing regarding the rink's acoustics: "Freeman paneled the walls with knotty pine and put in a rock maple floor. It was beautiful! He put in a drop ceiling so the sound would not bounce all over the place." Pyche, who skated there as a boy, claims that back in the early days they used a product called Tectum. "It was made from crushed and or compressed seaweed. Acoustically it was very good."[1]

Freeman was also the executive of the Dance-Tone Record Company, which specialized in music releases for roller rinks. The music on the Dance-Tone label (Revere, Massachusetts) featured various Hammond organ solos with artists like Phil Reed and pianist Frank Picher, who was the Bal-A-Roue's organist at the time. In August 1951 Freeman sold the building to Raoul E. Bernier, a Portsmouth, Rhode Island, rink operator and retired naval captain who held on to it for twenty years, eventually passing it to George Pyche. "I suspect he was too old to run it any longer. We ran it until 1983 when our lease was up,"[2] recalls Pyche.

The rink was a gathering place where adults, children, and teenagers congregated. At the time the Bal-A-Roue was considered to be one of the "finest rinks in the country." Always staffed with professional instructors, the venue held many professional competitions. The Bal-A-Roue accommodated adult professional skaters, but amateurs were welcome; it offered a "beginners night" every Wednesday.[3] "Ladies night" on Mondays and Thursdays was perfect for those who were on the prowl for a hot date, including sailors who were stationed nearby at the Charlestown Shipyard. In an issue of the Bal-A-Roue newsletter it claims, "These are two very popular nights with the ladies, and the men should not overlook these nights as the ladies always turn out in good numbers."[4]

The rink was equipped with a soda bar, ice cream freezers, a grill, lockers, and rentals. It also offered dance lessons and parties. If rink food wasn't your bag, just across the street from the Rollaway was Kemps Hamburgers, where generations of skaters indulged in eighteen-cent burgers and other delicacies. The hard working staff at the Bal-A-Roue prided themselves at maintaining a spotless rink, making certain that substances like chewing gum stayed off of the pristine oval hardwood floor.

When Hanley took up the sport at age fourteen, the roller skating phenomenon of the 1950s was fairly prominent. A 1951 Bal-A-Roue newsletter states that when the "Bal-A-Roue first opened roller skating was America's seventh rated participant sport. But, today as for the past four years, it is the first and favorite sport of more Americans."[5] In the beginning Hanley was not the greatest skater, often succumbing to constant ridicule from the regular rink kids who poked fun at his fumbles on the floor. According to Hanley, he couldn't skate very well: "One Sunday afternoon they were harassing me. They could all skate well, except for me."[6]

Determined to practice as much as possible, Hanley used to get into the rink for free through his high school bench mate Howard Hughes. Hughes had a close friend whose job it was to take tickets at the door. Fortunately for Hanley, his strict father thought the Bal-A-Roue was a good clean operation, gaining his full approval. With free admission and a green light from his dad, the young Bill was off and skating. Eventually Hanley became skilled at the sport, learning to do elaborate turns, jumps, and regular dances. Hanley recalls: "I got in for nothing so I was going six to seven nights a week. This allowed me to get better and I eventually became a great skater. This is when I fell in love with the organ music."[7] While frequenting the rink after school and on the weekends, Hanley became enamored with its powerful sound system and superb acoustics. Hanley remembers:

> It was played loud with a total of nine Hammond B-40 tone cabinets that held two 20-watt amplifiers and four 12-inch electro-dynamic loud speakers each. The speakers were up real high in the rink, sitting on the floor of a room that was used as a space for the former boat manufacturing operation. Having them elevated made it sound great. All of this in a rink that had excellent acoustics! I would go hear other, bad, sound systems and wonder why something couldn't sound as good as that roller rink![8]

The system at the rink left Hanley curious about how to achieve better quality sound. When old enough to tag along, Terry joined in: "Bill used to drag me along. He loved the organ. We serviced the sound system for the organ. We took them apart and cleaned them, fixed them and put them back together."[9]

When Hanley was seventeen, he and Terry somehow ended up with one of the broken-down organs from the Bal-A-Roue. It was a Hammond B-3 that lit up with lights when it was played, and it fascinated the young men. With permission from Hanley's father, the brothers and a few friends affixed four roller skates to the base of the organ and began rolling the heavy unit home. According to Hanley's sister Barbara: "The roller skates didn't work so someone got in the trunk of a car holding on to a rope while the others were pushing this thing down the street. I believe it was an old Studebaker Coupe going about five miles an hour."[10]

On October 18, 1952, veteran organist Benny Aucoin joined the staff at the Bal-A-Roue, where he performed in a "beautiful setting." While Aucoin sat behind the organ, he was silhouetted by "venetian blinds" while "the blue of the atmosphere" gave his space "a very homey touch."[11] It has been said that Aucoin was an immediate hit with the skaters, and Hanley agrees. Aucoin used to play in theaters during the silent-film era, but lost work because of the introduction of sound in film. The sound engineer still speaks of the remarkable organ playing by the well-known house organist: "Benny's hands danced over the keys of that Hammond, he was really amazing."[12] Up until this time Hanley had no identifiable connection to any particular music. Yet, the fanciful organ playing of Benny Aucoin seems to have had an enormous impact on him. Hanley claims it was organ music that influenced him to move into the sound business.

Live organ music at roller rinks was common during this time, and the organ of choice was often the Hammond B-3. According to Pyche, the Hammond was perfect for roller skating environments: "It was particularly good at rinks for some reason. It had plenty of volume and didn't slur the beats. It gave you true repetition, of what the organist was trying to put out."[13] Throughout the 1950s

Bill and his Medford friends gleefully skated to the magical waltz-like sounds that emanated from Aucoin's Hammond. Hanley recalls: "Aucoin's playing style was like no other. He also played a Novachord piano made with a 138-tube envelope generator, and a Clavioline, a small three-octave keyboard with an ancillary keyboard under it to his right."[14]

As Hanley's love of organ music continued to grow, he listened to artists like Milt Herth and his album *Hi-Jinks on the Hammond.* This record, Hanley admits, was one that he "wore out" as teenager. This long-lasting and significant love of jazz organ playing became highly relevant throughout Hanley's career.

The Bal-A-Roue turned off its recognizable glowing neon sign in the mid-1980s after seeing a decrease of interest in the sport. Now the rink is nothing more than a memory to the generations who enjoyed its pristine floor. A bank now occupies the location. Hanley's love for roller skating lasted well into his mid-forties. The sound engineer still has his old skates from that bygone era.

Hanley's passion for roller skating developed into an even greater love of dancing, something he still enjoys to this day. Attending Mrs. Putnam's School of Dance to learn ballroom dancing enhanced the young man's skills. After family dinners Hanley could be found dancing with his sisters and mother in the kitchen for practice. Barbara remembers, "At first Bill didn't want any part of it, but once he started skating he loved the rhythm."[15] Hanley also ventured into Boston to explore dancing at a venue called Steuben's the Cave. There was never a cover charge.

The Cave opened in 1945 and for twenty years was Boston's premier night-spot. Hanley reveled in the nightly floorshows, big band, jazz band, and swing band appearances—but not the sound. According to Hanley: "I actually used a fake I.D. to get in. I remember seeing Tito Puente perform there. You could not hear a thing at the Cave, it had a poor sound system and everyone on stage used only one microphone."[16]

SCHOOL AND TRAINING

Hanley had an extremely difficult time in grade school. When he attended the Immaculate Conception Elementary School in Malden, many of the nuns were his father's former teachers. Hanley's father was an exemplary student, so naturally the same was expected of him, setting him up for failure. Hanley recalls: "My father was so smart he had two double promotions and jumped grades. I was a dud in school and he was a super student."[17]

Something of an outcast, Hanley was often disconnected from the rest of the class. Most every day he gazed out the classroom window, uninterested in the day's lessons. When instructor Sister Corita gave up fighting for his attention, she handed the young boy paper cutouts to paint and draw for window decorations. Hanley remembers: "The nuns just gave me busy work. They didn't know about A.D.D. back then, there was no diagnosis."[18] According to classmate Bill Appleyard, Hanley was odd and really stood out in class. One day Hanley proved himself by way of a broken radio that one of the nuns owned. Appleyard recalls:

> I think it was in sixth or seventh grade and Bill was around thirteen. The nun had a secondhand radio that someone had given her and it had a lot of static. She asked the class if someone could take it home and have their father fix it. Hanley raised his hand and said he could. Of course the kids in the classroom were behind their hands laughing at him thinking he was going to mess this all up! He brought it back the next day working like a charm. We were all amazed. Hanley had other things on his mind.[19]

Around fourteen, Hanley embarked on a whole new development of learning at the Melvin V. Weldon Vocational High School in Medford. At the time, vocational school was an inevitable and logical choice for someone like Hanley and other academically underachieving students. The Radio & Television Electronics vocational program was exactly what he needed. Becoming quick friends with his vocational school teacher and mentor, Thomas A. Rawson, he blossomed here. Hanley successfully graduated from Weldon in 1955 with a focus on radio, electronics, and television, excelling in all these subjects.

THE BOSTON GARDEN

On his days off, Hanley spent as much time as possible studying the sounds and acoustics at local venues. He noticed that larger spaces like the Boston Garden suffered from subpar sound. Barney Noonan, head electrician at the Garden, was a friend of Thomas Rawson, Hanley's vocational school teacher. One day Rawson made a call to the Garden and spoke to Noonan about a kid in class who was into sound systems. Before long, Bill was taking the trolley from Medford to Sullivan Station in the Charlestown neighborhood of Boston. Here he caught a train on the Charlestown Elevated, known as the "El," that transported him into Boston's North Station.

Hanley had free access to the Garden and its sound room during a time when the facility only saw touring acts and performances like the Ice Follies, Ice Capades, Ringling Bros. and Barnum & Bailey Circus, and other sporting events. Even at his young age, Hanley noticed acoustical deficiencies with the modest-sized sound system utilized by the Garden. For the young man with an interest in sound, he felt the four Altec A7 500 speakers and 360-degree circular cluster hanging above the 11,000-seat arena were insufficient at best. Hanley reflects:

> I used to go in and hang out with Barney and this guy named George Mac-Lennan who was a stagehand. I saw all the shows that were coming in and spent time listening to the sound, which was not very good. I heard better sound coming from the Bal-A-Roue and Jack Boston's record player than at the Garden. Later on I tried to sell them the idea of good sound, but they were not interested because of the cost involved. I wanted to get good intelligibility for the crowds, for the games, and crowd control. They wouldn't spend five dollars to rent a speaker from me.[20]

Still too young for a work permit before high school, Hanley cleaned lockers at the Tufts University gym. The gymnasium was located next to the Tufts Office of Engineering. Hanley disliked the job; nevertheless, it allowed him to exchange ideas and stories with the engineering students who were involved with antenna and rocket studies. Inspired, the young vocational student could not wait to graduate and finally get into the field of electronics and sound.

Hanley's skill of networking at such a young age would help his business in years to come. While at the Garden he mingled with the production staff, including longtime Boston Garden and Fenway Park organist John Kiley. By the late sixties Kiley called on Hanley's sound company to produce a sound system for his Hammond X-66 organ at Fenway Park, and for $40,000, the sound company did just that. Engineer Harold Cohen, who was in charge of wiring the project, claims that this was one of the first centrally located bi-amplified stadium sound systems in the country.

Cohen claims the grid from where the sound emanated is still visible in center field: "It was a two-way system with Hanley's 410 cabinets being used. Being on the ground they were ground coupled acoustically so consequently very low bass was produced. The system was powered by Crown DC 300s, which were at the time some of the first high-powered transistor amplifiers on the market. We had the system in for about eighteen years."[21]

DEMAMBRO SOUND

After graduating high school in 1955, Hanley began to look for a job. Desperately wanting to use the knowledge he gained in school, an obvious position would have been with a local electronics or sound company. With not much to choose from, DeMambro Sound of Boston was the only sound company in the area. DeMambro was a wholesale electronics dealer that had a sound installation department. In addition, the company sold high-fidelity radio and television parts.

Radio Shack was another possible local choice for Hanley. By the mid-1950s the Massachusetts electronics wholesaler specialized in high-fidelity equipment and had been a resource of electronics for Hanley early on. Known as the "Shack," it was one of the oldest of its kind in the Northeast. A 1956 *Boston Globe* article reports: "Radio Shack is now thirty four years old, venerable for the still young electronics field, in fact one of the oldest and largest shops of its kind in the world. Its customers include musicians, factories, schools, government agencies, churches, foreign governments—even the Shah of Iran has ordered from its catalog."[22]

Of the two, Hanley felt he would fit in with a company like DeMambro because they installed PA systems for schools and churches. Unfortunately, they wanted nothing to do with the young sound engineer, refusing to give him a position with the company. Owner Joe DeMambro was considered something of a local entrepreneur. As a boy during the 1920s, DeMambro tinkered with "crystal set" devices, which he assembled and sold to radio fans that were "bewitched by the magic of the airways."

Known for having a keen business sense, DeMambro realized that radio and sound had a bright future. During the depressed 1930s his business was struggling until he landed a break investing his last $350. Partnering with a banker friend, DeMambro developed his home shop business into a multimillion-dollar firm. After the war his climb to success grew even greater. With his wife by his side, and an increasing credit line from his bank, DeMambro was able to grow his company into a local empire. Expanding his wholesale and retail operations meant that he was able to hire more technicians, warehouse workers, truck drivers, and salesmen.

By 1953 DeMambro was settled into a prime location at 1095 Commonwealth Avenue. With twelve branches of showrooms and warehouses strategically placed throughout New England, his company was able to efficiently provide electronic parts and components to industrial and commercial customers. His "store to door" service was time saving and customers liked that. DeMambro became known for his sound division, which installed "internal communication systems" and "recording equipment" that was used in various schools and hospitals.[23]

LAB FOR ELECTRONICS

Rejected by DeMambro, Hanley searched for whatever work he could find. With his mind set on other ventures, going to college seemed less and less of an option. After a series of various other jobs, a unique opportunity fell into Hanley's lap. Because of his father's friendship with a then junior Boston politician, Thomas Phillip "Tip" O'Neill Jr., the emerging sound engineer landed an apprenticeship at the Laboratory for Electronics (LFE) in Waltham. LFE was affiliated with the Massachusetts Institute of Technology.

LFE was Hanley's first real electronics job after graduating high school; he worked there for five years. His time at the company taught him the fundamentals of the trade that he would use throughout his career. According to Hanley it was somewhat intimidating: "There were a bunch of heavyweights there, MIT boys, who were really sharp."[24] Unaffected by egos, it didn't stop the ambitious young man from moving forward.

Because of his skill set Hanley quickly excelled through the various departments. Beginning with soldering school, which was a breeze for the young electronics wizard, he then moved on to the harnessing department, where he dealt with wires. This progression was followed by moving on to the printed circuit board department and eventually all the way up to the engineering building. "I learned a lot at LFE. I was building the first forty-pound radar set models and making printed component circuit boards. I worked with other engineers during this time."[25] Hanley claims that this experience offered him sufficient training that allowed him to excel within the emerging field of electronics and sound.

After LFE, in 1957 Hanley began part-time employment at the Ad Cole Corporation. Addison Cole, former president of LFE, brought Hanley on as one of the company's earliest employees. Ad Cole was contracted by the MIT Instrumentation Laboratory to develop aeronautical instrumentation. Hanley was hired as an engineering aide working on rocketry projects like gyro table controls. This is when Hanley first became aware of the McIntosh amplifier (MA 30 or 35) used to power the table controls. While employed at Ad Cole, during his free time, Hanley was also working to get his business up and running.

With considerable arm-twisting, Hanley's father attempted to steer his son into more conservative and practical types of trades. But Hanley would not have it and could not understand his father's indifference toward his passion for sound. No matter how much his father tried, he could not stop his stalwart son from becoming a sound engineer.

At the time no one would hire Hanley regardless of his expertise in sound, so it became his personal quest to become the sound engineer he always imagined:

No one cared for better quality sound. I was begging for work. They would say "I have my sound system so why should I bother with you?" At Carnegie Hall they had a 35-watt Stromberg amplifier in the closet. When you walked in the stage door and took a left up the stairs there it was, their sound system. It was ridiculous! I had to fight all over the country to get my sound system into places. But I was not going to give up at this point, I thought what I was doing was worthwhile.[26]

Immersed in anything to read that was audio related, for industry information Hanley most often referred to Dr. Leo Beranek's 1954 book *Acoustics*. Filled with information on microphones, loudspeakers, speaker enclosures, and room acoustics, this was Hanley's go-to reference on acoustical concepts and theory. Beranek, a professor, acoustics expert, and co-founder of the Boston based acoustical consulting firm Bolt, Beranek and Newman, understands why a young sound engineer like Hanley used his book so closely: "My book on acoustics was one of the first to talk about loudspeakers, microphones, amplifiers, and rooms in some detail. If Bill Hanley wanted to learn something about what was available and why it would give you the best sound quality, he would have got it from my book."[27]

Even though schooling was discouraging, Hanley still had a desire to attend college to complete a degree in engineering. After all of his training at LFE and Ad Cole, this seemed like a good idea. But he needed prerequisite courses missed in high school in order to qualify for admission. So Hanley enrolled in English and math courses. However, at the same time his newly established sound business began to grow, and this pushed further studies aside.

Chapter 6

430 SALEM STREET

FINDING A HOME FOR HANLEY SOUND

In 1954 Bill Hanley established Hanley Sound Inc. out of his basement at 56 Farragut Avenue in Medford. An article claims, "His original office was in the basement of his parents' home."[1] Huddled in his makeshift laboratory, the persistent young sound engineer was looking for light within the dark ages of sound reinforcement: "I took over my father's bench which he didn't like and I also had a big bench in the cellar of my house. This was really the beginning of Hanley Sound. It was where I first started my radio and television repair company."[2] Before Hanley Sound had an actual physical commercial space to work from, the sound engineer often stored his increasing surplus of equipment at his parents' home.

As Hanley's business grew he rented out one side of a two-bay garage located on Salem Street in Medford from a family friend he called Mr. McDonough. The garage was located right around the corner from his Farragut Avenue home, making it convenient at dinnertime. The building, a cement block structure, sat next to a doctor's office down the back of a long driveway. Here Hanley and his friends Phil Evans, brothers Tom and Dick Hawko, and Phil and Ed Robinson helped Hanley build out the space by adding insulation, portable heaters, a floor, wall structures, shelves, and workbenches. The material that Hanley used here was hauled from nearby South Border Road in Medford. Most of it came from the dismantling of a military installation. According to Terry, he and his brother were "great improvisers" when it came to repurposing.

In the fall of 1957 Hanley met Ed Robinson through mutual high school acquaintances. Becoming fast friends, Hanley asked Robinson to help with his new operation. Robinson claims: "A buddy of mine said you need to meet this guy, he is really into electronics. At the time Hanley was called 'Wild Bill.' He was kind of an off-the-wall, strange character at times. Eventually, I became interested in what he was doing and began to help him out with some of his projects."[3]

The unheated garage could not withstand the cold winters, so Hanley and Robinson borrowed a gas-fired salamander heater from his uncle John's store in Tewksbury. According to Robinson: "It was freezing cold! We needed it so we could survive the winters in there!"[4] The heaters did the job but they smelled and that forced Hanley to always keep the door open.

A sophomore at Medford High School, Robinson's younger brother Phil, was only fifteen years old when he began as a helper at Hanley Sound in early 1958. According to Phil: "My older brother Ed worked for Bill, and I just hung around. I was attracted by all of this stuff going on with electronics."[5] In those days Hanley didn't leave the garage much unless he was out installing equipment. He often fixed radios and televisions in the space in order to maintain a revenue stream.

Between late 1958 and early 1959 Hanley enlisted for a short stint in the Army Reserve. Putting his business briefly on hold, he attended boot camp at Fort Dix in New Jersey, then power generator school at Fort Gordon in Georgia. Eventually he was stationed in Boston, where he could resume his sound business.

The shift from his home to the garage marks the official beginning of Hanley Sound. Always keeping an eye out for possible expansion opportunities Hanley noticed some vacant space across the street from the garage. What looked to be an ideal spot was the vacant former Star Department Store, and next to it was an occupied fish store. The department store space was originally made up of two storefronts, offering ample room for the equipment bursting out of his garage. According to Hanley, "In the early days the department store housed a Kennedy Butter and Egg and Hood Milk store before it became one space."[6] Acting swiftly, Hanley signed papers and moved his operation across Salem Street into the empty department store around late 1960. According to Ed: "It had been vacant for some time, and I recall Bill working out some sort of deal. The building needed a lot of work."[7]

The transition was not as smooth as Hanley would have liked. He fought tooth and nail with his skeptical father to convince him the move was a wise one. According to Patricia, "When Bill bought the building, I thought my father was going to have a heart attack."[8]

By 1961 Hanley's move was under way. Little by little he and others renovated the space. Phil and Ed, who assisted, were due to leave soon for military service. Phil remembers: "It was kind of run down at the time and had not really been kept up. I am guessing the building was from the 1930s."[9] Having very little heat, the first winter at Hanley Sound was a cold and damp one.

Until Hanley's office was built, he worked mostly out of a mobile studio he bought from Faircast Systems, a public address sound company located in Manchester-by-the-Sea, Massachusetts. One journalist writes, "His office is a trailer fashioned into an actual radio station except for a transmitter."[10] The mobile studio was parked in a lot behind the Salem Street location. If Hanley ended up working into the early-morning hours, he slept on a cot at the rear of the vehicle surrounded by his electronic and audio equipment. Later the studio was destroyed, succumbing to an act of arson. According to Phil: "After the fire was out, the melted telephone still rang. We all got a kick out of that!"[11]

Once settled at his new location, Hanley waited patiently for the adjacent fish store to become vacant, which it eventually did by 1964. An old man owned it who didn't like Hanley, hippies, or anyone other than his fish-buying customers. Hanley recalls, "I ultimately ended up buying all of the 430 Salem Street locations when they became available to me."[12] Still thick with the stench of fish, the store made up the third bay of Hanley Sound.

The company was evolving and now had three separate sections to accommodate a diversity of functions. Once construction was final, there was enough space for offices, workbenches, electronics, and fabrication areas replete with a table saw. There was a lot of pre-construction necessary before all of these areas were cohesively functional. In 1964 *Boston Evening Globe* writer Robert Glynn recalls Hanley's shop in an article called "Young Men, Big Job." According to Glynn: "His equipment is now stored in five Medford garages, a cellar and in two storehouses in Newport, Rhode Island, and in New York. Hanley acquired the property that contained three stores months ago and he and his men and boys have rebuilt its front along modernistic lines and are working on the interior."[13]

One of Hanley Sound's earliest employees, David Roberts remembers when he arrived in 1964: "Everyone was sort of camped out in the first two bays. It was very primitive. The tin metal ceiling from the first two bays was still there. There were remnants of walls and counters. The office was heaped with stuff. There were cables, and amplifiers, helter-skelter all about. Bill had a desk somewhere in the corner."[14] Consensus from those who remember is that the interior of Hanley Sound was organized chaos. Although there are very few pictures of Hanley Sound from this time, *Boston Globe* feature writer and music reviewer Nathan Cobb recalls its original condition back then: "Bill Hanley Sound Inc. is housed in an unidentified brick and wood one-story building on Salem St. near Medford Square. The interior is a patchwork of styles surrounding what may once have been an attempt to present normal office décor. But no more: people and equipment are chicken-wired together into a cluttered collage that would do justice to any attic."[15]

Hanley also leased part of the basement under his store location, but the space was small. He and his crew set about making more headroom and storage. Amid the drilling, welding, and construction that went on at 430 Salem Street, nothing can compete with the story of blowing up a boulder in the basement. This needed to be done before the concrete floor could be poured.

While digging out and leveling the floor, Hanley and his crew came across a large 9x4 ft. blue granite boulder. Due to its size the rock needed to be dynamited in order to crack and remove it. Hanley remembers: "We were at water table by the time we dug this out. One of my crew members brought in a friend who was a dynamite expert."[16] Knowing the explosion could attract unwanted attention, Hanley turned up the music in the shop to try match the decibel level of the dynamite. In earthquake-style magnitude, the explosion shook all of Haines Square. Some say it even jolted the sandwiches out of people's hands at Mueller's restaurant. "Later people were asking me if I had felt the earthquake,"[17] laughs Hanley.

After the demolition Bill employed Terry, who was smaller and more agile, to help with the backbreaking job of removing the rock. Terry and others schlepped the entire boulder out by bucket brigade, all of the dirt and rock ending up in the backyard. "It was a pretty big job. I did most of it myself with some friends from school. We took the dirt out and it took a REALLY long time!"[18] recalls Terry.

A large area behind the building served as a parking lot for Hanley's many trucks and trailers, most likely violating some city code. These violations were ignored because his father was a Medford police officer. Hanley Sound was not zoned for industrial work, so the sound engineer often kept the shades of his shop windows down. "No one understood what we were doing anyway so it didn't matter,"[19] adds Hanley.

A Medford community leader and three-time war veteran, Major Tom Convery recalls being suspicious of Hanley's business in the neighborhood: "I would ask myself. Why were all the curtains drawn all the time? Were they keeping books back there? There was a sign on the window of his shop that said Hanley Sound Inc. We didn't know what—SOUND—meant? What does that mean? The guy selling fudge across the street from Bill's shop was running books and we knew that."[20]

THE VOICE OF THE THEATER SPEAKER

Hanley built his first speaker cabinets in 1957, and by the mid-1960s he was building his own sound systems. What didn't exist he made himself. His reputation within the business of sound was growing nationally. Having a number of large music and political accounts under his belt by then, Hanley Sound was venturing into a new and exciting field of concert sound reinforcement. With his business expanding, Hanley acquired additional spaces. The more money he made, the more equipment he could buy to serve a diversified clientele.

The Voice of the Theater speaker (VOT) by Altec Lansing served as an instrumental component to the live sounds of jazz, folk, and rock and roll. At the advent of the sound reinforcement industry, VOT was the speaker that laid the foundation of live outdoor sound. As audiences got bigger, sound required more power; a speaker that could deliver was needed. It was a logical and obvious choice for Hanley, and was available and powerful. These impressive speakers could throw big sound at large events, and for many first-generation sound engineers like Hanley it was all he could get his hands on.

The VOT speakers were used in movie houses across America during the 1950s through the 1970s. Resting on casters so they could be easily rolled around, they sat behind the movie screens in medium to large theaters. The sound blew through the tiny holes of a perforated movie screen. Many of these theaters were going out of business as television was tightening its grip on American viewing audiences. It was perfect timing for the growing sound company. Buying secondhand equipment saved Hanley lots of money and allowed the sound engineer an inventory of extremely efficient and great-sounding speakers.

Hanley Sound employee Ray Fournier remembers: "Bill checked the papers for the theater sellouts and auctions and bought up all of these systems. He was able to set them up on scaffolding outdoors. No one had ever done this before. Usually people just set up horns outside and they sounded awful."[21]

Beginning in 1945 the Altec VOT speaker was offered in different sizes to accommodate varied movie house dimensions. The series included the A1, A2, A4, A5, A6, A7, and A8 speakers. The A2 and A4, which Hanley used most, were fairly common in early rock and roll sound reinforcement. Depending on the style or model number they could be giant in scale, some even towering over ten feet high, making them difficult to move. If you were lucky enough to acquire an A2 of this series, one had to be prepared to haul over 1,300 pounds of speaker. Terry remembers: "I used to go out and pick the speakers up with my friends. We carried them up stairways, and it was hard work. We did some serious lugging."[22]

According to Hanley: "We ended up with a lot of the Altec 210s and 410s because they were in the theaters. The A4s were at least 9x7x4 feet and weighed around 800 pounds. The bass cabinets alone were 8x3x4 feet. These were big and heavy and the A2s weighed even more!"[23]

The A4 speakers were equipped with two 515 low-frequency drivers, a ten-cell multi-cell horn, along with a 288 compression driver. They had side wings that were two feet long for increased low frequency horizontal directivity. Hanley often modified and removed the wings for concert events. The speakers provided fairly high sound pressure levels despite using the limited amplifier power available at that time. Hanley observes: "They were filled with tar to dampen them for resonance. The different cells aimed the sound in different areas and they were great for what we were using them for. They got 500 Hz of passive crossover which was fairly standard."[24]

By the 1960s, Altec's powerful movie theater speaker became the preeminent standard for concert sound reinforcement. If Hanley was to make an impact on the music scene it was necessary to stockpile these speakers, which he did. Early on he spent nearly every dollar earned buying as many as possible. Eventually the speakers took up almost every inch of his office space.

With ready access to this equipment, Hanley Sound was able to support multiple tours, venues, and festivals supplying ample volume to growing audiences. The consistent application of the VOT speaker emerged as his ideas on building systems of sound for large outdoor events began. Altec's Voice of the Theatre speakers were used for many of Hanley's early shows and their extensive application should be considered one of his early innovations.

Chapter 7

JAZZ!

THE BOSTON ARTS FESTIVAL

Besides the Bal-A-Roue, one of the first places Bill Hanley claims he saw and heard a high-quality sound system was at the Boston Arts Festival (BAF) held on Boston's Public Garden in 1957. According to a *New York Times* article, the BAF was "a program of art exhibits, opera, ballet, orchestral music, poetry, and drama."[1] Here Hanley absorbed sights and sounds as he walked the isles of green and white canvas canopies that accommodated the fine arts exhibition.

In the public garden there was a temporary outdoor cantilevered stage at the edge of the pond for performances, which were held each evening at 8:30 p.m. To support the concerts and recorded music, high-powered loudspeakers were strung up carrying the music throughout. Hanley remembers: "There were two sound systems there. One was a distributed system along the walkway where the art was displayed. The other was for the stage event that was designed by David Klepper."[2] Klepper, who had joined the local acoustical design firm Bolt, Beranek, and Newman (BBN) as a sound consultant in 1957, was an early inspiration for Hanley. The acoustical designs of Klepper at the BAF gave the young sound engineer ideas for building better sound systems: "They were mixing from a Scott amplifier in a room they built way in back of the audience. This of course was not an ideal position, I thought."[3] Klepper can see why Hanley made this observation:

> Although the rear of the audience was not ideal, it was better than at the side of the stage, which had been typical of theatre sound up to then. The amplifier was Scott, but the controls were custom-built by Lake Systems using Altec parts. Each mike had its own control and the controls were like a map of the stage, showing where each mike was located. This allowed the operator to have the best possible control.[4]

The sound was unlike anything Hanley had ever heard before; it was bigger and better than he ever imagined: "Klepper was using great big 400-pound loudspeakers up on top of the roof of the theater! It was great!"[5] According to Klepper, the firm was consciously trying to create high-fidelity sound in an outdoor setting: "We at BBN positioned these systems at the top center of the stage, trying to preserve directional realism. The amplified sound arrived at the listener's ear shortly after the live sound on the stage. Later, this concept was expanded to have left, center, and right channels, but this was not in the budget for the Boston Arts Festival."[6]

While Hanley was examining the Klepper system, he ran into Michael Wynne-Willson and Alex Troy, owners of Faircast Systems, a public address sound company located in Manchester-by-the-Sea. They were making announcements throughout the public gardens that day. Willson was something of a local entrepreneur who owned what he claimed was the only "Sponsored Sound System in the United States." Willson was an Englishman who settled in Hamilton, Massachusetts. His company offered a unique live, outdoor commercial advertising venture referred to as "sound casting." According to Hanley, of the two men, "Troy was more of the engineer and Willson facilitated the dog and pony show."[7] Described in a *Boston Globe* article, the Faircast System was a "sound and master of ceremonies service designed to meet the needs of those who operate and manage church, local and state fairs, conventions, and agricultural shows. Thus, because of the sponsors, there isn't any charge for the use of the system."[8]

Hanley noticed that Willson's air-conditioned, soundproof, mobile radio studio was affixed with a low-powered transmitter—a signal, as Hanley recalls, that could be picked up within a quarter of a mile of the broadcast. The unit had over thirty loudspeakers and was equipped with several microphones, perfectly sufficient for a sound engineer. It could broadcast recorded and taped music, news, and public address announcements. This impressed Hanley enormously.

Faircast Systems was one of the early sound companies to be hired for sound reinforcement at the 1957 Newport Jazz Festival; however, their approach proved to be unsuccessful. By now Hanley realized that the Jazz Festival needed help: "Faircast Systems was trying to make a buck selling airtime and playing background music. The trailer had a radio studio and his speakers were all around the area blasting advertising to the festival people. I had heard he got a job at Newport using old University loudspeakers and that wasn't good."[9]

Meanwhile, Hanley heard that Willson's company was going out of business. By the early 1960s Hanley had bought the trailer he was once so enamored by at the BAF. This multipurpose mobile trailer was an instrumental resource for Hanley, as it gave him a base of operation during his early Newport years. Later it fell victim to an act of arson.

For Hanley, breaking into sound was difficult. The type of service he was offering was new, and no one seemed to take the young engineer seriously. Convincing promoters, managers, and others that his innovative methods in sound reinforcement would be a better solution was a struggle for most of his career.

GEORGE WEIN'S STORYVILLE

Since its inception in 1954, the Newport Folk and Jazz Festivals in Newport, Rhode Island, have grown to monumental proportions. To this day they are considered among the most highly anticipated annual cultural musical events in the United States. Newport Festival impresario George Wein had his foot in the door more than a decade before visionary promoter Michael Lang conceived of the Woodstock Music and Arts Fair in 1969. Like Hanley in sound, Wein should be considered the "Father of the American Music Festival." As one author writes, Wein is "Short, stout and balding, a caricature of the cigar-chomping promoter, in fact he plays very much against that type (and doesn't smoke cigars)."[10] Wein—not only a consummate fan of jazz music—is acknowledged among his peers as an accomplished musician for his unique piano style.

More than sixty years ago, Wein took a chance at putting together a festival in Newport, and succeeded. Wein grew up in Brookline, Massachusetts. As a young man, he denounced his family business against his parents' wishes, to be closer to his musical ambition, jazz piano. Wein nurtured his passion into his thirties and eventually opened the well-known Boston jazz venue Storyville. In 1959 he married Joyce Alexander, a then rare interracial marriage. This long lasting union, of forty-six years, ended when Joyce passed away in 2005. Known to be an "outside of the box" type of individual, Wein is primarily responsible for the longevity and success of the Newport Festivals.

Before Hanley became involved with the Newport Jazz Festivals (NJF), you could find Wein at Storyville. It opened in late September 1950 and was located at the Copley Square Hotel on the corner of Exeter and Huntington. Lasting only around six weeks at the Copley location, in February 1951 Storyville relocated downstairs in the Hotel Buckminster at the corners of Beacon and Brookline, close to Fenway Park. Here Wein had a modest sound system that he had bought for $600. It consisted of a little amplifier with two speakers that hung next to the bandstand.

Faced with new obstacles, Storyville moved back to the Copley Square Hotel by 1953, then in 1959 to the Hotel Bradford on Tremont. By now Wein was busy with his Newport Folk and Jazz Festivals as well as the Castle Hill Concert Series in Ipswich. In 1961 Storyville closed. According to Wein, without the venue there would have never been a NJF: "I won't say that Newport killed Storyville . . . but there's no question that Newport hastened the process. My Club, which had served as a training ground, was withering on the vine; I decided to let it fade out."[11] After dissolving the club, Wein further established himself in Newport, embarking on a legacy in jazz that still survives: "So jazz moved south, down the coast."[12]

NEWPORT JAZZ: BEDEVILED WITH SOUND PROBLEMS

On a cold winter evening in 1953, descendants of the founders of the Lorillard Tobacco Company, and wealthy summer residents of Newport, Elaine and Louis Lorillard approached George Wein with an idea. While at Storyville, Lorillard convinced Wein to facilitate a jazz concert in the resort town. According to Wein, the Lorillards were "insistent on the idea," so he gave in.[13] Unsure about hosting an event at the proposed Newport Casino, Wein felt that Newport's cobblestoned streets and historic waterfront seemed "trapped in time."[14] However, the casino, which held one of the first US National Lawn Tennis Championships in 1881, had some potential. According to then Storyville and Newport publicist John Sdoucos: "I remember when Elaine and Louis Lorillard came to Storyville that year to talk to George about coming down to Newport. George said, 'You know I have been walking around with this idea for two years now.' George wanted to do an outdoor music festival with the popular music of the time, and that happened to be jazz."[15]

In his book *Myself Among Others: A Life in Music*, Wein reflects on the Newport Casino location, where the first festival was held in 1954: "I had seen my share of wealth and society, but nothing could have prepared me for this. The heyday of the Gilded Age may have faded, but Newport still seemed to glow with its aura. The entire resort community harkened back to the proud excesses of a bygone era. It was not a particularly jazzy place."[16]

With $20,000 of Lorillard financing and having no official festival rulebook to refer to, Wein set out into the uncharted territory of putting on a music festival. The inaugural event was to be set up on the lawn of the casino in an athletic complex and recreation center built in 1880 in the shadow of the Bellevue Avenue Historic District. The First Annual American Jazz Festival hosted a who's who of jazz musicians that included Billie Holiday and Dizzy Gillespie. Wein's legacy begins in Newport during the summer of 1954, on that mid-July weekend.

The two-day festival saw over 11,000 in attendance. Although its first year was a huge success, the sound was a big problem throughout the casino grounds. According to Wein: "Outdoor sound and lighting was a total mystery in 1954. We needed to project music with high fidelity to an open air audience in a large field. The equipment available in 1954 was far from adequate, but we made the best of what we could get."[17] For the event, Wein hired Hsio Wen Shih, an MIT-trained architect, to build the stage. According to Wein, Shih designed an "attractive sheltered stage constructed of thick heavyweight cardboard."[18]

Reporting for the *Providence Journal*, William Keogh declared Wein's festival a huge success: "To say the whole affair was a success is a considerable understatement. The first Newport Jazz Festival was a sensation, and possibly as much a surprise to its sponsors as to anyone else."[19] However, others gave it a lashing regarding its sound quality. Journalist Howard Taubman of the *New York Times* wrote that "there were some bugs in the festival, and that the loudspeaker system was disconcerting."[20] Hanley recalls Sid Stone Radio Labs of Boston, a purveyor of public address equipment at the time, supported the sound that first year.

Clearly, the technology to support live outdoor music was less than suitable for Wein's event. The *New York Times* continued in its less than favorable review: "The loudspeaker system will have to be improved tremendously so that everyone in the arena hears substantially the same thing. Sitting at the side last weekend, one heard a huge noise coming out of a speaker beamed directly at this section of the audience as well as the more remote sounds from the stage almost as an echo."[21]

Those who were in attendance in 1954 questioned whether a "jazz festival should be held outdoors at all." Some complained that the music became "diffused" making it difficult for performers like Ella Fitzgerald, Dizzy Gillespie, George Shearing, and Billie Holiday to make "contact with the listeners."[22] Problems like this plagued future Newport Jazz Festivals until Hanley Sound became involved in 1960.

Beyond anyone's expectations, the festival exploded in size and popularity, catching everyone off guard, especially the Newport Casino Board. When Wein began planning for its follow-up, he faced resistance from the board and the Newport City Council. Only weeks away, in June 1955, he settled for Freebody Park, an open-air municipal sports field located behind the casino grounds. "Freebody Park didn't have the cachet of the Newport Casino, but it was only a block away ... close enough for jazz,"[23] adds Wein.

Over 27,000 attended the 1955 Jazz Festival—but even though in a new location, it still suffered from sound problems. Hoping to avoid previous issues, Wein hired an outfit called Lang and Taylor out of Waltham. Henry C. Lang, a boutique hi-fi speaker and sound equipment designer, engineered the festival sound system that year. The loudspeakers for some of the workshops and receptions held at a large estate owned by the Lorillards (Belcourt) were a combination of University loudspeakers with amplifiers and testing equipment by H.H. Scott, out of Cambridge. According to the *New York Times*: "A band-shell was erected, Henry Lang, who is well known in hi-fi circles, was selected to take charge of the amplifying system. In passing, it might be noted that the acoustics tonight were excellent, far superior in fidelity to most comparable events. But apparently nobody has been able to lick the feed-back problem."[24]

Wein reflects on this time of unsuitable sound reinforcement in his book, where he references a *Newport Daily News* article describing the morning setup of the 1955 Jazz Festival. The newspaper analysis touts the new arrangement by Lang and Taylor for outdoor sound: "More than forty direct driver loudspeakers will be used in the high fidelity system. They will be driven by six 70-watt laboratory amplifiers giving a total 420-watts of audio output. The speakers are all banked on top of the upward-sloping shell roof. Engineers tested the system last night in a twenty-five mile per hour wind and got a uniform 8 decibel reading even in the most distant parts of the arena."[25]

According to Wein, the journalist's reflections were slightly fabricated: "It's funny to read this description in retrospect, since the sound that year was a catastrophe. But there was nothing funny about it at the time; it almost ruined the festival."[26] It would take several attempts by Wein to get Newport's sound right.

Intermittent rain dampened the 1956 Jazz Festival at Freebody Park. Yet as time passed, Wein's Newport Jazz Festival continued to see an influx of festivalgoers deemed too great for the small seaport resort town. Hanley recalls the film *High Society* about a jazz musician and a Newport socialite, a film that really seemed to help boost attendance at Newport in the coming years.

In 1957 Faircast Systems, the same company that Hanley saw at the Boston Arts Festival, took on responsibility for sound reinforcement at Newport and failed. Word was out now that the NJF was the anticipated event of the summer, yet sound quality was still an issue. Fred Flukes, a friend of Hanley's father who worked for the *Boston Globe*, knew Bill loved audio. One day he called Hanley's home and said he had read that the NJF was having sound problems. Seeing an opportunity, Hanley began to hunt down the right people in an attempt to get his foot firmly into Wein's door.

In 1957 Hanley ventured out to Wein's Storyville to see if the festival producer would give him a chance. Instead, he ran into his assistant, promotions and pressman Charlie Bourgeois. Bourgeois got his start at Storyville. Wein remembers: "Charlie's taste in music was as impeccable as his sartorial sense. Charlie Bourgeois has been a surrogate member of our family for more than fifty years."[27] Unimpressed with what Hanley had to offer, Bourgeois brushed him off, claiming they were "all set for sound." Bourgeois sent the young sound engineer home mildly discouraged. Hanley had to try and figure out another way.

After the sound disasters at Newport, sound engineer Myles Rosenthal from New York was hired for the job in 1958. Hiring him was a logical choice for

Wein since Rosenthal was already providing sound for large concerts at the New York's Forest Hills Tennis Stadium. How Wein heard of Rosenthal is unknown, but Hanley assumes he must have seen some of his work at that venue. According to Hanley: "Myles used good equipment. It was all the big theater stuff, JBL speakers and McIntosh amplifiers."[28]

In the early summer of 1958, Hanley made it down to the 14,000-seat tennis venue in Queens, curious to hear what Rosenthal was doing for sound reinforcement. During his visit Hanley learned that Rosenthal got into Newport via a bet: "Myles had a contract with two partners who were doing snow removal for Kennedy and LaGuardia airports. They were the promoters of the concert series. Myles had somehow proved that he could have high quality outdoor sound to George over a bet he made with him and the other promoters. The promoters thought you just couldn't have good sound outside."[29]

Hanley drove down to Freebody Park one day in early July 1958. Determined, he walked over to Rosenthal's mixing position, located stage left in a small room, and reintroduced himself. According to Hanley, Myles was actually a nice guy. But his mixing position was less than ideal and Hanley remembers it being difficult to see or hear properly out of the room's modest space that had only one glass window: "'Do you want to mix kid?' Rosenthal asks me. I emphatically said 'YES!' I was all over it!"[30] Hanley pronounces that from this day on, he was now "on the inside" and never looked back.

Occasionally Wein came in and helped mix from this location as well, seemingly only when Hanley was at the board. This was probably because he didn't trust the young sound engineer. "When I would fuck up, he charged into the room and yelled at me!"[31] laughs Hanley. It was this serendipitous opportunity that allowed Hanley to begin his long career of mixing sound at Newport, an effort that was initiated by simply volunteering his time and being persistent.

After the 1958 festival Hanley volunteered to do additional smaller gigs: "After George saw what I was doing he became impressed, he was really happy."[32] Then, in January 1959 Wein provided the sound engineer with another shot, this time located at the Bradford Hotel in Boston. This was one of the first independent jobs outside of the Newport Festival that Hanley did for the producer. It was the last location for Storyville, which closed in 1961. Here Wein continued to evaluate Hanleys skills as the venue's sound engineer. Hanley recalls, "While I was working and going back to school part time, George let me do a couple of gigs for him at the Bradford Roof. I wasn't there very long. I believe he was testing me out and it went well."[33]

Hanley's first event at the Bradford featured American jazz flautist Herbie Mann and piano legend George Shearing. Wein claimed it was the "biggest success ever to hit the Boston jazz scene."[34] As Hanley mixed the performance amid a full house, all ears were on the musicians.

In between Hanley's pursuit of the Newport Festivals, he ventured out to cultivate new work, and the West Coast seemed like a good start. Concert promoter and jazz radio broadcaster Jim Lyons was a natural choice. Lyons had launched his Monterey Jazz Festival in Monterey, California on October 3, 1958. In an attempt to build new relationships, Hanley flew out to speak to Lyons, only to find that he was too late. The bid for the 1959 festival had already been established by the time he got there. A sound engineer by the name of Jim Meagher of Meagher Electronics in Monterey provided sound reinforcement for that festival. There are leaders in the sound industry today who apprenticed under Meagher. Many of them claim that he was Hanley's West Coast parallel.

NEWPORT BECOMES A REALITY

George Wein brought sound engineer Myles Rosenthal back to the NJF in early July 1959. Hanley was invited, too, but this time with a little bit more responsibility. Still, it wasn't smooth sailing once he was allowed onto the festival grounds. Up until this time, his musical background was mostly in organ music. Nevertheless, although Hanley did not fully understand the complexity of jazz, he did realize more could be done with the sound at Freebody Park. Moreover, he noticed Rosenthal's lackadaisical and somewhat inadequate approach to sound reinforcement application.

Hanley could see that the sound was getting lost in Freebody Park's open-air field because Rosenthal had only four theater speaker systems set up in the audience. These were mounted on twelve-foot towers, with only two 200-watt McIntosh amplifiers, and two McIntosh mixers. Drawing from what he learned in Leo Beranek's book *Acoustics*, the young sound engineer could now analyze and give a professional opinion on what was good (or bad) sound application at Newport. His immediate thought was that things needed to change.

Forever socializing, Rosenthal often left Hanley alone for lengthy periods of time over the festival weekend. At the console, Hanley sensed that Rosenthal really didn't understand that much about the sound system or his role and responsibility as a sound engineer: "Myles played the social scene but I didn't mind at all. The recording companies were controlling everything and wanted the right microphones. One time they sent me six feeds to two McIntosh mixers, which I have never seen since. Then the next thing you know I am fully in charge of the sound system in this little room! This is where it all started."[35] Impressed, Wein asked Hanley to assist Rosenthal the following weekend with the inaugural 1959 Newport Folk Festival on July 11 at Freebody Park. He graciously obliged.

Later that summer, on August 21–23, Wein also produced the First Annual Boston Jazz Festival at Fenway Park. Holding a concert event at the baseball stadium posed many challenges for stage and speaker placement. According to Hanley, the stage was placed on the infield, between first and second bases, facing the grandstand on the first base side: "I heard Myles was having a difficult time with the acoustics bouncing all over and rolling back in on the grandstand. It was somewhat of a disaster."[36]

By the time the 1960 NJF was being planned, Hanley learned that Rosenthal was not coming back to Freebody Park. Having been involved with the festival for a few years now, Hanley knew Wein's working style and his ways with the production staff. Eventually Wein became convinced that Hanley could do a better job, and invited Hanley Sound in as the contracted sound company for the seventh annual 1960 festival.

On Saturday July 4, with Rosenthal out and Hanley officially in, his long-awaited solo performance in sound reinforcement was cut short by a riot. Over 12,000 angry college-age fans that had driven great distances to hear their favorite jazz performers were turned away at the gate of the sold-out festival. Tensions flared; bricks, rocks, and tree limbs were used as weapons, some of which smashed into the windows of people's summer homes.[37] Car windows were broken, some vehicles even overturned during the chaotic drunken student melee.

Local officials responded with high-pressure hoses and tear gas aimed at the furious mob. According to the *Chicago Tribune*: "The rioting began while the Saturday night performance was in progress before an audience of 15,000 in a walled, open air arena in the city's Freebody Park. Thousands of persons who had hoped to buy tickets had been turned away for lack of room."[38] As a result of this "jazz riot," Wein was met with resistance from the Newport City Council, forcing him to cancel the scheduled Sunday and Monday performances. The festival did not return to Freebody Park until 1962.

With the festival at Newport on hiatus, Hanley continued to support additional events for the festival producer in the following months. On August 26–27, 1960, he provided sound reinforcement for the Boston Jazz Festival at an East Coast theme park known as Pleasure Island located in Wakefield, Massachusetts. Performers including Nina Simone, Dave Brubeck, Art Blakey, and Duke Ellington entertained audiences. For the event, Hanley built and modified eight Rebel 5 speaker cabinets that were designed by hi-fi audio pioneer Paul Klipsch. The modifications were made in his friend Tom Hawko's basement using Tom's father's table saw. These speakers were the ones used at the local armory Hanley's father had introduced him to years earlier.

FRONT OF HOUSE

Even though Rosenthal had let Hanley mix, being off to the side and so far away from the stage didn't make sense. Hanley had made suggestions about working from the front of the house to Wein and Rosenthal, but it fell on deaf ears. According to Hanley: "Working from the front of house was a progressive idea at the time. But no one listened to me."[39] Nevertheless, Hanley could still see that Wein was genuinely concerned about the quality of sound at Newport.

Bob Jones, Newport's longtime technical director, reflects on Wein's attention to detail about sound quality: "George never let us sit around while someone was performing. We didn't have sound checks back then, but when people started playing we were instructed to listen and he would do the same. He would get off of the stage, go to the back of the facility and walk one end to the other, to make sure people could hear correctly."[40]

In order to properly check the sound quality and levels at Newport in 1959, Hanley had to physically leave the room he and Rosenthal worked in. The sound engineer knew that this position was wrong, claiming: "Trying to listen from inside a room was terrible! It didn't work."[41] Hanley realized that if he could mix near and/or in front of the stage, he could have better control of the sound: "It was probably at the Pleasure Island Jazz Festival where George first noticed me mix sound from a front of house (FOH) position. I was already doing this at events prior to this, but George wouldn't have it, he was still testing me. After he saw what I could do at Pleasure Island, it led to me getting the green light to move my mixing location from that small room to front of house. Eventually George fired all of the recording companies coming in and hired me to do it all from front of house."[42]

Mixing sound from FOH was a Hanley innovation. Sitting out in front of a stage in a venue and mixing was nearly unheard-of prior to the late 1950s, especially in a theater or big band ballroom setting. Even at venues like the Boston Garden and Madison Square Garden, sound was most commonly controlled from backstage. You often would see the person in charge of the sound running backstage making adjustments, then running back out front.

Without anyone in a FOH position there was minimal control over the sound situation, and Hanley realized this. Terry Hanley claims: "This was where the lighting equipment was, like the dimmers etc. In the old days the lighting person was backstage. The only people who were ever out front were the people running the spotlights."[43] Promoters and house managers, unwilling to take a financial hit, didn't want to lose paying customers; they prioritized making sure every seat was sold—leaving sound quality as an afterthought. Eventually Hanley turned this around by "demanding" to be out in front of the stage, now commonly known in the audio engineering world as FOH, or front of house positioning.

Despite Hanley's hard work at Newport during these early years, he claims that he was either unpaid, or worked for next to nothing. Yet the sound engineer didn't care, as the NJF gave him an opportunity to let his ideas develop. Hanley reflects that he didn't have much of a business sense at the time, and for him the NJF experience was invaluable: "I had a habit of either undercutting myself or often volunteering in order to get the job. What I was doing in sound reinforcement was such a new business that there was really no way to know how to measure the value of what I was doing. It was hard to know the worth of it. There really was no marketplace to build off of."[44]

Realizing a long-term permanent appointment at Newport was ideal, Hanley felt he had to further convince Wein that his sound systems would work well. If successful, it meant a steady and certain future as the premier sound company at one of the most important festivals in the United States. For the young sound engineer, his time spent with Wein and Rosenthal was experiential learning that he could apply to future events.

METROPOLITAN BOSTON ARTS CENTER

In early 1959 Hanley read that the newly designed Metropolitan Boston Arts Center Theater was being constructed on the Charles River near Soldiers Field Road. Designed by MIT graduate Walter W. Bird, the four-story, steel-framed roof structure was the first of its kind in the country. The impressive roof of the 2,000-seat theater was constructed of an "air inflated detachable vinyl coated nylon," spreading over 145 ft. in diameter.[45] Designed with a slant for better acoustics, the dome looked more like a "tilted cookie" rather than a traditional theater.

A 1959 *Boston Globe* article claims: "The white nylon roof was manufactured by Birdair of Buffalo, New York, in two layers, a top and a bottom. Each layer consists of 108 panels seamed together. The two layers are connected with an outside zipper, which goes around the perimeter. When the roof was raised, air was pumped into it through a galvanized metal pipe two feet in diameter."[46] The roof was kept at a constant pressure, and inflated all the time. Although strong enough to withstand seventy-mile-per-hour winds, the venue proved to be acoustically inadequate.

As the structure was being built, theater management had been promoting a production of three "Shakespearean in the round" plays led by Sir John Gielgud and Margaret Leighton.[47] All involved were anticipating a huge success in this new venue. But when the productions began, they found that the sound became trapped inside the inflated roof. According to Hanley, "it bounced all over the place." With no sound baffling, actors and audience members heard a torrent of echoes during performances; the quality of the sound was intolerable.

By the end of 1960, the theater's operating manager William Hunt sought emergency assistance to help with the sound issues. Building a local reputation, Hanley who was only twenty-three years old at the time, was called in to help: "The sound was really bad in that tent structure. I recall building eight Rebel speakers, running them 70-volt which worked great."[48] Phil and Ed Robinson remember this being a big job for Hanley Sound at the time. Phil recalls: "I remember when they asked us to come in. They were having problems. This was a massive tent structure with a balloon roof. The reverberations were terrible."[49] According to Ed: "I used to get up on stage before the show and test the sound system for Bill. There were a lot of physical problems with the facility."[50]

While Phil was setting up equipment, he remembers, Hanley was constantly trying to balance the sound in the difficult venue: "Bill ended up using this frequency-shifting device to help prevent some of the feedback. This worked really well. Once he figured it out you could hear clearly everywhere in that theater. It was incredible and they were really happy."[51]

Hanley amusedly remembers a rather large RCA 77DX microphone hanging by a wire from a steel cable that nearly struck Sir John Gielgud's head: "He ended up grabbing it and nonchalantly incorporating it into his performance! This was for his production of *Much Ado About Nothing*."[52] According to Ed, Hanley got along well with the actors: "After the show ended I used to drive his car home while he went off to party with the cast!"[53]

Hanley was only involved with a few productions at the theater. Later his mentor Leo Beranek, and his firm Bolt, Beranek, and Newman were invited to analyze the facility's unique acoustical problems. A contemporaneous *Boston Globe* article states: "The firm of acoustical engineers is speaking of the scheduled improvements in amplification at the Arts Center Theater." According to the members of the group, some of their proposed recommendations were two simple innovations: "First, the actors will be near the microphones, and there will be a grid of microphones located 10 ft. above the players. The mikes will be visible, but that cannot be avoided."[54]

Hanley adds that one logical solution was to make the theater a permanent structure rather than a summer open-air venue, an idea that never came to fruition. At a public hearing, the Metropolitan District Commission was advised to improve the theater because of its bevy of problems, "So that people can listen in comfort and actors can act in comfort."[55] After only one season, the short-lived theater met its fate and was dismantled. Unable to capture an audience, the project was unsuccessful and ran out of money.

CASTLE HILL

In 1961 the fate of the NJF was uncertain. The riot that occurred the year before had left the festival in debt for over $110,000. Unable to secure a license from the Newport City Council, and no backers for its eighth season, George Wein and Newport Festival president and co-founder Louis Lorillard focused their efforts on a summer concert program at the Castle Hill Mansion in Ipswich, Massachusetts.

Wein was optimistic, since annual music events had been held there already. In a May 6, 1961, *Billboard* article, Wein states: "We are determined that Castle Hill shall become America's summer center for the performing arts. We intend to make it a showcase for the greatest talents in classical music and drama, as well as attracting the great names in jazz, folk and gospel music."[56] In the summer of 1961 Wein produced a series of folk, jazz, classical, and pop concerts on the rolling lawn at Castle Hill, requiring Hanley's assistance.

The twentieth-century country Versailles-esque Crane estate overlooked a salt marsh bordering Ipswich Bay. The location provided a picturesque venue for concert attendees, replete with beautiful gardens and views of sand dunes. The garden area of the venue could seat 1,700, while the amphitheater could comfortably hold 4,000. The mansion was the former summer home of the early twentieth-century plumbing magnate Richard T. Crane Jr.

The popular folk group the Kingston Trio was scheduled to perform on August 4–5. According to Hanley, an event like this was another opportunity to prove himself to Wein as a skilled sound engineer. Arriving at dawn on Friday August 4, 1961, the Hanley and Robinson brothers set up a sound system at the base of the sloping lawn overlooking the bay.

By now, Hanley realized that raising his speakers high in the air provided better results. In order to achieve this, he used construction grade scaffolding, with wooden planks resting on metal poles. Construction workers who build, repair, or clean buildings commonly use this type of structure. However, Hanley's deployment of scaffolding in music production may be a first. Once erected, the structure was purposefully elevated slightly higher than it needed to be so a block and tackle could be affixed to a board across the top. The block and tackle mechanism consisted of ropes and one or more pulley blocks. The challenge was getting the speakers high enough without the scaffolding tipping over.

Once the speaker box was strapped in, Hanley and his crew pulled and hoisted the 1,300-pound speakers to the top. Wooden planks were then placed under the dangling speaker across the midsection of the scaffolding. Once in place, the speaker was then precariously lowered onto the resting planks. There were concerns about the structures' safety, but according to Phil it was because no one had ever seen speakers on scaffolding before: "I recall going up in a rented truck with the scaffolding and sound equipment. The speakers were miserable to get up. They were very bulky and very heavy. It took four or five people to get them up onto that scaffolding. Raising speakers up in the air like this was something new."[57]

The equipment for the event consisted of two Altec 210 speaker cabinets, two McIntosh 275 power amplifiers, two 105 multi-cell horns, and two RCA BN6 mixers. Terry was preparing the stage area and other equipment setup: "I remember preparing for the Kingston Trio and it was blazing hot. After getting the speakers up on scaffolding I began setting the mixers up and the microphones. I had a pretty good ear so Bill asked me to mix the event. I recall George Wein testing us for future gigs."[58]

The 1961 concert series was a huge success, with crowds of six to seven thousand pouring into the area. The Kingston Trio alone brought in 3,500 attendees and boasted the largest crowd in history on the grounds up until that time. Coincidentally, the group broke up a week later, canceling a performance on Cape Cod. According to those who were in attendance on that hot August weekend in 1961, the quality of Hanley's sound system was impeccable.

Chapter 8

NEWPORT NEEDS A HOME

Throughout Hanley's career he always had to persuade prospective clients of a better solution for sound reinforcement, and festival producer George Wein was no different. The young sound engineer needed to be persistent if he was to break into this newly established field. But his convictions about quality sound, which had run deep ever since he was a young boy, would soon begin paying off.

By 1960 Hanley was entirely on his own as the established sound company at Newport. Jazz was Hanley's introduction into the application of live concert and festival sound, and Wein's festivals at Freebody Park proved to be a testing ground for his ideas. These formative years of the Newport Jazz and Folk Festivals are responsible for creating the blueprint of his developing festival business during the 1960s and 1970s.

With the difficulties Wein had faced during the 1960 NJF, the producer decided to skip the 1961 season altogether. However, in 1961 promoters Sid Bernstein and John Drew presented "Music at Newport" at the open-air field on June 30–July 3. The production was a financial disaster. Wein resurrected the NJF in 1962 at Freebody Park once again.

Both a blessing and a curse, the 1962–64 Newport Festivals were growing in size and popularity but the surrounding community was becoming increasingly frustrated. The volume levels, the influx of people, and automobile congestion generated many complaints; the need to relocate became evident. By 1965 the Jazz Festival would be moved to a new location accompanied by better sound, a change in staff, and a new stage.

ASTRA ARTS LIMITED AND THE WENGER WAGON

In 1959 Newport Festival board member and music manager Albert Grossman had introduced lighting designer Chip Monck to the Newport production staff. Wein was impressed. After several seasons of working the Jazz and Folk

Festivals, Monck offered to build Wein a new, more advanced stage for the upcoming 1964 season. With a torrent of complaints coming in from area homeowners and businesses, Wein thought it was a good idea.

Hanley recalls that when he assumed the role as sound engineer at Newport in 1960, "A modest thirty-foot stage was positioned at the end of Freebody Park with its back to the residential area near the casino tennis courts."[1] Up until then the stage was built by a contractor. During the rest of the year it was conveniently stored under the concrete grandstand of the facility. According to Hanley: "Before Chip came on the Newport scene, Aloysius Petruccelli (Al Pet) handled lighting and stage management. He was a heavy in the Boston theater and production scene."[2] Monck recalls, "The previous stage was inadequate," claiming his newly proposed stage was unique because "it allowed us to put lights exactly where we wanted them; the former did not."[3]

Monck met sound engineer Bob Kiernan in March 1964 at the New York venue Basin Street East. Monck asked Kiernan to help with the production. During the following months Monck, Kiernan, and Izzy Siedman, a friend of Wein, designed and built the stage at a location in upstate New York. According to Kiernan, Monck was able to persuade the festival producer that a new and better stage was needed: "George paid for it and we built the stage in the summer with Izzy at his father-in-law's farm in upstate New York."[4]

According to Hanley, Chip's new design looked more like an "Erector Set" complete with sound and lights: "It was made up of many different pieces. As I recall it took a lot of time to put together."[5] Kiernan reflects that the construct was portable and durable: "It was built out of aluminum I-beams and structural extruded aluminum. It was unique because it did not involve welding, we could just bolt everything together. I would not call it an Erector Set because Chip put in a lot of time in planning it. At the time there was not much going on in the industry with regard to outdoor stages."[6]

Chip's proposal caught Hanley's immediate attention and he thought that Kiernan's company Creative Theatrical Services, known as Kiernan Sound, was joining Monck's business Chip Stage Lighting. The new and brief partnership resulted in Astra Arts Limited, which included Siedman. Hanley believed the lighting designer was offering Wein a complete concert production with Kiernan replacing his services and this worried him: "Up until this time, Chip was lighting and I was sound. It felt like he was trying to sell George on a complete package of stage, light, and sound. It could have pushed me out of Newport entirely."[7] A merger like this could have meant the end of Hanley Sound at Newport.

Monck recalls this differently, claiming, "Astra Arts Limited might have come out of a loud drunken conversation between Kiernan and me."[8] Summing it up more or less to a partnership founded on loose promises: "I don't recall how it came to be? Kiernan might have come in as a consultant, but I really had no intentions to do anything with audio or sound whatsoever. But we did build a new stage. Izzy was a friend of Wein's and a good builder. He and I built the stage at his father-in-law's New Paltz property and Kiernan was there."[9]

Worried and feeling threatened, Hanley was motivated to look for work outside of New England. "When Chip went ahead and built this Erector Set–style stage in 1964 at Freebody Park for George with Kiernan and not me, I realized it might be a good idea to spread my wings."[10]

The sound engineer knew he needed to get established in New York City as soon as possible, and by late 1964 he did. That year he purchased his first fully portable stage, the Wenger Wagon, referred to as the Showmobile, Mobile Stage and Canopy. This allowed him the opportunity to sell a nearly complete concert production package for his clients. Kiernan recalls: "I think Hanley's gut instinct was good after seeing what Chip and I were attempting to do for George. This got him in a direction that was smart for the time."[11]

In 1964 the Wenger Corporation was marketing its diverse music product line through its catalogs and direct mail. The "Showmobile" caught Hanley's eye as he thumbed through the catalog. For around $10,000 Hanley could offer additional production value to his clients. Not long after, Hanley made his way out to the company's Owatonna, Minnesota, headquarters to take a closer look. There he met with President Harry Wenger and his son Jerry to discuss their product. Jerry Wenger recalls when Hanley came to meet them in 1964:

I remember when Hanley came in and bought a Wagon. He explained to my dad and I that he was doing outdoor shows. We were glad to sell it to him. Hanley was probably one of the first guys to use our product with a complete sound system. Eventually we had about a dozen or so Hanley types that owned our Showmobile units. These were guys doing outdoor events. Soon after Stan Miller of Stanal Sound from Kearney, Nebraska, was using our equipment as well. We probably should have been more focused about guys like Hanley and Miller and had a product specifically tailored for their market. But we decided to not go in that direction, and focused elsewhere. We had a product that worked, but we never controlled that marketplace. They controlled ours. Hanley knew what his expertise was. He took our product and moved in that direction.[12]

In the mid-1950s Harry Wenger saw a need for an all-in-one roadworthy mobile portable stage after seeing many bands in the Midwest performing on hay wagons. His son Jerry recalls the impetus for his father's unique design:

In 1958 or '59 my father was so sure that everyone needed a wagon to perform on outside, that he built one of his first outdoor pre-Showmobile mobile stages. These were the very first ones ever built. In the Midwest everybody had hay wagons pulled by either a tractor or horses and bands would play on these. They were flat with a wooden bed on it, had four wheels, and usually loose or bales of hay along the front. If you had a concert in town, someone would say "Hey is your dad's hay wagon free? Could you bring it down here?" That's kind of how it was, and my dad felt that there was a need for one of these. So he got into building them.

The first one he built was a hay wagon with an improved deck on it. By the early 1960s my father began working with an industrial designer and that's when the Showmobile Mobile Stage was created. It was a total redesign, so you had the side panels that opened up, different types of floor systems, etc. We were building these and they were completed by 1963–64. We found out quickly that the people who bought them weren't the local high schools bands. They were local communities, counties, or cities. They used them for a variety of things like fairgrounds, town events, store openings, and political events. They were easy to use.[13]

Harry Wenger was a lot like Hanley. In 1946 Wenger, known as the "Music Man," was a local high school band director in Owatonna. Wenger was a gearhead and dedicated his inventiveness to developing music products that helped regional organizations, music associations, and students in school music programs across the nation.

The one thing that bothered Wenger was a lack of quality music equipment. So he set out on a course to make it better, and did. Today the company offers over 600 products all over the world serving the performing arts, production, broadcast, and entertainment marketplaces. Jerry Wenger, who took over his father's multi-million-dollar company in 1970 claims: "Like Hanley, my father was a tinkerer, and innovator of many things. If my father needed something, he invented and built it. He began our company in his basement, just as Hanley did in his."[14]

Even though Monck's Newport stage was successfully constructed for the 1964 season, Wein remained loyal to Hanley. The festival producer continued the engineer's contract for sound reinforcement services until 1967. According to Kiernan: "I never did the sound that year when Chip and I got together. There were really no sound companies that I was aware of. However, Hanley was already around."[15]

The short-lived Astra Arts Limited company trucked the stage in from upstate New York for the 1964 Jazz and Folk Festivals. By 1965 Wein moved his event to Festival Field, where Monck's design was used one last time. According to Monck: "It was used at Newport for two seasons and after that in August of 1965 it was moved to Wein's Down Beat Jazz Festival at Soldiers Field in Chicago, Illinois. George found the stage set-up not fast enough and a little too difficult to construct, so he decided to give it to the Smithsonian for a tax credit."[16]

Monck's stage saw many powerful performances in its lifetime, including Bob Dylan's epoch-making 1965 electric performance at Festival Field. By 1966 a new and permanent stage was contracted for the venue. Monck would finally make his exit after several seasons with Wein's production, claiming, "George was watching his pennies, and I had other things to explore."[17]

According to Hanley: "George had hired an architect by the name of Russell Brown to design the new stage at Festival Field. It was made of concrete and steel and it was massive! From what I recall it was over 100 ft. long."[18] Wein's new setup had an enormous backstage area that included dressing rooms, showers, lounges, and restrooms. Wein recalls: "It was hard to believe that the space had been an open field just six weeks prior. There were 200 stage lamps connected by nearly four hundred miles of electrical cables, and 16,000 ft. of underground sewer lines."[19] It took over 500 workers to finish construction at the site in time for the July event.

As leaders in the new and developing field of live music production, Monck and Hanley continued to run into each other throughout their careers. They were not the closest of friends in the early days, since they were both finding their own way. "Hanley and I never really interacted at Newport. I had far too much to do with my work. We really did not fully get to know each other until much later in our careers," says Monck.[20]

Later, Hanley and Monck would meet at various pop festivals (and venues), including Woodstock. It was Wein's gut instinct to put these two talented people together: "Chip Monck, and Bill Hanley, who handled our lighting and sound, respectively, would later form the nucleus of the technical team at Woodstock."[21]

The festival impresario had the intuitiveness to construct a team of talented technicians to facilitate his productions, with Hanley and Monck at the heart of them. As one article states: "Mr. Hanley is one of several young people working for Mr. Wein. 'Chip' Monck the lighting man, is another. This as Wein claims at the time 'reflects his faith in young people. They won't let you down if they have anything to offer.'"[22]

FROM FREEBODY PARK TO FESTIVAL FIELD

Over time George Wein realized that Freebody Park was becoming more of a problem than it was worth. In his book, the producer reflects on the issue just after the closing of the 1964 Newport Folk Festival at the open air festival site: "It's hard to imagine we kept the festival operating at all, especially with the unprecedented influx of 70,000 people to Newport. The town was saturated with kids who had no place to stay. Congestion in the downtown area was omnipresent; all business in Newport essentially came to a halt."[23] With pressure and backlash from the Newport City Council, Wein set his sight on a new home for his growing venture.

Still, Wein looks back with fondness for his time at Freebody Park as cited in a book called *Backstory in Blue: Ellington at Newport '56*: "We had the greatest years in jazz in '62, '63, and '64."[24] The 1964 Jazz and Folk Festivals marked Hanley's final year at the site and he didn't mind. The sound engineer was becoming frustrated with the park and could tell that the new location would allow him to further expand his ideas: "Festival Field was a much more suitable location, unlike Freebody Park which was in a residential area."[25]

According to Wein, he christened the new site Festival Field days before the 1965 Jazz Festival.[26] Festival Field was approximately thirty-five acres in the northern area of Newport, near the State Highway, city dump, and sewage plant. "The new site on Cornell Highway, which later became known as Festival Field, is where the council approved the 1965 Jazz and Folk Festivals,"[27] adds Wein. This new location could seat around 12–15,000 and had enough room to park 2,500 cars.

Hanley was in full agreement as Wein proposed a permanent "shell-like performance area" for the outdoor venue. Complete with a raked roof and 70 ft. stage, the construction would be finished before the 1966 season. The position of the new stage, according to Wein, "Faced north to project the sound toward the local naval base property." He claimed the sound system would not be projecting sound "toward any nearby homes" this time around.[28] Wein had taken a ten-year lease on the new location.

Although the 1965 NJF was rain-soaked, it was still a success, and for the 15,000 who were in attendance, most if not all were elated. Hanley was busy providing sound for more than 120 musicians in over "seven beautifully performed programmed sessions." An unparalleled lineup of jazz icons included the Duke Ellington Orchestra, Count Basie and Quincy Jones, Dave Brubeck, Frank Sinatra, and Buddy Rich, who was said to have "drummed up a beautiful rain shower." A 1965 *Billboard* review of the first Jazz Festival at Festival Field, hailed the work of impresario Wein, Hanley, and Monck. The writer urged the public to make reservations for the festival for the following year:

> The crowd of 15,000 sat in the downpour and cheered Buddy on his unbelievable effort. For an event like this, which is staged outdoors, it is necessary to make elaborate plans for sound and lighting. Chip Stage Lighting of New

York created the very special lighting moods, and Hanley's Sound Equipment Company of Medford, Mass., projected the musical nuances in a most effective manner. The recording crew coordinated their sound with the public sound system and did a very fine job.[29]

For this festival, all of the recordings were executed and coordinated by Hanley Sound with the exception of Frank Sinatra, who performed on July 4. Many of the purists felt that booking Sinatra was in poor taste to the jazz tradition. A huge name by then, the singer arrived at the festival site in rock star fashion, by helicopter. Hanley and others claim the crooner's 168-foot yacht *Southern Breeze* sat moored in Newport Harbor with nineteen-year-old actress Mia Farrow onboard. A recording crew from Columbia Records attempted to record the event. Hanley recalls: "Sinatra performed with Count Basie and had his own sound people come in and take over the recording. Frank was parked in the harbor on his yacht and I never spoke to him about it."[30]

Wein had around $300,000 to invest on improvements for Festival Field that included sound reinforcement. In a 1966 *Boston Globe* article, Wein emphasizes the need for quality sound at his festival, not just for audiences but for musicians as well: "He is already proud of his sound amplifying system, the work of Bill Hanley of Medford. It is the need for amplification, Wein explains, that dictated the choice of concert as opposed to staged performances. It is difficult to plan a [sound] system that works not only for the audience but also enables the performers to hear each other."[31]

Wein's festivals were seasonal, making the winter months much slower at Hanley Sound. Although the company was growing, there were still many accounts that owed him money. On February 1, 1967, Hanley Sound employee Ray Fournier, was interviewed about the company finances. Fournier relates that Hanley was out of state and would be for several days, negotiating contracts. He claimed that sales had "expanded and as a result the sound company was becoming more diversified" in its offering of sound equipment.

In the financial report for 1966, the company had four employees, its net worth was estimated at $40,000, assets at $65,000, and sales, around $75,000.

Over the years, the company has devoted most of its investment, and profits realized in fixed assets like real estate, the latter has been renovated to a considerable extent during 1964 and 1965. Also, the company has been owed considerable amounts of monies beyond the stipulated payment date, as a result, this has effected the manner in which the company has been reported in meeting its obligations which as noted by above suppliers has been predominately slow.[32]

A lot had occurred during Hanley's tenure at Newport. In many ways, production there was the sound company's lifeblood. From 1960–67, season after season Hanley and his crew hauled the mobile studio to Freebody Park and Festival Field. As one journalist claims: "His office is a trailer fashioned into an actual radio station except for a transmitter. The trailer served as his base of operations to provide the audio for student record hops, and the Newport Jazz Festival for the past three years."[33]

During this time, there were no other reliable choices for sound reinforcement for live events of this size. If any, they mostly provided public address that was by no means suitable for the world-class music at Newport. As a result, Hanley Sound became the premier sound company for not only the Jazz and Folk Festivals but for other jazz festivals in Chicago, New York, and New Orleans. Wein's focus and investment in quality sound for his productions was strong, and Hanley was evidence of this. According to Wein, Hanley's significance as one of the firsts in the industry was indisputable: "Outdoor sound was in a very pioneering stage and the most adventurous guy I knew with sound was Bill Hanley. He understood what we were looking for and he gave us the best he had. It was really the beginning of it all because Hanley was the pioneer. You couldn't pick anybody else; there wasn't anyone around other than some professional sound companies that really weren't experimenting with outdoor sound."[34]

Hanley's last summer at Newport was in 1967. A final invoice was issued for $4,500 on July 16 from the Hanley Sound Equipment Company to the Newport Folk Festival for "Sound Reinforcement and Engineering." After its tenure at Festival Field, the Newport Festivals finally settled into their current location at Fort Adams State Park beginning in 1981. George Wein hired Hanley off and on for various touring jazz festival productions throughout the country well into the mid-1970s.

To this day, Hanley and Wein remain friends and look back on these years fondly. Hanley's invaluable experiences at Newport fostered many of his future innovations. Wein's festivals provided the funding, ideas, reasoning, and impetus for Hanley's career to move beyond jazz, to folk, and then on to the hair-raising sound-pressure levels of rock music.

As African American writer/jazz critic Albert Murray claims about jazz: "When you see a jazz musician playing, you are looking at a pioneer, you are looking at an explorer, you are looking at an experimenter, and a scientist. You are looking at all those things, because it is the creative process incarnate."[35] I suspect that Bill Hanley encompassed all of these ideals as well, but on the technical end of the musical spectrum.

Chapter 9

ORCHESTRA SOUND

THE METROPOLITAN OPERA

Large instrumental ensembles were a challenge for Hanley and his crew. Rock, jazz, and folk were at the forefront of his craft, but occasionally his work included orchestral sound. Symphonic orchestras almost always include large numbers of musicians, all of whom need to be miked and balanced. Wein and Hanley faced sound issues during a four-night production of the Metropolitan Opera at Newport. The event was part of a production for the Newport Opera Festival held on July 12, 1966.

Operatic sound was a new experience for the festival producer, who admitted: "It shows how little I really knew about opera at the time."[1] Most opera purists wanted a more natural acoustical sound, something Hanley and Wein quickly realized when they were met with a frustrated response from the audience. Wein recalls: "When the first movement of *La Boheme* began, I walked over to Bill. Bring out that chorus, I advised him. He amplified the chorus, and the heavens just exploded with these voices. Within a matter of minutes, opera lovers from the audience and the Met staff came storming back to us, absolutely furious."[2] Once the sound engineer got his bearings, his powerful system handled the sounds of the Metropolitan Opera fairly well.

It was not unlike Wein to occasionally blow up at Hanley if the sound wasn't right, or if there were any technical difficulties. In 1973 at a Louis Armstrong celebration, the producer was freaking out because Hanley's system let off an extremely loud squeal. Hanley recalls:

> I was doing Satchmo's 73rd birthday party at the Singer Bowl in Queens, NY. It was a rush job, and I had just built a new console, checked it out on the bench, and it looked OK. In the middle of the introductions, all of a sudden, it squealed like crazy. Here we are, in front of 20,000 people, and Wein is on stage, screaming at me, 'Don't you know what you're doing after all these years?!'"[3]

As the story goes, Wein was screaming at the sound engineer in front of the crew and surrounding audience. A nervous Hanley could not figure out if the problem was coming from the console: "It sounded like it was going into oscillation. It wouldn't snap on, it kind of rose up in level. Here I am, banging on the console!"[4] Eventually the problem was solved when Hanley noticed an individual inconspicuously standing by the mike box. According to Hanley it was someone from the recording company. "Here was this jerk . . . plugging his signal generator into the mike inputs. Luckily, someone caught him. Was I embarrassed!"[5] Hanley adds that the venue's unique space made the sound more intense: "The Singer Bowl was this concrete rectangular set of stands, with no roof or anything. If you slapped your hands it lasted, and you heard it for nine seconds."[6]

ARTHUR FIEDLER

Throughout his career Hanley and his crew of engineers had numerous conflicts with orchestra conductors like Arthur Fiedler, who had a fifty-year run with the Boston Pops Symphony Orchestra. The story is that Fiedler was adamantly against having his orchestra miked, not fully understanding Hanley's concept of close miking as many instruments as possible. According to Harold Cohen, Fiedler did not like sound people at all: "If you are not close miking everything then what the conductor is doing is really not being mixed effectively. If you put a proper mix on the instruments you have more control. Fiedler didn't like us intruding on his work."[7] According to Hanley, Fiedler used to get mad at him because he wanted to put up so many microphones, resulting in many arguments.

Hanley claims that he needed as much direct sound as he could get, adding reverb artificially. That way, everyone would hear the entire ensemble. "When you have a solo flute playing, you drop 30dB in level, and the people 300 feet away are now in the ambient noise. I used to think that was poor. The conductors at commencements asked for one mike for a thirty-five- or forty-piece orchestra, but now they often go along with multiple-microphone techniques."[8]

Once Fiedler realized the control Hanley had over the sound from the stage, he relented. Hanley recalls, "It was very difficult to try to convince him, but he realized that the outcome was much better."[9] The sound engineer's technique of using multiple microphones came from working with some of the big bands at Newport. It was right around his NJF, Pleasure Island, and Bradford Hotel experiences when he began using twelve to fifteen microphones for the larger multiple-pieced acts. This was an approach that was unusual and unheard-of during the early to mid-1960s.

Working with conductors and bandleaders like Count Basie and Duke Ellington wasn't always easy either. In 1963–64 both bandleaders were reluctant to use Hanley's onstage microphone arrangement.

Once George pushed out all of the recording companies and I was in charge of it, I got everybody close miked. One time I put three microphones on a grand piano and we were ready to go. I had five or six microphones set up in the front row, then the trombone row, the trumpet row and then the drums. Then they start playing and I can't hear the piano. I said "What the hell is going on?" They had their band boys pull the microphones off of their piano and you couldn't hear a bloody thing! They didn't understand what I was trying to do. As soon as I would turn my back, all of a sudden I was no longer picking up their instrument.[10]

It seems that the jazz bandleaders naively thought that Hanley's microphones would override their principal performers. Still baffled by the occurrence, Hanley reflects: "They didn't understand that when you are in a club this was fine. But when you are out in the open air playing in front of ten to twenty thousand people with no microphone, you are a dead issue. I had to really force the issue!"[11] Hanley never had the opportunity to explain the technical issues with the bandleaders, claiming, "they were often scooted away, at which point someone else told them."[12]

THREE RIVERS NEW ORLEANS MUSIC FESTIVAL

On August 24, 1970, trumpeter and bandleader Al Hirt performed at Pittsburgh's first annual Three Rivers New Orleans Music Festival at Three Rivers Stadium. Despite low audience numbers (about 7,000 in attendance), and impending rain, the *Pittsburgh Press* still considered this the best "jazz show ever staged outside of Newport."[13] Rain did not halt the production and the sound system was noticeably great that evening. Lead sound engineer Sam Boroda was in charge. According to reviewer Lenny Litman, the sound system was tops: "It's impossible to say too much about the sound. The full-bodied quality came from a $150,000 set-up brought in by Hanley Sound of Boston, the firm, which provided the same system for the Woodstock Festival. The company sent in four engineers and two technicians with fifteen microphones and fifty speakers."[14] The Hanley crew arranged the sound system around home plate. Hanley claims that providing sound for a symphony orchestra at a sports stadium, especially when there are empty seats and no audience to absorb the sound, often posed acoustical challenges.

THE PITTSBURGH SYMPHONY AND MADISON SQUARE GARDEN

Although Hanley is not best known for his orchestral work, reviews of the Pittsburgh Symphony and a performance at Madison Square Garden establish his company as having expertise in this area at an early time. In 1971, on a June night at Three Rivers Stadium, during a Pittsburgh Symphony concert, Hanley and Boroda were preparing the sound system for some very important orchestral work featuring conductor Donald Johanos and American pianist Van Cliburn. According to Boroda, "Bill left me in charge and I set up and engineered the entire concert."[15]

As one reviewer reflects: "A special sound system was installed at the stadium so that every listener will get the full vibrant musical message of the evening."[16] The result of Hanley's experiences working with orchestras and symphonies paid off as the review of the event continued its high praise: "Regarding the complex sound system, anyone who can shade a Van Cliburn pianissimo and the soundtrack of a war through the same speaker system—and make the effect natural in the acoustics of an outdoor, cavernous sports palace—is a genius."[17] Hanley and his team's work pleased thousands in attendance that evening in Pittsburgh.

In December 1971 president of Columbia Records Clive Davis hosted a black-tie concert event called Columbia's Festival of Stars at Madison Square Garden, which Boroda mixed. Amid a crowd of 18,000, artists such as Percy Faith, Peter Nero, Vikki Carr, and Johnny Mathis performed on a "circular, raised stage located smack-dab-in-the middle of the Garden Floor . . . with some sixty musicians."[18] The lights were dimmed, as the stage revolved making a full rotation every seven minutes.

According to *Cash Box*, the artists put the orchestra through a "hoedown,"[19] adding that Carr was especially magnificent that evening: "The feminine change-of-pace was a delight, and Vikki's low-cut brown gown drew whistles. But it was her voice that drew the applause. Vikki belts with the best of the belters and can tear your heart out. . . ."[20] Like his boss, Boroda recalls classical musicians pushing the microphones directed at them out of the away: "During the Vikki Carr performance at Madison Square Garden, a violinist pushed her microphone to the side! I did not care. I walked right up on stage and pushed it back and asked her not to do it again!"[21] Even so, the sound that evening was exemplary. *Cash Box* adds: "The Hanley Sound System was magnificent. So clear was the sound that one could even pick up an occasional imbalance in the orchestra during a number or two."[22]

THE CLEVELAND ORCHESTRA

Hanley shipped $90,000 worth of sound equipment to Cleveland Stadium for a rehearsal with sixteen members of the Cleveland Orchestra and resident conductor Louis Lane in May 1972. He eventually used a whopping $250,000 worth of equipment brought in from Medford for the performance. An article in the *Cleveland Plain Dealer* describes Hanley's impressive orchestral setup, as well as the acoustical challenges he faced while providing sound in an outdoor orchestral setting: "The rehearsal was for the benefit of sound engineer Bill Hanley, who admittedly faced a difficult task. The Stadium's acoustics are the worst in the league."[23]

Similar to working with Arthur Fiedler, Hanley was used to working with worried conductors. Yet Lane seemed to be pleased with the sound engineer's work and overall effort. According to journalist Dan Coughlin, "Conductors are finicky people, but Louis Lane seemed satisfied with the reproduction." Citing Hanley's work, the fastidious conductor claimed: "This isn't a concert hall. You can't have the intimacy of a concert hall with 80,000 people. But the sound system will reproduce it effectively, everybody will be able to hear."[24] Hanley recalls this event being even more difficult than Pittsburgh. The sound engineer's equipment included thirty-six speakers that were set up in a circle around the playing field. All 100 members of the orchestra performed on a stage built right on the fifty-yard line of the stadium. Hanley and his crew hung acoustical drapes throughout the stadium that effectively absorbed the sound.

THE CINCINNATI SYMPHONY ORCHESTRA AT RIVERFRONT STADIUM

On September 10, 1972, Hanley provided sound reinforcement for the Cincinnati Symphony Orchestra. This time Al Hirt and his accompanying group "Short Legs and the Saints" were a highlighted act for the third annual event. A few days later Hanley received a personal letter from Cincinnati Symphony production manager John Gidwitz thanking him for his outstanding work. The letter contained two reviews of the event.

September 12th 1972

Dear Bill:

It was great working with you, many thanks for all of your efforts. I enclosed copies of the reviews from both papers, since both of them take time to mention how good the sound system was. Looking forward to doing it again next year!

Sincerely,

John Gidwitz[25]

After the concert the positive reviews regarding the quality of sound kept coming, some claiming, "We're a little amazed ourselves at the sound."[26] According to *Cincinnati Enquirer* music critic Gail Stockholm, Hanley's system made an impression: "A new sound system in the stadium designed by William Hanley picked up the gorgeous tones of the soloists and the full range of the orchestra remarkably well."[27]

When the Symphony Jazz Quintet performed Frank Proto's arrangement of the song "Aquarius" from *Hair*, Hanley and his technicians needed to make quick and specific adjustments and could be seen scurrying around the stage, adjusting wires, microphones, and speakers that were arrayed in a curve in front of the orchestra. Regardless, Stockholm felt, "On the whole, the fourteen speakers (each the size of a large television console) picked up the orchestra's sound beautifully, with a good sense of presence and almost no audible distortion."[28] It was obvious that by now Hanley had mastered his craft for symphonic sound reinforcement.

Chapter 10

FOLK AND DISTORTION

THE NEWPORT AND PHILADELPHIA FOLK FESTIVALS

Inspired by a Storyville packed house performance in 1958 by "Queen of Folk" Odetta, George Wein set his eyes on a festival to showcase this style of music. At the time, Wein was not aware of the folk movement occurring within the coffeehouses of Cambridge, Massachusetts, or nationally. It was here where college-age students were enthusiastically absorbing the sounds of Joan Baez and others. However, in 1958 there was the Berkeley Folk Music Festival in Berkeley, California. Directed by Barry Olivier, these festivals ran annually until 1970 and were considered very important, especially during the folk music revival.

Rather than embedding a limited folk segment into his already established Jazz Festival, Wein called upon Odetta's manager Albert Grossman in mid-1959 to assist with launching a full-on folk festival at Newport. According to Wein, "The first Newport Folk Festival was roughly modeled after its jazz counterpart, with both evening and afternoon concerts and reserved seating."[1] With their planning in place in July 1959, Wein launched the first annual Newport Folk Music Festival (NFF) a week following the Newport Jazz Festival. The producer invited Hanley to assist Myles Rosenthal on this inaugural occasion.

By 1960 Rosenthal was gone, leaving Hanley in charge, where he remained until 1967. Hanley's influence on festival sound reinforcement begins at Newport. In American popular music, Wein's festivals provided a model for others to come. Following is a chart indicating which year and location Hanley Sound was officially involved at Newport.

As Hanley helped mix sound that first year in 1959, he witnessed a variety of traditional and new folk talent. Dozens of musicians performed like Odetta, the New Lost City Ramblers, Earl Scruggs, and the Kingston Trio. Last, there was Pete Seeger, who in many ways is considered the harbinger of the new folk revival, captivating young folk enthusiast audiences with his banjo playing and storytelling prowess. Wein recalls being overwhelmed by the various

	Newport Jazz Festival	Newport Folk Festival
1957	Hanley assisting, FP	No festival
1958	Hanley assisting, FP	No festival
1959	Hanley assisting, FP	1959 Hanley assisting, FP
1960	Cut short, riot	Hanley, FP
1961	*	No festival
1962	Hanley, FP	No festival
1963	Hanley, FP	Hanley, FP
1964	Hanley, FP	Hanley, FP
1965	Hanley, FF	Hanley, FF
1966	Hanley, FF	Hanley, FF
1967	Hanley, FF	Hanley, FF

* "Music at Newport" produced by Sid Bernstein and John Drew. No Hanley.
FP = Freebody Park
FF = Festival Field

"sounds and styles" unified by the early folk movement occurring at Newport: "An unspoken feeling was in the air, a sense that folk music was approaching a threshold. The music at the Newport Folk Festival that summer constituted more than a revival; it was a transformation."[2]

The rise of the NFF and the accompanying folk revival were at their height around the time Hanley was launching his new technological applications in sound. He was there for many folk festivals around the country, but none were as significant as the NFF. According to then proprietor of the Folklore Center in Greenwich Village, "Izzy" Israel Young, NFF had "done for folk music what John Hammond's 'Spirituals to Swing' concert of 1939 did for jazz." Comparing Wein's NFF to the 1938–39 Carnegie Hall concerts put on by Hammond seems accurate. Hammond, a record producer, music critic, and civil rights activist, put on the show showcasing a timeline of spirituals to big swing bands. The events involved African American artists and integrated audiences.

Somewhat praising Wein for providing a forum for traditional folk music aimed at broader audiences, Young claimed, "Folk music is no longer an esoteric art belonging to the Southern Mountains and other assorted cultural pockets in America."[3] With audiences growing each year, Wein's NFF helped launch the careers of many folk artists.

By the end of the 1960s the folk festival movement and accompanying audiences grew beyond Hanley's expectations. In a 1970 article in *Sing Out!* writer Stephen Calt reflects on the flurry of folk festivals across the nation: "Folk music festivals have since proliferated into the early years of the twenty first century in the United States, eventually numbering many hundreds scattered across the country. In addition, more all-inclusive festivals have been held, such as Newport and Philadelphia."[4]

The First Annual Philadelphia Folk Festival (PFF) had a successful beginning in 1962 at the Wilson Farm in suburban Paoli, Pennsylvania. It eventually moved to Old Pool Farm in Harleysville because of growing audiences.[5] After learning of Hanley's exemplary work at Newport, the festival sponsor, a non-profit organization called the "Philadelphia Folksong Society," asked Hanley for help with the second annual PFF held in September 1963. According to a 1965 *Billboard* article, "In its three years, the Philadelphia Folk Festival has attracted the top names in folk-dom, including Theodore Bikel, Mississippi John Hurt, Doc Watson, Judy Collins, Pete Seeger, Dave Van Ronk and The New Lost City Ramblers."[6] The roster of performers for the 1965 PFF was nothing to scoff at. On September 10–11, 1965, the headliners included Theodore Bikel, Judy Collins, Jean Ritchie, Tom Paxton, Phil Ochs, and the Beers Family, for all of whom Hanley or one of his top engineers mixed sound.

For decades now PFF, also referred to as Folk Fest by locals, has featured a wide assortment of acts and performances, building its ideals around a sense of community. Hanley remained the festival's primary sound engineer until the early 1980s, mixing sound for an elite and diverse roster of traditional and contemporary folk artists. Hanley worked for the PFF considerably longer than any other festival, claiming that it was "as significant as Newport" with regard to his development as a sound engineer. Even though his sound company had a fairly positive experience at the PFF, from time to time Hanley struggled to convey his ideas to the festival leadership: "The event itself was a pleasure to work, yet as always, difficult for the festival organizers to agree to my innovative ideas."[7]

At this festival the sound engineer should be credited for his groundbreaking use of large-screen video projection. By the early 1970s he was displaying black-and-white performance images of musicians close up so those in the back could see. This turned out to be a huge success. Hanley also used this technique at the Fillmore East during an historic Janis Joplin performance in the late 1960s. Frustrated, the sound engineer claims he "tried for years to get the festival to use these video screens," eventually agreeing to do it for free in order to "convince them."[8] Finally, those in charge relented when they saw the audience's positive reaction to Hanley's innovation.

Ever since, a video screen with an accompanying live-feed, close-up projection of the festival stage has been used. According to Steve Twomey of the *Philadelphia Inquirer*, "The festival was helped out by introducing a closed-circuit TV system that featured two large screens on each side of the stage which enabled ... the audience to examine the styles of those on stage."[9] One other significant event happened at the festival. In 1970 Bill met his future wife

Rhoda Rosenberg. Back then they could be found dancing together on the side of the stage.

As the years rolled on, the young and focused sound engineer's repertoire included some of the biggest names in jazz, folk, and rock and roll. Yet, mingling with so many well-known performers meant nothing to him. Hanley was not a groupie; in fact he kept his distance and had tremendous respect for the musicians and their music. The music showcased at the Newport and Philadelphia Folk Festivals was diverse, highlighting blues and country artists. The variety of music introduced to Hanley's fine-tuned ears when mixing sound for different genres was often a challenge. Newport and Philadelphia allowed him the opportunity to further his vision of providing the best quality sound he could produce for the time. After all, he was following his dream; he had come far from that young boy who fixed his neighbors' televisions and listened quietly to that crystal radio set in his childhood bedroom.

PETER, PAUL AND HANLEY

Peter Yarrow of Peter, Paul and Mary first met Hanley at the NFF in 1963. According to the folk singer, Newport was his first and best experience with the "warmest sound that anyone was producing."[10] Yarrow was quite the sound "aficionado" during the early Newport years. Having some audio sensibility, he knew what was needed for good sound quality during performances.

By 1965 Hanley's philosophy, knowledge of sound reinforcement, and "ears" were far superior to anyone working in the field. In one situation, somewhat unusual for the day, Yarrow recalls Hanley using condenser microphones: "Unlike others, Bill would hang condenser microphones around the performers, which was extraordinary. I believe this provided the depth and richness in the sound that you would want to hear."[11] In Murray Lerner's 1967 documentary film *Festival*, Yarrow, Hanley, and Bob Dylan can be seen on the '65 NFF stage conducting a sound check, also very rare for the time.

By 1966 Yarrow and Hanley had formed a tight working relationship at Newport. As journalist Neale Adams wrote: "Mr. Yarrow said he felt the 'sound has been better this year than ever before. Getting the proper sound is a subjective thing. To bring something up you need to bring something down.'"[12] Yarrow's commitment to Newport's production was quite evident then. Working the console instead of being on stage in between acts, Yarrow often assisted Hanley during the festival. Adams continued: "The entertainer said he is working on the technical end this year instead of appearing on stage because he is concerned with making Newport come off."[13] In elementary fashion, Yarrow describes the function of the sound system to the rather uneducated journalist: "The console is a table full of electronic gadgets which control the dozen or so microphones on stage and sends the sound they get to the speakers. The sound can be modified by the console, if necessary."[14]

Yarrow was on the NFF Board of Directors working closely with Wein, whom he admits was "listening very closely" to what Hanley was doing: "It was beyond the general run-of-the-mill sound that was happening at the time, in any indoor or outdoor facilities. There were many critical ears evaluating the sound and Hanley's work at Newport including George Wein and Albert Grossman."[15] Yarrow was as passionate, if not as particular, about the quality of sound as Hanley was, acquiring his knowledge in the recording studio beginning in 1962, where he developed his "ears."

Hanley's significance, then, was not just with the application of technology, but also the natural gift of sensory audibility. Being able to hear with precision placed him ahead of other engineers—hearing beyond the sound itself. With this natural gift Hanley allowed musicians and audiences to create an emotional picture simply by being able to hear the nuances of the sound, and according to Yarrow, this is "ninety percent of what affects people at a large concert." The folk singer continues to acknowledge Hanley's work as transparent because you could actually hear the highs and the lows. "Quite often there was too much pumping of low/mid-range sound that was too aggressive; with Bill Hanley you never had any of that. Of course this was perfect for folk music as it was purely acoustical."[16]

Still friends, Hanley and Yarrow have shared many experiences together beyond Newport. These include some of the largest anti–Vietnam War demonstrations. Hanley's involvement with Yarrow created a bridge into more national politically driven events. When demonstration organizers needed to get the "message" to people during the civil rights and anti–Vietnam War movements, the demand for Hanley's expertise increased. Within just a few short years, the folk artists that Hanley mixed for at the Newport and Philadelphia Folk Festivals, were present at Washington, DC, and New York protest marches.

PULLING THE PLUG ON PURE FOLK

Thirty-seven years before Bob Dylan went "electric" at the 1965 NFF, lawyer, performer, and folklorist Bascom Lamar Lunsford organized the 1928 Mountain Dance and Folk Festival in Asheville, North Carolina. Lunsford, known as the "Minstrel of the Appalachians," in many ways gave birth to the modern folk festival as we know it in the United States. A purist, Lunsford did not like outside artists coming to play at his festival, trying to keep it as locally authentic as possible. According to some, he was extremely apprehensive about the long-haired, bearded travelers who came to the festivals "during the early-McCarthy-era-1950s."[17] Passionate about maintaining the integrity and authenticity of his festival, Lunsford often "scoured the mountains of Western North Carolina" looking for new talent.

Uniquely, Lunsford kept his festival free of electric instruments and sound reinforcement. When electric instruments did manage to creep in, it was so they could be heard above the loudness of the clog stepping. Eventually the "shoe tapping" was banned to cut down on the noise so audiences and dancers could hear the music. "The ban on amplified instruments was enforced; the one on toe and heeltaps was not."[18]

Loyal Jones and John M. Forbes's book on Lunsford's career, *Minstrel of the Appalachians: The Story of Bascom Lamar Lunsford*, details Lunsford's disdain for electric instruments. "In reference to electronic instruments, Festival chairman Jerry Israel said, 'we want musicians, not electricians.' The banning of amplifiers was possibly an attempt to get back to more traditional forms, which was part of the motivation for doing away with shoe taps."[19] Lunsford stepped down from running the festival in 1965 after having a stroke. That same year, on a July weekend, the Newport Folk Festival provided a stage for an electrifying performance by folk artist Bob Dylan, forever changing the landscape of the art form and the way Hanley applied sound reinforcement.

For whatever nuance you lose when plugging in like Dylan, you gain something that is not there in an acoustic performance. Lunsford, Wein, Hanley, or any other folk purist had difficulty grasping that the act of "going electric" was a form of creative expression. Bob Dylan's performance at Newport is what Hanley defines as the artist "plugging into Niagara Falls." According to Hanley, "When I did the Beatles and when Dylan went electric is when I first noticed electricity becoming an extension of the musician's message."[20] In Hanley's experience he was affected by the change of folk into rock, regardless of whatever the driving force was behind it.

For someone who strives for clarity in sound, Hanley may not have understood Dylan's creative approach. Perhaps Newport Folk was not the appropriate stage for the folk singer to introduce this new sound; nevertheless, rock music was evolving. If there is any truth that music reflects our emotional connection to society, culture, and politics, then Dylan's performance was a precursor of that evolution of folk into folk-rock. The times were a-changin'. According to author Elijah Wald, in his book *Dylan Goes Electric!: Newport, Seeger, Dylan, and the Night that Split the Sixties*: "The 1960s were a period of dramatic upheaval, and 1965 marked a significant divide. The weekend Dylan walked onstage with his Stratocaster, President Johnson announced he was doubling the military draft and committing the United States to victory in Vietnam."[21]

As rock and roll music changed the acoustical impact of sound, it influenced Hanley's techniques in sound reinforcement application. With every stomach-rattling, bass-aggressive, mid-range, searing top-end sound that rock music offered, the engineer had to adapt. Hanley knew he had to change and did so successfully until faced with something different—distortion. According to Hanley's role model, acoustics expert Leo Beranek: "Distortion is when a sound is not clear and distinct. There can be two kinds. One is when not all the frequencies are in place. Such as not enough bass, or not enough highs. The other is when the sound has a sort of gravely quality to it. This is a sound that is not pleasant."[22]

According to a *Crawdaddy!* magazine article, shifting from an acoustic to electric guitar produces a distinctly different sound: "An electric guitar is not just a louder version of a conventional one; further, although it is possible to get sustained vibrato, reverberation, and other effects through an amplifier, to use these effects tastefully and to make a sound that is not simply a mass of muddy noise is a skill that must be acquired."[23] At Newport Dylan tested this skill on an audience that were anticipating a familiar repertoire of folk songs.

What Hanley experienced during Dylan's electric set on the night of July 25, 1965, at Newport, was something his ears were unaccustomed to. Wald writes that Dylan's newly formed band "crashed into a raw Chicago boogie and, straining to be heard over the loudest music ever to hit Newport, he snarled his opening line: 'I ain't gonna work on Maggie's farm no more!'"[24]

Those in attendance at Dylan's performance that weekend claimed it forever changed the face of contemporary music, giving birth to folk-rock. According to Wein, "The young figurehead of the folk movement, the songwriter who had given us 'Blowin' in the Wind' and 'With God on Our Side,' had remade himself a rock star."[25] What Dylan presented to the Newport audience that July evening unleashed a torrent of emotions affecting not just folk purists but also those who were in control of the sound.

According to Hanley: "I thought, all of a sudden we have a situation that's different. Dylan is now making distortion as part of his material."[26] From a spectator's perspective, Yarrow recalls what hit you was a "physical assault of sound waves." This was one of the first occurrences Hanley ever recalls hearing distortion of that magnitude: "Dylan was overdriving my amplifier!"[27] It was a pivotal moment that not only permanently changed the world of folk music, but the ways the sound engineer conceptualized and applied the reinforcement of sound moving ahead.

THE STAGE IS SET

George Wein was happy; the 1965 Newport Folk Festival was now nicely settled into a new location at Festival Field. It offered an improved stage designed and lit by Chip Monck, accompanied by an improved Hanley sound system. By now Hanley provided sound reinforcement not just for Newport's main performances but also for its smaller workshops. According to Newport technical director Bob Jones: "Hanley showed up no more than two or three days in advance. He was the one who said, 'okay, this is how these speakers will go up' and 'these need to go into a certain direction.'"[28] Jones recalls that finding sufficient help for the festival setup in those early days was difficult: "We were in the position where we did not have riggers and stagehands, just local people helping us. The only person to rely upon who knew how to do this was Bill Hanley. He was not only the sound company, he was the guy who told us literally how to put this thing together."[29]

Regarding Hanley's sound system that year, *Broadside* journalist Robert J. Lurtsema claims, "You won't need a hearing aid." That's because the sound engineer paid close attention to the quality of sound being produced and positioning was everything. For example, on the main stage of the festival, Hanley's placement was as follows: On either side of the stage there were six fifteen-inch Altec Lansing woofers and tweeters mounted on tall platform scaffolding (three on each). Powered by, four MA 275, 150-watt McIntosh amplifiers that gave the system enough power. The microphones the engineer used were specially modified Shure 545's that contained transformer packages and connectors different from the original factory units. Impressed by the sound, the *Broadside* writer acknowledged that Hanley's system had enough fidelity to "hear the pick hit the strings all the way from the back row."[30]

An equally impressive production feature, Monck's new stage consisted of "Forty-one circuits, seventy-one separate instruments and 8,750 feet of cable" and with "line-of-sight execution" that can make the stage "one big spotlight, narrow in for a soloist, fade on a blue cool, or splash in an upbeat red."[31] These two important production elements accommodated Dylan's legendary performance.

Little did Hanley know that the 1965 NFF would be yet another landmark event in his long career. The happenings of this weekend have stayed with the engineer ever since. Nothing seems as controversial in the history of rock music as Dylan's performance on that Sunday evening of July 25, 1965. Even though the festival featured traditional folk music, electric instruments had been played here prior. Just two years earlier in 1963, bluesman John Lee Hooker had played a blazing electric set. During the 1965 festival, artists like Muddy Waters, the Butterfield Blues Band, and the Chambers Brothers all used electric instruments before Dylan. Bob Jones remembers: "There were other electric bands at Newport that year. Paul Butterfield played that very afternoon. But no one complained, not a soul. We also had all these blues bands from Chicago with no complaints."[32]

Hanley recalls Yarrow insisting that a sound check be engaged that day so Dylan and his band could remember where they placed their volume levels. Yarrow claimed the sound setup was not prepared to handle what "Dylan was trying to achieve." Hanley had overdriven his sound system and blown speakers before and it wasn't pleasant. The sound check gave Hanley some idea of what was to follow. Anticipating a potential disaster, the engineer was worried that the wooden chairs the audience sat in could reflect the sound back on to the stage, making it hard for Dylan to hear what his guitar amplifier was doing. Hanley was correct. According to author Elijah Wald: "I don't think there is any problem with amplifying an electric band on a really good system for amplifying an acoustic band. But the question is how high can you drive that system? This is the change you have with Dylan. You not only have an electric band, but you have people on the controls saying this needs to feel really loud."[33]

With Hanley's sound system ready, stage lights dimmed, audience seated, Yarrow stepped into the center spotlight and announced "Ladies and gentlemen, the person who's coming up now ... has a limited amount of time ... his name is Bob Dylan!"[34] Dylan emerged, and was greeted with an incessant roar. The folk-singer-turned-instant-rock-star rolled into "Maggie's Farm" while members of the Butterfield Blues Band, which included lead guitarist Mike Bloomfield, elaborated the song. As the electric number pronounced itself into the salty Newport air, the "new sound" held the folk purist audience captive.

It was very loud for 1965. Hanley, up until then, had not heard anything comparable. The distortion levels coming from Dylan's band sent Hanley's crew scrambling to try and figure out its origin. Perplexed by the situation, the sound engineer thought it was a malfunction in his system claiming: "I didn't know really what to do. I couldn't tell where the distortion was coming from. I was really worried. I could not differentiate between the distortion of his amplifier and the distortion of my power amplifier. I was maxed out at the limit of my loudspeakers!"[35]

In the background, bellowing screams of joy and disdain could be heard. Hanley remembers, "I was afraid that Dylan would turn up his guitar amplifier throughout the performance because of the crowd."[36] For those closest to the stage area it was hard to hear and understand Dylan's lyrics. As Hanley explains in Wald's book: "The stage was raked, slanting toward the audience and driving the roar of the instrument amplifiers directly into the faces of the people up front."[37]

Dylan's voice echoed over Newport's Festival Field through Hanley's speakers that were placed on either side of the stage. Those who were backstage and at the foot of the stage were blown away by the instrument levels. Most were unable to make out a word of what Dylan was singing.[38] For those positioned further out in the crowd, the sound was much clearer. According to Wald: "People were upset for a lot of different reasons. Some were upset because it was simply so loud. There were others who felt the sound was bad. Others claim they were unable to hear Dylan sing. Yet some thought it sounded great. I believe it depended a lot on where you were standing."[39]

In Joe Boyd's book *White Bicycles: Making Music in the 1960s*, he gives the reader his interpretation of the historic chain of events. Boyd was an occasional stage manager who worked for Wein at the time. Boyd was summoned backstage while watching Dylan's performance that evening where he was ordered by festival leaders to tell Hanley or whomever was at the console to turn it down. Boyd recalls: "Someone tapped me on the shoulder, 'they're looking for you backstage.' Alan Lomax, Pete Seeger, and Theo Bikel were standing by the stairs, furious. 'You got to turn the sound down, it's far too loud.' I told them I couldn't control the sound levels from backstage."[40] With no walkie-talkie communication system in place, the stage manager maneuvered through the excited audience over to Hanley and his crew. In order to gain access to the sound console, which was located a few yards from the stage, Boyd climbed up some stacked milk crates located by the lighting trailer, and hopped over the fence.[41]

According to Boyd, it was Paul Rothchild, Albert Grossman, and Peter Yarrow who were at the console with no mention of Hanley whatsoever. They were "grinning like cats" when I asked them to turn down the volume. Yarrow replied to Boyd, "Tell Alan the board is adequately represented at the sound controls and the board member here thinks the sound level is just right."[42] As Boyd left to deliver the message, he recalls Yarrow extending his middle finger emphasizing an additional note to be delivered to Lomax. Boyd remembers it being an extremely "tense" and "confrontational" atmosphere.[43]

Hanley recalls that at some point that evening Lomax, Seeger, Bikel, and Boyd hurried back over to his mixing console begging him to turn down the volume while the crowd both booed and cheered. It was common practice that Hanley's crew and Newport notables like Peter Yarrow and Albert Grossman "hung out" at the mixing location in those days. "Yarrow and others were always looming around my equipment then," adds Hanley.[44]

With so many different scenarios, it's hard to pinpoint who exactly was on the knobs when Dylan was playing. Wald suggests: "My impression is that it was Peter Yarrow who was in charge of both the sound check and performance. I imagine Hanley was trying his best to give Yarrow what he wanted while trying not to compromise his system."[45]

Pete Seeger disliked distortion of any kind. Is it true that the folk singer threatened to take an axe to Hanley's cables? Or is it a myth? In the 2005 film *No Direction Home* Seeger recalls saying: "Get that distortion out of his voice . . . It's terrible. If I had an axe, I'd chop the microphone cable right now!"[46] simply because he felt that Dylan's lyrics should be heard clearly. The pacifist never got a chance to use his axe.

After 1963, when Seeger became involved in organizing the NFF, his vision was for a musical gathering rather than a series of concerts like at Wein's Jazz Festival. For Seeger it was more about sharing the music rather than performing music for the audience. In fact, many like Seeger felt that Dylan's performance was more like a concert rather than an act of sharing. Wald adds: "For Pete the workshops represented this. People learning from genuine, authentic, southern rural masters and then they would go off and play on their own. In Pete's mind the folk festival was supposed to be a participatory event rather than a concert."[47]

Yes, Seeger was filled with antipathy as many of the folk purists were then, and rightly so. This was after all a folk festival. It wasn't just the NFF leadership who were angered, but also loyal fans. Yarrow adds: "The unhappiness of the crowd that day was more or less a statement of Dylan not just going electric, but going commercial. It was the infancy of handling those kinds of instruments and the integration of them into acoustic sound."[48]

Wein claims the face of folk music changed that evening: "No longer was there the semblance of a pure folk community that resisted corruption by outside forces. The young idealistic folk fans, which had valiantly resisted the mainstream tastes of their friends, no longer had to hold out. Rock and roll was no longer taboo; if Dylan could cross that line so could they."[49]

It's believed that neither Seeger, Wein, Bikel, or Lomax were present at Dylan's initial sound check. Even though Hanley, Yarrow, and a select few knew that Dylan was going to play some semblance of electric music on that evening, no one could ever predict its outcome or impact on the future of popular music. What happened on the Newport Folk stage was the catalyst of a new style of music often labeled as folk-rock. Yet, for all intents and purposes, what is folk-rock but a version of rock and roll? Whatever this music evolved to be, to a sound engineer like Hanley it was becoming far more distorted and louder than ever before.

For many years others have taken credit for Hanley's work in the industry. This happens most commonly about events like Woodstock and Newport. "Sound by: Hanley Sound Inc." can be seen on the inside credit page of every official Newport program he worked for. Hanley was not in the limelight for sure, yet it was his sound system, his crew, and most importantly his ideas behind the sound application at Newport. To be clear, it was he and his crew of engineers who were most often adjusting the levels at the console, hanging speakers, running cables, and setting up microphones. Bob Jones reflects: "I would have heard of any complaints about the sound if there were any, but we never had any under Hanley's reign as our sound company."[50]

The emerging counterculture at Newport that year set the stage for what was to come. It fostered many of the performers who emerged as stars at future festivals like Woodstock four years later. Artists like Richie Havens, Arlo Guthrie, Joan Baez, and others launched their successful careers from the Newport stage.

Newport was not only a venue to hear music but also a place where people went to be with each other. This makes the festival a forerunner of Woodstock. The first day of Woodstock was all acoustic, with almost half of the performers Newport alumni. Friday evening even ended with a set by Joan Baez that included a singalong of "We Shall Overcome," just like she did so many times at the folk festival.

DYLAN MOVES ON

Bob Dylan's tour continued after his Newport performance. Only a few shows remained in the United States, then he moved on to Australia and Europe. On August 28, 1965, the singer was interviewed after a performance at Forest Hills Tennis Stadium in Queens, New York. Here folk purists continued to vehemently boo the musician. According to *Billboard*, "Two musical worlds—folk and rock and roll—collided Saturday night at Forest Hills Music Festival here, but Bob Dylan provided something for both elements."[51] When asked if the combination of rock and roll's electronic sound, coupled with folk music's meaningful lyrics, summed up what he was doing at the time, he claimed: "Yes. It's very complicated to play with electricity. Most people who don't like rock and roll can't relate to other people"[52]—a statement with which Hanley or any other sound engineer might have been in full agreement.

For the remainder of the 1965–66 tour, Dylan realized he needed a sufficient sound system to travel with. At the time only a few artists actually traveled with sound gear. With limited options, a musician either sang through their guitar amp or the public address system of the venue, and Dylan was no exception. Nearly the same age as Hanley, first-generation sound engineer Richard Alderson came to Bob Dylan's rescue. Yet he too had to survive the perils of managing the performer's new sound. According to Alderson: "When Dylan decided to go on a world tour half of which would be electric, he played three or four gigs in the United States; they were all disasters one way or another. I recall Forest Hills being one of them."[53]

While Hanley was busy working at Newport, Alderson was working for Bob Dylan's manager Albert Grossman. Grossman managed Peter, Paul and Mary and several other acts. He hired Alderson for the remainder of the tour beginning in February 1966 in Australia. According to Alderson: "I had been doing ad-hoc sound reinforcement for Albert's bevy of talent like Peter, Paul and Mary and Country Joe and the Fish. Then I was asked to come along and help Dylan with the rest of his tour."[54]

Distortion was also an issue for Alderson at times: "I was the first to do live rock and roll for Dylan on this tour, they had never seen anything like it before. But my gear wasn't suitable for rock and roll, it was more suitable for the acoustic stuff. Most of the time distortion wasn't an issue, but if Dylan was pushing really hard it didn't help."[55] Alderson claims that Hanley had an entirely different problem, mainly because he was dealing with a completely "different" venue. Alderson continues that he wasn't a "big venue" guy like Hanley, indicating, "These were small theaters and venues which sat only 2,500 people." On that tour Alderson recalls the "biggest house" he had to deal with was Royal Albert Hall, seating only 4,000 people. "Hanley was dealing with huge amounts of people. It was a different sound setup."[56]

Alderson built two road systems for the European leg of Dylan's electric tour. "We left one of the systems out in Australia and I don't know what happened to it. But when I got back to New York, I immediately put a system together for Stockholm in only two days."[57] The engineer's equipment on this tour included a combination of Sennheiser and Electro-Voice microphones, Klipsch La Scalla speakers, McIntosh amplifiers, Altec Lansing tube mixers, and additional stage monitors. Alderson recalls: "I miked everyone with only one mix that was all the same feed going everywhere. I usually mixed from the hall and walked around. I was dealing with Mickey Jones who was the loudest drummer at the time. I was dealing with a performer who was singing a dictionary full of words in front of all of this. No one had ever heard anything like it."[58]

RICHARD ALDERSON

A self-proclaimed hi-fi hobbyist in high school, Richard Alderson left his Lakewood, Ohio, home in 1955 for New York City at eighteen years of age. According to Alderson: "I had no intention of ever becoming an audio or recording engineer. However I loved audio, I loved recording, and I loved music. I was a hi-fi nut!"[59] When it came time for him to make a living in the big city, he got a job at the original Sam Goody on Ninth Avenue. Here the young audio enthusiast helped the record store run their hi-fi department. Moving on, Alderson found himself installing hi-fi systems for the elite on Madison Avenue. Shortly thereafter he met audio pioneer, and the wealthiest man in America at the time, Sherman Fairchild: "I had done some work for him and he became so impressed that I became his personal assistant for a while."[60]

Around the end of the decade Alderson was twenty-two years old and independently running his own service: "By 1961–62 I was a professional sound engineer. Although I didn't have any schooling, I was a quick learner."[61] Having minimal equipment on hand, Alderson mostly rented his sound systems to accommodate his client's needs. "Peter, Paul and Mary had gear which I used. I had my own mikes, but I mostly rented gear. I didn't feel the need to have my own equipment; I didn't want to accumulate gear."[62]

Like Hanley, Alderson worked his way around the circuit, providing sound for the folk and jazz clubs around Greenwich Village. In early 1962 Alderson acquired a six-month stint as house sound engineer at the Village Gate with lighting guru Chip Monck. It was during this time that he inherited a job working for the very popular performer Harry Belafonte. At the time, if a performer was savvy enough and had a budget, they might have traveled with their own sound system. This is what Belafonte did.

A rarity then, the singer toured with an elaborate production that included not only a sound reinforcement system but a sound engineer. Alderson remembers: "I got the job from these two techno guys from Manhattan and rebuilt Harry's whole setup. I could tell these guys had no music sensibility but I did. I remember Harry having a full production when I started. As far as I know no one in the world carried live sound and an engineer with them, Harry was the first."[63]

After 1966 Alderson slowed down his efforts in sound reinforcement, focusing more on recording. In 1969 he decided to leave the United States. He packed up all his belongings and drove to southern Mexico, where he lived until 1975. Richard Alderson is a forgotten character not just within the Dylan history books but in the history of sound reinforcement and recording. He too should be recognized as significant to the field. "I used to think that the reason I was forgotten was on purpose. But I suspect I am forgotten about in the history books because I left the country when Nixon was in office."[64] Although working within the same circle of colleagues, Alderson and Hanley's paths never crossed.

NEWPORT FOLK '66

According to George Wein, the 1966 NFF offered the typical array of "authentic" artists. However, it seemed that Bob Dylan had left an indelible impression, as the festival that year offered a "small but noteworthy concession to the folk-rock trend" by presenting electric bands like the Blues Project and the Lovin' Spoonful.[65] Hanley was prepared, having survived the previous year's distorted debacle.

Up until now Hanley and his crew traveled to Newport in a station wagon, with an attached trailer. It usually contained numerous speakers, microphones, and amplifiers. If needed, he and others made a few round trips back to Medford with the occasional unfortunate breakdown along the way, as engineer Harold Cohen recalls: "It was always a bumpy ride. The trailer was known for blowing a tire or two."[66]

At the end of the 1966 Newport Festival season, Hanley had been with the organization for close to ten years. The following would be his last. At twenty-nine years of age, his philosophy on sound reinforcement was the same then as it is now, stated in "Woofers Tweeters and Hanley," where Hanley claimed: "I want to make people feel like they are sitting in the conductor's seat. I want to create a sound so that everybody feels he is emotionally involved in the happening on stage—the musical happenings."[67]

Hanley's philosophy was to get the sound heard accurately and everywhere at large outdoor venues whenever and wherever he provided sound reinforcement. Two of his most important concerns were a) not "blasting" the people sitting directly in front of the speakers and b) where to properly "aim" the sound. Claiming that he wanted the sound to "flow into the people," he positioned his speakers to do just that. According to Hanley, if you were in the audience while he was mixing, the sound in front actually traveled "just over your head" with just enough "dropping down in each section."[68]

Most sound engineers agree that wind is an issue when providing sound outdoors. For Hanley, especially in open-air spaces such as Freebody Park, Festival Field, or even later at Woodstock, it was a constant problem. The sound engineer claims that you can compensate for this by where you "aim the speakers." Nevertheless it was difficult for Hanley to get the sound loud enough at Newport. Dynamic range is an issue too, Hanley explains: "The system must make a whisper sound like a whisper and still be audible; a yell like a yell and still not jolt everyone out of their seats."[69]

On a hot July day in 1966, while Hanley was working on the wire configurations at Festival Field, the sound engineer went into great detail about his equipment design for the event.

At the heart of the Hanley System are two very large speakers to either side of the festival stage. On a platform almost 30 feet high are the "woofers," 8 ft. tall boxes with amplifiers, which produce the low sounds. On top of these boxes are the smaller horn shaped "tweeters" which produce the high notes. The whole affair weighs a ton-and-a-half and has to be put in place by a crane. The speakers produce sound which ranges from 30 to 18,000 cycles per second, or equivalent to a very good high fidelity record.[70]

These were pioneering days, and the challenges Hanley faced went beyond having enough equipment or even poor weather conditions. Simple things like keeping jazz musician Dizzy Gillespie's trumpet directed into the microphone, miking and managing the sound levels for Frank Sinatra or the soft-spoken Barbra Streisand, or even juggling the power of the Duke Ellington Orchestra. All of these were problems to be solved. Additionally, Hanley was learning how to adapt to the newness of distortion and fuzz coming out of the amplifiers of the Paul Butterfield Blues Band, Muddy Waters, and Bob Dylan. It was a time of true innovation in sound reinforcement, and through Hanley's method of trial and error, he helped define what we now know about effective sound reinforcement application.

According to journalist Neale Adams: "Mr. Wein had words of high praise for his young protégé. 'I think Bill has come a long way. He is a product of the festivals.'"[71] Wein himself was a perfectionist, and rightly so. As one 1966 article states, regarding the new Festival Field sound system: "We are not satisfied with the sound yet. But now with a permanent stage in place, Mr. Hanley can think about a permanent sound system."[72] At the time Wein suggested that "two towers" could be replaced with one "single speaker system" on the top of the stage or maybe leaving the two towers and adding "uppermost speakers" to the stage area, something the festival producer proposed as a "current thought" to Hanley's established system configuration. "The sound is excellent—but not perfect yet. I'm going to make him do a lot of work this year," Wein explained.[73]

During the Newport Festival years, Hanley claims, he was fully prepared to install a permanent system at an "overall cost of $75,000 to $100,000," suspecting that most of the problems would be "solved" by doing so. Engagements like these are a further testament to how Hanley was always experimenting at Newport, his laboratory of sound. Former Hanley Sound engineer Karl Atkeson speaks of Hanley's attention to the quality of equipment used during the early days at Newport:

Bill used amplifiers that had a frequency response of ten cycles to one hundred kilohertz/kilocycles. But there was no finer gear available. You could buy a driver that was guaranteed to almost last forever. If someone inadvertently put a bass signal on the woofer it would survive. But Bill wouldn't buy that stuff. Bill bought the aluminum foil diaphragm devices that were so light and crisp. They took nearly no power to operate and responded wonderfully right up to 20 kilohertz, beyond human hearing, whereas other gear was not as good. We had safety measures where we only used special amplifiers to use the high frequency devices that were fragile and delicate.[74]

On one occasion Hanley gave his new employee Ray Fournier a shot at mixing. Having a previous career in studio recording, Fournier was more than suitable for the job. But an appointment like this did not sit well with Wein, because Ray is blind. For Hanley this made perfect sense, but according to Fournier it was difficult to get George to believe in his abilities. Fournier recalls: "George didn't think I could do anything because I was blind. This had happened to me before, people judging me because of my blindness. That was the one thing about Bill. He never questioned my ability. He was the easiest guy I ever worked with. Not a mean guy at all even when situations got difficult."[75]

The summer of 1967 marked Hanley's final year with the NFF. Even though he continued to work with Wein on other productions, they decided to part ways. On Sunday July 16, 1967, folk singer Arlo Guthrie introduced his anthem "Alice's Restaurant" to the Newport audience. This was actually one of Harold Cohen's first experiences as a sound engineer for the company. He explains: "One of the most memorable moments for me that year was when Arlo Guthrie performed 'Alice's Restaurant' for the first time that weekend. Arlo had everyone on stage singing it with him including Bill—he was up there!"[76]

There was a lot on the horizon for Hanley and his company after Newport. With upcoming festivals and tours planned, Hanley kept moving forward. Even today, the sound engineer looks back on his time at Newport with fondness. "These were great years for my company and my knowledge as a sound engineer. I'll never forget them."[77]

Chapter 11

THE INAUGURATION OF PRESIDENT LYNDON B. JOHNSON

PRESIDENTIAL SOUND

While listening to the January 20, 1961, inaugural proceedings of President John F. Kennedy, Hanley felt inspired that he could do a better job with the sound. As a result, he pursued more political accounts, feeling that the equipment commonly used was unsuitable. In June 1962, Hanley was hired for the Commonwealth of Massachusetts Democratic and Republican state conventions. The Republican convention was held on June 15 at the Worcester Memorial Auditorium. It was a unique space for installing sound reinforcement, complete with a 116-foot-wide proscenium. It was here that Hanley multi-miked the convention floor, which resulted in superb sound for all of the delegates. "From what I recall I applied additional modifications to my microphones for this event."[1]

Two years later in July 1964, Hanley made an unsuccessful pitch to supply the public address portion of the Democratic National Convention at Boardwalk Hall in Atlantic City, NJ. According to Hanley: "I flew down to Atlantic City. They didn't know me there, and after being told to wait, the contract was awarded to someone else. I know I could have improved on the setup."[2]

Interested in working for government projects, the young, confident, and self-proclaimed "audio perfectionist" soon showed the world what he could do. In the winter of 1965 Hanley got his shot when Johnson's Inauguration Committee awarded his firm the official contract, as a low bidder, over two competing sound companies. The others were an established Washington company and DeMambro Sound of Boston—the same firm Hanley had applied to when just out of high school.

According to Hanley winning this bid was an unlikely opportunity since "Bidding on a sound contract in Washington was rare at the time," and because

"Up until 1961 when Kennedy was inaugurated only one company had provided sound since the introduction of sound systems in Washington."[3]

DeMambro's clients at the time included Archbishop Cardinal Cushing of Boston and the Democratic Party. By late 1960 Kennedy and Cardinal Cushing were putting pressure on the Architect of the Capitol to change its sound service vendor for Kennedy's inauguration. They had DeMambro in mind. Hanley believes that Cushing, Kennedy, and DeMambro had a close relationship. Hanley recalls, "Joe DeMambro provided sound for most of the Catholic schools and churches in Boston. Cardinal Cushing wanted him for the job."[4] Kennedy and Cushing's attempt failed, as far as Hanley remembers. He believes an "RCA distributor" most likely provided the sound for Kennedy's 1961 inauguration, with DeMambro supporting the inaugural parade.

After winning the Johnson inaugural bid, Hanley immediately went out and bought more equipment. News spread within the Medford community about the sound engineer's upcoming gig in Washington, leaving many family, friends, and colleagues filled with pride and elation. Throngs of local newspaper journalists approached Hanley to learn more about his preparation. In a 1964 *Boston Evening Globe* article, Hanley spoke candidly about the work that needed to be done: "My job is to make it possible for the dignitaries and the anticipated 100,000 spectators to hear and understand every word that is spoken during the historic ceremony."[5]

And so in 1964, the twenty-seven-year-old, highly determined Hanley and his crew were preparing for one of their biggest jobs ever: The Inauguration of President Lyndon B. Johnson and Vice President Hubert Humphrey at the Capitol building in Washington, DC, on January 20, 1965. The sound engineer's local newspaper the *Medford Mercury* claimed, "Hanley has had to get special security clearance for the work since he will be on hand beneath the Capitol Rotunda personally supervising the amplification for the Presidential platform."[6]

At the time, the small Medford firm had only six workers, all of whom were less than twenty-two years of age. The *Mercury* reported, "He will be accompanied by his crew of young and enthusiastic full and part time employees, typical of the quick and intelligent young men taking over this modern age of electronics."[7] Still under construction, the "Hanley Sound Equipment Company" had only been founded a few years prior. While settling into his 430 Salem Street location, Hanley was immersed in his audio work within the confines of his trailer parked in a lot behind the business. By the mid-1960s the sound company was just beginning to gain a reputation for its high-quality work.

Even though Hanley had been juggling other large-scale events during this time, none of them created as much of an impact on his career as the inauguration of a president. By changing some of the original audio specifications and introducing new types of amplifiers, microphones, and speakers for the Washington event, the sound engineer left a heady impression around the Capitol. The current Washington, DC, standard of reinforcing presidential sound is indebted to Hanley's influence, although very little has been written about it.

Hanley's friend and part-time crew member Phil Robinson had been on leave for a term in the Navy beginning in 1961. When he was discharged in December 1964, he was curious to see how his old boss Bill and his business were doing and decided to pay him a visit. Robinson arrived at Hanley's shop and saw major changes: "Hanley had just won the bid for the inauguration. His business was a bigger operation. There was a lot more equipment and people. There was more activity with cars and trucks coming and going."[8] Hanley was excited to see him, welcoming his old friend back with open arms. Immediately he transitioned into assisting Hanley.

By now Hanley had his hands full and was relying on the latest communication technology to keep in touch with his sound company. By using a Bell Boy Pager he could be notified if someone needed him at the office. According to Hanley: "It was small enough to keep in my jacket. If someone wanted to reach me I would receive a signal, and then call into a service center. That person would then pass on the caller's message to me. It was a really handy device."[9] Robinson had never seen one before: "I don't think they were that commonly used."[10] Hanley was forever into new technology, even having a car phone that he kept in the trunk of his car.

After Christmas of 1964 many of the Hanley Sound crew began taking measurements and trucking equipment to the Capitol. It was a huge undertaking for the small firm. Bill and Terry handled the logistics of getting the equipment down to Washington. Hanley and some of his crew flew into the Capitol approximately two weeks prior to take wire measurements for around forty loudspeakers. These were to be installed on the Capitol and the Senate and House wings. This was no easy task, because the sound engineer and others needed to string at least five to six miles of wiring to the speakers from his control room, which sat beneath the Capitol's Rotunda.[11] Always prepared for the worst, the "inauguration sound crew" constructed a backup system in case the first did not function properly.

At the time Hanley was confident in his quality of service for the event, as he explained to the press: "The number of full-time concerns providing audio equipment throughout the nation can be counted on your fingers."[12] Courage to take chances like this is what drove his new business and crew into a new area of sound application. Without a talented staff of engineers and crew people, Hanley would have faced great difficulty initiating an undertaking as complex as an inauguration.

With only two weeks to prepare, Hanley's team worked tirelessly. The talented and technologically savvy individuals involved included Ken Sommers and Bill Kelly, who were electronic technicians. There were also Phil Robinson and Ken Jeffrey, a friend of Terry Hanley's, who was a young helper and still a student at Malden Catholic High School. All were residents of Medford and very young. At only nineteen, Terry was one of the firm's lead engineers, and often one of the first to go out on jobs for his older brother. Last, there was Hanley Sound foreman, David Roberts, a mechanical designer who had received specialized electronics training in the US Air Force. Due to his experience Roberts rounded out the crew nicely.

Some recall the drive down to the Capitol as harrowing. According to Roberts: "We were in this box truck. We drove down in a blizzard by the skin of our teeth!"[13] Once everyone was assembled in Washington, DC, Roberts was put in charge of renting an apartment for the crew in what he referred to as a rather "seedy" area near the Capitol. He claims, "There were no sheets for the beds so I ended up looking for some in the middle of the night at a local shop."[14] Only half a mile from the Capitol, the apartment offered a place close by to rest and warm up after long cold days.

The sound specifications for Johnson's inauguration were based on those of the previous inauguration of President Kennedy. Acoustical consultant Lewis S. Goodfriend had authored the specifications that were to be the outline of equipment Hanley Sound was required to provide. Hanley recalls, "Lewis S. Goodfriend was hired by the Architect of the Capitol, J. George Stewart, to write these specifications."[15]

While carefully reviewing the supplied specs back in Medford, a number of acoustical concerns arose for Hanley. Yet, if any changes were to be made, he would need clearance from Walter Rubel, assistant to Stewart in Washington. Hanley remembers that Stewart's office was somewhat rigid to work with; they were in charge of the entire physical plant, Capitol building, and inauguration.

One of the concerns that Hanley noticed was that the intended requirement for speaker placement already had television lights placed in that location. "After seeing Mr. Stewart's pictures, I realized they were wrong. The pictures were of lights and not loudspeakers! Whoever wrote the initial specification screwed this up. Since he was the Architect of the Capitol I brought this concern to him."[16]

In Hanley's view, the system outlined and intended for Johnson's inauguration was not going to be sufficient. Known for using quality equipment, Hanley immediately changed the specs for amplifiers from Altec (or Ducane) to McIntosh. After learning of the changes, Goodfriend claimed he was horrified when he realized Hanley was bringing in his Newport sound system to Washington,

DC. Changes like these sent fear and anxiety through the organizers' minds, since they were accustomed to using a complete, integrated sound system.

To Goodfriend and others, the Altec system was easier to specify because its components were designed to work together. He felt that by using a "pop music" sound system, one compiled from different manufacturers, made it inadequate for the inaugural proceedings. According to Goodfriend: "One major concern that came into my mind was the use of McIntosh amplifiers. The spec was based around Altec Lansing equipment so I had to stop and think for a moment because these were primarily used for home hi-fidelity. It wasn't a single system, it was a collection of equipment and this made me feel very uneasy."[17]

Goodfriend thinks that Hanley landed the job for a couple of reasons. First and most obvious, Hanley was the low bidder, and second was his reputation with handling large crowds. Goodfriend observes: "Interconnectability is probably the biggest problem in noncommercial systems. If the Architect of the Capitol reviewed what Hanley's equipment was beforehand, he might not have got the job."[18]

At the last minute, President Johnson decided to bring in the Mormon Tabernacle Choir, their first-ever performance at a presidential swearing-in ceremony. While on the phone, Rubel asked Hanley if he could make additional improvements to accommodate the choir. Knowing he could, Hanley confirmed with an emphatic "Yes, of course I can!" Because Rubel's systems specs were inadequate for a 350-member choir, additional alterations had to be made. Terry who mixed for the choir, recalls some of these modifications: "We had to convince them to let us use our McIntosh amplifiers, which were the best at the time. The amplifiers were modified so we could use them like a rack mount. We put meters on them so we could monitor the output based on their specifications."[19]

The choir arrived at Dulles International Airport from Salt Lake City on Tuesday the 19th. Bused to the Capitol grounds, choir director Richard P. Condie met with Terry and Bill around 5:00 p.m. for a short rehearsal in the bitter cold. As the choir warmed up, the director and sound engineer discussed details about miking. On inauguration day Terry fine-tuned equipment as choir members checked in, finding their places on the platform. All around the engineer, news reporters were abuzz on telephones and typewriters. Television anchors and camera operators were positioned in makeshift buildings. A helicopter flew overhead while watchful guards remained at their stations throughout the grounds.

For the choir, Hanley and his crew used around sixteen Shure 545 (Unidyne III) modified microphones constructed at Hanley Sound. These particular microphones were the predecessors to the SM-57. According to Shure: "The world renowned Mormon Tabernacle Choir used Shure Unidyne III microphones to carry their voices and songs to the huge crowd. Some 15 to 20 Unidynes were stationed throughout the Choir as part of the extensive sound system that covered every facet of the inaugural ceremonies."[20] During the performance Terry was situated in front of a rack of five RCA five-channel mixers totaling twenty-five channels. To cap it off, Hanley also used Altec Voice of the Theater 210s, three units in each, located on the left and right scaffolding towers.

Due to Rubel's concerns about the performance of Hanley's Newport gear, the sound engineer hauled two separate sound systems to Washington. According to Hanley: "Rubel didn't want me to replace his specs because he didn't trust me, so I had to buy all the equipment he had specifications for. If I didn't, I would not have got paid."[21] The only exception to the specs was Hanley's choice in power amplifiers (McIntosh) to which he and Goodfriend had already agreed. The engineer needed to further convince the inaugural leadership that the rest of his incongruous sound system could work seamlessly with additional power and clarity. To prove himself even further, the sound engineer simultaneously A-B tested (one system against the other) his Shure, Voice of the Theater, and McIntosh system against theirs.

As they began the test in the bitter cold, a light snow fell on the overworked and exhausted crew. Terry had just finished chipping ice off both systems when they were turned on and sound roared into the frigid Washington air. The University spun aluminum loudspeakers were no match against Hanley's Newport system.

In disbelief, assuming Hanley was cheating, Goodfriend flew into the control room under the Rotunda. The A-B speaker check had revealed Hanley's system was better than he had imagined. To anyone listening it was evident that Hanley's sound system far surpassed theirs, and begrudgingly Rubel agreed to use the Hanley sound system. "This was the first time they had high-quality speakers for the inauguration," recalls Terry.[22] Bill adds: "They had a twelve-inch woofer and a re-entrant trumpet in the center up on scaffolding towers, and it was poor, so they went with mine. I put up six 210s, three on each side, and blew them away."[23]

Anticipating results, most of the crew were sitting on the front steps of the Supreme Court building where the installation was located. Roberts said: "I remember setting up for that A-B test. We could distinctly tell when our system was on and when their system was on. Let's put it this way, clearly Washington, DC was schooled in the old bathtub University WLC stuff, and it sounded like crap against our equipment."[24]

The microphone of choice used at a presidential platform up until then had been Electro-Voice. Included in the A-B test was Hanley's preferred microphone, Shure. Hanley recalls: "We had two A-B tests done, one for the microphones using Electro-Voice, and Shure. I was using a shock-mounted version of the SM57. We decided that Shure had a better sound as far as the microphones went. It was because of their reliability and performance."[25]

Because of Hanley, and the good work of his crew, Shure microphones have been used at presidential inaugurations ever since. Such uses are a tribute to his impact on the famous microphone company. Shure historian and director of application engineering Michael Pettersen explains: "There was never any consistency as to which microphone had been used until the Johnson administration. Eventually they realized that the president needed to have consistent sound on the microphone. Bill had a part in this, and the president has only used Shure SM57 microphones ever since. Bill was really a pioneer."[26]

During much of this time, Secret Service agents shadowed Hanley and his team. According to Hanley this made everyone on his crew seem nervous and tense and that made him nervous and tense. "They were keeping a close eye on us, it was after Kennedy's assassination," adds Roberts.[27] Hanley preferred to walk around and listen to his sound system while in use, a method of tuning the sound. This made the assigned Secret Service agents very uncomfortable. The sound engineer was told that he was required to work his sound system from a remote location.

Out and about, just as the newly elected president was going to speak, the sound engineer was abducted by four Secret Service agents and carried against his will into the Rotunda of the Capitol. Dark suit and tie flapping in the wind, Hanley was questioned about his credentials:

> We had set everything up for the news crew so I walked out to see if the NBC pool feed was good. After this I went back to where I was stationed under the crypt located beneath the dome two or three floors. When I arrived they would not let me back into the room. It seems that I was issued the wrong kind of pass. The Secret Service had their grip under each of my arms and around each of my legs! Rubel was going crazy because they wouldn't let me back in and he wanted me around to make sure everything ran smooth![28]

According to Robinson: "Bill wanted to go out and see and hear the quality of sound himself so he could tune it. He went out one door and they wouldn't let him back in. He had quite a blow up about this and threatened to pull the plug! It eventually got smoothed over. I think this made the Architect of the Capitol very nervous."[29]

On that cold January day in 1965, Hanley and his crew brought large-scale high-fidelity sound to one of our nation's greatest events. His contribution to the audio industry was now marked in history. When the ceremony began, the United States Marine Band performed its scheduled numbers, after which the choir sang "Give Me Your Tired, Your Poor" and "Battle Hymn of the Republic."

Following the swearing-in, the choir successfully closed out the program with "This Is My Country." Many claim President Johnson was smiling with great pleasure regarding the quality of sound. The following month Hanley received a letter from J. George Stewart praising the performance of his sound company:

February 11, 1965

Dear Mr. Hanley:

You were the successful low bidder for a contract to provide and operate a temporary speech reinforcement system for the Presidential Inaugural Ceremonies held on the East Front Plaza of the United States Capitol on January 20, 1965.

This contract was subsequently amended to include a music reinforcement system capable of amplifying and distributing the voices of the Salt Lake City Tabernacle Choir, an organization of about 350 members. Your diligence, cooperation and superb professional performance contributed much to the success of the official Inaugural Ceremonies, and to the splendid reproduction of the Choir's renditions.

I am personally grateful to you and your organization for the excellent services performed under most unusual circumstances.

Yours very truly,

J. George Stewart

Architect of the Capitol[30]

Roberts recalls being within ten feet of President Johnson, and how the entire event went "flawlessly" with no hiccups or disasters: "It sounded great. Bill had worked his way up to Johnson's Inauguration. I remember there was a long hard fight for that job and Bill thought if we can get this job, we could do anything."[31] With the Big Apple on the horizon and other projects brewing, the following year, the sound company took on even bigger jobs.

HANLEY AND SHURE

The March 1965 edition of Shure Microphone's newsletter *Sound Scope* highlights the significance of the company's involvement with the inauguration. The title on the cover boasts "Shure Records a Page in History," with no mention anywhere of the sound company who introduced the product to the White

House. Before Johnson was sworn in as 36th President of the United States, Hanley positioned four Shure Unidyne III microphones on the rostrum so the world could hear every word being said during the event.

There is mention of the actual sound system in the newsletter, yet no word of who deployed it. It claimed: "The public address system, praised by radio and television commentators on the scene—enabled everyone in the audience to hear every word. The Shure Unidyne III professional microphone had the honorable distinction of being the only microphones used as the source of sound for the extensive radio and television broadcasting pools."[32] To this day Hanley has no idea why the microphone company chose not credit him as the catalyst of this major development with their product.

The relationship between innovators like Hanley and the available technology of the day is important. Hanley's influence on the Shure microphone company is a prime example of how the engineer's creativity shaped the development of their products. For example, in 1959 Shure introduced a series of studio microphones (SM) that evolved into the SM57 (1965), followed by the SM58 (1966). The SM57 (Unidyne III), known for its natural clean sound, was the model microphone that Hanley used at Johnson's inauguration in 1965. The SM57 microphone resulted in the SM58, known for being rugged and having great sound quality. Eventually the SM58 transformed into a common and popular microphone for many rock and roll musicians for years to come.[33]

According to Michael Pettersen: "We brought these microphones out because we thought they would be great for speech and public speaking. We thought they would be big in the TV studio environment too."[34] Unbeknownst to Shure, sound engineers like Hanley would use Shure products to mike live rock and roll music. Microphones that were typically being applied in studio environments were now being used on stage and in multiple ways, yet another Hanley influence.

In an effort to get the sound loud enough while having more control, Hanley was one of the first to mike all instruments including amps, drums, and vocals. Hanley remembers, "Instead of miking just the singer, I decided it was more effective to mike all of the instruments for more control over the sound."[35] British sound engineer Dinky Dawson, of Dawson Sound, acknowledges this shift during an early visit to the United States. After seeing instruments miked at the Fillmore East in late 1960s Dawson recalls, "It was here that I noticed all these guys like Hanley who were starting to put microphones on all of the instruments."[36]

According to Pettersen: "Hanley was the one to figure out that the SM57 and SM58 were great for that. Shure was not hip to that at the moment, but Bill was. His work on the Beatles' tour is how they started using the Shure Microphone. Bill really helped us learn how to use our products."[37] From the onset Hanley

Sound primarily used Shure microphones at a time when there were other options. In a field where different mikes were used for different instruments, it seemed logical for Hanley to standardize its application for everything.

MAKING MODIFICATIONS

In the early days the Hanley brothers were all too aware of the utilization of microphones with twist-lock connectors common in electrical industrial wiring. But these posed drawbacks because they forced engineers like Hanley to make an inventory of separate cables. According to Terry: "One of the earliest we used was a standard high impedance Amphenol butt connector. Some of these gave us trouble so we looked for other solutions, as they were unbalanced and prone to humming. Eventually we then went to a broadcast standard Cannon type 'P' (or three-pin), but they were too expensive. Then we found a better solution with an improved Cannon (three-prong) XL, and then to an XLR connector."[38] Realizing it was easier to have one style of microphone with a cohesive cable and connector (XLR three-pin connector) system, the engineers at Hanley Sound discovered a way to innovate and modify the Shure microphone.

Hanley solved this problem by purchasing microphone parts directly from Shure. He and his crew figured out that if you bought the microphone cartridge and the transformer handle (from the more expensive model) and attached them, you'd have a microphone that could be plugged into any microphone cable. Terry observes, "We came up with the portable master microphone cable, which people were not doing at the time. We made our own from scratch."[39]

After consistently ordering these parts in lots of fifty, Shure questioned why Hanley wasn't buying complete microphones, and only ordering specific parts. According to Harold Cohen: "After six months they were producing them the way we did. This was a major change for them and a major change in the industry."[40] As a result, Shure "re-tweaked" the original 545 and 565 microphones that had the four-pin female (high or low impedance) Amphenol style connectors on them. According to Pettersen: "It was Bill and others saying 'You know you should be using this XLR connector.' So we eventually dropped the Amphenol connector."[41]

After the Johnson inauguration, things were shaping up back in Medford and Hanley found himself in need of an office manager. Never a good businessman, Hanley desperately needed someone savvy enough to run the office, organize his staff, and cultivate more jobs. As one article states, "When he and his crew return from Washington, Hanley intends to, first, hire a secretary and then complete remodeling the new home for his expanding business at 430 Salem Street."[42] The aforementioned "secretary" ended up being the feisty, headstrong and organized Judi Bernstein. Famously known for maintaining a tight ship while keeping Hanley in line and on time, Bernstein remained by the sound engineer's side for many years to come.

Chapter 12

HANLEY SOUND NEEDS A MANAGER

JUDI BERNSTEIN

In 1964 twenty-eight-year-old Brooklyn native Judi Bernstein was booking acts for visionary talent agent Frank Barsalona of Premiere Talent in New York. "Prior to coming to Hanley Sound in 1966 I had worked in the business for some time."[1] The highly successful Premiere Talent Agency was one of the first to book rock acts, earning it a firm grip on the developing industry. A 1968 *Cash Box* article explains: "Barsalona had worked heavily on booking the first Beatles tour, and realized that the new music, new rock, would be the music of the future. Not just on disks, but in the college market (which was on a folk rock kick then), in nightclubs, concert halls, movies and other media."[2]

At Premiere, Bernstein handled many of the up-and-coming British acts of the day. In charge of music productions she booked smaller acts into colleges. *Cash Box* reported: "Premier has established itself enough to sign the cream of the then-top English touring groups: Herman's Hermits, The Animals, Freddie and The Dreamers, and Wayne Fontana and the Mindbenders among others. Nine agents are currently employed and a West Coast office is in the works."[3] Collectively, these experiences contributed to Bernstein's twelve-year tenure as Bill Hanley's right-hand woman.

While on a business trip to meet promoters in New England, Bernstein ran into Newport publicist John Sdoucos. The two already knew each other because of her work at Premiere. While in conversation Sdoucos convinced Bernstein to manage acts instead of booking them. After serious consideration, in just a few months, Bernstein left Premier and moved to Boston to work for Sdoucos. Bernstein claims she was never really satisfied with her position at Premiere: "John convinced me to come up to Boston, so I packed up everything, my dog and me and that was it. I only got to do the small acts at Premier. When the big acts came from England, the heavies in the office would book them, so I never had a chance."[4] According to Sdoucos: "I got Judi coming out of an agency in

New York. She had to get out of there I guess. I told her to come on over and you can work for me. This is where this relationship developed. This is how she met Bill Hanley."[5]

New England jazz promoter and producer John Sdoucos has had a long relationship with Hanley going back to the beginning of the Newport Jazz Festival. Sdoucos traveled all of New England representing Newport, having begun his career in 1954 as a publicist for George Wein's club, Storyville, and then later at the NJF. "I covered all of New England at the time; remember, we weren't worldwide yet. We had to really kick butt and that's what I did. I ran the road; I hit the radio stations, newspapers, and television."[6]

Over the course of a long, fruitful relationship with Wein, Sdoucos worked on many festivals with Hanley, including Newport Folk, Ohio Jazz, and Kool Jazz. Hanley's relationship with Sdoucos worked out well for the sound engineer. During this time the Newport publicist introduced Hanley to the Boston group Barry and the Remains, indirectly charting a path toward working with the Beatles.

In mid-1965 Hanley was feeling good about the quality work he had done for Johnson's inauguration. Finally settled into the shop in Medford, Hanley Sound was kept busy working at Newport and supporting various college acts. Often overwhelmed, the sound engineer needed someone at the office who could keep it all together. That fall, while providing sound reinforcement at one of Barry and the Remains' local college shows, Hanley met Bernstein for the first time. She recalls: "Somewhere along the line Bill was doing sound and stage for the colleges with the Wenger Wagon and I was working for John at that time."[7] The two spoke briefly about his needs for the office but were not to meet again until October 14, 1966, during a Mamas and the Papas performance at the Commonwealth Armory in Boston.

As the group performed, both Hanley and Bernstein firmed up details at the sound console. Linda J. Greenhouse, writer for the *Harvard Crimson*, reported that the Mamas and the Papas had a "fine upbeat performance," claiming "you go to rock and roll concerts like a fan goes to, game after game. Sometimes, when your group looks as good on stage as they sound on records, you win."[8] This statement is a testament to Hanley's work that evening.

When Sdoucos's business slowed down, Hanley's business was on the upswing, further establishing his need for a business manager. Bernstein offered to assist Hanley on a part-time basis: "Bill asked me to come help him out, so I found myself working for both of them at the same time—it was crazy! Bill certainly needed a manager then; it was the beginning of the beginning!"[9] After she realized the new and exciting work Hanley was involved in, she joined Hanley Sound full time, claiming, "working for Bill felt cutting edge."[10]

The addition of the savvy young New York booking agent was a smart business decision for Hanley. What she brought to his sound company was experience, organization, and a fluid compatibility with musicians and managers. Bernstein spoke show-business language and communicated it well. She brought a Rolodex of her own contacts, which didn't hurt: "While with Hanley Sound I referred all of my contacts from Premier to Hanley."[11] These were contacts Hanley needed in order for his business to flourish.

Anything that Hanley Sound was involved in from 1967 to 1977 most definitely went through Judi Bernstein. Eventually she became a legend of her own at the sound company and in the industry as well. Never having an official title, Bernstein was always referred to as the company's "office manager." Being the only woman in a shop full of men, Bernstein often felt the pressures of being a female in charge. Even so, the fiery manager was feisty and in control, putting anyone in their place if they slighted her.

Bernstein managed the business office, in-house technical staff, handled the bookings, contractual negotiations, scheduled crew and equipment, and oversaw sound installations in Boston and New York. In addition to her multitude of responsibilities, she arranged necessary permits, scouted venues, and almost anything else that needed to be done. Her position with the sound company was something she referred to as more or less a "juggling act." According to Bernstein, "I booked the jobs, set up the jobs, told the guys where they were going, and made sure they got there; I also took care of the licenses."[12]

Some of Judi's most important work was assisting with large demonstrations in Washington, DC, as well as many important music festivals, including the Woodstock Music and Art Fair, Newport Jazz and Folk Festivals, New Orleans Jazz and Heritage Festival, Miami Pop, and the Ann Arbor Blues Festival. Bernstein was most proud of her work as the on-site technical director for a Canadian Festival called the Festival Express in 1970, often referred to as the "Trans Continental Pop Festival," a musical train tour across Canada carrying major artists like Janis Joplin, the Grateful Dead, the Band, and Buddy Guy. Her responsibilities included managing the crews for live sound, stage, lighting, and audio recording. The audio recordings for the event eventually made it into the famous DVD documentary of the tour.

Keeping in touch with Hanley became more difficult as the jobs got more detailed and frequent. The summer months brought additional festivals and outdoor events, and for those who worked at the sound company it was not uncommon for their jobs to require round-the-clock availability. Hanley Sound crew member David Roberts recalls Bernstein's significance to the business: "Judi became important to us. She knew where everything was going, where everybody was, and had a way of curtailing people at the other end of the phone.

She both smoothed and ruffled feathers if necessary. She became, in my view, indispensable. Everyone went through Judi. Very few people had access to Bill."[13]

Trucks were always coming and going (some occasionally breaking down) at Hanley Sound. In the rear of the 430 Salem Street building, stage crews and engineers loaded and serviced the tractor-trailers, converted Frito-Lay vans, and portable Wenger Wagon stages. Out of all of the Hanley Sound trucks, none was as famous as the forty-foot tractor-trailer known as "Yellow Bird." According to Bernstein: "I remember Mary Hanley [Bill's mother] writing a check for fifty thousand dollars for the Peterbilt truck we purchased. Her hands were shaking when she was writing that check!"[14]

Back then Hanley was known to have a pocketful of pre-purchased airline tickets. This allowed him to chase down promoters and acts across the country. Conveniently, he could alter the hand-written flight number at will. Bernstein recalls: "I remember driving to the airport with him and hiding under the seat! He drove so fast to get on that plane!"[15] Often the sound engineer didn't know where he was going, randomly showing up somewhere based on a lead. "I would call the office and ask Judi for updates, 'Where am I going? Who am I seeing and what's it all about?' Then she called the promoter and informed them that I was at the airport and needed to be picked up."[16]

Equipped with only a desk phone, Bernstein remembers constantly trying to track down the busy sound engineer: "Half the time I didn't know where Bill was. Half the time I couldn't find him. He may be back from the coast at the end of the week, but then again he may stop in Detroit, and he usually gets hung up in New York. He had no phone!"[17]

One of Hanley's methods of securing a job was by walking into a venue, looking around briefly, and clapping three times. Then he would leave and go on to the next venue. This impressed promoters who might need his service. Bernstein recalls: "This was all he needed to do. It's done much differently today. After the promoter had a great show and great sound, they often asked when Bill was available next. They kept calling for us to do more work."[18]

Bernstein was also instrumental in attaching a rider agreement between bands, promoters, and the sound company. This was a contract issued between the sound company and performer that distinctly specified certain sound, stage, and light requirements that needed to be in place for the artist prior to the show.

Often a group who was set to perform at any given event was left to the dangers of an inadequate sound system. Bands like the Beach Boys and Ike and Tina Turner soon realized the quality a Hanley Sound system had to offer. So the experienced manager convinced them to attach a rider to their contracts. Bernstein was one of the first people in the music business to implement this type of contractual negotiation. "It never existed before. I was the first to do this, just so that we would be the sole provider of sound for that act."[19] By the end of the 1960s, musicians and promoters expected quality sound and such riders came to be commonplace throughout the music business.

After Woodstock, Hanley became something of a popular public figure. Different people floated in and out of the office on a daily basis just to meet him. It was unique for a sound engineer to reach celebrity status. It seemed that a growing fan base was emerging of people who recognized his talents in sound. "Sound by Hanley" was often displayed on handbills, programs, albums, and theater marquees along with the top-billed acts. Bernstein remembers: "Bill was the first, so everyone wanted to work with him. He often stayed with friends he had all over the country. He was well liked and everyone wanted to know him, he was a popular guy."[20]

After leaving Hanley Sound in the late 1970s, Bernstein went on to become the founding editor of *Pro Sound News*, a popular trade journal for the sound industry still in print today. She opened a business with her husband, Hanley Sound engineer Harold Cohen, and at one point worked for Terry Hanley for a short while.

Those who remember Bill Hanley for his quality work readily acknowledge Bernstein for hers. The strong-willed office manager will always be remembered for her contributions and influence during her time at Hanley Sound. "I gave Bill Hanley ten years of my life. It was a great ride. I couldn't think of anything else I would've rather done. There wasn't any time to think at all. It was new and interesting and happening. Bill was pioneering the path, and there was no one there to say he was right or wrong, he was doing it."[21] Judi Bernstein-Cohen passed away after a brief illness on May 26, 2015, and is survived by her husband of forty-four years, Harold Cohen.

Chapter 13

GO-GO DANCERS, BATMAN, AND THE BEACH BOYS

SOUNDBLAST! '66

By the end of 1965 Hanley and his crew were getting a taste of what it was like to pull off large-scale concerts. Supporting continuous performances and festivals further established the sound company as a leader in the new and developing field of sound reinforcement. In 1966 Hanley Sound assisted the Beatles and Beach Boys tours with some interesting events in between. It would be a busy year for Hanley and his crew.

Concerts held inside stadiums always posed acoustical problems for the engineer. This was something Hanley faced increasingly while providing sound for the Beatles on their August tour. But even prior to these performances, in June 1966 Hanley experienced similar issues.

A concert called Soundblast! '66, held at Yankee Stadium on Friday, June 10, was replete with sound issues. The large-scale event advertised some of the "biggest names in pop," with groups like the Byrds and the Beach Boys headlining the show. The diverse lineup featured Ray Charles, the Cowsills, the McCoys, the Marvelettes, Jerry Butler, Stevie Wonder, the Gentrys, and the Jimo Tamos Orchestra. According to *Village Voice* writer Richard Goldstein, "The acts spanned the spectrum of the American rock scene (an increasingly expansive spectrum it is)."[1] Hanley Sound provided full stage and sound production for the eclectic lineup.

"Go-go 66," as Hanley refers to it, was strange and confusing and a complete disaster. As Goldstein relates, "It should have been a gratifying combination of jazz-folk-gospel-acid-dance-rock," but the sound was "lost somewhere in the mammoth infield of the stadium," adding that the "end product sounded like a cross between a Jehovah's Witness revival and a drag race."[2] Poorly promoted, Soundblast! '66 sold a mere 9,000 tickets out of the 70,000 seats. The heavy rain probably didn't help either. According to Hanley: "It was a disaster and I never got paid for it. We were lied to by promoter Jon Jaymes."[3] The event promoters, TAJ Enterprises, were made up of "three young producers all less than twenty-seven years of age."[4] One of the youths was the shifty Jon Jaymes, who was notorious for his sly ways. Jaymes was known to be a "pathological liar," according to Hanley.

The concert got off to a late start at 9:00 p.m. The show commenced with sixty-six teen-aged go-go dancers who "rode out of the right field bull-pen on bicycles."[5] As the evening continued, Soundblast! '66 became intensely chaotic as fistfights and fireworks exploded in the grandstands. The infield was brimming with roaming photographers and people without proper passes. Goldstein reported: "Autograph hunters roamed wild in the outfield, and a horde of press photographers knelt at the foot of the stage in homage. The sound was lost in the pseudo-event."[6]

Hanley had faced real problems at other baseball stadiums and Soundblast! '66 proved no different. For those who were performing, it was even worse. In Goldstein's article, McCoys band member Randy Zehringer described the challenge of playing in a large open area as "awful and terrible, it's like singing in the middle of a freeway with all the noise and open space. You can't hear the audience reaction and you can't hear your own sound."[7]

The *Village Voice* article's account of the sound quality of the Byrds' performance at Soundblast! '66 was no better: "Finally connected to their amplifiers by electric umbilical cords, they began to play. But the sound wasn't worth the amps. The group seemed incapable of sustaining effective harmony in person, and their ambiguous raga–rhythms lost themselves in a haze of echo and feedback."[8] Empty stadium seats only increased the severity of the problem. "Most of the stadium's seats were unused, and the sound of trickling applause echoing off empty wood seemed anything but frenetic."[9] Hanley recalls that the sound "reverberated" all over. In a letter sent to Hanley dated June 20, 1966, from TAJ Enterprises, one of its promoters Joseph Talbot wrote:

DEAR MR. HANLEY:

I WOULD LIKE TO TAKE THIS OPPORTUNITY TO EXPRESS OUR APPRECIATION FOR THE OUSTANDNG QUALITY OF THE SOUND SYSTEM WHICH YOU PROVIDED FOR OUR SHOW SOUNDBLAST '66 IN YANKEE STADIUM ON JUNE 10TH.

I AM WELL AWARE OF THE TECHNICAL DIFFICULTIES INVOLVED IN SETTING UP A COMPLETE SOUND SYSTEM IN A 70,000 SEAT BALL PARK CAPABLE OF PICKING UP SUCH DIFFERENT PERFORMERS AS THE BEACH BOYS, RAY CHARLES, STEVIE WONDER, THE BYRDS AND THE MCCOYS- NOT TO MENTION A 17 PIECE ORCHESTRA.

WE FEEL THAT YOU DID AN EXCEPTIONAL JOB UNDER DIFFI-
CULT CIRCUMSTANCES AND WE LOOK FORWARD TO USING YOUR
SERVICE AGAIN IN THE FUTURE.

VERY SINCERELY,
JOHN T. TALBOT
PRESIDENT

In the not-distant future, the application of sound reinforcement transitioned from stadium environments to that of fields, farms, and raceways. Eventually, events in large baseball stadiums like this resurfaced in the 1970s when
sound reinforcement technology became more advanced and systemized as
the large-scale festival market narrowed.

GASSED WITH THE STONES

On October 28, 1965, Hanley had been hired to provide sound reinforcement
for a Rolling Stones press conference at the New York Hilton. The following year
on Friday June 24, 1966, the band kicked off their North American tour in Lynn,
Massachusetts, and the sound company was called on for their expertise once
again. The first date of the tour was at the Lynn Manning Bowl High School
Football Stadium, not that far from Bill's home in Medford.

Before the Rolling Stones took to the stage, three bands performed: local
band the Mods had won the opening spot, followed by the McCoys and the
Standells, who were known for their hit "Dirty Water." As a warm, hard rain
soaked the 5,000 fans attending, local radio DJ Arnie "Woo Woo" Ginsburg
finally introduced the Rolling Stones. In a mad fury hundreds rushed forward, breaking through wooden barriers and surging toward Hanley's portable
Wenger Wagon stage. Nervous local police reacted by firing off tear gas, and
the Stones jumped into their limos and took off.

According to the *Daily Evening Item*, "The mob broke through police lines
and charged towards the performers' platform."[10] Crew member Ray Fournier
recalls he and Hanley fighting off the biting effects of the tear gas: "There was
chaos in the audience and the tear gas didn't help. I don't think the Stones
played much of a set."[11] The band noticed folding chairs being hurled their
way, placed their instruments down, and took off after playing only half a set.
Hanley remembers: "It rained and there was tear gas everywhere. The crowd
was very angry and throwing wooden chairs all around the field."[12]

An article titled, "Rolling Stones Concert a Gasser!" describes the chaos
that evening: "After ten songs, rain begins to fall harder and the Stones rush

from stage as youths follow. Lynn cops drop tear gas bombs to quell mob. The event makes world headlines. The Beatles consider a concert here in August, but choose Suffolk Downs instead. Stones' concert tickets were five bucks."[13] Mick Jagger recalls: "It was a bit of an outdoor crazy. It wasn't well-secured. A few people got a bit drunk. There were a few cops and that was the end of it."[14] Hanley met the Rolling Stones again three years later in December 1969 for a series of historic shows at Madison Square Garden.

THE BATMAN CONCERT

The next morning, June 25, eyes still stinging, Hanley drove to New York for a concert at Shea Stadium. The sound engineer was heading to an event known as the "Batman Concert," yet another financial disaster for all involved. And if it could get any worse, it was a brutally hot day. A rather odd musical spectacle, television superhero Batman, played by actor Adam West, headlined this one. Frank Gorshin, who played the Riddler, was included in the lineup of performers.

There were rumors that the event would be canceled due to poor advance sales. Acts that were slated to perform defected early after hearing the news. As production crews threatened to dismantle platforms an hour before the show, West came to the rescue. Generously, the caped crusader stepped up and promised to pay the frustrated workers, but Hanley wasn't included. An article called "Batman Concert Zonks" stated in June 1966, "The real hero was sound-man Bill Hanley of Hanley Sound who took a financial bath with equipment, engineers and mixers who worked the evening show at Hanley's expense."[15]

The acoustics in Shea Stadium were notoriously difficult to manage, and with no more than 3,000 in attendance, it made the situation even worse. According to New York Times writer Robert Sherman, the lack of attendance was noticeable: "And so the show, such as it was, went on before a total audience that couldn't have exceeded 3,000. That's not a bad crowd in Carnegie Hall, but dispersed through the three extensive tiers of Shea Stadium, it seemed a handful."[16]

In a rented Cadillac, Batman circled the bases of Shea Stadium, while Hanley sat stationed in the middle of the field behind his console, surrounded by his speakers. Sherman jokes that "The Batmobile was in for repairs, no doubt."[17] The television superhero's grand entrance was the highlight for loyal fans, bringing the sparse crowd to their feet.

West and Gorshin took to the stage continuing with their quintessential "hero/villain routine" and also a few solo vocal performances. Sherman adds, "Gorshin's impersonations were for the most part lost in the wide open spaces, but Adam West revealed a mellow, resonant, crooning voice that set the kids cheering all over again."[18] The Batman concert demonstrated that nothing could have saved Shea Stadium from poor sound, not even if the caped crusader was Hanley himself. "His [Hanley's] satisfaction came from coming up with a "great sound" in the difficult Shea Stadium,"[19] one writer claims.

Hanley was beginning to see his hard work pay off, and 1966 would prove to be an interesting and busy year for his company. Now established in New York, the sound engineer was often in Greenwich Village, working tirelessly to establish himself throughout the Big Apple.

THE BEACH BOYS TOUR OF 1966

In 1966 Hanley Sound was planning to go on the road with two of the most well-known groups of the decade: the Beatles and the Beach Boys. According to Terry: "These were the first big tours we did, and I remember us doing them simultaneously. I recall the Beach Boys and the Beatles were two of the most popular groups in the United States"[20]—and he is right. In May of that year the Beach Boys had released their now iconic *Pet Sounds*, and in August the Beatles their epic *Revolver*. It was a significant time for the company and the music scene.

Brian Wilson may have had the final say in securing Hanley Sound for the performances, even though by 1966 the musician had retired from touring altogether. According to Hanley it was most likely the "red bearded, and golf cap wearing"[21] Beach Boy, Mike Love who initially inquired about a possible agreement: "Love probably heard my sound system first, possibly somewhere in the Midwest, then again at Soundblast! '66."[22]

Love could have heard Hanley's sound system in many places that year. The band had performed at various East Coast locations, including the Crane Estate in Ipswich on April 30 and at the Massachusetts Institute of Technology in Cambridge on May 6. Hanley Sound was an obvious choice by promoters given that the company was local to these particular venues.

The Beach Boys were plagued with bad sound in that year. On May 7 at Providence College in Rhode Island, the band played through the primitive PA system of the auditorium. As journalist Steve Gilkenson wrote at the time: "If there are any complaints to be made about the Beach Boys' half of the show, it has to be with their sound system. Occasional high pitched feedback from the amplifiers was very annoying."[23]

It was either Mike Love or Brian Wilson who demanded a Hanley sound system for the 1966 Beach Boys tour beginning in August. Hanley recalls: "I went out to Love's beach house in Los Angeles where we talked out the details.

We eventually made it onto their rider, which is the first time in the music business that this ever happened."[24] Road manager Dick Duryea handled the remainder of the details once the system arrived on the road.

A close friend to the band, "Balsa Bill" Yerkes, recalls the buzz around the group when they decided to bring in the Medford sound company. Yerkes first heard the news about the new sound system on July 30, 1966, in Atlantic City. "That April they were putting everything through the Fender Showman amps; by November you could see the plywood stage monitors Hanley had built."[25] According to Yerkes, it was a big deal at the time to have a separately designated vocal PA system: "I recall the band talking about using the same system as the Newport Jazz Festival. They were very proud to have a PA when prior their vocals went through the second channel of their amps."[26]

If you were in the audience, Hanley's crude unpainted plywood custom monitors could be seen on stage during portions of that tour. While their recent hit "Good Vibrations" dominated the airwaves, the band performed the complex song with clarity through this relatively new advancement. "The Beach Boys were heavy vocally compared to most other rock and roll bands. They had so many rich layers of harmonies going on and would often have trouble getting a good vocal mix. Hanley's system made it much better. What a difference in the sound," adds Yerkes.[27]

Plans came on quickly, and with very little prep time the Hanley Sound crew moved tons of equipment to various venues overnight. This tour was uniquely difficult because of logistics. Getting from one venue to another, plane tickets, car rentals, angry promoters, and overall communication was a challenge. For a portion of the tour the Hanley sound system was loaded onto a DC3, a plane that the Beach Boys chartered. Crew member David Roberts recalls, "Trying to get the Voice of the Theater speakers into the cargo bay of the plane by half an inch was a challenge."[28]

Once loaded, Roberts and other crew members waited for hours for a phone call from Judi Bernstein. It was a hot August late evening at Chicago's O'Hare Airport and some of the exhausted crew slept behind the freight terminal rolling-staircase. The following day Roberts awoke to the aroma of rotting strawberries that were baking in the sun on pallets awaiting shipment "I remember waking up very early to a purple, pink dawn. It was high summer. The air was filled with this sweet smell. I will never forget it."[29]

Eventually the call came in: two tickets were waiting at the Air Canada terminal, and off to Toronto Roberts and the other crew members went. According to Roberts, "I was broke and exhausted."[30] After several shows, the Beach Boys decided to let the sound company go. Because of the numerous tour dates involved, it was becoming a costly endeavor. At the time, bandleader Mike Love felt that transporting and deploying Hanley's system every night didn't make much sense, especially when they played so many shows in various locations across the country. Hanley claims: "Mike Love didn't want to use us any longer because the group was playing so many dates. It ended up costing them something like $10,000 to ship our equipment around on their private plane and set it up."[31]

Extensive touring services—involving stage, sound, and lighting—were still in their infancy in 1966. This system of concert production had not yet been streamlined and made efficient. For those dedicated crew members involved with the rigors of early touring, it was a shared and common credo that the show must always go on. Those who were part of Hanley's crew should be considered some of the first touring rock and roll "roadies" in the developing music business.

Chapter 14

HANLEY AND THE FAMILY BAND

THE COWSILLS

By 1966 Bill Hanley had forged many relationships with managers and band members due to his good work in sound reinforcement. One of these relationships was with William "Bud" Cowsill, father and manager of the band the Cowsills. After hearing the quality of Hanley's sound system somewhere in Rhode Island, Cowsill asked the sound engineer to join them for upcoming shows. Band member and son Paul Cowsill remembers, "Bill and my dad probably ended up talking to each other in Newport at some point."[1] As the Cowsills grew in fame, Bud Cowsill hired Hanley Sound for future concerts and tours.

The Cowsills were made up of Barbara (mom) and her six children ages eight to nineteen years old. Formerly a chief petty officer in the US Navy, "Bud" Cowsill was known to be a stern manager. According to the *New York Times*: "The whole Cowsill thing was Dad's idea. He founded the group four years ago when he retired from a twenty year stint in the Navy and decided that his singing, drumming and guitar playing family was ready for more than charity shows and family jam sessions."[2] The band's harmonies, catchy lyrics, and young vibrancy made them a good fit for Hanley to practice his sound application techniques. In retrospect, the Cowsills were an extremely important band in the history of the company.

Not only was the band musically talented; the Cowsills were the primary inspiration of the 1970s television show *The Partridge Family*. Yet the family band was far more complex than those who lip-synced and pretended to perform in a televised environment. When assembled they performed with striking harmonies coupled with instrumental fluency; they could really play. But their performance capabilities were often overlooked because they were perceived as more of an affably happy and approachable type of group. Hook-laden and complex hits like "Hair" and "The Rain, the Park and Other Things" proved this stereotype inaccurate. As journalist Alfred Aronowitz said of the group: "The

Cowsills' contribution is not to music, but to show business, momism and the American family. By that criterion they are great. As a matter of fact they are irresistible."[3]

Hanley was the first sound engineer the Cowsills had ever met. When he rolled his truck into their shows, it made a big impression on the band. According to Paul Cowsill: "Suddenly Bill came in and the speakers were as big as our bus; this had to have been around 1965 or 1966. Sometimes I helped set up his gear."[4] Later in life Paul became a sound engineer himself, attributing this career choice to Hanley: "It was all asses and elbows with Bill Hanley. It seemed like he loved doing the sound thing. It was so new at the time that he really is a pioneer. Bill was always sweating and working his butt off and I became a sound guy because of him, I am sure of it."[5]

It was essential that the Cowsills be able to hear each other on stage. However, they often struggled to do so. Prior to Hanley's introduction, the band played through Shure Vocal Masters, a typical and commonly used PA for any singing group of the day. Paul, who was the acting stage manager and sound person, recalls the insufficient setup: "A column of eight speakers, that was it, that was the sound for the Cowsills before Bill!"[6]

When band member and brother Bob saw what Hanley brought in for equipment, he was amazed at the sheer size of it. As he recalls, their modest Shure column speakers were as "thin" as his arm: "We were coming off these toothpick speakers, and we didn't even know what good sound was. I doubt we even knew what the word monitor was."[7]

While on stage the band often turned their column speakers inward for monitor sound. At one of their performances in 1966 Hanley heard the band performing and realized that there were no highs or lows coming out of their speakers. Paul remembers: "We had turned the two end columns in so we could hear ourselves. The other two speaker/columns were facing the audience. All of these vocals coming out of the mid-range like a big batch of mashed potatoes must have really bothered Hanley."[8]

The band went through a major transformation after Hanley solved their sound problems. According to Bob Cowsill, Hanley had his work cut out for him with six young and boisterous singers to corral: "Bill Hanley's greatest contribution to the whole rock scene was that he was the first with the stage monitor. When Bill's equipment came in, it was heaven on stage for us. It seems so elementary but it wasn't, it was revolutionary is what it was. He saw it and solved the problem."[9] Eventually father Bud asked Hanley to record the group's live album, *The Cowsills in Concert*, released in 1969. *Cash Box* reported: "Plans have been made to have the Cowsills record their first live album during an appearance this week at the University of Illinois. The LP, recorded for MGM under the supervision of the Hanley Sound Co., will include about an hour's worth of the performance featuring several original songs and a medley of the Cowsills' pop hits."[10]

Hanley Sound was engaged in full production touring by 1968. It was a time when there were very few, if any, sound companies supporting full concert production touring. While the music business was still in its infancy, the Medford firm was already deploying its crews on the road for various concert tours. It was backbreaking work. Gig after gig, night after night, Hanley and his team unloaded trucks and semi-tractor trailers filled with heavy equipment. These rigs often contained speakers, amplifiers, horns, cabinets, microphones, stands, and cables.

Harold Cohen was on tour with the Cowsills in 1968 and recalls the efforts of early concert touring with the family band: "We did not carry their equipment for them for the most part, just our own. But on that tour it was just me and a truck driver and it was hard work."[11] For each town Cohen and the Cowsills pulled into, they were met with screaming teenybopper fans that fawned over the group. On one occasion while Cohen accompanied the performers in their station wagon, he recalls hordes of overly excited kids strewn all over the hood of the vehicle.

The Cowsills' 1968 tour was an extremely grueling one for Cohen. They traveled five hundred miles or more in between dates, often including two and three shows per day: "I was traveling closely with them in 1968. I traveled the Midwest and East Coast. We did state fairs in Ohio, Wisconsin, New Jersey, Pennsylvania, and some shows down south."[12]

Depending on the venues—mostly fairgrounds that held auto races, demolition derbies, and rodeos—the engineer would often try to set up before noon. Known for being organized, Cohen placed the speakers and other expensive equipment in a specific way so the setup and breakdown went more efficiently, keeping a close eye on it always. "It was usually a 4 pm show. We could not leave our equipment because the rodeo cowboy people were thieves. We had to guard our equipment with our lives, I learned that early on."[13]

Bud Cowsill became quite fond of Cohen. "Bud treated me well. I was a skinny kid but all muscle. We were always in a strange and new town and some of these fairs had great food, but it was hard to eat because my hours were rough. Often I went without eating. Bud always saw to it that I had some food and a quart of milk at my console every night before the performance. If not, I wouldn't have eaten until midnight."[14] In 1969 Bud invited Cohen to help the band with a commercial for the American Dairy Association in Newport.

Chapter 15

HANLEY AND THE REMAINS

BOSTON BEFORE THE REMAINS

During the early to mid-1960s, most venues in Boston were small, and music was often performed in church halls, community centers, ballrooms, and college gymnasiums.[1] Rock and roll was not yet found in large arenas or coliseums. However, the Boston Arena, now Matthews Arena, located at Northeastern University, was an exception. Although mostly used for ice shows and hockey games, Hanley provided sound reinforcement for political events and many other performances at this location.

A Who's Who of the entertainment industry performed here, including acts like Chubby Checker, Jerry Lee Lewis, the Supremes, and the "first-ever appearance" of the Motor Town Revue, starring Marvin Gaye, on November 2, 1962.[2] Other locales like the Lynn Manning Bowl (football field), Suffolk Downs (racetrack), Rindge Tech, and the Walpole Skating Rink were hotspots frequented by top performers.[3] Brett Milano in his book *The Sound of Our Town: A History of Boston Rock and Roll*, writes: "Rock wasn't yet the stuff of smoky nightclubs and packed arenas; it was a show that your parents could drop you off to see in broad daylight. A popular local band would be able to jam a thousand kids into one of these clubs on a Saturday afternoon."[4]

According to Boston booking agent John Sdoucos, "There were no other places to go other than the gin mills." He claims, "The Boston Garden was for basketball and hockey," adding: "When I tried to book Joan Baez into Symphony Hall, they said 'my goodness we can't book Joan Baez in here, she has dirty feet.' Joan used to walk around with no shoes on."[5]

Hanley recalls jazz music as the popular sound in the early 1960s, especially with the college crowds who were attending the Newport Festivals. In Cambridge and Boston, folk music performed in coffeehouses dominated the scene. By 1964 that scene was about to change, as a new generation of rock and roll–loving college students embraced a grittier sound. The music of the

Beatles and the Rolling Stones had a great impact on many towns across the United States, including Boston. With so many colleges in the Boston area, a thriving and diversified garage rock music movement emerged. Groups like Barry and the Remains, the Lost, and the Barbarians, all hoping for stardom, played their hearts out night after night. Radio play was an important part of this equation as well. Local AM radio station WBZ offered an all-rock format along with DJ Arnie "Woo Woo" Ginsburg (WMEX), integrating many Boston bands into their popular radio schedule.

LADIES AND GENTLEMEN—THE REMAINS!

In the fall of 1964 four Boston University students exploded onto the Boston rock and roll scene, taking everyone by surprise. That band was Barry and the Remains. Of all the groups to perform in the area, the Remains are an important part of Boston's rock and roll history. Having this group on Hanley's roster of clients catapulted his company into a whole new realm of touring, accompanied by new and improved methods of sound reinforcement application.

The Remains were a mid-1960s original garage rock band with R&B, soul, and British Invasion influences. As Milano claims, "Nowadays you'd call the Remains a garage band, but that term didn't exist in 1965."[6] The band is considered one of the most influential and loudest from this time, and to this day they still have a cult following even though they were only around for a couple of years. The group consisted of Barry Tashian (guitar/vocals), Bill Briggs (keyboards/vocals), Vern Miller (bass/vocals), and Rudolph "Chip" Damiani (drums). All were highly accomplished, professional musicians. By 1966 the band landed the break of a lifetime as the opening act for the Beatles on their final tour of the United States. Drummer "Chip" decided not to go on the tour, and left the band; N.D. Smart became the band's new percussionist.

Highly popular with the student population, the group had a number of local Top Five hits, including "Why Do I Cry," which climbed all the way to #3 on the Boston pop music charts. One could find the Remains playing at nearby haunts such as the Rathskellar, Where It's At, and the Banjo Room or an occasional frat party if you were a hip teen-ager during the years of 1964–65.

The band's dorm room at Boston University's Miles Standish Hall sat conveniently across the street from owner Gene Brezniak's Lounge Bar, which later became known as the Rathskellar, or the Rat. The popular club was located in the heart of Kenmore Square. When not playing a gig, the band often practiced in the basement of the dormitory building. Those who were in Boston in November 1964 might remember the Remains playing their weekly Wednesday night gig—known as Remains Night—in the back room of the venue.

Selling out almost every week, Brezniak became motivated to open up the basement of the club because of the large crowds the band brought in. On certain nights when the Remains performed, a line of fans could be seen outside the building, up and around the corner, then onto the Brookline Avenue Bridge over the Massachusetts Turnpike. At times the line of people snaked all the way to the far reaches of Fenway Park.[7] According to Miller: "We turned up our amps, pounded the drums and played uncivilized rock and roll while the audience drank beer, danced, and perspired. It was dark, damp, noisy, smelly, and fun. No matter how wild the crowd got, the music never stopped."[8]

The band's booking agency HT Productions was located nearby at 739 Boylston Street, across from the Lennox Hotel in Kenmore Square. Sharing the office were agency owners and Newport friends of Bill Hanley, John Sdoucos, and Freddie Taylor. If you wanted to meet with any of these individuals, they could be found four flights up by freight elevator. The office was a happening spot for up-and-coming bands and aspiring booking agents of the day, including the now music mogul giant Don Law. Law worked at HT Productions as a college agent while a student at Boston University. He co-managed Barry and the Remains and was instrumental in securing the group a record contract with Epic.

Bill and Terry Hanley met the Remains through their relationship with Sdoucos in early 1965. Bill recalls: "John at HT Productions managed Barry and the Remains. He was doing college shows. We hooked up with the Remains through him."[9] According to Sdoucos, his agency was instrumental in keeping the band noticed by exclusively using Hanley Sound: "We kept the Remains alive and a lot of other acts. They would hang around the office. We were booking the hell out of the Remains. They were rockin' n rollin', and could cover anybody's tunes and wrote some great tunes themselves. They were a great band and all the colleges wanted the Remains and Hanley was our go-to sound company."[10]

The sound company was crucial in helping the group with some of their earliest college circuit gigs. At Boston University's Union Forum, Brandeis University in Waltham, and at the University of Massachusetts Terry was often sent out to support the band. Terry recalls: "Barry and the Remains were my act. I did them all the time in town. I worked with them when they were nobody, until they got their record deal."[11] The band's reputation as raucously loud should not be confused with sloppy performance or poor musicianship. They were known for their unique presentation because of how tight and disciplined they were on stage.

Since the emerging college circuit was growing, it meant that sound reinforcement at different college facilities posed problems. Usually a PA at a university consisted of one speaker centrally located near a scoreboard and an amplifier and a couple of microphones hanging from the ceiling on the stage. For a band like the Remains this kind of setup was grossly inadequate.

Sound quality became very important to the band. As their popularity grew, so did the need to enhance the audible experience for their expanding fan base. In early 1965 members of the band ventured out to Hanley Sound in Medford to see what the engineer had to offer. According to Briggs: "We met Bill in the back room of Hanley Sound. I remember his warehouse was full of stuff! It was controlled chaos, stocked to the ceiling! It was wires, speakers, amplifiers, and boxes just stacked from floor to ceiling! Hanley rented a system to us and left it in the club for a week."[12] Briggs happened to be a sound perfectionist, adding: "If the sound wasn't right, I wasn't happy. Hanley Sound was reasonable so we brought them on to do the bigger gigs."[13]

By fall 1965 the Remains popularity skyrocketed, forcing them to take a year leave of absence from Boston University. This decision allowed the band to play the many college circuit shows booked by Sdoucos. The booking agent and acting manager scheduled the band as much as he could on weekends at New England colleges. At one show at the University of Massachusetts, the Remains performed using a Hanley sound system with Bo Diddley and the Shirelles in front of a crowd of around 4,000 people.[14] At the time Sdoucos was managing local act, the Barbarians. The popular Barbarians were known for their one-handed drummer. According to Hanley Sound manager Judi Bernstein: "Yes it was unusual, the drummer a.k.a. 'Moulty' of the Barbarians had only one hand!"[15]

Sdoucos was looking for other outlets for his acts, so when he learned about the large cultural activities budgets of area colleges, he applied his focus there. In Fred Goodman's book *The Mansion on the Hill: Dylan, Young, Geffen, Springsteen, and the Head-On Collision of Rock and Commerce*, Goodman and Sdoucos recall breaking into the college circuit: "The schools had tremendous budgets, he recalls. The University of Massachusetts had eighty to one hundred grand. I said, "Hey, we can do groups like The Coasters, The Drifters, Hank Ballard and the Midnighters?"[16]

Soon Sdoucos realized most if not all of the colleges had poor sound systems. According to Miller: "One of the things that was happening at that time was the college market was opening up. The colleges were having rock and roll shows because they had fairly good entertainment budgets. We played a lot of college shows. We played the Rathskellar and the Banjo Room in Cambridge but that was a different kind of thing. College-age people patronized most of the clubs we played at."[17]

Hanley's influence on the Remains helped the band stand out from others. Miller remembers: "Bill and Terry had these big old grey Altec A7 cabinets.

These were just humungous to us. These were the Voice of the Theater speakers and they were loud."[18] Sdoucos recalls that the Remains were "loud but not in a negative sense. The sound was big from a little four-piece band. They had a big, ballsy, gutsy sound."[19] He claims Hanley had a lot to do with that.

The sound engineer's expertise played a major role on not just the quality but the volume of the Remains' sound. By doing so the band stood out for their booming performances then, even though now your average contemporary barroom blues band is considerably louder. According to Bill Briggs, "Loud was a different frame of reference then; we were *very* loud."[20]

In a 1967 *Crawdaddy!* magazine article, rock music critic Jon Landau emphasizes this distinctive characteristic about the band when using a Hanley sound system. Landau had witnessed them perform early in their career and became inspired by their sound. A musician himself, Landau looked to Sdoucos for representation for his band, Jellyroll. It was in 1964 at one of Sdoucos's "subsequent college packages at Brandeis University" where Landau "saw and flipped for Barry and the Remains."[21]

> When they were introduced, they ran on stage, plugged into two Fender guitar amps through which they were running all their equipment and two microphones, and smiled. Four soft syncopated chords, and they broke into their first song at a volume, which was for me beyond belief. It was there when people started yelling for them to turn the volume down and Barry just stood there grinning and said, "Hey, this is our volume," and then broke into some ear-splitting hard rock piece. It was in the embryonic stages, but it was all there.[22]

An enormous fan, Landau saw the band play again only two weeks later at Boston University and noticed how tight they had become: "They were playing to a much better audience, and now you could see it all happen. The sound, the music, the feeling was there."[23] Barry Tashian recalls that the Remains had attempted to fix their sound prior to the sound company's influence but without luck. In Milano's book Tashian says that, when the band first began playing gigs at the Rathskellar in 1964, they "bought some speakers, these huge metal horns that were meant for public address systems or football games."[24]

Before Hanley, the group the Kingsmen hugely influenced the Remains' sound. Miller recalls: "We did a fair amount of shows with the Kingsmen; they were using these bullhorns. So we went out and bought these bullhorns. That was our PA system and some sort of amplifier we amplified them with. They were pretty tinny. This was pre-Hanley."[25] According to Tashian, eventually Hanley made some major changes to their sound: "Either Terry or Bill showed up with all of these Voice of the Theater speakers, power amplifiers, and groovy microphones. So we were kind of beefed up."[26]

When the Remains were preparing for a gig at a college or auditorium, the Hanley brothers usually showed up ahead of schedule with a trailer full of equipment. Tashian explains: "It was all very minimal. They might have put a mike on the snare drum and definitely the vocals, but it wasn't at the stage yet with big mixing consoles in the back of the hall with EQ and everything."[27] However, Hanley often experimented with different things. Tashian continues: "At one show the engineer set up long skinny microphones explaining to the band that they were television microphones."[28] Eventually the sound engineer made the bandleader a twelve-inch two-way coaxial (Altec Lansing) loud-speaker cabinet. Unique for the time, it had hidden spring-loaded handles on either side with wheels on the bottom. Tashian remembers, "This was the most favorite one that I ever had."[29]

Don Law, a high school friend of Tashian and a Boston University student, alerted music business executives about the Remains' unique sound. In January 1965 the band auditioned with Epic Records, leading to an eventual recording deal. Epic released four singles, including their popular hit "Why Do I Cry" in March 1965. Based on their local success and feeling confident, the band relocated to New York by the spring of 1966. Tashian reflects in his book *Ticket to Ride: The Extraordinary Diary of The Beatles' Last Tour* that "NYC was the place to be."[30]

At this time the group left Sdoucos, eventually connecting with publicist John Kurland, who became the band's official manager. According to Miller: "We loved John Sdoucos but we really wanted to broaden our horizons. It's just he wasn't breaking anybody nationally. We were ready to expand. We were signed to Epic Records and had two or three singles out already. It was time to make a change."[31]

New York was good to the band. They gained success through appearances on Ed Sullivan (1965) and *Hullabaloo* (1966) television shows. Their self-titled album was released in September 1966. The Remains disbanded just before its distribution.

The Remains were with Hanley Sound for over a year and a half before they moved to New York. Hanley still supported the band with local gigs in and around the New England area throughout the early part of 1966. The sound engineer encountered them again, this time with the Beatles in August 1966.

How the Remains landed an opening spot on the Beatles' tour is another story. In 1965, Bob Bonis with the General Artist Corporation (GAC) booking agency, and soon-to-be manager John Kurland had attended a Remains performance in Boston. According to Tashian: "They came up to see us play at a place in Kenmore Square called Where It's At. And it was just during that time when we were playing really strong shows, which we did that night too."[32] Later Bonis, who was already booking English acts, became interested in co-managing the band with Kurland. As luck had it, Bonis became instrumental in booking the Beatles' final tour. Bonis approached the Boston band in late spring of 1966 and asked if they were interested in opening for the group that summer.

Having the opportunity to support the Beatles gave the Remains the instant recognition and stardom they were hoping for. Miller explains: "It was ironic; we had just made a pact that we weren't going to open for anyone anymore. We were going to headline our own shows. We wanted to do it on our own. No sooner than we made that pact, it took us three seconds to break it."[33]

The wide-eyed bass player recalls the technical negotiations about sound reinforcement that occurred between Kurland, Bonis, and Beatles manager Brian Epstein. According to Miller, the agreement was that "the Beatles could use our PA system, if we could use their amps for the entire tour."[34] Once both management teams finalized plans, the Remains were locked in to play twenty minutes of original music as an opening act. Soon they joined a roster of three other groups for that summer tour, which included Bobby Hebb, the Ronettes, and the Cyrkle.

During the summer of 1966, Hanley Sound was running twenty-four hours a day with its overworked employees sleeping on the floor. Their focus was to get the equipment to the specified gig and set up on time. Most Hanley Sound employees agree that you did what you had to do to get the job done, often under intense pressure.

Tours in 1966 usually ran four to twelve weeks, depending on the contractual agreement. Sometimes there were shorter two- or three-week tours depending on if they were national or regional. Often Hanley crews pulled one sound system out of a theater in Boston at 1:00 a.m. for a show the following night. When Hanley heard that the Remains were opening for the Beatles in Chicago in August, he knew he had to drive out to see them. Only this time he had his eyes on providing sound reinforcement for the biggest band in history—the Beatles.

AFTER THE REMAINS: THE PSYCHEDELIC SUPERMARKET

In 1967 Hanley was renting out his sound systems and installing them at most of the local venues in Boston and in New York. At clubs like the Psychedelic Supermarket, the Catacombs, the Unicorn Coffee House, the Boston Tea Party, and Jazz Workshop, Hanley met the demands of increased volume. But, Hanley did not have enough money to staff each one with a permanent soundman. If there were a technical issue, he either tended to the problem himself or sent someone out. According to Hanley: "I put in a sound system at the Jazz Workshop for Fred Taylor. It was a Western Electric mixer with a McIntosh amplifier. I also had a system in the Boston Tea Party for Ray Riepen, and the Psychedelic Supermarket for George Popadopolis. George also owned the Unicorn Coffee House at 815 Boylston Street."[35]

A new psychedelic sound called the "Bosstown Sound" was emerging in Boston. More or less a marketing scheme manufactured by record producer Alan Lorber, what was occurring in Boston in 1967 was similar to what was happening in cities nationally, especially San Francisco. A concert with the Velvet Underground at a place called New England Life Hall at 225 Clarendon Street in Boston was one of Hanley's first shows of this period.

By the mid-1960s Boston promoter Popadopolis began booking rock and electric blues bands in addition to the standard folk music at his coffee house the Unicorn. The club was considered at the time to be more of a commercial coffeehouse than its competitor, the traditional Club 47. According to the Boston College student newspaper the *Heights Review*: "The Unicorn is equipped with a powerful sound system needed to handle these groups. The progress of the Unicorn seems to follow fairly closely that of the mass student culture, and is adapting, as near as its physical structure will allow, to current tastes."[36] The Unicorn already had a Hanley sound system in it. Liking what Hanley was doing, Popadopolis called on the sound engineer in mid-September 1967 for his new venture abutting Boston University. That venue was the Psychedelic Supermarket, located at 590 Commonwealth Avenue outside of Kenmore Square.

The Supermarket was difficult to locate. The structure, a converted parking garage, was hidden behind Commonwealth Avenue. If an individual were to attend a show they needed to venture down an alley to access the venue. By night the lower floor of the garage was converted to a concert space seating around 300. The sound was known to reverberate wildly within the cold concrete environment.

The Supermarket was preparing for a series of shows by the English group Cream on September 8–16. In late October 1967 Chuck Berry performed there. Popadopolis requested a Hanley sound system to rent, as the concrete space was becoming more and more of an acoustical challenge. But the venue's sound quality remained a problem. *Harvard Crimson* journalist James R. Beniger referred to the Supermarket as a "damp basement garage" with album covers lining walls, "The Electric Prunes, Surrealistic Pillow, Fresh Cream, The Grateful Dead—all call out in glaring psychedelic script." Beniger continues that prior to Berry's performance, "No more than a dozen people sit at tables near the stage—mostly teeny-bopper couples with happy-colored beads and sad faces."

Hanley's sound system screamed psychedelic music: "I'm so glad, I'm so glad, I'm so glad."[37] Described as having no windows, low ceilings, and being an overall cold environment, it was clear the Supermarket's days were numbered.

Before removing his sound system from the garage entirely, Hanley assisted Popadopolis with a series of Grateful Dead performances on December 8–9. According to Hanley: "I helped George and supplied him with some Voice of the Theater speakers, Altec horns with University woofers, Shure 545's, and McIntosh amplifiers. But the sound was just horrible, the venue had terrible acoustics."[38] Eventually Popadopolis merged his other venture, the folk-based Unicorn Coffee House, with the Psychedelic Supermarket. He officially closed its doors by the end of the decade.

Popadopolis was considered by many to be an extremely shifty individual, often cheating fans and bands of their money. It was reported: "Groups canceled contracts and left because they would be paid less for long stands. The exposure was supposed to make up for the lesser pay!! One out of two bands would leave a gig after one set for various reasons and regular clubgoers remember him raising ticket prices from $4.50 to $5.50 when he knew that a show was going to sell out."[39]

The Boston Tea Party's then-manager Don Law claims Popadopolis was the kingpin concert venue owner in the area with the Supermarket and Unicorn under his reign: "George had a reputation for not being very honest. His favorite thing was if you had a line out on the street he would get on a stool and change the ticket price while everyone was standing in line."[40]

Chapter 16

THE BEATLES FACE A FEVERISH PITCH

PUBLIC ADDRESS SYSTEMS IN BASEBALL STADIUMS

Prior to the Beatles invading America in 1964, the quality of live sound in the United States or abroad at any given venue was arbitrary and unpredictable. Often musicians relied on an instrument's amplifier to compete with louder acoustic instruments, for instance, a guitarist in a big band might overwhelm a horn section. In some cases vocalists plugged directly into a guitar amplifier, utilizing it as a PA. But most often a band used the primitive PA system the venue offered. If they were lucky there might be a small speaker cabinet on stage—or two for larger PAs—powered by a low-wattage amplifier and a few vocal microphones.

Plain and simple, there just wasn't enough power if you had a sizeable audience to reach. After the advent of high-powered, high-fidelity equipment in the 1950s and 60s, we see distinct changes occurring in how rock and roll music was reinforced with sound. In the 1960s and '70s, the requirement of better sound systems becomes even more apparent as the music business evolved and good sound became expected.

Back then the Beatles did not travel with a PA system in the United States, and their performances in enormous stadiums gives us a good example of the requirement for additional power in larger settings. In many ways they were the band that set a standard for how good concert sound should be, by showing how bad it was at these venues. According to Ringo Starr: "Now we are playing stadiums! There were all those people and just a tiny PA system—they couldn't get a bigger one. We always used the house PA."[1] The need to get louder was mostly evident during the "Beatlemania" phase, which began in 1963 and continued through their final tour in 1966. That fandom even lasted through the 1970s, although the group had disbanded by then. When Hanley met the Beatles on their last go-around in 1966, his friends from Boston, Barry and the

Remains, were the opening act for the Fab Four, performing in front of some of the largest audiences of the day.

Sound reinforcement for baseball stadiums was a challenge for Hanley. The Beatles performances Hanley supported occurred in the same year he did Soundblast! '66 at Yankee Stadium and the Batman Concert at Shea Stadium. Close friend of the Beatles Tony Bramwell reflects on the issues the band faced with stadium sound in his book *Magical Mystery Tours: My Life with The Beatles:* "In those days, the PA systems were adequate except in stadiums, where they were hopeless. The whole industry from recording to playing to marketing was finding its feet, still so basic that it is almost unrecognizable by today's standards."[2]

Road manager for the Remains, and official photographer for the Beatles tour, Ed Freeman considered the sound "pathetic" at the venues with a dedicated PA system, referring to it as "unintelligible noise." Freeman recalls: "Looking back, what strikes me most is what an earth-shaking event that tour appeared to be at the time, and yet how modest a production it was by current rock standards."[3]

PA systems like these were in place for announcements, not music, and most had a terrible delay. In effect, uncontrolled delay for a musician poses real problems, since they need to sing in time with the music. It was quite distracting when, every time music was played on stage, the band would hear it seconds after. "It was kind of like playing in the Grand Canyon or something,"[4] recalls Remains bassist Vern Miller. According to Tashian: "What I heard was an out-of-sync echo from the stands. All I could do was close my ears and plow through the songs. We were singing on stage and behind second base, and the sound was coming out in the stands 150 ft. away, two, three seconds later."[5]

GLENN D. WHITE

To better understand the complexities that Hanley and public address operators were dealing with, it's helpful to examine the Beatles' acoustical problems prior to 1966. Facing this complexity, acoustician and public address operator Glenn D. White was stunned on August 21, 1964, when the Beatles performed at the Seattle Center Coliseum in front of over 14,000 screaming fans. According to White, the sound system was only audible when the screaming died down. When the group belted out a few words it was followed by another wave of screaming lasting for the entire twenty-two-minute set. Before the Beatles arrived in Seattle they sent two individuals to the venue to see if White could enhance the public address system. They wanted him to get the system as loud as possible in order to have the band heard over the frenzied roar.

In order to accommodate the band, White scrounged around the complex for additional equipment to add on to what was already being used. In an article in *Tape Op* magazine, White confirms a couple of key points: 1) the sound department at the Seattle Center Coliseum was solely in charge of mixing the performance, and 2) the type of equipment he used for sound reinforcement. White recalls: "All of our amplifiers in those days were 80-watt tube amplifiers. So I added three Altec/Lansing 300 Hz 2-cell horns (model 203, on 288 C Altec drivers). Each horn had its own 80-watt amp, and the woofers were JBL 15" models."[6] White speaks in detail about the placement of his speakers for the Beatles' Seattle Coliseum performance:

> I put two of those, one on top of the other, facing one of them back to cover the far end of the auditorium, the other one facing down a little bit to cover the far end of the main floor. Then we had some 800 Hz. horns that were also Altecs. They were 805's, two rows of 4 cells each, and I had one of them pointed right down on the stage so that the Beatles could hear themselves and also cover the audience close to the stage. The speaker cluster was right above the stage.[7]

White used Electro-Voice 655C omnidirectional microphones for the group, with one on Ringo's drum kit. According to White, the drummer was amazed when he could actually hear the response from the sound system as he hit his stick against the microphone.

Between the band's performance at Carnegie Hall on February 12 and the final date of this first US invasion in September 1964, the screaming fans got louder. Beatlemania was now catching on, which meant fans versus live sound posed a serious issue for the band. In Washington "The Beatles had worked the crowd like a politician," but in New York "They remained aloof, frustrated by the audience's apparent refusal to listen."[8]

Not being able to hear clearly embittered the band, especially John Lennon, who reached his breaking point during their performance in New York. Bob Spitz explains in his book, *The Beatles: The Biography*, that Lennon's impatience eventually surfaced right after the band's seventh song: "He stepped forward, looked at the audience sternly . . . and yelled, 'Shut up!' The indifference to the music became, apparently, too much to bear."[9] After lasting thirty-four minutes on stage, the Beatles left the venue.

Spitz continues to detail how poor sound troubled the group in 1964, again angering Lennon to the point of frustration. After the band's "standard half-hour" set at a particular gig, Lennon claimed: "It was ridiculous! We couldn't hear ourselves sing."[10] Often they couldn't tell the key they were in during any given number. Succumbing to the monstrous roar, the Beatles gave in, and "barely even attempted to strum the chords at times."[11]

Aware of the Beatles ongoing sound issues, Hanley attempted to help them just as they were taking America by storm. The young sound engineer describes this in "Wired for Politics: Audio Enthusiast Lives for Sounds," published in 1964 in the *Boston Globe*: "Regardless of what you might think of their voices, or their songs, their timing and execution is very professional. I tried unsuccessfully to go trouping with them, and handle the sound."[12]

From 1964–66, the group trudged through it all despite continuing obstacles of intelligibility. While performing live, hiding behind their bobbing heads and innocent smiles, no one would have known at all they were having problems hearing each other.

DUKE MEWBORN

Even though technical issues plagued the Beatles throughout their entire touring career, one exception was a performance at Atlanta Stadium on Wednesday, August 18, 1965. Sound engineer F. B. "Duke" Mewborn handled the sound reinforcement for this performance. He worked for an Atlanta hi-fi and PA rental company called Baker Audio.

By now the Beatles were used to not being able to hear themselves while performing. Mewborn was aware of these problems and decided to provide the band with stage monitors aimed directly at the group during their performance. Mewborn recalls: "We were afraid of echo so we put a monitor speaker which was right in front of the stage down on the ground on a chair facing them. This allowed the Beatles to successfully hear both their voices and instruments and they loved it!"[13] The sound engineer did not use the stadium system speakers, although he did use the stadium electronics. The stage positioned in the center of the playing field had two stacks of Altec A7 speakers on both sides clustered at first and third base. They were powered by four Altec 1570 amplifiers supplying 175-watts of sound.[14] Mewborn used cardioid mikes on the band claiming, "They had a restricted pattern" that rejected the "ambient sound coming from the sides or below."[15]

Chuck Gunderson writes in his book, *Some Fun Tonight!*, "The group gave a rousing concert bolstered by the great sound quality." During the set Paul McCartney shouted his approval, "It's loud, isn't it? Great!"[16] In a note to the sound company written after the performance, band manager Brian Epstein referred to the PA as "excellent" followed by "Without question, it proved to be the most effective of all on the US tour in 1965."[17]

Epstein asked Mewborn to go on tour with the band, but the sound engineer politely declined claiming, "I had a business to run. Being a roadie, that's a different life."[18] Mewborn estimates the screaming fans produced decibel levels reaching one hundred: "We made it work, and we managed to overcome it the best we could."[19]

Chapter 17

BEATLES BY CHANCE

CHICAGO

In 1966 the hysteria of Beatlemania was still evident across the United States but slightly waning; even so, sound was still an issue. Pockets of fans remained hungry to see the four boys from Liverpool, England, perform live. Tony Bramwell reflects on the group's reluctance to tour the States that year: "Despite saying that they would never go abroad again, another American tour had been booked for August 1966. The Beatles were bored and restless and said they wished they didn't have to go."[1]

The Beatles had become "prisoners of their own fame," as witnessed first-hand by members of the Remains. Tashian observes, "It appeared that they had entered a dangerous position, for a while, Beatlemania was at its peak, they had ignited the flames of controversy, and were seen by many in the US as problematic."[2] In March of that year, during an interview in the UK, Lennon claimed the Beatles had become more popular than Jesus Christ. By the time the group finished the rest of their US tour that summer, tempers flared within Christian communities over that controversial statement.

The entire tour covered fourteen US cities in August 1966. On some occasions the band had two performances per booking, one in mid-afternoon and another at early evening. This made nineteen performances in total. Tashian describes the tour as some of the most memorable days of his life. It began on August 12, 1966, at the International Amphitheater in Chicago, and ended August 29 at Candlestick Park in San Francisco. Hanley provided sound reinforcement for several of the performances.

The Remains were slated to open in Chicago, where the British group had played prior in September 1964. Nervously, the Boston band played two shows, one at 3:00 p.m. and the other at 7:30 p.m. in front of more than 13,000 frenetic audience members. Both bands intended to use their own equipment and expected to play through the sound system available at the amphitheater. This was before any knowledge that Hanley Sound was to accompany them on the tour.

Hanley had wanted to tour with the Beatles since 1964; now he had a chance. He trucked his sound equipment from Medford across the country to Chicago to see what would happen. He showed up unexpectedly, as some of the Remains recall. According to keyboardist Bill Briggs: "All of the sudden Bill shows up with his trailer truck unannounced, and begins setting up his equipment. I don't think the Beatles knew and when they saw it they said 'What's this?'"[3] It's unknown to the Boston band how Hanley arranged to provide sound reinforcement in Chicago. A business decision like this would have been left to the group to discuss.

Remains bassist Vern Miller recalls: "I don't have any knowledge of the Remains contacting Hanley in advance. I don't remember that conversation. The impression I got was that it was presented as a gift to the band."[4] With no evidence of a deal being struck or contract signed between band and sound company, the ambitious sound engineer most likely didn't ask for any money up front. According to Briggs, "It is quite possible that since the sound was known to be so bad in Chicago, that our manager John Kurland contacted Bill on the side without the Beatles knowing about it."[5]

Hanley's intuition and impulse to come to Chicago was on target, since the quality of sound at the Amphitheater was notoriously inadequate. Tashian, who kept a detailed journal regarding the experience, recalls: "Hanley pulled his truck right into the Amphitheater, unloaded, and set up their state-of-the-art sound equipment right beside the in-house PA system. What a joke. The in-house stuff was so archaic next to Bill's powerful amps, good mic's, and Altec Voice of the Theater speakers."[6]

When the time came for a sound check, there was a squabble between the union, Beatles manager Brian Epstein, and Remains manager John Kurland. When the union realized the band wasn't using the house system they were upset. Tashian recalls: "They told us we had to use the house system, which was ancient. It was pretty bad compared to Bill's equipment."[7] Miller remembers: "Beatles manager Brian Epstein turns to our manager John Kurland and says, 'I see our sound system is here.' Kurland recognized Hanley and says to Brian 'No that's not your sound system, that's ours!'"[8]

As clear as day, Epstein could hear the difference in quality of Hanley's sound system compared to the house PA. According to Tashian, "Brian looked at the two sound systems and decided that the Beatles should go with OUR system," adding "It was the Hanley Sound system, our sound system, that the Beatles used that night."[9]

Hanley anticipated that providing sound for the raucously loud audiences would be a challenge. Judith Sims, writer for *Teen Set* magazine, was sitting behind the stage during this performance. Sims detailed the audible chaos that ensued during the Beatles' show, claiming the "roar was deafening" and that her "ears were ringing." She continues, observing the difficulty for the band: "It's a wonder the Beatles can see or hear at all after going through that so often."[10] Even though there were challenges, the Hanley sound system impressed Epstein. He invited the sound company for additional dates during eastern and midwestern portions of the tour. Tashian remembers: "I told Hanley we had no budget to pay him. We could only pay a roadie or two to help Beatles' road manager Mal Evans. Bill said 'I don't care I am going to go out anyway!'"[11] If it was fate that landed the Remains on that tour, then it was quality sound that gave Hanley his chance.

The sound engineer was both focused and overwhelmed during these concerts. Known to be friendly and outgoing, Hanley was a geeky sort of guy often referred to as the "mad scientist of sound." Wearing a white collared shirt buttoned all the way to the neck, he looked the part. As the tips of his wavy hair danced in the ball field wind, he tweaked the dials of his sound machine into overdrive.

The mercury rectifier tubes within the four RCA 600W battleship amplifiers intensified, glowing bluish purple. According to Hanley, he pushed the military-grade equipment to the limit: "They weighed two or three hundred pounds each and didn't have great frequency response. You had to drive the inputs with 100W, but it was not enough."[12] Briggs recalls Hanley's system as science fiction then: "There he was behind these huge glowing tubes, twisting dials. It made me think of the sounds of electricity as they were generated in Dr. Frankenstein's laboratory. He was a mad scientist! I had never seen anything like this before! He had a big smile on his face with his headphones on."[13]

Some of the electronics on this tour came from a junkyard Hanley frequented called Eli's, officially known as Eli Heffron and Sons on Elm Street in Cambridge. The facility boasted a "Surplus of Electronic Equipment with Inventory Changing Weekly." It was here where the inquisitive sound engineer scrounged a surplus of used electronics and test equipment in an effort to make a better sound system. Others in the area, including audio buffs and MIT students, also scavenged Eli's for components to build audio systems. One other junkyard close by offered Hanley additional equipment, including the amplifiers he used for the Beatles sound system.

Hanley remembers: "You couldn't hear yourself think during these shows, 46,000 teenage girls screaming at the top of their lungs, 120+dB ambient noise. It was unbelievable!"[14] Sound reinforcement at this level was still new to the audio community, and it was Hanley putting all the pieces together right before everyone's eyes and ears. But not every show had screaming girls. For example, southern fans were a bit more polite, which allowed clearer and audible experience for the group. At a typical southern show there were lulls in the screaming as the Beatles tore into their popular catalog.

After Chicago, the Beatles moved on to Detroit, Cleveland, Washington, DC, Philadelphia, Toronto, Boston, Memphis, Cincinnati, St. Louis, and New York, then on to the West Coast to finish the tour. The band and production entourage traveled to most of these dates on the tour by either bus or chartered plane. Hanley's sound equipment was trucked from show to show, and depending on the distance between each date, he simply could not support them all.

Moving equipment posed a logistical problem for the overall production. Yet, on a few occasions the sound engineer did accompany the Beatles and the Remains on their chartered plane. Epstein was worried that there might not be enough room for the band's equipment on the plane. According to Miller, "There was a big problem with figuring out how we were going to get our amplifiers from city to city."[15] Often after a show you could find Tashian helping the road guys load guitars and amplifiers off of a truck and into the plane's cargo hold.

Hanley left the Beatles after their performance at John F. Kennedy Stadium in Philadelphia on August 16. The band went on to perform in Toronto. The sound engineer and crew needed to make it back home in time to prepare and support the Beatles' show at the Suffolk Downs racetrack in East Boston on August 18. When the Beatles arrived at Boston's Logan International Airport at 3:45 that afternoon, Hanley and his team were setting up the sound system yet again.

The local promoters who put on the event, Gerald Roberts and Frank Connelly, knew Hanley well. The two had booked many shows at the Carousel Theater in Framingham and often hired Hanley Sound. Suffolk Downs had the capacity to hold up to 50,000 people and featured a 16,000-seat concrete grandstand. Although expecting to exceed the audience limit, the promoters were only able to sell 25,000 tickets. Later on, Hanley Sound supported bands like Aerosmith and the Jackson Five at this same location.

Boston Globe writer Ernie Santosuosso writes about the system Hanley was using for the Beatles: "I was up front some four feet away from the stage. I should mention that I was stationed face to face with a set of ominous-looking speakers. I didn't miss a note. In fact I may hear again by New Year's. The sounds emanating from the speakers and zeroing in on me cleaned my sinuses more efficiently than any antihistamine."[16]

In an effort to keep the Suffolk Downs' dirt as pristine as possible—and keep fans from charging the stage—organizers had the Beatles' temporary wooden stage set up in the center of the racetrack area. This created a substantial distance between the band and their screaming fans and some of the fans didn't like that. In epic fashion, the Beatles ripped through a thirty-minute set with clearer sound than on other tour dates. The concert took place only a few short miles from Bill's hometown of Medford, and is further evidence that he had come far in a relatively short period of time.

The following day the Beatles left for their date in Memphis, which Hanley was not part of. The sound engineer bid his farewell to Epstein and the band. He would see them next at Crosley Field in Cincinnati and then at Shea Stadium in New York. It was evident that a dark cloud loomed over this tour. Lennon's claim that the Beatles were more popular than Jesus Christ had created an enormous negative stir while they were in the States. Even though Lennon at a Chicago press conference, emphasized it was taken out of context, it made people upset.

According to Miller, when they left for the South the Beatles were under heavy protection: "I remember being in Memphis on a bus from the airport to the Coliseum surrounded by police and secret service. This was the Bible belt!"[17] At one stop in Washington, DC, the Beatles were picketed by a chapter of the Ku Klux Klan nailing a Beatles LP to a wooden cross, and vowing "vengeance" toward the band. Throughout the country, various conservative groups held public burnings of Beatles records.[18]

Chapter 18

SHEA STADIUM SOUND

THE BEATLES' SOUND AT SHEA STADIUM 1965

Because of supply and demand, it was logical for the Beatles to play big venues like William A. Shea Municipal Stadium in Queens, New York. Here they could perform for large crowds in one shot, and promoters loved that. Even with their specially designed Vox "Super Beatle" AC 100-watt amplifiers turned up to ten—combined with the house PA—it could not be heard over the 50,000 enthusiastic (mostly female) Beatles fans.

The sound company that supported the Beatles at Shea on August 15, 1965, is unknown. With no evidence of the group touring with a sound company or engineer until Hanley did, some suspect that road manager Mal Evans occasionally acted as soundman. Handling the band's instruments, from retuning guitars to setting up drums, Evans's possible role in mixing the sound is highly plausible.

When the band took to the Shea stage at 9:02 p.m., a rocketship roar of screams deafened them. According to George Harrison: "Shea Stadium was an enormous place. In those days, people were still playing the Astoria Cinema in Finsbury Park. This was the first time that one of those stadiums was used for a rock concert."[1]

Roy Clair, founder of Clair Global, suggests a New York outfit called Neagle Sound provided sound for the sold-out performance. But there is no evidence to back this claim.[2] After analyzing photos of equipment used, Hanley claims it wasn't much: "Whoever was the sound company, they used what look like eight-inch Temple speaker columns about four feet long."[3] *New York Times* writer Murray Schumach incorrectly identifies these speakers as amplifiers: "Little more than the pulsation of electric guitars and thump of drums reached the stands although more than fifty 100-watt amplifiers had been set up along the base paths of the baseball diamond."[4]

According to experts in the audio field, the vocal mix was not played through the house PA but rather through an array of Electro-Voice LR4 column speakers positioned about the field aimed toward the hysterical fans.[5] According to Hanley, "The delay, combined with those field speakers and the screaming house speakers, would have been atrocious."[6] Regardless of who was deploying sound reinforcement, it was no match for the "wall of piercing sound that blared from the stands."[7]

A Shea Stadium production sheet cites Fred Bosch as sound engineer and Bob Fine as sound mixer for the event.[8] Both were positioned behind the elevated stage platform at stage right. Bosch and Fine were registered with the Audio Engineering Society at the time. It's well established that they handled only the audio recording portion of this show, since it was being documented and filmed for television release. But it's possible that Bosch could have provided full-on sound reinforcement for the event. Since 1951 he had worked extensively in the film and television industry.[9] Bosch was known for his involvement in creating the six-track sound system for Cinerama.[10] A former field engineer for the Altec Service Corporation (specifically for theater sound systems), Bosch would have been a knowledgeable and suitable candidate for the job.

Bob Fine may have made Bosch a good teammate, because he had a recording studio in New York called Fine Sound. After his death in 1982, an Audio Engineering Society memoriam observed: "He then founded Fine Sound in New York, and those studios became the birthplace of many of the early and vital stereo LP disks that featured a high degree of separation. These records in many ways, helped materially speed the consumer interest and acceptance of stereo."[11]

Recording in a stadium, let alone one with a fairly consistent tumultuous roar, meant that Bosch and Fine probably could not hear very well. The *New York Times* reported that the Shea performance was riddled with acoustic issues: "Fans drowned out almost all the singing. The Beatles were situated on a stand at second base."[12] The problems were so severe that Beatles producer Sir George Martin had to work on additional overdubs and retakes on some of the band's performance tracks early the following year.[13]

The roar of shrieking fans must have overwhelmed the hired sound company. A year later in 1966, Hanley had a similar experience at the same venue. According to Tashian: "It would've taken an act of God to get the Beatles to be heard at Shea. All of that screaming at once, sounded like a rocketship taking off."[14]

HAROLD COHEN AND SHEA STADIUM 1966

Engineer Harold Cohen joined Hanley Sound during the summer of 1966 at age twenty-two, while still in college. Terry Hanley introduced the young engineer to the Medford firm while Hanley was preparing to go on the road with the Beatles. According to Cohen: "I figured I could fit the job into my weekend schedule. I was interested in stage lighting, with sound becoming a natural progression for me. I realized quickly that the company was involved in some fairly innovative things."[15]

Cohen claims you weren't always "compensated" for being young and smart, relating that he was fired after the first month: "It was over my pay. As it turned out I was the highest paid employee in the company and knew the least. So I was let go and went to work for a professional photographer back home."[16] Eventually the Hanley brothers came to an agreement with Cohen regarding his compensation.

At the time Cohen arrived, the company's clients were accustomed to the quality Hanley's equipment was producing. Cohen claims, "Most of them never heard anything like it before." Cohen liked this and came to share the same convictions about sound as Hanley:

I designed my own apprenticeship there and I felt as though by learning from Bill and Terry I was getting the best. Bill used the best equipment and although he wanted power, he also wanted it to sound natural at no cost. Bill really focused on producing high quality audio back in those days. He wanted to represent the performer (or politician) in their best light always. What we were doing was so unique. We often got billing alongside the acts at places like Madison Square Garden, on outdoor billboards, and other concerts. We were a feature that promoters could hang their hat on because they felt we were a superior product. Bill had to definitely convince people to use his equipment and his ways, but he really provided quality sound with integrity. We were one of the only sound companies that were doing live sound on a whole different level than others. Bill really set the standard of high-quality audio.[17]

Cohen became one of Hanley Sound's longest-serving employees and a unique character within the generations of Hanley Sound crew members. He was a reliable, straight-laced sort of fellow with short hair. This meant that Hanley could use him for political events. Cohen, like Hanley, worked for both political parties and over time engineered sound for the Kennedy, Nixon/Agnew, McGovern, and Kerry campaigns. Yet nothing would compare to working with the Beatles.

By the end of August, Hanley and his crew had provided sound reinforcement for the Beatles and supporting acts in Chicago, Detroit, Cleveland, Washington, DC, Philadelphia, Boston, Cincinnati, and the band's final New York performance. Soon after the modest 12,000-person show at Crosley Field on August 21 in Cincinnati, Hanley and crew departed the tour. Hanley had worked at Crosley Field before and knew the promoter, Dino Santangelo, very well. When the band got off the stage, Hanley and his crew took their time striking equipment in the extreme afternoon humidity, anticipating a long journey home. Hanley told band manager Brian Epstein he would meet the group in New York on August 22 for their scheduled press conference.

After Cincinnati the Beatles traveled 300 miles to Busch Stadium in St. Louis, Missouri, for a gig that evening. Performing in two different cities on the same day was unheard of then, but due to rainy weather the day before, the band's dates had been consolidated. Hanley later heard about their rain-soaked Busch Stadium performance in St. Louis: "It's a good thing I didn't go. I heard it rained fairly heavily, and that would not have been good for my equipment."[18]

Hanley was right; St. Louis was one of the most difficult of all the dates on the Beatles' tour. There was poor weather and the band was again victim to another horrific PA system. The show received mixed reviews. Robert K. Sanford, a journalist for the *St. Louis Dispatch* wrote, "Fans at Busch Stadium got plenty of volume for their money," stating that the "song lyrics were difficult to understand" and that the "rain did not dampen the echo qualities of the stadium."[19] Music for the event was channeled over 200 speakers within the stadium structure. Jack Goggin, public address system operator for Busch Stadium, recalls, "The system works well when performers speak distinctly, but distinct enunciation is not a notable ingredient in rock-n-roll music."[20]

On August 22 Hanley began setting up for the Beatles' press conference at the Warwick Hotel in New York. Earlier that day the band had flown in from their show in St. Louis, arriving close to 4:00 a.m. The Beatles gave two press conferences and Hanley recalls each being a "zoo." The following night the group performed at Shea Stadium. For all the venues plagued with sound problems on this tour, Shea received more attention than any other. In hindsight it was a really big deal to play this venue. New York is the media capital of the world and the Beatles had had their performance at Shea filmed and recorded the previous year. Tashian recalls: "New York was the big time. I remember the lights were so bright at Shea Stadium. There was a lot of excitement in the air."[21]

The 1966 tour had not been as successful for the Beatles as those previous. It was burdened with bad weather, political, and sound problems. Even though it wore on the band, from a fan perspective you might not have noticed.

However, Remains keyboardist Bill Briggs did. Observing the issues the Beatles had on stage he recalls: "The pitch of the audience overshadowed the music and the Beatles knew that and played accordingly. They kind of played fast and loose."[22]

Chicago Sun Times journalist Glenna Syse reflects on the rabid behavior of the fans during the 1966 Shea Stadium show: "You have to see it to believe it. . . . It is not the kind of thing you believe by hearing. These conclusions are not sociological they are medical." The journalist then goes on in a comically written review: "When you plug over 10,000 young female larynxes into the Beatles' circuit, you produce a vibration that causes a disease called labyrinthitis, which is an inflammation of the inner ear that sometimes results in loss of balance. It is an ailment that seems to affect only those age 15."[23] Tashian remembers that "Beatlemania" had reached a point where, at many of their concerts, the words of their songs could not be heard above the screaming fans and "Shea Stadium was no exception."[24] This is something to which Hanley attests.

The enthusiastic, anxious, and noisy 1966 Shea Stadium crowd was 40,000 or more. Hanley's total system power consisted of only 600-watts, with no monitors. Hanley recalls using monitors on a few dates on the tour but not for all. However, when monitors were applied, Tashian claims it made a huge difference: "When the stage was on second base at these ball parks, Hanley placed a monitor in the middle and flipped it around so it was facing the stage. Miracle of miracles we could hear ourselves!"[25] Most automobiles we travel in now have sound systems that offer more power.

These were the early days of concert sound and you could not simply purchase what you needed at a music store. The Shea Stadium configuration was comprised of the following: A dozen or more "monster" Altec 210 low-frequency horns, and Altec 203B two-cell multi-cells with 290 drivers. According to Hanley: "The 288 drivers had better high-frequency response but a 288 didn't last the night. We were using the 290s, even though the forty voice coils were an inconvenience."[26] For the event Hanley used a 600W RMS amplifier built by RCA weighing around 300 pounds, something that the engineer claims was surplus from a battleship: "I recall they were using these same RCA amplifiers for the public address system at the Indianapolis Motor Speedway.[27] For microphones the sound engineer used Shure 546s, which he placed on the amps and drums. Hanley claims he even had a small mixing board with EQ on the input, which was custom-built back at the shop in Medford: "The mixing console was a custom-built unit, using sixteen channels of Langevin modules."[28]

For previous events, held at baseball stadiums, Hanley kept his speakers low to the ground. He spread them out around the bases of the field in order to keep the reverberation contained and the sound from bouncing all over. At Shea the configuration was slightly different; he kept his speakers in two separate clusters at stage left and stage right for more concentrated sound pressure levels. According to Hanley, "I had the band on the pitcher's mound, the speakers on the first and third base line, and I made this big circle of sound, all facing up, so the speakers didn't cross."[29]

Tashian remembers: "There were four or five speakers spread out over the infield facing the seats. Bill did what he could. I think he even put a mike on the drums and maybe some of the instrument amps in some cases."[30] The band's guitars, basses, and keyboards were fed through solid-state Vox Super Beatle amplifiers. Tashian continues: "They gave us two in tandem to each player. We just cranked them up and expected the whole ballpark to hear perfectly, which was kind of a joke."[31]

Hanley recalls that during this time of his career, "I spent all the money I made on new equipment so I could do a better job."[32] Spurred by Bob Dylan's transformative electric performance at Newport just a year prior, by 1966 the sound engineer was scouting "larger venues" for work. "I realized electricity was an extension of the musician" adding, "After Dylan went electric I had learned more about it and put together an enormous amount of gear."[33]

In the space of one year, venues and sound requirements seemed to get bigger and more demanding for Hanley. "Sound had changed from being high fidelity (faithfulness of reproduction). Since the Beatles, it's become a battle of levels. With the Beatles, the on-stage levels were getting higher and higher, and the supporting sound system had to grow just to keep up."[34] In a *Pro Sound News* article, Hanley describes his passion for quality sound at bigger events: "I was hung up on fidelity; I was the first guy to really go bananas with it and try to bring it outside to the masses. A great deal of joy that happens for people at concerts is from the intensity and fidelity."[35]

New York music producer and promoter Sid Bernstein was instrumental in hiring Hanley for the 1966 Shea Stadium performance. According to Hanley, he had been trying to work with Bernstein for some time: "We got the Beatles job through Sid Bernstein who I was chasing in New York for one to two years already."[36] The impresario was fond of the sound engineer's work even though the Beatles 1966 Shea Stadium performance was mostly inaudible. Bernstein claims in his book *It's Sid Bernstein Calling . . . : The Amazing Story of the Promoter Who Made Entertainment History*: "I called Bill Hanley, an audio genius who worked out of Boston. Bill set up a sound system that proved to be a vast improvement over the first [1965] Shea concert."[37]

For this Shea performance there were at least 10,000 less in the audience compared to the previous year. Jonathan Gould claims in his book *Can't Buy Me Love: The Beatles, Britain, and America*, "Elsewhere, at some of the big ballparks—including Shea Stadium in New York—there were rows of empty seats."[38] Managing sound in a baseball stadium if there were rows of empty seats was a problem. Unoccupied seating meant less sound absorption, but in this case Hanley did not think it mattered much.

Harold Cohen, who had just joined the crew in 1966, observes: "We could not overcome the intensity and volume coming from the audience. The equipment in our inventory was not designed to handle those levels."[39] Hanley's attempt to contain and amplify the sound at Shea was a challenge any sound engineer would have faced. He recalls the event, as "sheer insanity" with a sound so loud that "you didn't even know there was a concert going on." Still amazed, he remembers: "You couldn't do your job very well because you couldn't even hear yourself think, you couldn't hear any sound coming from the system. It's like it wasn't even on!"[40]

Hanley was slammed with what he claims was 135 dB of intense screaming that day and his system couldn't come close to that. He emphasizes that he used "everything he had" during shows like this and nothing more could have been done short of a miracle. *New York Times* journalist Paul Montgomery claimed, "As usual, the noise was deafening, the music all but inaudible, the hysteria high."[41]

As a sound engineer, Hanley could have never imagined the challenges he faced on this tour. He was and is revered for managing quality sound, even though the Shea performance is not a good example. For his efforts, he should always be remembered as the Beatles' first sound reinforcement engineer. Remains bassist Vern Miller recalls: "Using Hanley's equipment felt like the big time. Bill Hanley was really smart and really knew his stuff inside and out."[42]

Shea Stadium was the last date on the Beatles tour for which Hanley provided sound reinforcement. This entire experience was something the sound engineer will never forget. "At the end, Beatles manager Brian Epstein begged me to come with them to England, but I respectfully declined the offer. I regret it now, but I told them I was too busy, which was true."[43]

Hanley's sound system served as an important tool for all of the musicians on this tour. Working with the Beatles was a great résumé builder for Hanley. His innovativeness demonstrated to the audio engineering community that there was a need for additional power for an emerging live-music business. Hanley led by example through an attempt to solve an unsolvable problem. Due to this, Tashian's appreciation of Hanley has grown over the years. The bandleader is thankful for the sound engineer's generosity and tenacity: "At the time I must admit I didn't know that Bill was special. Or that what he was doing was so unique. I was very self-absorbed in the music. It wasn't until years later that I realized Bill saw a chance to go and set up a sound system that might be noticed by the Beatles and he went for it."[44]

THE END AND BEGINNING OF AN ERA

Hanley attended the Beatles' final concert at Candlestick Park in San Francisco on August 29, 1966. This event was the final Beatles concert on this fourteen-date tour, and as one writer claims, "The Beatles belted out a final version of Little Richard's "Long Tall Sally" and with that were gone for good."[45] The sound engineer showed up without his tractor-trailer filled with sound equipment this time. McCune Sound handled production for this performance.

In 1966 the Beatles' changing persona aligned with an evolving American culture. By the end of this final commercial tour, the Beatles' music was noticeably shifting in a different direction. Their "paisley patterned" shirts poking through their grey military style suits is another indication of this change. It was a different look from the smart-looking, grey collarless garb they wore for their 1964 invasion tour.

It was a rapidly changing scene as 1966 rolled into 1967. San Francisco's neighborhood of Haight-Ashbury became a gathering place for a new community and a sharing-based hippie culture that was born out of the 1950s Beat Generation. Four months after the Beatles' final performance at Candlestick Park, on January 14, 1967 a mass gathering celebration of over 20,000 called the "Human Be-In" occurred at Golden Gate Park. The event sparked a new movement of individuals who rejected the materialistic values of modern life, characterized by a change of culture, social activism, music, and festivals. Unbeknownst to the busy sound engineer, the emergence of the "Summer of Love" and the music that represented it propelled him into a new realm of rock and roll sound reinforcement. The psychedelic era had truly begun.

As the Beatles' tour ended, a torch was about to be passed. The West Coast sounds of psychedelic and folk rock music moved eastward and Hanley was prepared. In the coming year the sound engineer witnessed the beginning of this shift in music and culture. The festival movement was on the horizon as was his inevitable transformation into—The Father of Festival Sound.

Hanley now had the West Coast in view after his company had a rigorous year of touring with the Beatles and Beach Boys. Hanley bid unsuccessfully for sound reinforcement at two Bay Area venues, Chet Helms's the Family Dog and Bill Graham's Fillmore West. Hanley remembers: "Chet told me he was 'all set' and Graham wasn't at all interested as they were using companies who were closer to them. It was difficult trying to convince these promoters that I could offer them a better solution."[46]

Chapter 19

THE COMPETITION

This chapter provides background on some of the sound reinforcement companies across the United States that eventually formed Hanley Sound's regional competition. Far removed from Hanley's Medford home base, parts of the West Coast, East Coast, and Midwest had their share of up-and-coming first-generation sound companies. However, most of these firms were isolated from one another.

MCCUNE SOUND SERVICES

In attendance at the Beatles' final performance at Candlestick Park on August 29, 1966, Hanley looked on at the modest sound system being used. West Coast sound company McCune took on the baseball stadium challenge. Hired by local promoters Tempo Productions, the McCune logbook entry for the show said, "Bring everything you can find!"[1]

Abe Jacob, former McCune Sound crew member, comments on the facilitation of sound for this event: "The no frills system consisted of Altec A7-500 loudspeakers (twelve spread along the infield lines) two Altec 849 Studio Monitors, four stage monitors, and four Altec 1567A 4-channel microphone mixers."[2] Jacob, who went on to become sound engineer for the Mamas and the Papas, claims: "The vocal mikes we used at Candlestick Park were AKG D202. I later used the same mikes for Mama Cass. She was impressed that she was using the same mike as John Lennon."[3]

Jacob recalls the sound reinforcement landscape back then: "It was Hanley, McCune and Stan Miller in Nebraska. These were the main companies doing concert performance sound in the United States in the mid- to late 1960s. At McCune they were using Altec Lansing loudspeakers, and like Hanley, McCune slightly modified them."[4]

At the time only a few West Coast, Bay Area sound companies (and engineers) mirrored what Hanley was doing on the East Coast. Quality concert sound during the 1960s was as essential on the West Coast as anywhere else in the country. The festival scene was developing and vibrant performances had been occurring in many popular ballrooms like the Family Dog, the Winterland Ballroom, and the Fillmore West. Swanson and McCune Sound were the two companies most called upon for their services. They're still in business today.

Established in 1932, the San Francisco–based Harry McCune Sound Services Company (McCune Sound, Lighting and Video) is now known throughout the world as one of the earliest and most innovative in the sound reinforcement industry. Founded by former auto mechanic Harry McCune, the firm has seen enormous growth over several decades and three generations of leadership.

Early on in his career Harry McCune Sr. could be seen driving a sound truck replete with speaker horns on top in support of ethnic picnics for the local German, Irish, and Italian communities. During this time McCune built his own amplifier using a schematic from a book. When big bands passed through during the 1930s and '40s, McCune also rented equipment. Gradually the company began to support theaters, corporations, and live rock concerts.[5]

Harry McCune Sr. was a true inventor. He built one of the first answering machines, a device that contained a recording cylinder and a phonograph. When someone called his office it would trip the ring and drop the needle on the phonograph record. The record had McCune's voice saying, "I am unavailable. Please leave a message." This then started a little cylinder recorder so people could actually talk and leave a message. It became so popular that people called just so they could leave a message. Since his invention tied up the telephone exchange, the local fire department forced him to disconnect the device. According to longtime McCune employee Mike Neal: "Harry was very cheap. He didn't want to pay someone to stay in the office and answer the phone!"[6]

Harry McCune loved to race automobiles. During the 1950s and 1960s his area had a few racetracks. While in attendance he noticed the track workers were having difficulty communicating with each other during events. They had been using the antiquated intercom systems previously installed by the phone company. According to Neal, "The only intercoms that were around were the old carbon mikes that made your ears hurt, they also weren't very loud."[7] McCune saw an immediate need for something more advanced.

Harry solved the problem by placing a directional current on a loop that carried it. Located at each position of the track, little boxes housed a microphone input and a preamp that boosted the microphone level. The boxes were equipped with a headphone amp (tapped off of the audio loop) and a pair of noise-canceling headphones. This small yet significant improvement made it possible for the track employees to control the volume. "The race track people loved it, so Harry built a bunch of these,"[8] adds Neal.

Harry hired technician Bob Cohen to assist him with building these inter-coms. Neal relates, "As soon as the project ended, Bob quit and started Clear-Com."[9] McCune built things out of necessity, especially when rock and roll came into the picture. According to Neal, individuals like Bill Hanley and Harry McCune Sr. should be remembered because they started this small industry out of a "need":[10]

Harry Sr. was really inventive and designed and built what we call "Harry Boxes," which are still around. They were a big ol' steel Bud box that's got some real interesting circuiting inside. The staples of the McCune business were the old Altec multi cellular horns and 80-watt amplifiers. We used stock Voice of the Theatre speaker boxes.

From a West Coast perspective, I think Harry Jr. created the concept of a soundman and sound company. He traveled on the road with a system. I don't know anyone that was doing this before him. He cut his teeth on a lot of stuff in San Francisco. When I started with the company, rock and roll was a big thing out here and all of these rock and rollers were looking for louder, bigger, and better. McCune was about as loud as you could get so we worked with the Grateful Dead and Janis Joplin early on. We did the Monterey Pop Festival; guys like Abe Jacob and John Meyer came out of McCune.[11]

Over the years McCune Sound fostered and trained many leading engineers in the sound industry. Some of these are Harry McCune Jr., Mike Neal, Dan Healy, Abe Jacob, Ken DeLoria, and Bob Cavin. Healy, sound engineer for the Grateful Dead, worked for McCune Sound at one point in his career. When Healy was on the East Coast with the Dead, he rented equipment from Hanley Sound: "I was a longhaired rat infested hippy creep, most people wouldn't approach us with sound equipment, and Hanley didn't judge me."[12] According to Healy, there are many comparisons to be made between Hanley and McCune:

I worked at McCune for a while and had my own system there. So while I was technically an employee for McCune I still had my own equipment. It belonged to them but it was my design. That's where I met John Meyer and Abe Jacob. Abe was one of the first guys I knew. He was a theater guy; I went into rock and roll. Harry McCune used to make his own condenser micro-phones back in the day. He was a real guy and a mentor to me. He treated me like a son, and I loved him dearly. He did sound up until his nineties. He was the type of guy that created stuff because he needed it. He would work out and invent ways of doing stuff just so he could manifest a certain sound system. It was a means to an end for him. He was less of a mad scientist and more of a big picture innovator guy. I think Harry McCune Sr. should be considered a pioneer as well. I would also rate Hanley up there with McCune, like Harry, Bill Hanley was a leader and a friend.[13]

Leading the pack of the McCune alumni is John Meyer, who has over forty patents associated with his designs and is considered a genius in the business. Meyer remembers: "McCune was pioneering sound for PA. This was why I wanted to work for them. On the West Coast they were the biggest supplier. There were lots of other little companies around doing events."[14] Designing touring systems by the mid-1960s, McCune Sound was already serving a diversified clientele. Over time the company has become known by its long list of important shows, including the Monterey Pop Festival and the Beatles' final performance at Candlestick Park on August 29, 1966.

SWANSON'S SOUND SERVICE COMPANY

Founded by Art Swanson in 1926, the Swanson Sound Service Company, located in Oakland, California, was another early force in the business of sound reinforcement. Swanson Sound began as a typical public address, local event, and rental shop. Swanson had panel trucks that served as advertising billboards for the community. The trucks were outfitted with large audio horns mounted to their roofs. Early on they had a searchlight division, a component to the business that was later phased out.[15]

Don Neilson, a longtime employee, bought out the business during the mid-1950s. Many say that Neilson elevated Swanson Sound to a new level of innovation. For example, according to Maureen Droney of *MIX* magazine, Swanson was one of the first companies to use "Bi-amplification, with active electronic crossovers feeding separate amplifiers for woofers and tweeters, an innovative concept for the time."[16]

Neilson eventually grew the business into one that could offer permanent large touring sound systems and custom design installations. Grateful Dead sound engineer Dan Healy reflects on the company: "In the San Francisco Bay Area there were basically two old school sound companies. They were McCune and Swanson, who was in the East Bay. Swanson was the very first company that I rented sound equipment from."[17]

Swanson served many influential acts during its time. Meyer recalls: "Swanson got some good gigs and they built stuff. But they were pretty much dedicated to doing rock and roll. No Broadway shows or anything like that."[18] The most influential work of the company can be seen during the psychedelic era of the 1960s, when they did sound for Jefferson Airplane, the Grateful Dead,

Crosby Stills and Nash, Joni Mitchell, Bill Graham's Fillmore West, and the Winterland Ballroom.

Roy Clair, president of the worldwide sound company Clair Global, recalls Swanson's significance to the sound industry: "In the beginning it was Bill Hanley, Harry McCune, Bob Kiernan, Stan Miller, and Swanson. Swanson Sound was the biggest in Oakland. Because of what the Clair Brothers did with Jefferson Airplane, Swanson called us to do their groups back east, like Iron Butterfly."[19]

MEAGHER ELECTRONICS

There were only a few places in North America during the 1950s where you could experience creative innovation and problem solving for high-fidelity sound. One location was in Medford, the other in Monterey. Electronics and sound innovator Jim Meagher of Meagher Electronics was another trailblazer in sound reinforcement application. Born on May 14, 1919, Meagher had a head start on Hanley by at least ten years. Although they never met, they did share similarities. Very little has been written about Jim Meagher until now.

Meagher grew up in the Pacific Grove area of Monterey, California. When he was in the first grade, the young audio and electronics enthusiast worked in his parent's shed on three- and four-tube radios. Often, even at age eight, Meagher loaded his Radio Flyer Wagon with his neighbor's inoperable phonographs and radios for repair. Like Hanley, he had a real passion for ham radio and Morse code. Meagher's father thought his son would take an interest in his dry goods store, a venture in which Meagher was uninterested.

Meagher graduated from Pacific Grove High School in 1937 (the year Hanley was born) and Hartnell College in 1939. During World War II he worked as an electrical technician for Lockheed Aircraft. After the war, Meagher left his father's business and began working at a local music shop in Monterey called Abinantes. There he met Carmen, his future wife, when she brought her record player in for repair.

It seemed that the two were a perfect match. Carmen took interest in Meagher's business. Eventually the sound engineer taught his future wife how to solder, presenting her with a temporary engagement ring of the same crude material. They were married in 1947. Two years later, Meagher opened his electronics store with Carmen.

Of its many locations, Meagher Electronics operated from Webster Street in Monterey longest of any. There was a very large warehouse where the audio engineer stored hundreds of phonographs and wooden console radios that customers left behind. Meagher's son Robert claims, "They had been left by customers who decided not to pick them up because they didn't want to pay the repair estimate charge."[20]

Meagher Electronics was a multifaceted business, advertising itself as an Altec dealer, radio and television repair, and recording studio. Meagher was one of the first Altec representatives and dealers on the West Coast, if not in the country. The audio engineer provided public address systems for music, conferences, and public events, including the Big Sur Folk Festival, the Monterey Jazz Festival, Feast of Lanterns, and the Merienda.

Later, Meagher built his own recording studio at the shop. It was here that folk singers Joan Baez, Mimi Farina, and her husband Richard Farina recorded early demos of their music. According to Tychobrahe senior sound engineer and former Meagher employee Jim Gamble: "We had Joan Baez in the studio playing the piano at one point! Altec speaker amps are what Meagher used for listening in the control room. He had an AMPEX machine, which for the first time had three tracks! Count them, three tracks!! Tape of course."[21] Former employee Ken Lopez remembers: "Jim had figured out what Motown was using and had linked three AMPEX tape recorders together to get three channels. It was cutting-edge stuff in 1963–64."[22] Sound engineer Dan Healy recalls Meagher's business: "Jim had a really nice recording studio. It wasn't fancy but the audio quality was staggeringly good. He had mike pre-amps he made himself."[23]

Meagher was a night owl. Jim Gamble remembers hanging out late at night with the sound engineer as he worked away and talked politics: "Meagher was a Democratic thinker in a rather Republican area."[24] Meagher often felt alone in his progressive views.

Gamble assisted Meagher with organizing the cluttered shop. The sound engineer was known for hoarding vintage radios and other electronics. Gamble recalls: "The shop was a mess. We put in shelves and put things in order like mike cables, etc."[25]

Unlike Hanley, Meagher did not tour. He traveled locally taking Voice of the Theater speaker systems into the field. In his 1956 Plymouth station wagon, a monster of a car, you could easily fit two huge Altec A7s. Robert recalls, "My father installed sound at an amazing number of fairgrounds, churches, auditoriums, hotels, and also homes with hi-fi speaker systems in the walls."[26]

At a very early time in sound reinforcement history, Meagher was innovating, and like Hanley he had a deep passion for sound. Robert remembers: "His most precise instrument was his ear. He could listen to a sound like a word and then tell you the different frequency components of it. He had a great ear."[27]

In 1959 Meagher served as the hired sound company for the Monterey Jazz Festival. It was a gig that lasted for several years but for which Meagher rarely gets credit. His son Karl claims that this did not bother his father: "McCune ended up taking over the Jazz Festival. Jimmy Lyons of the Monterey Jazz Festival [MJF] always got the credit before my dad. Fame wasn't important to him."[28]

In 1958 Hanley visited Lyons looking for work. Lyons was a "well known D.J." who "spearheaded the MJF for three memorable days on October 4–6" of that same year. A 1958 article titled "Sound Recording and Reinforcing at The Monterey Jazz Festival" states that Ampex recording engineers provided the sound reinforcement. The crew filled a panel truck, a small moving van, and two station wagons with $35,000 worth of gear. The engineers traveled to the festival grounds "towing two large theater speaker systems on trailer wheels, plus a third such system knocked down and packed inside."[29] It took two full days to set up for the event. Ten years later the Monterey Fairgrounds venue and its associated attributes and location was used for the Monterey Pop Festival (1967).

Meagher, passionate about the sound he was producing, was known to endlessly tweak the horns of his Voice of the Theater speakers until they were just right. According to Robert and Karl, their father always made sure that he had the "best quality equipment" to do the job. Meagher thoroughly researched the acoustics of the environments he worked in and was very careful about miking.

Meagher brought his studio techniques to the stage, using fifteen to sixteen mikes for a full-sized orchestra. Robert recalls: "My father always had two to four overheads at all times for orchestras. It would be six or more if it were a choral thing. We would be up in the rafters dropping stuff."[30]

Meagher was always willing to offer his expert opinion, educating many performers on proper microphone use. *Chicago Tribune* writer Lillian Roxon claimed that Meagher's sound system at the Big Sur Folk Festival was "So clean and pure that even people a mile away on either side, those who maybe could not get in, can hear as perfectly as those inside the grounds."[31] Jim Gamble reflects on working with Meagher back then: "He was a highly eccentric guy. He had all Altec A7s, Altec amplifiers and mixers. He had all kinds of really nice mikes. At Big Sur he was there bringing in the big Altec long throw horns to hit people up on the highway. He was a genius in his time."[32]

Like those who worked and learned under the direction of Hanley, over the years several West Coast sound industry innovators passed through the doors of Meagher Electronics. Some of these are Jim Gamble (Tychobrahe), Dan Healy (Grateful Dead), and Ken Lopez (JBL). Lopez, while walking home from high school one day during the mid-1960s, stumbled upon the engineer's shop. "Like me, a lot of other people floated through his shop and got their start in sound reinforcement. He was teaching people, like Hanley was on the East Coast."[33] Lopez, who has had a successful career as a music educator, audio engineer, and as former vice president of JBL, considers Meagher a mentor.

Harry McCune of McCune Sound also rented equipment from Meagher. Karl remembers: "I think McCune learned from my father. He talked to him. McCune had the power and the business. He was the Henry Ford of sound installations, making lots of them."[34] According to those close to Meagher, he didn't care much about his competition. He selflessly taught others even though he knew they would borrow his innovations and offer a similar product. Robert claims his father "created his competition."[35]

Like many during this time of innovation, Meagher faded out of the business around the mid-1970s. According to Karl: "My father was making music systems when others were making public address systems. He was selling people on this idea. It's possible that his successes at Big Sur, Monterey Pop, and Monterey Jazz are the reason people wanted to come."[36] James Albert Meagher died in 1997 at seventy-seven years of age. He owned Meagher Electronics Co. for almost five decades.

STAN MILLER

By 1964 Hanley had the largest collection of sound equipment around. It was necessary because he was busy feeding closed-circuit broadcast television, installing permanent sound fixtures, and supporting sound reinforcement at Newport and other festivals in Ohio and Pennsylvania. The sound engineer found himself consumed with work back in Medford. Hanley faced minimal competition in the sound reinforcement business at this time. Whatever the case, he was leading the pack. According to Hanley, pickings were slim back then, claiming, "It was all controlled by the manufacturers and distributors."[37] If there was one competitor in the United States comparable to Hanley, it was Stan Miller. Miller was close to Hanley in age and blazed his own trail in sound reinforcement.

At the 1964 New York World's Fair, Miller ran into Hanley while he was hauling equipment in his station wagon during a tour for the group the Young Americans. According to Miller: "I eventually got my first experience with international touring with them. We went to Australia, Thailand, and Japan."[38] Miller was impressed with Hanley's operation at the time: "In comes Hanley with this big truck! It was incredible. I thought to myself, I need a big truck!"[39]

He remembers being impressed by the size of Hanley's operation: "I saw Bill with his big trailers, which no one had at the time. I believe this was the first time anyone was using this sort of touring equipment. Bill Hanley inspired me to do the same."[40] Reflecting back, Miller claims, "Guys like me and Hanley had to pull ourselves up by our bootstraps," adding: "We didn't have money to go buy something in the early days and it didn't exist anyway. So you figured out how to do something because you couldn't buy it off the shelf."[41]

Born on October 25, 1940, Miller grew up in Holdrege, Nebraska. As a child, Miller was fascinated by the sounds coming out of his parents' rather expensive Silvertone radio situated in the middle of his living room. In junior high his interest in audio was piqued when a band instructor took him to hear a hi-fi system in the neighboring town's high school gym. He was hooked. Like Hanley, Miller ran popular sock hops, toting his turntables and a modest self-built amplifier to his school gymnasium. "I hooked them up, and played records for the high school kids."

Miller was acquiring a reputation as a hi-fi buff and was known throughout the community as "Stereo Stan." According to Miller, "People came to me and asked if I could fix their hi-fi equipment."[42] A radio announcer in high school in the mid-1950s, Miller worked at a 500-watt day-time station in Holdrege.

By age twenty-one Miller was enrolled at Kearney State Teachers College in Nebraska. It was here during the early 1960s that he noticed an influx of groups traveling and playing in college gymnasiums. A college friend of Miller's asked if he could help by hooking up some of his sound equipment in the gym for a show. He enthusiastically accepted. By now Miller had built several Heathkit amplifiers and loudspeaker boxes with Altec components. Lugging his sound system from his mobile home across from the college, his methods and ideas eventually became popular.

Miller additionally managed the college theater and radio station. Here he noticed acts like the Smothers Brothers and the Carpenters traveling through without their own sound systems, but the equipment the college offered did not suffice. According to Miller: "The college had four Atlas rear entrant horns at each corner of the gym, and a Bogen amplifier with one microphone input. It was for calling basketball games. If a band came they would just plug their microphone into their amplifiers, and that's it."[43]

According to Miller, Variety Theater International, a company out of Duluth, Minnesota, was touring Midwest colleges: "When the Christy Minstrels came in, the organizers were determined that they could make more profit if they offered the group their own sound system."[44] This is what inspired Miller to go out on his own. "I kept building more speaker boxes, more amplifiers, and used to haul my stuff around in a station wagon."[45]

With more advanced equipment, by 1962 Miller was supporting additional college performances. The sound engineer frequented Altec Lansing seminars, and opened a commercial Altec franchise as well as a commercial audio store. It was around this time he established the Stanal Sound Company in Kearney, catering to pro and consumer clients. Miller continued providing sound reinforcement for more shows, expanding his business, and designing equipment.

In 1967 he began an impressively long fifty-year relationship as chief mixing engineer for artist Neil Diamond. "I was on a bus and truck tour with Peaches and Herb, and Neil shared the bill with them. It was a little theater in Vermillion, South Dakota, 300, maybe 400 people, and Neil's road manager said 'geez that was good. Can you do a show with us in a week or two?'"[46] Miller never looked back.

At seventy-nine, Miller is probably one of the oldest sound engineers in the business who up until 2018 was still actively touring. According to *FOH* magazine: "Along the way he worked with legends including John Denver, Johnny Cash, Sonny & Cher and Bob Dylan, among others. He was audio designer/consultant for . . . Pink Floyd's *The Wall* concerts in 1980–1981."[47]

Like Hanley, Miller is revered as an innovator in the industry and is considered by his peers one of the firsts in sound reinforcement. Miller is known for advancing sound reinforcement technology into the digital domain. "Bill Hanley and I are Parnelli award winners, which they only give to old people. He was certainly doing things on the analog stage a very long time ago. He was a driving force and inspiration for people like me."[48]

KIERNAN SOUND

By 1964 Hanley had set his sights on New York. During his short stint in the Big Apple there were very few sound companies established there, and these were mostly outfits renting and designing theatrical sound for Broadway. According to Hanley the key players in sound in New York at the time were Masque Sound, Sound Associates, and Kiernan Sound.[49] Shannon Slaton in her book *Mixing a Musical: Broadway Theatrical Sound Mixing Techniques*, claims Masque Sound was established in 1936 by three stagehands and became well known for their work on Broadway; ten years later (1946) Sound Associates began.[50]

During the mid-1960s, Bob Kiernan was making a mark for himself in the early sound reinforcement business in New York. Eventually, he became the leader of midtown Manhattan's premiere sound company, Creative Theatrical Services, known as Kiernan Sound. If Hanley had any competition in the Big Apple at all, it was Bob Kiernan. Kiernan remembers: "Hanley was around. There were really no other sound companies that I was aware of. He had a lot more equipment."[51]

Growing up in the Bronx, Kiernan got hooked on stage technology early on at Cardinal Hayes High School. He recalls, "I never had any interest in being on stage, being behind the scenes really intrigued me."[52] After high school he maintained several jobs, one of which was at a summer theater, earning ten dollars a week plus room and board. This position allowed Kiernan to hone his skills in stage, lights, scenery, and sound technologies.

In 1959, while a junior at Manhattan College, he realized he wanted to be in show business rather than his original plan of teaching and coaching. After doing a series of shows for Manhattan, Kiernan dropped out, and continued to work the college concert circuit. Initially more of a technical stage and lighting guy, Kiernan's interest in sound surfaced later, when it became an increasingly vital part of his freelance work. Although he was hired to provide full production for his clients, he became drawn to the idea of being a primary supplier of sound.

In 1962, Kiernan met the manager of artists Al Hirt and the Brothers Four. When Hirt was slated to play Basin Street East (BSE), a popular nightclub located at 137 East 48th Street, the trumpeter's manager called Kiernan for assistance. By April 1963 Kiernan became the permanent sound engineer at BSE. Not your average club, BSE showcased a Who's Who of performers of the day, among them Bill Cosby, Woody Allen, Ella Fitzgerald, Count Basie, Benny Goodman, and Barbra Streisand.

At capacity the room sat four hundred and saw a full crowd most every night, with shows at 8:00 p.m., midnight, and 2:00 a.m. Kiernan sat in the back, encased in a little booth where he controlled his sound system. He recalls what he was using for equipment back then: "It was an Altec system with a small rack that had an Altec 1567A mixer and a standard stock small bass enclosure with some horns." Regarding the acoustics of the space, Kiernan notes: "It was a great room. You really didn't have to mike an entire orchestra individually. Considering the room size, it was more of a live mix than an amplified one."[53] The sound engineer, who mostly worked alone, poked his head out of the sound booth window to listen for some level of accuracy. With one hand controlling a follow spot aimed at the performer, his other adjusted sound levels.

Kiernan first met lighting and production designer Chip Monck during a Miriam Makeba performance at the nightclub on March 11, 1964. Makeba, a South African vocalist and civil rights activist, hired Monck as her production person and had been booked for a benefit for the Student Aid Association of South Africans. Kiernan and Monck quickly hit it off, sharing some of the same philosophies in sound and lighting.

Monck was employed at the Village Gate jazz club, where he was doing similar duties. According to Monck: "Bob was doing about the same as I was at the time, except he was both audio and lighting. BSE had no obstructions, and he was just over the seated audience heads. His follow spot always left this telltale shadow on the back wall. He was good operator, and an important figure in the industry."[54] Monck had been scheduled to do a small tour with Makeba but was committed to something else, and asked Kiernan if he could assist. Eagerly accepting the offer, the sound engineer talked to the management at BSE, and subbed himself out for a few dates.

Even though he had a large amount of equipment by 1964, Kiernan Sound was not yet fully established. "I had some equipment. I toured with the equipment but I was not yet 'Kiernan Sound.'"[55] That changed soon when the Ford Motor Company contacted the sound engineer to do a series of free college concerts across the country. The company was promoting its Ford Mustang, and it was this gig that propelled him into the field of sound reinforcement even further. Kiernan reflects: "I am not sure how they found me. Possibly one of the artists I had done shows with? A marketing guy named Gene Gilbert who was the founder of youth marketing had a contract with the Ford Motor Company and brought me on board."[56]

Gilbert ended up hiring Hilly Krystal, later the founder of the New York club CBGB. A year later, in 1966, Krystal reached out to Kiernan to provide full production, lighting, and sound for his Central Park Music Festival. This is when Kiernan Sound was born. According to Kiernan: "Out of the Gene Gilbert gig came a call from Hilly, who was a partner with promoter Ron Delsener; both were founders of the Central Park Music Festival. When I set up for that first season, I remember having to get together with some lawyers, as I realized that it was time to incorporate."[57]

Kiernan Sound settled in a rented space on the fifth floor of a building in Soho. "This is really when I had a company, where I kept the books (as badly as I did it) and tried to get paid for my jobs. Officially it was 'Creative Theatrical Services,' but others called it 'Kiernan Sound.'"[58]

Keirnan became known in the business by setting a standard of pricing for his services. According to Roy Clair of the Clair Brothers, Kiernan was the first to make any money on a tour: "Here's the rest of us and we're sitting there going, oh my gosh! I am almost sure that he charged Tom Jones $1,000 a concert! I think that is the most anyone received until then. The most we made was $400 to $500!"[59] Later Kiernan's business expanded, and he moved to Las Vegas in the 1970s. By the end of his career, for almost ten years he was Frank Sinatra's exclusive sound engineer.

In Hanley's view there was still plenty of room to grow his company and client base in New York: "There was hardly anyone supporting this bustling live music scene. I had to be there. It was the entertainment capital of the world."[60] In order to acquire additional work in the city he doggedly frequented nightclubs, placed ads in the Yellow Pages, the Red Book, and chased down promoters like Sid Bernstein, New Jersey promoter John Scher, and Teddy Powell, another well-known rock promoter. Hanley also had a close friend, Mike Revel, assisting him in cultivating new work. "I met Mike Revel at Soundblast!

'66 at Yankee Stadium. He was trying to bring in more business for me while I was in NYC. Mike took over my apartment at 888 8th Avenue when I finally left NYC."[61]

CHARLIE WATKINS

British audio engineer Charlie Watkins, known as "The Father of British PA," is another important figure in early sound reinforcement. His company Watkins Electric Music (WEM), founded with his brother Reg in 1949, was originally a record shop located in Tooting Market, London. By 1951 the brothers had moved to Balham, where they also began selling instruments.

The British audio engineer is known in the industry for his innovations in sound. Designs such as his "Watkins Copicat" Tape Echo Unit and his invention the, WEM "Slave" PA System, have become standards in the industry. Second-generation UK sound engineer Stuart "Dinky" Dawson of Dawson Sound found a mentor in Watkins early on. According to a *Pro Sound News* article, Dawson relied on WEM technology and guidance from the inventor. After coming back from touring the United States in the late 1960s, he went to Watkins for input. "As soon as I headed over to see Charlie, it was love at first sight."[62]

Known for loudness, a WEM PA system was first used at the Windsor Festival in Balloon Meadow at the Royal Windsor Racecourse in 1967 and was capable of at least 1,000-watts of power.[63] According to Watkins: "The invention of the 'Slave' PA System is possibly my most rewarding development. They started referring to me as the 'Father of British PA.' Suddenly artists could be HEARD, with my massive PA system with a starting power in excess of 1,000-watts."[64] However, with bands and performers like the Crazy World of Arthur Brown, Peter Green's Fleetwood Mac, Cream, Ten Years After, Jeff Beck, and John Mayall and the Bluesbreakers on the bill for the three-day festival, the system was no match for loud rock and roll.

From the 10,000 who attended, some say the system was not powerful enough to sustain and was marred with technical issues, leading to a lot downtime between sets.[65] But others claim that complaints from nearby neighborhoods compelled the engineer to turn the system down, making the audience and performers unhappy. Although for the time the Watkins WEM system was capable of great volume, it lacked clarity as Tychobrahe crew member David Pelletier recalls: "The Who showed up with their WEM System and it was loud, that's all it was."[66]

Like Hanley, Watkins supported many notable bands and festivals during the 1960s. Among these were Jimi Hendrix, Led Zeppelin, the Rolling Stones, the Byrds, Elton John, Ten Years After, and Pink Floyd. Watkins is known for his work at the Windsor Festival, the Rolling Stones in the Park, and the 1969–70 Isle of Wight Festivals, among others.

While working with so many British acts that toured in America, Hanley couldn't help but acknowledge the WEM PA systems they were using while traveling through. According to Hanley, he had never met or heard of WEM creator Charlie Watkins, but he did see the equipment from time to time and was not impressed: "I saw a lot of the WEM equipment with bands like the Who and Ten Years After. In my view there was too much distortion and high SPL with Mr. Watkins's design. Every time I saw it I froze. The name Charlie Watkins never came up until I saw all of his speaker columns."[67] It is unclear whether Watkins was aware of Hanley or anyone else within the elite group of first-generation American sound engineers.

Chapter 20

BIG APPLE! BIG SOUND!

CHIP MONCK

Hanley arrived in New York City in 1964. By now he had already been with the Newport Festivals for close to eight years. According to Hanley, it was around this time that producer George Wein was thinking of replacing him with Chip Monck and Bob Kiernan's Astra Arts Limited. Monck's home turf was New York, and although he and Hanley had a working relationship, there remained an air of friendly competition between the two. Even though Monck and Hanley traveled in similar circles, it was still a struggle for the sound engineer to get settled in New York. "It took Chip and I awhile to become close friends,"[1] says Hanley.

Born on March 5, 1939, lighting designer Edward Herbert Beresford "Chip" Monck grew up not too far from Hanley in Wellesley Hills, Massachusetts. Throughout his youth Monck had a growing interest in welding and machinery, later evolving into learning theatrical lighting at Wellesley College. At one time Monck worked for Harvard University's theater company. By 1959 Monck was established in New York with his Chip Stage and Light Company, working at the Village Gate, and later for George Wein's Newport Jazz and Folk Festivals. Referred to as "The Gate," the Greenwich Village jazz club was where Monck for a time applied his lighting techniques.

To support the many well-known comedians, jazz, and folk artists who rolled through, Monck repaired and borrowed equipment from the Altman Lighting Company in Yonkers. Founded in 1953, the company has been a leader in stage, film, concert, and architectural lighting. Monck quickly became friends with its owner Charles Altman. Conveniently, Monck rented the basement apartment under the popular venue. According to Monck it wasn't a bad deal: "I got $1.87 for each show, coming to a grand total of $16.83 a week, but I could eat and drink there, too."[2] Monck claims that Bob Dylan wrote the folk opus "A Hard Rain's a-Gonna Fall" in his apartment on his electric typewriter.

Monck later broadened his talents on many tours, festivals, and at other venues, including the Monterey International Pop Music Festival, the Byrds at the Hollywood Bowl, and major Rolling Stones tours. By the late 1960s he and Hanley were working with each other at the Fillmore East and festivals like Denver Pop and Woodstock. Monck recalls: "Hanley, as far as I know, and as far as I go back (at age seventy-nine), was the first person you would consider to use in a major venue. What he did was build the skeleton of the audio industry. Much like what I did in lighting."[3] Monck and Hanley remain friends to this day.

THE CHAMBERS BROTHERS

Hanley's sound influenced many groups, but none like the four talented brothers from Carthage, Mississippi—the Chambers Brothers. The family band consisted of George (bass), Willie (guitar), Lester (harmonica), and Joe (guitar). All were self-taught musicians. The Chambers Brothers were known as a crossover band offering a multilayered sound of gospel, folk, funk, rhythm and blues, pop, and rock and roll. Their style, as New York Times writer Mike Jahn claims, was a "smoothly rocking" and "exciting" one. "They have the bits of soul but the happy rolling feel of gospel music. At their best moments they work a careful formula."[4]

Raised from an impoverished sharecropping family, the brothers began on the gospel circuit, quickly adapting to the emerging folk scene. Eventually they settled in Los Angeles, regularly performing at the Ash Grove in front of an expanding white audience. Soon their fan base grew to prefer their developing pop and rock and roll sound to gospel.

A big break for the band came at the Newport Folk Festival in 1965, when jazz, blues, and folk singer Barbara Dane brought them on as her guest. Despite the hoopla surrounding Bob Dylan's performance that weekend, the Brothers impressed the audience first, opening the festival with an electric set of blues and soul of their own. This is where Hanley and the Chamber Brothers first met. George Chambers recalls: "We were there primarily as a folk group. Back in those days we were still doing a lot of gospel. Our sound kept developing from one place to the other. Our sound slowly made a turn. I was learning my instrument at the time."[5]

After that weekend, Hanley and the Chambers Brothers became friends. Bill even found the brothers a place to stay near his home in Medford. According to Joe Chambers: "If I can remember correctly we lived at Bill's house or somewhere near his home for a few nights. We were folk at the time, and with the folk music scene people would often take you into their homes. Bill was a nice, gentle, humble, soft-spoken man who was serious about sound."[6]

Lester Chambers recalls the sound engineer at Newport running on and off the stage and into the audience carefully checking the sound: "Hanley helped us make a great transition from acoustic to electric. He knew where to place speakers so everyone could hear perfectly. They were angling to the right, left, upwards, and forwards. I had also never seen monitors before this. I could hear myself breathing! It was phenomenal!"[7] As the group's repertoire expanded, electric instruments eventually became an extension of their sound.

After signing with Columbia Records, the Brothers landed in New York City in 1965. It was at a club called the Downtown where Hanley further assisted them with their unique sound. The group was booked for a few weeks, and that gave the sound engineer ample time to figure out some of their technical difficulties. Very few clubs then had a sound system sufficient for a rock and roll group like the Chambers Brothers. "It was like dying on stage. Every group depended and relied on a club to have the sound system,"[8] adds Lester.

Compared to other groups, the Chambers Brothers had a relatively small stage setup. The group carried very little equipment with them. This was perfect for the smaller venues they often played. According to Willie: "Back in those days there was one microphone and everyone sang around it. There were no monitors, no side fills; there was none of that."[9] Cousin and acting soundman for the group was Julius; he and Hanley hit it off nicely. "I was everything at that time. I was manager, road manager, sound, and lights. When we went on the road I always did sound. I remember seeing Bill at Newport but it was when we were working at the Downtown in New York where he really helped us."[10]

The band was having difficulty hearing each other on stage. Dealing with noisy feedback caused frustration, resulting in arguments. Similar to the family band the Cowsills, Hanley noticed the Chambers Brothers were also using an insufficient Shure Vocal Master PA setup. Julius remembers: "He studied us and asked us a lot of questions about what we wanted. There were hardly any decent PA systems at that time. I told him we needed to hear ourselves and we need a PA that can playback to us."[11] Hanley explained to the brothers that he had a PA but needed a day or two to bring his equipment down to the club. Once set up, it took a few tries before the engineer got the sound just right. Willie recalls: "He brought down the system and a couple of speakers. This was fine and now we could be heard, but we still couldn't hear ourselves. Then he brought another couple of speakers for monitors, which was totally unusual for the time, no one was doing that."[12]

Enhancing the Chambers Brothers' sound allowed them to perform to their fullest capacity. This development played an essential role in their success in New York and abroad. Lester observes: "The second time I was like OH MY GOD!! DID YOU HEAR THAT? I COULD HEAR THE SINGING EVERY-WHERE!! I could hear all the instruments and it was just fabulous. This had been a real struggle for us. Being able to hear well was impossible up until Bill Hanley."[13] According to Willie: "People weren't used to hearing us that clear, that bright, and that loud. It was amazing and thanks to Hanley; he made us very successful."[14]

Lester Chambers became known for a unique scream so loud that it could blow out a sound system. "But we never did blow up Bill's system!"[15] adds Joe. Roy Clair of the Clair Brothers claims that when the Chambers Brothers screamed the word "TIME!" from their hit song "Time Has Come Today," it blew out his compression drivers: "The Chambers Brothers taught us a lot with their music. That 'TIME!' took more SPL (sound pressure level) than any sound or music from any group! When they all yelled, 'TIME!' they used to laugh because they knew they were blowing up peoples' sound systems."[16]

Over time Hanley taught Julius various techniques, mentoring the cousin in the art of sound. "I learned a lot from Hanley. He was laid back, and easy going. I began to realize certain things we needed and I asked him for them. He always had good sound equipment."[17] Often the Chambers Brothers traveled from venue to venue with a Hanley Sound System of their very own. According to Joe, "He changed our perspective of sound immensely and should be remembered as one of the most innovative sound engineers of the period."[18]

SOUND IN THE CITY

In 1965 Hanley expanded his efforts in New York. Since most of the entertainment happenings in the country were occurring here, it made perfect sense. According to Terry: "NYC was where the action was. The music business was in New York, not in Boston."[19] In New York, Hanley hustled for work, just as he had done back in Massachusetts. But convincing concert promoters and band managers that his methods in live sound were the answer proved frustrating for the sound engineer.

Hanley rented a one-bedroom kitchenette-style apartment in a brand new, postwar, twenty-story building at 888 8th Avenue for a few hundred dollars per month. The modest space served as an office and a place to crash for his tired employees and helped give Hanley Sound a proper home base in the Big Apple.

David Roberts, who had worked for the company on the Lyndon B. Johnson inauguration in January 1965, was there when Hanley first acquired the apartment. "When he got his apartment, I went down for a party. It had parquet floors, white walls, and that's it. This was the first time I was exposed to a NYC rock and roll type party. I remember sleeping on the floor."[20] According to Hanley Sound crew member Bill Robar: "One time I was working with Terry putting in a system somewhere and stayed at Hanley's apartment. The place was tiny but it had a color TV in it and I remember it was one of the first times I had ever seen one! They weren't that common yet."[21]

To store the equipment he was amassing, the sound engineer rented out a nondescript 2,500-square-foot storefront on the Lower East Side of Manhattan on 5th Street between 2nd and 3rd Avenues. This was not far from the Anderson and Fillmore East theaters where Hanley would eventually provide sound systems. According to Hanley, the location was ideal because it was "street level and convenient."[22]

Hanley Sound employee Billy Pratt often frequented the space. Pratt exclusively handled many of the Tri-state area shows including the Chambers Brothers. "Hanley had a little storefront right up the street behind the Fillmore where he kept equipment. It was filled with large speaker boxes, horns, and other stuff. I had a key, went there, grabbed whatever equipment I needed and delivered it to the stage door behind the Fillmore."[23] Unassuming to the passer-by, the storefront windows had wire mesh over them and a steel grate that could be pulled down to protect all the expensive sound equipment.

Still in high school at the time, Terry kept things in order back at Hanley Sound in Medford. On weekends, days off, and during summer vacation, he drove the company truck, set up gear, and oversaw the day-to-day activities at the shop. Often, he would have to leave the office for a service call to New York. Terry remembers: "I jumped in my station wagon for a NYC service call, hit the gas and flew out of there. In certain circles I was called the 'Wildman' for doing this."[24] When the New York work was finished, he turned around and headed home that same day. Terry, like his brother Bill, was a straight shooter; neither brother drank or did drugs. "There was no time for it. I was so busy working out of Boston. I was handling most of the major US tours that we did, like the Cowsills and Jefferson Airplane."[25]

Hanley had sound system installations at various New York clubs including the Bitter End, the Apollo Theater, Ondine's, the River Boat, the Cheetah Club, the Electric Circus, the Generation, Steve Paul's the Scene, the Fillmore East, and Café au Go Go. According to Terry: "We did people like the Blues Project and Jesse Colin Young and the Youngbloods at the Café au Go Go. We had a subgroup of guys who worked out of NYC that were associated with us at the time."[26]

At Steve Paul's the Scene, a basement club on 46th Street off 8th Avenue, Hanley installed a system for the 5,000-square-foot space. In 1967 owner Steve Paul also had a popular primetime nationally syndicated television show called *The Steve Paul Scene* featuring many top performers of the day. On Labor Day weekend, Paul brought Hanley in as a consultant for a "two-hour color TV special dealing with contemporary music." According to *Cash Box*, "Sound consultant for the show was Bill Hanley, who lists the Newport Festivals and the James Brown Show among his credits."[27]

In late 1967 New York club owner Barry Imhof called Hanley and asked him to install a sound system for his new venue, the Generation, in the former location of the old Greenwich Village folk club the Village Barn at 52 W. 8th Street. The Generation was slated to open on April 2, 1968. According to *Billboard*: "Bill Hanley with offices in New York and Boston, installed sound equipment in such clubs as the Café au Go Go, the Bitter End, and the Electric Circus. His firm also specializes in outdoor sound jobs and has done audio for the Newport Festivals. Hanley Sound, which has a staff of eight sound engineers in Boston alone, usually rents equipment out to clubs like the Generation."[28]

On the day of Dr. Martin Luther King's assassination, April 4, 1968, Janis Joplin paid tribute with a performance at the Generation. Bill Hanley and Jimi Hendrix looked on from the audience. Back then Hendrix frequented the venue, participating in many late-night jams and impromptu onstage collaborations with acts like Big Brother and the Holding Company, B.B. King, and Sly and the Family Stone. Later that year, Hendrix and his manager Michael Jeffery bought the failing club and built the famous Electric Lady Studios.

There were a few other venues the sound engineer unsuccessfully tried to get his equipment into. Among these were the Roseland Ballroom and a jazz club called the Village Vanguard, located on 7th Avenue in Greenwich Village. Known as a hot spot for Beat poets and folk musicians of the day, this was where Hanley had his first interaction with political activist Abbie Hoffman. Their next encounter was at Woodstock. "Abbie threw me off the stage at the Vanguard during an anti–Vietnam War benefit. He was crazy!"[29] remembers Hanley.

Advertising and word of mouth was how Hanley was getting most of his gigs in New York, and it was paying off. In 1967 Café au Go Go manager Howard Solomon called Hanley for help. The club owner was in dire need of a quality sound system for the emerging folk rock scene and needed his assistance. Nothing proved to be more valuable than this connection. It eventually led to Hanley Sound establishing itself at Bill Graham's Fillmore East in 1968.

Chapter 21

DOWN IN THE VILLAGE

GREENWICH VILLAGE

Bill Hanley's career encompassed many genres of music, although he is mostly recognized for his work in rock and roll. The mid-1960s transformative migration of folk music into rock was a significant event within the career of the sound engineer. It required a skilled individual with the ability to adapt to changing sounds. The evolving culture and politics behind the music of this time created a need for additional technology, something Hanley realized sooner than later. As cultural historian Morris Dickstein observes in his book *Gates of Eden: American Culture in the Sixties*: "The subsequent evolution of folk into rock showed that the complication of awareness and technique that we call modernism couldn't be abandoned after all. Instead the new music kept developing the complexity and subtlety of its resources."[1] It was the ebb and flow of multifaceted musical trends like these that guided the sound engineer's career.

Hanley was able to establish himself in Greenwich Village, located on the west side of lower Manhattan, by 1966–67. The poet and folk singer movement of the mid-1960s that the area became famous for was rapidly developing. People flocked to the Village because of its music scene, its rebellious Beat Generation history and its being the hub of bohemian culture, art, poetry, and politics. According to folksinger Melanie Safka: "It was fabulous! Music was everywhere! People were examining where the different sources of music came from. It was like a near renaissance on earth during those times in the Village. Every corner had somebody singing and writing. The whole scene became a rebirth and people were curious."[2]

But even before Hanley arrived in the Village, distinct changes were occurring. According to Blues Project guitarist Danny Kalb: "The sixty-five scene had elements of folk, which was transformative politically, socially and sexually. It had many elements that were not Woodstock yet. It started to go into that direction by 1967."[3] There was still a strong coffeehouse presence in the Village.

Yet some claim its authenticity had shifted, as record company executives lurked in the shadows seeking out the next Dylanesque superstar. "The surplus of singer songwriters suffocated folk music, which was slowly dying of ennui,"[4] claims Suze Rotolo, artist and former inspirational girlfriend of Bob Dylan, in her book *A Freewheelin' Time: A Memoir of Greenwich Village in the Sixties.*

Folk musician Tom Rush recalls the New York music milieu being very different from the one in Boston in 1965–66. Rush worked with Hanley on several occasions throughout his early career. "The Boston club scene was more of an amateur one, and people were doing it more for the love of the music. In New York there was much more of a 'C'mon let's get matching shirts and go on the road' kind of approach. The people in New York were trying to build careers. It was a bit more formal."[5]

By late 1966 an intense wave of creativity and talent had come and gone in the Village. Rotolo reflects on what the atmosphere was like then, claiming, New York was getting "grittier and more dangerous" and the streets at night were "menacing" and "people were acting crazier."[6] Rotolo also recalls folksingers "Paul Clayton and Phil Ochs were erratic and broken," and the friends she "cared about" were all high on drugs.[7]

As the counterculture was finding its new self, darker times were imminent. More than half of the country was opposed to the developing war in Vietnam. Collective waves of protest involving hundreds of thousands surfaced across the United States within the next few years. The sounds of folk rock were emerging, and for those who were there this change seemed inevitable. "The counterculture was imploding; chaos lurked along the edges,"[8] warned Rotolo.

From 1965–67 a developing West Coast folk rock movement influenced what was happening on the East Coast. It included bands like Moby Grape, Grateful Dead, Jefferson Airplane, and Buffalo Springfield. Most important were the Byrds who, with their transformative reinterpretation of Bob Dylan's catalog, created a highly influential and memorable folk rock sound. Their hit of Bob Dylan's "Mr. Tambourine Man" (and album of the same name) exploded onto the scene, engaging national audiences. In 1965 it reached number one on the UK Singles and *Billboard* Hot 100 charts.

The mid-1960s British Invasion—which included prominent bands like the Beatles and the Animals—had a significant influence on popular music. By the mid- to late 1960s, groups like Pentangle and Fairport Convention created a unique style of British electric folk music. If you mix all of this with the newly electric Bob Dylan, the Byrds, and Beatles tours of America, the result is a creative evolution that stirred within artists who began their careers as blues, bluegrass, and folk musicians. Performers like the Youngbloods, the Blues Project, Simon and Garfunkel, and the Lovin' Spoonful who had been immersed in the folk revival just a few years earlier began to change their sound.

Many artists other than Bob Dylan launched their careers in the Village. And for a while it seemed like an infinite amount of incredible first-class talent wandered the streets in lower Manhattan. Venues like the Bitter End, Café au Go Go, and later the Bottom Line were showcases for many of the performers. According to Steve Katz, founding member of the Blues Project and Blood, Sweat and Tears, "From 1965 to 1968, it was like a circus in the streets of the Village."[9] If the Village was the center of music for the 1960s bohemian counterculture, then Hanley was most certainly its soundman.

By 1967 the Village was the nexus of folk rock, and it needed quality sound. Hanley could not have arrived at a better time. According to Hanley Sound crew member Karl Atkeson: "We did all the NYC clubs, and boy it was fun. All the big performers played in New York. I particularly enjoyed working in the Village. I was there a lot and we had equipment stored at a garage nearby. If we did an installation at a club the gear stayed there."[10] Among the many rooms that showcased this new energy referred to as "folk rock," none was as special as the Café au Go Go.

THE CAFÉ AU GO GO

The 375 seat Café au Go Go was located in the heart of Greenwich Village. Hidden in the basement of 152 Bleecker Street, the venue opened its doors on February 7, 1964. The club was a hotbed for some of the best rock and roll performances the city had to offer. Howard Solomon's Café au Go Go preceded Bill Graham's Fillmore East by four years. And even though Graham's Second Avenue theater absorbs most of the limelight for being New York's premier rock showcase, the au Go Go was something unique.

Unlike what was occurring within the ballrooms of the West Coast (and England) the au Go Go was not a venue for elaborate stage shows and psychedelic backdrops. It was a somewhat uncomfortable club, suitable enough for fashionably dressed enthusiasts who yearned to witness the next best thing in music. According to Hanley, "The Café au Go Go was the center of high-energy music other than folk."[11] The au Go Go was too large to be considered a coffeehouse, but only held between three and four hundred people. If at capacity, depending on where you were standing, you were only offered a distant side view.[12]

Owner Howard Solomon was short in stature and smart witted. Many remember him as being kind and having high energy. According to Blues Project drummer Roy Blumenfeld: "There were times when he would actually chase the IRS out of the Café au Go Go wielding a bat. Howard was a gem, if you could get past the baseball bat."[13] Solomon and his wife Elly lived around the corner on MacDougal Street, allowing the couple to come in and out of the club intermittently. A somewhat peculiar straight-laced atmosphere, the au Go Go served ice cream, ice cream sodas, and coffee instead of alcohol, making it accessible to underage music fans. *New York Times* journalist Jesse McKinley wrote: "Mr. Solomon's basement club played host to dozens of influential music stars. Renowned for its fine brick-wall acoustics and groovy coffee concoctions."[14]

Before the rock music explosion, almost every night the Café au Go Go was bustling with jazz, folk, and comedy performances. Comedy legends like George Carlin and Lenny Bruce (who was arrested on obscenity charges at the Café) entertained audiences with their controversial material. But by 1965 Solomon was having difficulty filling the room as the Village folk scene was beginning to recede. In order to survive, the feisty manager began to book more electric rock and blues acts. Among many others, Richie Havens, Judy Collins, Jimi Hendrix, the Blues Project, Odetta, Country Joe and the Fish, Van Morrison, Howlin' Wolf, the Chambers Brothers, Blood, Sweat and Tears, and the Youngbloods were booked by Solomon to perform at his club. The Café was the first New York venue where Joni Mitchell and the Grateful Dead performed.

At street level, the entryway to the au Go Go sat below a small awning. A short flight of stairs brought you down to the club's main entrance. As you entered, past the area of admittance, the coatroom became visible and then the sound and light control booth. An adjacent space to the right offered an area for food preparation as well as Solomon's office. The au Go Go's low ceiling contributed to the cramped feel.

According to Hanley, "It was a narrow subterranean like structure that was about seventy to seventy-five feet long and twenty to twenty-five feet wide."[15] The au Go Go stage and Hanley's speakers were slightly to the left, midway down the elongated space. Crowded by clusters of seats and tables, the stage was backed by a long brick wall. This arrangement generated less than ideal acoustics for those who performed near or around it. When there was too much amplification, the sound system occasionally produced feedback. Hanley remembers, "It was a difficult room to handle."[16]

Aside from managing and owning the venue, Solomon was in charge of renting the entire building, which included a small theatre above the au Go Go. The Garrick Theater, referred to as the "upstairs," offered prime rehearsal space for many of the café's visiting bands. The structure of the building gave the au Go Go its unique architectural footprint.

Hanley recalls: "A realty company or landlord owned the building. Howard was keeping everything on the hush-hush because he didn't own it. There was a lot of empty space and empty rooms on the upper levels. To the groups he brought in, Howard acted nonchalantly as the director of the building."[17]

While Hanley was in New York, he received a phone call from Solomon about renting a sound system. "Howard must have seen my ad in the phonebook."[18] Agreeing on a mutual time to discuss the club owner's needs, the sound engineer brought his heavy-duty equipment to the venue, eventually renting Solomon his system. In order to keep costs down, the money-conscious owner promised Hanley that he would introduce him to everyone in the rock and folk world for a discounted price. This was a logical agreement that eventually exposed Hanley's system to up-and-coming groups during many rock and roll shows at the club.

Solomon was very supportive of Hanley's services. For house bands like the Blues Project, who had a long-term engagement at the club, using a quality sound system was a way for both them and Hanley to advertise their talents. In this way Solomon was an important character in the sound engineer's career. "Howard was keeping me in the center of the music scene and was very helpful to me. This was my main place in getting started in the rock world. I was meeting acts here and people in the music business. Howard was promoting me."[19]

Robert Sontag, Café au Go Go employee and occasional sound and lighting engineer, ran the sound system for Solomon when Hanley was away. Sontag, twenty-one years old at the time, went out on tours with some of the au Go Go acts, and saw the quality and performance of Hanley's equipment. "Since I went on tour I could see the sound in California, and it was OK at the ballrooms, but Hanley's sound system was superior especially outdoors at live events. Anything Hanley gave us was better than what was in place at the time."[20]

Hanley's au Go Go sound system was comprised of two seventy-five-watt McIntosh amplifiers on one chassis. The cabinets were two fifteen-inch JBL speakers accompanied by two eight multicellular Altec horns, driven by Altec 290s, a common club configuration that he used at the time. According to Sontag: "Most of the equipment consisted of a preamp (multi-input) a graphic equalizer, and an amplifier. If I remember right, the pre-amp was an Altec."[21] When working other events, Hanley left about half a dozen Shure 565 microphones behind, more than sufficient for miking this size room.

In March 1967 musician Frank Zappa and the Mothers of Invention arrived in New York from Los Angeles. Hanley claims that Solomon leased space to Zappa for about six months. Here, the West Coast performer played his experimental avant-garde music nightly (through September) upstairs from the au Go Go at the Garrick Theater.

According to Hanley, the Mothers of Invention's unique sound seemed to get lost among the hip Village audience. For those who were there, including the sound engineer, many said it sounded more like noise. *New York Times* writer Dan Sullivan wrote that Zappa's sound was "Hostile—both in its headachy volume and in the lyrics that you can make out amid the roar. . . . If they are interested in attracting a wider audience, it might be suggested that they consider the uses of silence, as well as volume, to attract and hold an audience's attention. The listener finds himself paying more attention to what is on his mind than to what is in his ears."[22]

Zappa did a fair amount of recording himself. When he noticed Hanley recording many of the au Go Go performances, he became interested. Amazed at the level of quality the sound engineer was achieving, Zappa begged Hanley to leave his two Scully four-track recorders at the Garrick Theater rehearsal space. The engineer was reluctant, thinking he would never see his equipment again. He told the performer he would rather not. Although Hanley recorded a number of albums at the au Go Go, he rarely received proper credit for his work. The *Times* reported, "The club was the site for numerous recordings, both official and bootlegged, particularly of the blues performers who favored the club."[23]

On certain occasions Solomon asked Hanley to engineer and record ventures outside the Café au Go Go. Solomon managed folk singer Fred Neil, who in 1968 was booked for a weekend performance at a Woodstock, New York, café called the Purple Elephant. This rare gig was open to the public and music business people. Albert Grossman, members of the Band, Michael Lang, and many others attended.[24] The live event recorded by Hanley and produced by Solomon was used to finish out Neil's obligations to Capitol Records. According to Solomon: "I enjoyed producing this album and Fred was happy doing it and played better than he did in many a year. He was well received as you can hear from the enthusiasm of the audience who relished his every word."[25]

Hanley remembers, "I drove up from Boston to record and provide sound for Solomon who was managing Fred at the time."[26] Released in 1971, this was the last album of Neil's career. One side of *The Other Side of This Life* consisted of studio recordings; the other was the "live" side recorded by Hanley. Most of what Hanley recorded that evening ended up on the album. Solomon recalls Hanley's level of professionalism then: "*The Other Side of This Life* was a great experience. It was a joy to make and easy, thanks to Capitol coordinating with Bill Hanley of Hanley Sound."[27]

As Hanley's sound systems surfaced at various venues throughout New York, club owners were catching wind of his work. According to Hanley, Solomon often competed with Fred Weintraub, owner of the folk club the Bitter End, located just across the street from the au Go Go: "The quality of my work was becoming known around the area and I was soon installing a system in the thirty-by-thirty-foot space at the Bitter End for Fred Weintraub."[28] From 1966 to about 1969, music spilled out of Hanley's sound systems and into the streets of the Village.

THE ONDINE DISCOTHEQUE

Another midtown Manhattan club tucked away in a basement was the Ondine Discotheque at 308 E. 59th Street. This posh and trendy Upper East Side
Warhol hangout was equipped with a sound system that consisted of an Altec 100-watt amplifier, a pair of Altec column speakers, a five-channel Altec
1567a mixer, and three Altec condenser 21c microphones.[29] According to its
then soundman Jim Reeves, it was a "very sophisticated highly state of the art
system for its time."[30]

West Coast band the Doors were slated to play their first New York appearance at the club in November 1966. According to author Rich Weidman, "The
Doors signed for a one-month engagement at the trendy Ondine Discotheque,
then considered one of the hippest New York night clubs and a favorite place
for out-of-town bands to come play residencies."[31] Hanley remembers getting
a call by club owner Bradley Pierce to support the series of shows. "I suspect
that the sound system they had was not sufficient enough so they called me
in for consultation."[32] Anticipating a raucously loud event, the club's manager
most likely requested Hanley's expertise and additional equipment. Reeves
remembers Hanley, but is unsure why the manager called him in: "I suspect that
since the Doors were anticipated to be loud, Brad would have called Hanley
just in case."[33]

Reeves reflects on the sound technology that some of the New York clubs
had at the time: "In the latter half of the '60s, when the music revolution was
beginning, the sound systems in Manhattan clubs were not equipped to handle
the heightened energy of the acts that were to follow. Live PA sound systems
were just in their infancy then. It was new and exciting."[34]

Hanley remembers the Doors' first out-of-town gig as a memorable one: "It
was one of the first gigs for them on the East Coast and it was loud. No one
knew what to expect."[35] According to *Billboard*, the Doors performed with a
"bombastic roar of sound" giving the band almost the same "intensified sound
in person as on record."[36]

THE YOUNGBLOODS AND THE BLUES PROJECT

Back in Boston during the early to mid-1960s, the folk scene bustled at the Club
47 in Cambridge, while the Newport Folk Festival was happening concurrently
not far away. Radio airplay was another way for performers to showcase their
talent, and in Boston that occurred on a show called *Hootenanny* hosted by
deejay Jeff "Jefferson" Kaye (and later by deejay Ron Landry). Every Sunday
night WBZ took a break from its typical Top 40 rotation and played folk music.

At the time, *Hootenanny* was an easy way for Bostonians to become exposed to this style of music.

Artists like Jesse Colin Young of the Youngbloods benefited greatly from airplay in Boston. Young got his start in 1964 in Boston as a folk singer at Club 47. This was his first gig featuring songs from his debut album *Soul of the City Boy* on Capitol Records. His follow-up effort, *Youngblood*, was released in 1965. After he began playing with Jerry Corbitt and Lowell "Banana" Levinger in Cambridge, they assumed the name the Youngbloods. They soon gained a reputation on the club circuit.

From Boston the Youngbloods moved to New York. While playing at some of the area folk venues like Gerde's Folk City in the Village, they had difficulty hearing each other on stage. With no stage monitors and lackluster Shure columns for a PA, they were in need of a quality sound reinforcement system. According to Young, "The sound quality at some of these clubs was pitiful . . . until Hanley arrived."[37]

Before they met Hanley at the Café au Go Go in 1966, the band had difficulty developing the signature sound we have come to know. Founding member Lowell Levinger recalls: "Sound was always abysmal. Then we started playing at the au Go Go when I realized there is this new guy in town, Bill Hanley. Anywhere there was going to be a Hanley Sound system you knew the people in the back were going to be able to hear it, like the people out in the front."[38]

By early 1966 the Youngbloods became the house band at the Café au Go Go, earning around twenty dollars a night. Levinger remembers that even though Solomon was very supportive, he was not "overly generous" monetarily. A typical daily routine for the band, was to get up around 10:00 a.m., head down to the au Go Go, rehearse until mid-afternoon, go home to eat, then head back to play. "We were lucky enough to share duties and alternate as the house band with the Blues Project. When people came through who needed some backup, we would support them. And for the big shots like Cream and Muddy Waters, we opened for them."[39]

Fortunately, Hanley was there with the right sound system for the newly established folk rockers. According to Young: "We had a McIntosh tube amplifier and four cabinets with two on each side of the railroad car, which is what the au Go Go was shaped like. Hanley twisted the speakers off axis to cover the length and the depth of the room. It was so superior to what we had been doing. This was the first time I could hear myself. It was truly amazing."[40]

Like many groups at the time, Hanley was the first sound engineer the Youngbloods had ever encountered. The group spent many hours rehearsing with a Hanley sound system, and eventually brought it on the road with them. The sound engineer rented it to them for around $350 a month. While on an eight-city promo tour for RCA, the Youngbloods and their entourage of amplified equipment did not go unnoticed. A November 26, 1966, *Billboard* article titled "Do the Acts or Instruments Capture the Fancy of the Teens?" cites the band as toting "31 vari-shaped carrying cases containing over $10,000 worth of amplified apparatus weighing in at 1,942 pounds."[41]

While on tour, the group hauled Hanley's equipment as well as their instruments in a rented Hertz van. Bill Barko of B.C.L. Sound Laboratories designed the group's instrument sound system. Hanley reinforced the vocals with his. According to *Billboard*:

William Hanley, a Boston audio designer, worked out the vocal amplification equipment for the group. This section of the combo's armamentarium centers around a custom unit utilizing a McIntosh 275 which Young claims can dish out 150 "clean watts" with a combined RCA pre-amplifier-mixer that incorporates a Langevin studio equalization unit. Four Telefunken mikes feed into this, the group's heaviest piece of equipment. Two 6 ft. tall speaker columns on each side of the stage encasing six eight-inch Lansings provide the woofer and mid-range audio speaker assembly that flanks the group while two giant tweeter assemblies round out the Youngbloods' nearly unbelievable electronic entourage. Again these speaker enclosures were each custom-tailored for the group by Hanley.[42]

A 1967 *New York Times* article described the "sound" of the Youngbloods as "eclectic but individual," combining a diverse strain of "American folk music and jazz" with a "moderate dose of electricity and studio technique," and called the band's sound "dense, clean and exhilarating."[43] Clearly Hanley's sound system made it possible for the group to achieve the iconic music the Youngbloods are known for while on tour or in the studio.

The group's first two albums, *The Youngbloods* and *Earth Music* (1967), were rehearsed at the au Go Go. By the time the group reached the studio to record, they were accustomed to Hanley's system, a developing sound that evolved and naturally translated onto these early albums. According to Young, moving from Hanley's system and into the studio was seamless: "We just went in and played it and it sounded good on a record. Without Bill's sound I don't think it would've happened. Bill was essential to the music in our band. Being able to hear myself above the music helped me hear the music around it."[44]

In 1967 the Youngbloods left for the West Coast. It was the Summer of Love and they played the developing "free scene" occurring at some of the ballrooms and emerging festivals. By comparison, the quality of sound on the West Coast was not as good as what the band had become accustomed to in New York. "When we got there it seemed California was in the dark ages, like at the Avalon Ballroom for example. It took quite a while for them to catch up,"[45] says Young.

The unique sound of the Youngbloods is best evidenced in their version of the counterculture classic "Get Together." The arrangement and sound of this recorded song owes a lot to Hanley's sound system. Young observes: "Bill was personable and such a sweet guy for a soundman. He was a good salesman and he was selling something beautiful."[46]

The New York–based Blues Project left a defining impression on the Greenwich Village scene. A *Crawdaddy!* magazine article about one of the group's au Go Go performances reports: "The stage lights flash blue, red, blue, the audience lights flicker and burst, and occasionally the music is totally lost amid the cascading torrents of sound. The five brilliant musicians who collectively call themselves the Blues Project have once again won the battle for men's minds."[47]

The group's guitarist Steve Katz, recalls Hanley as a "mild mannered, straight guy" with "short hair." To the self-proclaimed "stoned musician," Hanley's appearance must have worried him. Yet, the sound engineer's easy demeanor quickly comforted those who were around. "Hanley was starting out when we were starting out. He loved doing the sound and loved his equipment; all of that was pretty obvious. We also had no monitors and I think Bill was one of the first people that probably used monitors. The sound was good."[48] Katz would meet Hanley again, only this time as a member of Blood, Sweat and Tears on a much larger stage during the Woodstock Music and Art Fair.

Band manager Herb Gart was instrumental in helping form both the Youngbloods and the Blues Project. These two groups later emerged as house bands at the Café au Go Go. Gart, who arrived in Greenwich Village from his home in Philadelphia in 1962, claims Hanley had a direct impact on the development and sound of these two bands: "The timing was very, very good for someone like Bill Hanley, and his fees were fair."[49] Establishing a successful career in the music business, Gart managed Bill Cosby, Tim Hardin, Buffy Sainte-Marie, Pat Sky, Don McLean, and Mississippi John Hurt.

During these transformative days most groups required quality sound equipment, and even more so as the scene shifted from acoustic to electric instrumentation. Gart recalls, "Suddenly things dealing with sound and speakers, and all that one needs to go electric, and be heard, became an important issue."[50] The bands also used Hanley's equipment for outside gigs and tours. From the perspective of a band manager, Gart gained a sense of security using a Hanley Sound system while on the road: "With Hanley we were genuinely able to have good sound almost everywhere we went. Most groups had to go with whatever sound was in place where they were playing. Some had shitty sound at best. That changed to being able to have good sound at a show. It became invaluable. At that point that was a fairly new idea."[51]

According to Hanley, when he left the Café au Go Go in late 1968, Solomon owed him a lot of money: "Howard couldn't pay his rent and the landlord closed him up. He told me to sue the landlord for the money owed to me, which was more like $20,000. I was not smart enough to do that."[52]

From 1965 to 1969 the Café au Go Go presented live music almost every night. Throughout its existence the club often advertised its performances in the *Village Voice*. In June 1969 its final ad read, "For lease 150–154 Bleecker Street—World Famous Café au Go Go—Legal Occupancy Cabaret 285, Theater 200." After briefly changing ownership, the club permanently closed its doors in October 1969.

By this time Hanley had temporary and permanent sound systems at many New York venues, including the Electric Circus, the Apollo Theater, the Bitter End, Ondine's, the River Boat, the Cheetah Club, the Generation, Steve Paul's the Scene, and soon, the 20,000-seat Madison Square Garden. Although Solomon could not pay Hanley the money he was owed, he certainly did provide him with business contacts, one of which included impresario rock producer Bill Graham.

In its time, the Café au Go Go was the preeminent rock venue in New York and the rehearsal space for countless musicians. Many of the bands Hanley encountered here went on to perform at the Fillmore East. He would also see many of them at the soon-to-be Woodstock Festival, where the lineup was an au Go Go alumni reunion. In March 1968 a second wave of rock music exposition occurred at Bill Graham's Fillmore East on Second Avenue, where Hanley was ready and waiting with a superior sound system—one the promoter could not refuse.

Chapter 22

MORE MONITOR, PLEASE!

THE BUFFALO SPRINGFIELD

In 1967 Hanley Sound was known as the premiere sound company in New York. However, efforts to expand his services with West Coast producers Bill Graham and Chet Helms were unsuccessful. According to Hanley: "I was trying to convince them both. Chet Helms of the Family Dog concerts at the Avalon Ballroom and Bill Graham at the Fillmore West. They were not interested in me at all. As I recall, they had a terrible sound systems."[1]

It was the Summer of Love while Hanley walked the streets of San Francisco. A change was in the air. From Haight-Ashbury to Greenwich Village it seemed that both coasts were on the verge of connecting through a new representation of rock music. According to Jefferson Airplane bassist Jack Casady, music was evolving during this time and sound quality was becoming a topic of conversation: "The shift started to come when you had places like the Fillmore and the Avalon come into being, places that had regular events happening. You had promoters like Bill Graham and Chet Helms that wanted to put on better and better shows. Everyone realized that the sound had to be improved."[2]

While in California, Hanley visited the Hollywood Bowl. He had heard about its state-of-the-art sound system used to support many symphony orchestra concerts and rock shows including the Beatles in 1964–65. Head engineer George Velmer gave Hanley a tour of the 18,000-seat venue. Hanley reflects: "I was always really curious about their sound console and wanted to see it for myself. I wasn't too impressed with the setup. They were using RCA stuff and Electro-Voice microphones and some Altec. I heard Wally Heider did some recording there. The place reminded me a lot of the Hatch Shell in Boston."[3]

It worked out in the end for Hanley, when the demand for his services became greater. Eventually pop and rock festival-based mass gatherings began to emerge all across the country. The Miami Pop Festival, December 28–30, 1968, in Hallandale Florida is considered the first major East Coast rock festival in a long line of similar events to come. Organizers for Miami Pop called on Hanley for assistance.

Sound reinforcement technology for live performances in 1967 was still in a developmental state. Commonplace now, monitors used for live sound were unusual then. If there were monitors at all, they were positioned either stage left or stage right of the performers. For example, on the West Coast at the Monterey International Pop Festival, photos reveal speakers deployed by McCune Sound used only in a side fill position. Stage monitors (known as floor monitors) are commonly multiple speakers aimed toward performing musicians. Positioning is everything, and having monitors aimed in the direction of the performers allowed them to hear well. Still, most musicians and sound engineers were not educated about the efficient placement and use of this application.

Through the 1960s amplified bands with multiple singers often had problems hearing themselves on stage. What Hanley refers to as the "footlight stage monitor" was extremely useful in cases like this. The term "footlight" in theater production is a lighting device used to illuminate the front edge of the stage floor. This position is where Hanley first used his sound application, assisting bands like the Chambers Brothers, the Beach Boys, and the Cowsills. His innovation allowed them to hear each other despite intense sound levels while on stage.

A stage monitor system has the most impact when electric instruments are used with acoustic instruments and multiple vocals. This configuration usually requires its own amplification and equalization, mixed separately from the primary front of house (FOH) sound system. The contemporary stage monitor sound system is a combination of speakers/monitors, equalizers, amplifiers, and a designated sound engineer mixing the monitor sound.

In the 1967 book *Professional Rock and Roll*, in a chapter called "Sound System," contributing writer Chris Huston vaguely mentions the application of a stage monitor setup: "In huge arenas and halls I've always found it best to try and keep the sound 'together.' By this I mean that it's best to keep the speakers as near to each side of the band as possible. This accomplishes two things: the band stands more of a chance of hearing themselves; and all of that sound comes from one source."[4]

During Hanley's West Coast expedition to cultivate new work, he attempted to locate one of his sound systems rented by the influential Buffalo Springfield. When Hanley finally tracked down his property, he noticed members Neil Young, Stephen Stills, Richie Furay, Bruce Palmer, and Dewey Martin were having difficulty hearing each other while rehearsing. Fortunately, Hanley was there just in time to help.

After some troubleshooting, the sound engineer positioned four twelve-inch Rebel speakers on the floor angled upward 45 degrees toward the ears of Young and Stills. This combination of monitor placement and the use of directional microphones for vocals resulted in higher sound pressure levels. This gave the performers the ability to hear each other more efficiently while playing live. Hanley remembers: "I was curious as to why they still needed my sound system since they were not on tour. They explained to me when I arrived that they couldn't hear each other when they were performing. So I helped Neil Young and the rest of the band with my monitor setup and *voila!* they could suddenly hear each other sing! They were harmonizing, so this was important."[5]

By solving the band's problem, Hanley realized that most performing musicians shared similar difficulties. As amplification of instruments got louder, the sound levels on stage increased also, and it became more difficult to hear. Up until this point, Hanley's primary concern during a concert had been the audience. It had not yet been on the sound engineer's radar to help performers hear each other while on stage. His work with the Buffalo Springfield helped them achieve their signature sound. According to Ellen Sander of the *New York Times*, "In the war of sound that the pop revolution has become, the quietly beautiful Buffalo Springfield albums have been consistently overlooked and underrated."[6]

Before this innovation Hanley had occasionally used quality speakers for side fills upon request. "I only used side fills if the performer wanted it. I never thought to place speakers in front of the performer's ears. It never came to my mind up until then."[7] This simple innovation was an eye opener for Hanley, forever changing the way musicians hear during a live performance.

Even though Hanley's initial discovery and introduction of the center-positioned footlight monitor can be determined from around 1966–68, others were inconsistently applying the technology as well. First-generation sound engineers like Harry McCune Sr. introduced the innovation in 1961. His grandson, engineer Allan McCune, claims his father and grandfather applied the technique for a Judy Garland performance at the Civic Auditorium in San Francisco: "The first time that a stage monitor was used that we know of was for Judy Garland. The rehearsal was not going well. My father came up with the idea of pointing a speaker at her. He jumped in the truck and dashed back to the office, grabbed a speaker, brought it back, put it on the corner of the stage, took a feed off the main system, turned up the amp and, like magic, the artist was happy."[8] Occasions like this demonstrate how Hanley and others across the country refined and shaped its future application during major concert events. Over fifty years later, the monitor has become an industry standard through elaborate stage and in-ear monitor system designs and arrangements.

The Buffalo Springfield made three albums during a brief and highly successful career that ended with a final performance on May 5, 1968. Their skyrocket to fame exploded, as did others of that folk rock genre like Jefferson Airplane. These bands were sometimes criticized for not sounding like they did on their albums while playing live. Having the ability to translate what was recorded in the studio into an outdoor setting was often difficult, mostly because of the state of sound reinforcement technology. Ellen Sander claims, "Electronic rock was interesting when the effects underscored and complimented the songs." She continues by asking: "What has happened? Are the San Francisco groups now earning five figures for concert appearances, unable to duplicate the warmth and excitement they projected when playing for free in Golden Gate Park? Have they fallen victim to the elaborate Sergeant Pepper Syndrome?"[9]

Between 1967–68, Hanley assisted when Jefferson Airplane was performing on the East Coast. His company provided state-of-the-art sound reinforcement on several tour dates during what is considered to be the band's loudest and most experimental phase. Bay Area sound companies McCune and Swanson supported most if not all of the band's West Coast dates.

In late 1967 Jefferson Airplane tour manager John Morris called upon Hanley Sound to help at a dilapidated Lower East Side theater called the Anderson. It was here that Morris and a few others were responsible for getting the venue showworthy. Hanley's involvement with the theater opened a path toward a future working relationship with concert producer Bill Graham.

Chapter 23

A FRENCHMAN AND
THE SECOND AVENUE SHUFFLE

In early 1968 a young Frenchman named Sam Boroda, from Marseille, a port city in southern France, came to America to live with his brother and mother in Allston, Massachusetts. Boroda had studied electronics at ORT France, an institution of education and training that helped him become technically proficient. When looking for a job in Allston, Boroda was referred to Hanley Sound as a suitable company based on his experience. He walked into Hanley Sound in a black suit and tie not knowing a word of English.

According to Judi Bernstein: "I remember Sam's brother was interpreting for him when he came in. Eventually I taught him how to speak English."[1] Boroda recalls that it was at Hanley Sound where he learned to speak English: "Everyday I learned one word here and there. One day from the back of Hanley Sound I screamed out the F-word! And Judi said 'okay, now he is speaking English!'"[2] If it weren't for Bernstein, Boroda would not have had a job at Hanley Sound. "Judi liked me because I was Jewish."[3]

In the early days the clean-cut and proper engineer often wore a black suit and a tie. His first job was with the Cowsills, and according to Boroda at this performance Hanley could see his potential: "Bill gave me a chance and let me mix. I could not speak English but I could read his body language. He liked what I was doing so I continued on. At the end of the concert I was signing autographs for the little girls."[4]

At the time the sound company was preparing for several Janis Joplin performances. The first was at the Anderson Theater on February 17, 1968, followed by a show at Boston's Psychedelic Supermarket February 23 and 24, and finally at the Rhode Island School of Design in Providence on February 25. It was here that the engineer got his first real taste of American rock and roll while working alongside Terry Hanley in front of an "enthusiastic crowd of stoned art students."[5]

According to Boroda, it was something he'll never forget: "Janis blew me away! At first I did not know who she was. I asked who the heck is that? I could not speak a lick of English. I mostly assisted in the beginning of my time at Hanley Sound. I remember it being a foggy night. On the way home we had to spend the night at a gas station on the highway."[6]

Over time Boroda became known as the "finicky Frenchman," having an attitude and being somewhat difficult to work with. To those who remember him, most remain indifferent about his personality, yet all claim he was a genius as a sound engineer. According to crew member Scott Holden: "To me he seemed to be a magician with mixing and working the sound system in general. He had great instinct, great taste, and a great ear."[7]

Boroda eventually graduated to mixing big-name performers, while staying away from the politics surrounding festivals. "I had an ear and I didn't play around. Often people who worked for Hanley were stoned. I was handed a lot of drugs. I think Bill and I were the only two who didn't do anything."[8]

Sam Boroda stayed with the company for many years, eventually leaving during the mid-1970s. He went on to work with some of the biggest names in the industry, taking what he had learned from Hanley and applying that knowledge to the fields of sound recording and sound reinforcement. "There was a day when you dropped the name Hanley Sound and no one asked any questions. The name was known for the best you could get in sound. Sadly, years later it got stained."[9]

THE ANDERSON THEATER

In New York, Hanley's work at the Café au Go Go allowed him an opportunity to network and interact with some of the biggest talent in the business. As a result, famed music manager Albert Grossman, concert promoter Ron Delsener, and rock concert producer Bill Graham took notice. Hanley recalls: "Howard Solomon must have called Bill Graham and others and told them that I was the man to go to. This connection is what did it for me. Because of this I ended up at the Anderson Theater with John Morris and then on to the Fillmore East."[10]

Hanley already knew Grossman from the Newport Festivals, where he was on the Newport Folk Foundation board with George Wein. Grossman managed Bob Dylan, Peter, Paul and Mary, and others. He often encountered Hanley at the Bitter End and the Café au Go Go. According to the sound engineer, Grossman had a reputation for being crass: "Grossman was more or less a crude businessman and didn't take me very seriously."[11] With or without the assistance of someone like Grossman, Hanley powered on, eventually having permanent and temporary installations around the Big Apple, including a short stint at the Anderson Theater.

The East Village of Manhattan in 1967 was a seedy and dangerous place and not particularly ideal for hosting music. With so many smaller clubs like the Electric Circus, Steve Paul's the Scene, and the Café au Go Go offering premier talent, it created an interesting and unique fusion of hippie counterculture and the dregs of society. Before Hanley most of these venues in lower Manhattan had sound systems that could barely sustain the increasing power of loud rock music. Even large-scale concerts held at Madison Square Garden were getting panned in the newspapers for poor acoustics.

In the midst of all of this was a dilapidated structure on Manhattan's Lower East Side known as the Anderson Theater. Located at 66 Second Avenue, about two blocks down and across the street from the Village Theater (soon to be renamed the Fillmore East), the Anderson wrapped around the corner onto East 4th Street. Opened in 1926, the venue began as a Yiddish vaudeville playhouse eventually showing Yiddish films. "It was named after . . . play agent, Phyllis Anderson, who was the wife of author Robert Anderson. The theatre continued to run Yiddish programs until the 1960s."[12] By the mid- to late 1960s it was used as a venue for rock shows.

The Anderson had the capacity to seat around 2,000 people, which was ideal for mid-size musical performances.[13] Seeing potential, John Morris viewed this area of New York as an untapped market for music. He wanted to establish a scene similar to what was happening on the West Coast.

While Morris was in New York in December 1967, he connected with colleague Joshua White of the Joshua Light Show. Later they met Sandy Pearlman and Richard Meltzer at Stony Brook University. They were associated with *Crawdaddy!* magazine and interested in getting shows going at the Anderson. Not long after, Chip Monck jumped on board to help with production at the theater.[14] Morris recalls: "The Anderson was available and the right size. These two guys from *Crawdaddy!* magazine barely out of college came to us and asked if we could help put together concerts. So we said yes. They had the idea and we did the booking once we started it."[15] In January 1968 they began planning a series of shows for the theater. By February a series of musical performances (called the Crawdaddy Concert Series) were held that included Country Joe and the Fish, Jim Kweskin, and Soft White Underbelly.

An unstable and untrustworthy fellow named Tony Lech financed what was needed to get things going at the fledgling theater. According to some, Lech owned a number of gay bars in the area. With the right combination of financing and booking, the formula was in place for concerts at the Anderson. But with very little music business experience, Lech lacked the leadership the theater needed to survive, causing its eventual demise.

When Hanley was first introduced to Morris and White in 1967, they were working for a production outfit called Sensefex. Hanley was assisting them in a series of weeklong performances with the Grateful Dead and Jefferson Airplane at the O'Keefe Center in Toronto, July 31–August 5. These shows produced by Bill Graham were an attempt to recreate the San Francisco scene on the road.

After Toronto, Graham asked Morris to move to the coast and help him run the Fillmore West. Morris eventually became tour manager for Jefferson Airplane. When the Airplane toured the states, Morris hired Hanley. Later Morris decided to settle in New York to promote shows.

Hanley was an obvious choice when Morris needed quality sound for the concerts at the Anderson. Along with Morris, individuals like Monck and White were already working there when Hanley arrived. Morris staged the performances while White ran his spectacular light show. Engineer Chris Langhart arrived around the same time Hanley did. Langhart was a professor in the theatrical technical department at New York University. He lived close by on 3rd Street, and on his walk to work he often passed the Anderson. Langhart recalls: "I caught up with Bill Hanley at the Anderson because he was doing sound at that location. My close friend Joshua and I worked at Carnegie, and he was there with his light show and so was Chip. So I stumbled onto the whole lot of them at once."[16]

Hanley recalls: "I was called in because I was the only guy who had enough gear to do these shows. I remember White and Langhart were there with Monck, who had sort of been out in the field working and talking with Morris at the time. I remember Morris was Graham's representative on the East Coast."[17] Hanley and Morris forged a strong and lasting relationship that continued at the Fillmore East and the Woodstock Festival, where Morris was an emcee and production coordinator.

Planning and producing shows at the Anderson proved to be difficult for Morris. He is quoted in John Glatt's book *Rage & Roll: Bill Graham and the Selling of Rock*: "It was really a hairy operation and done on a shoestring budget. At the shows at the Anderson, everything that could possibly go wrong, did. So we had to fix it. We had tons of energy but very little money and an old disintegrating theater to work with. But we got the acts and drew people. It was East Village '68 anarchy."[18]

In a last-ditch effort to keep the venue afloat, Morris tried to persuade Bill Graham to partner with Lech. What the Anderson needed was the expertise of a seasoned producer, and Graham was a natural choice. According to Morris: "It was a constant campaign to get Bill into New York. I had found working with Bill was exciting and interesting, and selfishly I wanted him to come to New York and set up a theater that I could run."[19] Finally convinced, Graham came to New York to see what was going on.

He arrived at the Anderson during the winter of 1968 to watch Big Brother and the Holding Company perform their two inaugural, sold-out shows. These performances on February 17, 1968, mark the first time Janis Joplin and the group performed in New York. The performances were spectacular, according to the *Village Voice*: "New York's golden ears came out ringing from the Saturday evening performances of Janis Joplin at the Anderson Theatre on Second Avenue."[20]

Sam Andrew, guitarist for Big Brother and the Holding Company, describes these memorable days: "It was amazing to be there on the Lower East Side at that time in 1968. That theater was a Yiddish theater right through the 1920s and '30s. I later learned of all these famous plays that had happened there. Classic Yiddish plays. For Big Brother and the Holding Company it was a real immersion in that culture."[21] Other groups including Procol Harum and Moby Grape performed at the Anderson Theater during its short life span.

Bill Hanley and Janis Joplin had a friendly and professional relationship throughout her career. When the singer and band took up residence at the Chelsea Hotel on West 23rd Street, the sound engineer often drove Joplin home after her gigs. Hanley remembers Janis as a "good kid" who liked to "talk a lot!" Crew member Harold Cohen recalls: "One time I was helping Bill with Janis at the Electric Factory in Philadelphia. Bill said to me 'scream into the microphone as loud as you possibly can and set the limiter based on that.' On the first number she screamed so loud that she nailed the meter!"[22]

By December 1, 1968, Big Brother and the Holding Company would be no longer, having played their final performance at a Family Dog benefit in San Francisco. While at a party at actor Sidney Poitier's Central Park West residence, Joplin asked Hanley to go out on tour with her new group, the Kozmic Blues Band. Hanley remembers: "Sidney was close to George Wein so I got to know Sidney and was invited to his home in NYC. At the party Janis and I began to chat. She asked me to help her out."[23] Hanley accepted the offer.

In early February 1969, Joplin's manager Albert Grossman booked the Kozmic Blues Band for a "sound test" at what he referred to as "the most obscure venue we could find."[24] The concert was held at Franklin Pierce College for a Winter Carnival at the rural Rindge, New Hampshire, campus on February 8. The show was a rehearsal for the band's premiere performance at the Boston Music Hall on February 9. According to Franklin Pierce social committee chairman Henry Ellis, Joplin's college appearance happened by chance:

Why would Janis Joplin agree to perform at Franklin Pierce? Besides the money, there was a specific reason she fell into our lap. She had recently broken up with Big Brother and the Holding Company and had formed a new band. The new group was to perform for the first time in Boston on

February 9th. They would be using a new sound system and needed a place to test it out under concert conditions. Franklin Pierce became the place for this "sound test" and the deal was done. We added the Paul Butterfield Blues Band as the opening act and Richie Havens for a concert on Sunday afternoon to close the weekend. Our student activities fund had just been reduced by $32,500.00; $20,000 for Janis Joplin, $7,500.00 for Paul Butterfield, and $5,000.00 for Richie Havens, but we were confident it was money well spent. As to making Franklin Pierce history, that was the furthest thing from our minds.[25]

After the Boston show, the singer and band had two scheduled high-profile performances at the Fillmore East. A year later on August 12, 1970, Hanley Sound crews provided sound reinforcement for Joplin's final gig ever, at Harvard Stadium in Cambridge.

After viewing the success of Big Brother and the Holding Company at the Anderson in 1968, the business-savvy Bill Graham saw the monetary potential to be had. However, even though the response to the music at the theater was spectacular, the venue was in dire need of financial and physical help. Released from his initial speculations, the producer was now interested in the business endeavor and agreed to meet with financier Tony Lech about a potential partnership. The two met and conversed over a bowl of soup at a New York café called the Tin Angel, where Lech's stubbornness got the better of him.

Unwilling to accept Graham's proposed offer, the deal fell through. The house manager of the Anderson, Jerry Pompili, and personal friend of Lech's at the time recalls the encounter: "I was with Tony at the Tin Angel Restaurant during that meeting. After he went ape shit all over Graham I told him he made a big fucking mistake, which he did. Graham just packed up and walked out."[26] But the Anderson had proved to Graham that a mid-sized theater offering music in the area could work. Forever the entrepreneur, Graham continued to look for other possibilities and found them at the nearby Village Theater, soon to be known as the Fillmore East. The Village Theater was located in what is now known as the East Village.

When word got out that Graham was setting up shop in New York, the anxious Anderson Theater staff was ready to jump ship, including Hanley. Most if not all of these talented individuals crossed over to Graham's new venue the Fillmore East. According to Pompili: "When Bill Graham opened up the Fillmore East, they all jumped over and worked for him. They weren't stupid. Soon after, the entire tech crew went over to Bill Graham and left the Anderson, every single one of them."[27] Morris recalls, "At the Anderson we all formed the team that ended up at the Fillmore East."[28] Joshua White comments: "We all began at the Anderson Theater, including Hanley. The Anderson was essentially the same as the Fillmore minus Bill Graham. The concert industry before the Fillmore opened was still a little shady in the sense that it was not just about making money but also stealing money. We began at the Anderson because we had a passion."[29]

At this same time, Graham was changing locations of his West Coast venture, the Fillmore Auditorium. That Fillmore held around 1,000 people; the operation moved to the larger Carousel Ballroom, holding around 2,500. When Graham flew back to New York, he focused his efforts on making an impact with the Village Theater. Hanley wanted in; however, the sound engineer knew that Graham was difficult to work with, and had refused to do business with him months earlier. According to Hanley, "Graham played his cards close to his chest."[30] Like the Café au Go Go, the new Fillmore East in the East Village evolved into a larger stage for rock and roll exposition.

Chapter 24

THE VICTORY SPECIAL

THE NIXON/AGNEW WHISTLE STOP TOUR

In between Hanley's work in New York, he and his crew kept busy with jobs of various kinds. Even though rock and roll was at the top of the list, so were gigs of a political nature. Although nothing could match providing sound for the inauguration of a president, political work helped keep the business afloat. In late 1968, Hanley Sound was contracted for another presidential campaign: a Nixon/Agnew whistle-stop tour. This event marked a return of traditional public address to political campaigning, via the rails. The "Victory Special," as it was called, was a fifteen-car train that traveled on a two-day campaign from Cincinnati through Ohio, ending in Toledo. Hanley only worked on a portion of their campaign trail. According to Russell Freeburg of the *Chicago Tribune*, "The train filled with local Republican politicians, Nixon staffers, and newsmen, clattered across the rails and rolled into towns with shut-down passenger depots."[1]

With some of Hanley's staff on board overseeing the sound system, the presidential nominee "Rode and spoke from the rear car of the train, which carried a sign that read 'Nixon Victory Special' on the tailgate."[2] Through Hanley's speakers, Nixon "Bashed Vice President Humphrey as well as the Secretary of Agriculture and the Attorney General."[3]

Before the train left Cincinnati on October 22, 1968, engineer Bob Goddard, Bill, Terry, and a few others worked on getting the multiple rail cars wired and up to spec for the tour. It was an immense amount of work. After the train departed, Boroda, Goddard, Cohen, and others including Bob Carr helped facilitate the daily events. Cohen handled live press feeds to the three major networks and also radio stations. He worked exclusively with vice presidential candidate Spiro Agnew.

Bob Goddard, Fillmore East engineer and occasional independent contractor for Hanley Sound, recalls getting the job: "When I jumped on to help out I didn't have my toolbox, and I didn't have clean clothes. Soon I was crawling across railroad ties running wires between railroad cars. By the time the train left I was on it with the crew. I got to ride on a private railroad car, served with silver and china at a dining room table by two porters."[4]

Bob Carr, Hanley Sound business manager of sorts, had a short stint with the company. Boroda recalls that Carr was influential in acquiring political jobs for Hanley Sound:

Before I went onboard Hanley and others spent a tremendous amount of time wiring the train. Bob had a very big link in bringing this job to us. Bob and I were on the train. We installed the equipment in one of the last cars, in the lounge. We had speakers, which we used to flip open when we came to a station. We also broadcast the rally in each town to the press car so that everyone could hear. I ran the audio for this and it was all done live. In the back train, which was Nixon's private car, it was Bob, Nixon, his family, manager, his security and myself. One of the reasons I was able to get this job was because I was not an engineer hippy, so I made the cut. I looked more corporate in those days.[5]

While on the train, Hanley Sound employees were referred to as the "advance audio crew." For 247 miles Hanley's engineers deployed sound for Nixon, setting up and breaking down five to six times each day. For the most part, crowds were large and overwhelmingly friendly. However, signs of the times were evident as antiwar protesters gathered in Dayton chanting, "End the war!" and "We want peace!" as the Hanley crew looked on.[6]

Chapter 25

THE WHITE MAN'S APOLLO

THE FILLMORE EAST

The Gothic Revival–style Village Theater was located on Second Avenue in the Yiddish theater district of the East Village. It had a number of names and phases of existence over its long history. In 1925 it was a vaudeville playhouse known as the Commodore. During the 1930s it was bought by the Loews Corporation and turned into a movie theater and renamed the Loews Commodore. During the 1950s it transitioned to the Village Theater, where comedy and live performances surfaced once again. Even later, having seen better days, the theater sat mostly vacant. It fell further into disrepair until 1967 when the occasional rock show occurred, hosted and led by producer Gary Kurfirst.[1] In 1968 West Coast rock producer Bill Graham took over, breathing new life into the 2,645-seat theater. Now known as the Fillmore East, it became one of the most important rock and roll performance venues in popular music history.

John Morris directed Graham to the vacant theater, knowing it was available and that performances were happening there already. First, the producer made sure that the space offered the basic infrastructure needed to support rock shows. Graham took important production elements into account, like sufficient areas for dressing rooms, load-in accessibility for equipment, and sound and lighting.

The theater was a relic of the past and the interior revealed such details. In his book *Live at the Fillmore East and West: Getting Backstage and Personal with Rock's Greatest Legends*, John Glatt writes, "The building exuded the faded glamor of a 1920s movie palace, with a grand proscenium arch, red velvet walls, and painted murals with a gilt chandelier hanging above the balcony."[2] Optimistic about expanding his Fillmore productions to the East Coast, Graham, along with other investors, paid around $400,000 in total for the building.

Morris, who eventually became managing director, was in charge of moving things over to the Fillmore from the Anderson. "Chip did the lighting, Joshua

White did the light show, Bill Graham did the booking, and Hanley did the sound. It was sort of everybody thrown together. Chris Langhart was head of the Theater Department at NYU and brought in all of these students. We set a pay scale of $2.25, that's all we paid them."[3]

It took a lot of work to revive the theater from a state of disrepair as it suffered from extensive water damage. According to Graham: "We tore down some brick walls and changed the whole back of the theater so we could get the wiring through. We made a lot of changes in the stage area to make it effective so we could do a variety of productions."[4] It didn't help that in February 1968, a month before opening day, there was a major garbage strike in New York. Those who were attempting to help set up the new Fillmore for its opening had to fight through mountains of trash stacked high in the streets.

Langhart, who worked many of the Loews Theaters in New York at the time, got going on the Fillmore's air conditioning, which was always breaking down. "Starting the Fillmore was a major construction job. On the uptown side of the audience there was a heap of rubbish around ten feet high and half of the depth of the auditorium. It all had to be removed and the labor had to be found."[5] By March 8, 1968, the Fillmore East was open and some of the most memorable rock shows in history followed.

BILL GRAHAM AND THE FILLMORE CREW

What Hanley was to sound reinforcement, Graham was to contemporary rock and roll production. Whether or not the impresario rock producer realized it at the time, he needed Hanley's expertise. Both individuals are major influences within their professions. Beginning on the West Coast, Graham merged elements of theater production with live musical performance, and by doing so he crafted rock exposition as we know it today.

By bringing the Fillmore experience to the East Coast, Graham demonstrated to the rest of the country how quality sound, stage, lighting, and theater production could work for contemporary musical performance. If an "experience" was to be had at the Fillmores, it was Graham who wanted to spoonfeed it to you. According to an article in *Variety*, "N.Y. Rock 'n' Roll Scene Makes Great Leap Upward With Fillmore East Bow," Graham's arrival in New York established "The first successful attempt in Gotham to wrap up live music in a showcase befitting the idiom's best," claiming the producer had given "Significant meaning to what has come off as madness."[6]

This creative endeavor could not have been possible without the highly skilled technical masterminds behind the Fillmore East production crew. Many of these folks would act as key players in the 1969 Woodstock Festival. Morris reflects: "Out of the Fillmore, a whole new system of lighting and sound had emerged. So many things were developed at the Fillmore East that people take credit for."[7] Graham claims, in his book *Bill Graham Presents: My Life Inside Rock and Out*: "Did I do any physical work to get the Fillmore East ready? I yelled. I had some pretty amazing people in there doing the work."[8] As luck would have it, most of the technically savvy Fillmore employees drifted over from the School of the Arts, founded in 1965 at New York University (NYU) next door.

Considered one of the leading performing and media art schools in the country, NYU's departments focused on cinema studies, dance, theater design, and lighting. Student engineers and professors in light, sound, and production like Bob Goddard, John Chester, Bob See, Lee Osborne, Steve Gagne, and Chris Langhart were all keeping the nuts and bolts of Graham's machine together. Langhart, instructed many of the students who migrated over to the Fillmore. Osborne, a professor at NYU in the film school, also specialized in audio. Some of these individuals wound up doing a great deal of work for Hanley other than their positions at the Fillmore.

Fillmore East staff like Morris and White came from the Carnegie Mellon College of Engineering in Pittsburgh. Technical director Chip Monck, a seasoned professional by 1968, nicely rounded off the Fillmore production team. Visual artists and music business professionals like Amalie Rothschild, Jerry Pompili, Lee Mackler-Blumer, and many others provided outstanding work and artistic contributions at the theater.

It made perfect sense for someone like Hanley to be included in this impressive group that worked at the Fillmore East. These individuals helped build the foundation of contemporary concert production. It was a common goal for the Fillmore staff to make each show better than the night before. According to house manager Jerry Pompili, Bill Graham truly cared about the audience experience:

> Bill Graham referred to the Fillmore East as the "white man's Apollo." I still think the Fillmore East was the best regular rock venue that ever existed. Let's face it, we did two shows a night, three acts a show, which ran on time like clockwork. We had a hell of a staff of guys who went on to be big Hollywood directors, playwrights, and producers. One of the stage managers went up to manage the Apollo. The Fillmore East was a breeding ground for the thing that changed the way contemporary music was portrayed. Bill Graham came along and put in professional theater lighting and first-class sound and set the standard, which developed into an art form. It was the Fillmore East which was the breeding ground for all of this. People like Bill Hanley, Bob Goddard, Chip Monck, and Chris Langhart and all of these different tech guys, this

was their experimental playground, and boy they did some crazy shit. Bill Graham was concerned with the experience. In order for you to provide good sound and lights you've got to provide an atmosphere where the artists could perform. Bill Graham was not a promoter, he was a producer.[9]

Joe McDonald, whose band, Country Joe and the Fish, played the Fillmore East many times, recalls: "The staff at the Fillmore was very sophisticated, the crew and all that. It all came together as a perfect storm of technicians and talent."[10] Fillmore East photographer Amalie Rothschild explains: "The Fillmore placed new demands on lighting and sound technology. The tech crew rose to the challenge, creating systems that became standards in the industry."[11] With a higher-quality concert experience in place, expectations grew for both the paying customers and performers. This placed heavy demands on those who were supplying technological innovations in concert production, especially sound.

Members of the Fillmore East crew recall Hanley on stage surrounded by a cloud of wafting smoke as he frantically soldered wires in preparation for that night's performance. Morris remembers, "My first image of Hanley was on stage with a soldering gun, putting wires together for the monitors."[12] The record of Hanley's work at the Fillmore East marks a significant tipping point in his career. The relationships and networking opportunities that he established in New York eventually propelled him into a pop festival market that was about explode.

THE JEFFERSON AIRPLANE TOUR

On February 18, 1967, Hanley Sound had set up equipment for Jefferson Airplane at SUNY Stony Brook on Long Island. At the time, a major blizzard had crippled the East Coast, making it difficult to transport sound reinforcement to some of their tour dates. According to Hanley Sound crew member Bill Robar: "My truck got stuck in a snow bank trying to get to this show! It almost didn't go off because of me! At the time David Freese was mixing the sound for the group. He was a really nice guy. He later worked for us. I remember him telling about a really great female singer on the West Coast; her name was Janis Joplin. She had not played the East Coast yet."[13]

The Jefferson Airplane tour continued, playing a two-week run at the Café au Go Go February 21–March 6 1967. At the time, club owner Howard Solomon had his venue outfitted with Hanley's powerful sound system, and the band liked that.

In 1968 Terry Hanley was assuming many roles for his brother's company, often acting as driver, engineer, and crew. Terry recalls the rigors of early touring with Jefferson Airplane: "We drove through an ice storm from Boston to Connecticut to get to a show in Washington, DC, the next night. I was driving the semi and setting up the sound system with only one guy helping me!"[14] That "one guy" was Robar, who recalls touring all over the country in Hanley's station wagon with trailer in tow:

> These were all mostly college dates back then. At first we had all this equip-ment in a trailer we hauled. We would be going over New York bridges with all of this stuff; the wind would be blowing us all around! At some point on that '68 tour we shifted from the wagon to a tractor-trailer. Terry and I would be driving this big rig with no license or anything—it was a simpler time. Nei-ther one of us ever drove a tractor-trailer either. One time we got pulled over on the New York Thruway and we didn't have any of the proper paperwork. Law enforcement left the tractor-trailer there on the Thruway with all of the Jefferson Airplane equipment and brought Terry to the police station. I had to take the rig to the venue alone, because the show must go on of course![15]

Nineteen sixty-eight was an exceptionally tense time in the United States. It was the height of the Vietnam War, and the assassination of Dr. Martin Luther King on April 4 was on the minds of the band and sound crew. Regardless, the tour continued and on April 19 the band played at Franklin Marshall College in Lancaster, Pennsylvania, followed by a date in Westbury, New York on April 21.

The Airplane performed April 27 at North Dartmouth Park, Cambridge, Massachusetts, for "Spring Weekend," not far from Hanley's company in Med-ford. This was followed by two shows with Iron Butterfly at the Shapiro Athletic Center at Brandeis University in Waltham, Massachusetts on April 28. It was a rather easy commute for Hanley and his crew.[16]

On April 30 the band performed on Johnny Carson's *Tonight Show*. The group sent for Hanley Sound to assist, because Carson's television studio was not fully equipped to handle live rock and roll sound. According to Hanley Sound truck driver Dick Butler: "I remember one of our rental trucks heading down to NBC studios to help the band. The Airplane would not do the gig without Hanley's sound equipment."[17]

Overruled by the studio, the band did not use Hanley Sound for the per-formance. Their new manager Bill Thompson (replacing Bill Graham) recalls the band being a bit uncontrollable in those days: "The Green Room, where they hung out, smelled like Tijuana! Carson was pissed off at the band and he shot us some dirty looks. They had a kid's set for the band with carousel horses and circus accessories, even though the band was in their twenty's. Of course it was lip-synched because they didn't have the technology to perform live at the time."[18]

In 1968 Jefferson Airplane scheduled a series of performances in early May at the Fillmore East and in Central Park. Following this Hanley was asked to support the band's upcoming European tour. Hanley obliged and joined the group on their first major tour of Europe, slated for August and September 1968. The sound engineer shipped his equipment oversees via Belgium Airlines a week early. According to Jonathan Cott of *Rolling Stone*, "The Jefferson Airplane [and Hanley] arrived in London a week earlier, flying in tons of equipment."[19]

The group headlined the Isle of Wight Festival on August 31 and requested the use of Hanley's sound system instead of the WEM PA system deployed at the festival by engineer Charlie Watkins. Hanley remembers: "Some of my equipment was shipped over specifically for the Isle of Wight Festival. They had set up a separate stage for us and the Jefferson Airplane."[20]

On this tour, somewhere in Germany Hanley recalls a venue accidentally advertising HANLEY SOUND on its marquee as the performing act instead of Jefferson Airplane. "It was extremely funny! We were dealing with a language barrier and because of it I got head billing!"[21] laughs Hanley. It was a likely mishap, since the company often received top billing along with the performing artist. Back then it was common to see "Sound by Hanley" on concert ads, articles, posters, and marquees.

Chapter 26

IT'S LOUD!

THE FILLMORE SOUND SYSTEM

Most of the artists who performed at the Fillmore East recall the sound as magnificent and very loud. Morris Dickstein opines in *Gates of Eden: American Culture in the Sixties* that sound at the Fillmore was more of a physical experience: "One effect of the blasting level of noise was that the music seemed to come from within rather than without, and to be emanating from one's own guts and vitals. Gradually the eardrums became calloused and hardened, and the sound was muffled into reverie and fantasy."[1]

Mark Farner of the American rock band Grand Funk Railroad, known for emitting intense decibel levels, recalls being worried about the size of the Fillmore in relation to the group's power: "Man, it was great! When we took the stage I went Whoa! I thought, I wonder if we're gonna be too loud and it wasn't; man, it rocked! Jimi Hendrix was in my dressing room when I got off stage. I will never forget the sound system at the Fillmore East!"[2]

Lester Chambers of the Chambers Brothers, also known for being loud, comments: "The Fillmore East was a house of joy! There were lots of totally happy people who came to hear music. This is what Bill Graham presented to us. Along with Hanley's sound system, you could not beat it!"[3]

Before the Fillmore East opened, Hanley was always networking and hovering around the venue. The sound engineer and his crew hung out with the Fillmore staff at Ratner's, a Jewish dairy restaurant located next door. According to Hanley Sound manager Judi Bernstein: "The operative word is 'focused.' Bill was focused. He had to convince Bill Graham to install a better sound system, which was tough to do. He fought tooth and nail to get a sound system in there and I think in the end Graham respected what he did."[4]

Eventually Hanley's sound system was leased to Graham for the theater. According to Hanley: "Bill Graham was a tough nut to crack. But he finally rented my system. He was very hard to please. He paid me peanuts."[5] In order to determine what sort of system was best suited for the theater, Hanley and Cohen had to examine the space beforehand. Because of Hanley's tenacious conviction, he was able to install his system in the Fillmore for its March 8, 1968, opening, and the reviews were great. According to *Variety* rock critic Marty Bennett: "Friday night New York's rock and roll scene made its most striking theatrical stride forward in several years. . . . Sound qualities, credited to the . . . specialist Bill Hanley, are letter perfect. This is another area in which the majority of local rock and roll shows have failed."[6]

This was yet another fortunate time for Hanley's business. West Coast folk rock was now becoming East Coast heavy rhythm and blues, and hundreds of bands expressed this new type of rock and roll at the Fillmore. Most every night, high-decibel sound levels filled with fuzz and distortion blasted off the stage.

A New Jersey native, it was easy for Harold Cohen to go back and forth from his parents' home to New York so he could work on the Fillmore sound system. Cohen recalls: "The Fillmore system was a point source system: A rental system for almost its entire life at the venue. I used to work on it quite a bit."[7] Cohen didn't really fit into the look of the crowd that mingled at the venue. This was a good thing since Hanley used the engineer on many political campaigns. Cohen adds: "I was not a hippie at all. I remember one time getting a call backstage at the Fillmore from the White House, because I was also a White House staff sound engineer. They wanted me down in Washington the very next day. It was surreal."[8]

It was because of Graham's attentiveness to quality production that the careers of emerging rock acts like Janis Joplin, Jefferson Airplane, Santana, Jimi Hendrix, the Band, and Crosby, Stills, Nash, and Young took off. Slowly but surely Graham was beginning to take quality sound seriously at his theaters. In his book *Bill Graham Presents: My Life Inside Rock and Out*, Fillmore East technical director Chip Monck talks about Graham's attitude toward sound at the theater: "Graham had already started his bit of brilliance. Which was, 'I'll build the best sound system available. You don't bring your fucking sound in here and stack it on the side of the stage and make my theater look like a piece of trash! Bring your monitor system if you absolutely have to, but my house system is what you use.'"[9]

It was in the producer's best interest to have a stationary state-of-the-art sound system remain at his theater, versus having bands truck their equipment in each night for three acts, two shows. Such advances within the business and production standards of rock and roll catapulted Hanley to a new level of quality sound. The Fillmore East was considered by the *Village Voice* as being "One of the best sound boxes in the city with a superb acoustical system."[10] This is due to Hanley.

The evolving Fillmore sound system consisted of "Twenty-six speakers, including some manufactured for civil defense alerts, strategically placed around the theater, as well as a two-ton center cluster speaker system, suspended over the center of the stage in front of the proscenium using a series of fly weights especially designed by Chris Langhart."[11] The system had four McIntosh MC 275 amplifiers total. There were custom modified Altec 210s (for the low end) that had two fifteen-inch woofers in each. These were about 2-½ ft. x 3 ft. x 5 ft. A pair of these speakers could be seen stacked up on either side of the stage. Another pair was flown near the lip of the stage. Hanley adds, "They were three-foot-long horn loaded speakers, four total, two on each side with four facing the balcony."[12]

The flyweight rigging system supported not only this large heavy speaker cluster aimed at the balcony but a lighting bridge as well. Monck, whose booth was located at the right-hand stage cluster, recalls the functionality and efficiency of the sound system: "The speakers were all on a fly system and they would actually come up and down. We could bring them right down to the front of the apron to do the repairs on them and then send them back up. It was all counter weighted."[13]

Fillmore East sound engineer John Chester recalls: "The horns were Altec, and one ten-cell horn or maybe two ten-cell horns on each side and a similar complement for the balcony. The drivers were 288s for the three-way system and some 290s for mid-range. Some of these were designed for civil defense warning sirens."[14]

The first time Chester actually met Hanley was in the lobby of the theater, sitting next to a stack of giant Altec 515b woofers with some tissue paper and a bottle of Pliobond cement. "He was pasting tissue paper over the tears in the paper cones on the woofers. At that point he could not afford to buy new cones for all of them. Bill had bigger speakers and amplifiers than I had ever seen before."[15]

Hanley was at the Fillmore East frequently during those early days, and when he was not, engineers Goddard and Chester often mixed the sound. Eventually Graham bought Hanley's system from him for $35,000, and over time Goddard and Chester built on it and further modified it to suit the evolving needs of the theater.

Like most promoters at that time, Graham was known for not wanting to give up seats in the audience, and this left no proper placement for front of house (FOH) mixing. Unlike other venues, FOH at the Fillmore East meant that engineer and console were positioned in a small private theater box above the stage, just barely into the audience. The specific location was slightly above the level of the stage-left speaker cluster, with center stage approximately 30 degrees to the left. From a sound engineer's perspective, there were a few downsides to this configuration. Even though the "professional grade-mixing console" was out front, it still was not in an ideal place. It's important that the sound engineer be able to hear what the audience hears.[16]

Another challenge was that this mix location made it a great deal louder for the engineer. With various horns stacked on top of boxes, the sounds intended for the audience were blown into the engineer's immediate area. For the concerned sound engineer, this arrangement could make the mix unrepresentative of what it actually sounded like in the house. Chester remembers: "I had to learn to mix in that specific location. You either had to have someone who was downstairs run upstairs and tell you how it sounded, or you had to do it yourself. At least you could see what the musicians were doing, like when they kicked over a microphone near a guitar amp or drum set."[17]

INTERCOM INNOVATION

It was common for the staff at the Fillmore East to build what they needed if it did not exist. Monck, in charge of lighting, needed communication with his staff. And for Chester, it was essential that the Fillmore production crew be able to talk to one another during a performance:

> One requirement was for lighting because you have follow spot operators and a lighting director who has to be in constant contact with them. You have an engineer in the sound booth that has to be in contact with those on the stage. If something goes wrong on stage, the guy in the sound booth can communicate with that person. Without that, someone from the booth has to run down and fix the problem.[18]

Early on at the Fillmore, Chester realized an intercom system was essential, so he built one. He quickly constructed a system that had sufficient volume to be heard over the extreme noise levels in the house. Up until then only a few of the Fillmore staff had seen theatrical intercom systems, but nothing remotely useful for the sort of sound levels they encountered with rock and roll music. Chester remembers, "I saw them first being used in a high school in Syracuse, New York, in their theater drama department."[19]

The Fillmore intercom system ran off a central set of electronics, where there was a mixer and amplifier. Chester adds: "Chip had a system that worked in a similar fashion. Chip certainly knew that he needed one."[20] He remembers there was a lot of parallel inventing going on then. Up until then no one seemed to be building intercom systems specifically for rock music.

That first intercom system, we built out of Radio Shack headsets with carbon microphones stuck on little wire booms. It was durable and shrewd. At that point the headsets had a metal headband. So we got out the brazing torch and some steel wire and we brazed around the headband and extended it down like a microphone would. And at the end of that, we brazed on a radiator clamp and stuck a microphone through the radiator clamp and wired it up.[21]

Several modifications were subsequently made to this improvised version of the Fillmore intercom. The staff had received headphones from a company called Telex, a standard then, but they were not durable enough. Finally they settled on Beyer headsets. According to Chester, they were able to create a headset using the Beyer brand with some additional modifications including a noise-canceling mike. Chester explains in detail about these refinements:

They sounded better than the Telex headsets but the problem was they didn't have noise-canceling microphones in them. So we figured out how to turn the microphone into a noise-canceling one by opening it up and ripping out a little cotton pad that they stuck in the back of it and replacing it with a piece of foam. So now we had a headset with a noise-canceling microphone on it! It had good shields on the ears that kept out a lot of the noise and decent electronics. This was the first intercom system that could deal with that kind of noise level. These were used for many years. We built them out of necessity for loud rock and roll because the previous technology had lousy headsets and lousy microphones. The technology that was out there picked up an enormous amount of sound and turned it into something ugly and raucous sounding. When it blasted into your ear it was pretty painful![22]

Along with many innovations in concert production, a new culture of concertgoer was evolving at the Fillmore East, different from the organic nature of West Coast rock ballroom audiences. At the Fillmore, amid its 1920s movie house grandeur and rococo frieze detailed proscenium, those who bought tickets sat in designated seats and watched. The theatrical production components of the venue contributed to a new standard of "focused" concertgoing. In turn, this affected the performer, who now had to face an eager audience in a highly intimate setting.

Traveling sound engineers could not help but notice the difference in quality of sound and production at the Fillmore. When English bands like the Who and Fleetwood Mac played Graham's theater, they were awed by the quality of its production. According to Bob Pridden, sound engineer for the Who: "I remember meeting Bill Hanley at the Fillmore East. They were situated up in the balcony overlooking the stage and it was fine. I saw CSNY the first time when they played there. I sat up front and it sounded fantastic, really fantastic."[23]

While on a tour of North America with Fleetwood Mac, UK sound engineer Dinky Dawson worked a show at the Fillmore. This is when, as Dawson claims, he first used Hanley's multicore snake cable and mixed from a true FOH position. Dawson recalls being enamored by the theater setup: "I was absolutely amazed at this big old system with big horns. This was the first time I ever saw anyone mixing from a booth. It had large controls with knobs all over the place and all of these tubes. It was loud; it kicked butt and was crystal clear! Come to find out, some of it was the Hanley system."[24] In an article titled "In Profile: Dinky Dawson, Concert Sound Pioneer," Dawson claimed: "That was the first time I was able to mix from the audience. Before, we were always mixing from the back."[25] Dawson was inspired by what he saw at the Fillmore, adopting similar techniques that he eventually brought back to the UK.

RECORDING

Just underneath the orchestra pit, located in the basement, Hanley installed an eight-track tape machine and a twelve-channel console for recording performances. Thanks to the engineer, we can still hear these early Fillmore East shows today. This was the same system that engineer Lee Osborne, assisted by Eddie Kramer, used for recording the Woodstock Festival in 1969. Hanley recalls: "On the stage right side of the basement I put in two 280 Scully eight-track one-inch recording machines, along with an eight- or twelve-track tape deck. I was trying to make recordings to generate more revenue to serve the record companies. It was a scramble to get it done, and put it down there."[26]

However, the recording location wasn't always located in the basement of the Fillmore. When the theater opened, Hanley and some of the engineers snaked a wire attached to the mixing console all the way out of the theater and into an apartment located upstairs. According to Joshua Light Show employee and Fillmore East photographer Amalie Rothschild, "There was always somebody up there monitoring the tapes and changing the reels."[27] Hanley is unsure how long the recording setup lasted in the apartment, claiming, "I believe we got moved out of there pretty quickly and ended up in the basement."[28] According to Joshua White, "Because of Hanley we have so many amazing recordings from the Fillmore East."[29]

In late winter of 1968–69, NYU School of the Arts student and Fillmore East employee Billy Pratt met Hanley at the theater. Pratt ended up doing a lot of work for Hanley: "Hanley started giving me duties right away."[30] Driving

Hanley's straight box truck that spent more time in Manhattan than in Medford, the NYU student hauled equipment to gigs in various college auditoriums and high school gymnasiums around the Tri-state area. Pratt got his orders directly from Hanley at the theater, rather than from Judi Bernstein. "I did many shows with the Chambers Brothers, Sly and the Family Stone, Janis Joplin, and Tim Buckley. Janis called me 'Whitey' because of my blonde hair. We got to know each other. After a show I used to keep a bottle of gin in the mike box for her. She was really appreciative of that."[31]

GET BOSTON ON THE PHONE!

While supporting the Fillmore East and other New York venues, Hanley often traveled uptown to provide sound reinforcement for the preaching of the popular evangelist named Reverend Frederick Eikerenkoetter. Known as "Reverend Ike," his over-the-top religious proceedings occurred at the enormous Loews 175th Street Theater. He eventually purchased the building and renamed it the Palace Cathedral, known as "Reverend Ike's Prayer Tower" in New York. The flamboyant reverend preached the gospel to the masses through one of Hanley's sound systems.

The Reverend Ike empire was famous for its creative use of media. The reverend sought out Hanley Sound for television and radio audiences nationwide. He became known for his "blessing plan." This is when radio listeners and audience members alike gave him money in return for his personal blessings and promise of prosperity. Hanley observes: "I remember people standing in line to buy his prayer cloths to bless everyone with. There was a ten, fifty, and one hundred dollar line! This is how he bought the Loews 175th Street Theater for $500,000."[32]

At one point the Reverend's ministry had grown so large as to have 1,770 radio broadcasts and his "videotaped sermons" telecast in ten major market areas. It made him a multimillion-dollar entity.[33] Hanley recalls, "Everyone needs a soundman, even the church, but there was no one like Reverend Ike!"[34] Hanley and some of his crew members, like Rick Slattery, were with the reverend for a couple of years supporting his remote radio broadcasts and tent-style church meetings all across the country. Slattery recalls going on short tours with Reverend Ike: "He had us set up a sound system (voice only) and we went out with him. We were paid to record all of his live performances that were later broadcast on the radio. This guy had six Rolls Royces! In order to look the part, the first thing they did was buy me a jacket and tie!"[35] Fillmore colleague Lee Osborne was also there, assisting with mixdowns of the reverend's sermons.

According to Hanley: "The reverend was a likeable guy, just your typical bull-shit evangelical type of preacher. I remember his slogan, 'You can't lose with the stuff I use!'"[36]

With so much going on in the city at one time, it has been suggested that the Hanley Sound system at the Fillmore East was intermittently compromised by pieces of it going to other club systems. Since Hanley had so many events happening around New York, he and others pilfered from the Fillmore system, frustrating the staff. According to Goddard: "Hanley came in and raided our system for other gigs and venues; this drove us to buy the system from him. Bill Graham told Hanley to 'leave the system alone because I am paying you a weekly fee!'"[37]

Although Hanley's initial Fillmore system was groundbreaking, it did not help his relationship with Graham. According to Bernstein, it really pissed him off: "Graham was known for screaming, 'Where the Fuck is Hanley! Get Boston on the phone!' This only occurred when things went wrong with the sound system. When things went right with the sound system, Bill was treated like a fly on the wall by Graham."[38] According to Langhart, who offers a more pragmatic perspective on the situation, this was disruptive to the production at the theater:

> The continuum at the Fillmore, which developed later didn't really fit in with the Hanley scheme. Hanley was very interrupt-driven. If there was a gig at the Cathedral of Tomorrow and we needed two extra horns at the Fillmore he would rush up there to get them. By then we needed to have the same scheme at the Fillmore every week because of the type of acts we were getting. So we were becoming reliant on required equipment. We ended up buying out Hanley's equipment from him. It needed to be a staid, dependent thing.[39]

Chapter 27

SIGHT AND SOUND AT THE FILLMORE

THE JOSHUA LIGHT SHOW

Sound at the Fillmore East may have been spectacular, but so were the visuals of Joshua White's Joshua Light Show (JLS). The light artist, only twenty-six years old at the time, led a troupe that had six to eight members. White is a significant character during this period of concert visual entertainment.

Although Hanley's sound was out in front and the JLS behind the stage, these two entities needed to be able to work in tandem. When in sync, both production elements danced beautifully with each other. Nevertheless, there were regular conflicts over artistic differences. White recalls that he and Hanley had something of a "love hate relationship," jokingly referring to each other as frenemies. White adds: "I got crazy over what Bill was doing because we were trying to be more theatrical and do things on our screen and that had tight sound cues. Bill was unflappable. I never saw him lose his cool. He never panicked and no matter how much I screamed at him he handled it very well."[1]

Clearly Hanley and White were different. The sound engineer was a bit older and White remarks, "I was a Jew from New York and he was an Irish Catholic from Boston. It was always "FUCKING BILL HANLEY THIS or FUCKING BILL HANLEY THAT!"[2]

On March 8, 1968, Hanley drove his red-cab 45-ft. yellow tractor-trailer to the Fillmore East. It was opening night, with performers Janis Joplin, Albert King, and Tim Buckley. Prior to the event White told Graham he would set up the theater marquee to include the JOSHUA LIGHT SHOW name on it. White was anxious to see his name up in lights associated with so many popular acts. But Hanley parked his truck in front blocking the signage. According to White: "I of course had an ulterior motive here. I had cleaned the marquee up nicely. Hanley parked in front of the theater and blocked only my name, not Janis Joplin's or Albert King's, just mine! In the front of the theater was this

FUCKING YELLOW TRACTOR TRAILER!!! Of course I was mad at him but he is such a sweet guy."[3]

White's curtain of "kaleidoscoped" color is now an image synonymous with counterculture live performance. For almost every show JLS's colors dripped, shifted, and moved in organic fashion to the sounds coming off the Fillmore East stage. With so many performers launching their careers at the Fillmore, the light show became as important as the sound. Both elevated the experience. White observes: "Hanley was me, but in sound. He recognized, as I did, without intellectualizing it, that people have eyes and ears. And just because you put a band on stage, does not make the show. People needed to have something to look at that was at a higher level than what they looked at before, which was black curtains in an auditorium."[4]

The JLS visual presentation was built on four essential elements: first, the "projection of pure colored light through various handmade and modified devices"; second, what White refers to as a "wet show" using water dyes and colored oil combined in a glass clockface, displayed overhead through a projector; third, a technique called "lumia" (colored forms against a black background) inspired by the 1920s light artist Thomas Wilfred who had a profound impression on White as a boy; and fourth, "cinematic" in nature, most often comprised of "concrete imagery that included film footage, hand-etched film loops, commercial cinema."[5]

The JLS rear projection system sat twenty feet behind the Fillmore East stage. Weekly, White's show produced images via 1200-watt lights that were needed to project images onto a 20x30 ft. vinyl screen that faced the audience. Huge amounts of equipment were splayed throughout two elevated platforms.

According to journalist Barbara Bell in a *New York Times* 1969 article titled "You Don't Have to Be High," the system included "Overhead projectors, two banks of four-carousel slide projectors, three film projectors, dozens upon dozens of color wheels, motorized reflectors made of aluminum foil, broken mirrors, Mylar, hair dryers, oil colors, watercolors, alcohol, glycerin, dozens of clear glass clock crystals, and (two) crystal ashtrays."[6]

For most every performance the JLS name graced the marque of the Fillmore East alongside the names of performers who played to its colorful dance. Although White did not invent the light show, he should be recognized for elevating it into a whole new realm of visual artistry. The cinematic element within all of White's displays eventually developed into live closed-circuit video. This contribution was Hanley's and used for the first time in February 1969 for a Janis Joplin performance.

LIVE MUSIC SCREEN PROJECTION

The use of closed-circuit television techniques with live music is discussed in a 1970 article in *Rolling Stone*. Hanley's friend, rock writer Alfred Aronowitz, suggests that up until this point, technologies applied in this format mostly had appeared at sporting events, not at pop concerts or even festivals. Aronowitz explains, "That experiment with closed circuit television, still untested in the pop field, has already earned hundreds of thousands of dollars in sports."[7] After Aronowitz witnessed its use at various sporting events at the Felt Forum and Madison Square Garden, he observed, "I think it's about time music tried it." Little did the journalist know that a year earlier, Hanley had already done this successfully at the Fillmore East.

While mixing at Newport, Hanley was bothered by the fact that he had difficulty seeing musicians perform on the stage. The idea of projecting an image of the artist in a large format for the audience had actually been on the sound engineer's mind for quite some time. In 1962, while providing sound for a Boston medical convention, Hanley witnessed a color television projector being used for presentations about burn patients. According to Hanley: "Later on, around the mid- to late 1960s, an opportunity arose to buy two Schmidt Optic Projectors from the old Paramount Theater in Boston. The rest is history. I later used them during the early seventies at George Wein's new Jazz Festival at Shea Stadium where I built a 16x12 ft. video screen out of plywood painted white."[8]

In mid-February 1969 at the Fillmore, Hanley had a chance to develop closed-circuit projection for a Janis Joplin performance. This was one of the first occasions where this format was applied to live music. It was White who convinced Bill Graham to give Hanley's idea a shot. White remembers: "We put Bill's state-of-the art-video projector in the center of our light show platform. He had two guys videoing the band on stage. This made a little tiny black and white image, which we integrated into our light show. I would have never been able to afford to put that video in my show if not for Hanley's contribution."[9]

Jerry Pompili, house manager at the Fillmore, who witnessed this innovation, claims he had never seen anything like it: "I have no idea how Hanley got that image. It wasn't a very good image but he got a close-up image onto that screen. But I remember seeing it during the show and I was just like wow! What the hell is that? I had ever seen anything like it before!"[10]

Hanley explains: "We had the screen at the lighting platforms that were located back stage fifteen feet off the floor. The camera was located in the first and second row of the audience aimed directly at Janis. The projector was connected to Joshua's light platform. The screen area for Janis was around eight by ten [feet]. Janis was in black and white."[11]

Big Brother and the Holding Company guitarist Sam Andrew recalls looking back mid- performance and seeing the projected image and being amazed. "It was really funny to see the birth of that thing as it was happening right then. What Hanley did was immediate, a first-time kind of thing. So many things were changing particularly with the light shows especially with Hanley's contribution being part of it. It was amazing to contemplate."[12]

As Hanley's camera equipment projected the enlarged image of the singer, those in the back could now see with some clarity. In an excerpt from an article called, "Janis Joplin's Face, Seven Feet High–Groovy!" *Village Voice* writer Howard Smith was detailed about this remarkable innovation:

> Janis Joplin's concerts at the Fillmore East last week were the debut for a light show innovation using closed circuit television. On the big screen behind her, two cameras and two special projectors flashed a nine by seven foot image of Janis's face in perfect sync as she performed. It was great for the people in the back; they could watch her move and see what her face was doing at the same time. For those in front there was the novelty of seeing her face at two different angles. Especially interesting were close-ups of her lips and shots of Sam Andrew's hand playing guitar, although they made the same mistake most TV shows do and only showed the strumming right hand, not the fingering left one where it's all really happening. The Fillmore is trying to get their picture larger and a more intense image, until they do they will only be using it at concerts from time to time.[13]

In 1973 Hanley applied his idea for video projection at a Jazz Festival in Shea Stadium where he used a four-screen system for promoter George Wein. Here the engineer used two-by-three frames and quarter-inch painted (white) planks of plywood. Positioned in the field they were twelve by sixteen feet. By now Hanley had purchased a handful of Panasonic cameras and used his crew to run the live video feed.

Now a widely used format at contemporary music events, not everyone then was receptive of Hanley's innovation. Hanley recalls: "At the Philadelphia Folk Festival I had to do it for nothing until they were convinced. It was well received and we eventually used it every year. It was like the audience was in the living room with their favorite performer."[14]

In 1973 Pompili was managing Graham's San Francisco Winterland Ballroom. Dissatisfied with the light shows on the West Coast, Pompili reflects on Hanley's 1969 innovation at the Fillmore. With a little more convincing, this time Graham agreed to use it again. According to Pompili: "I remembered back in 1969 at the Big Brother show. Hanley projected a video image out onto the screen during the light show. So I went out and bought a video projector and some cameras, a mixer and a three-quarter-inch Beta deck and put it together. But the whole idea came from Bill Hanley."[15]

THE WIRELESS MICROPHONE

Hanley was always interested in new technologies, first using a wireless microphone in the early 1960s. Crew member Karl Atkeson remembers:

> It was either in 1963 or '64, when I drove Hanley's father's Buick to a Barbra Streisand performance at the Colonial Theater in Boston. She needed a Comrex wireless microphone for her show. I parked right in front of the door, and left the vehicle to make the delivery. When I came back there was a uniformed officer guarding my un-ticketed ride. He said, "Who are you?" I explained, and then pointed out his dad's police decal on the back windshield. He let me go! This was why we were able to park anywhere in Boston like we owned the place![16]

By 1968 the wireless microphone was still a relatively new concept in sound and not at all common for rock and roll performances. That year Hanley introduced a version of this device to the Fillmore East and at the Whitney Museum for an event on October 3. At the Whitney the sound engineer affixed singer Grace Slick with a wireless microphone before Jefferson Airplane was to play the black-tie fundraiser. Not accustomed to such technology, Slick's drunken voice could be heard through the PA system while she was backstage. By the time she came out many in the audience were taken aback and offended by her derogatory remarks. According to Annie Fisher of the *Village Voice*, Grace was, "Stalking around with a wireless mike, spattering her own, in her anti-properness."[17]

On November 30, 1968, Hanley introduced the microphone to blues artist Buddy Guy before his set at the Fillmore East. The guitarist was wary. This was a natural reaction for an artist known for leaving the stage and interacting with the audience. With the freedom of a wireless microphone, Guy now had the luxury to wander about and leave the stage while still being heard.

Thrilled and bewildered, the audience at this particular performance was looking around wondering where the sound was coming from. *New York Times* writer Mike Jahn explains, "He often comes into the audience, sits at your table, smokes your cigarettes and sings to you. At the Fillmore East he came into the audience Saturday night, followed by part of his five piece band, singing and playing a prolonged explosive finale."[18]

Joshua White, who was there to witness this event, claims that Hanley was always experimenting and spending money on new technology: "This was not your ordinary headset microphone, you held it in your hand. Buddy comes out, picks up the device and goes into the house. He goes upstairs into the balcony and you can still hear him singing! When was the last time a wireless microphone got this sort of reaction out of a crowd?"[19]

Chapter 28

GETTING PAID

STILL NOT TAKEN SERIOUSLY

Getting paid by the artist or promoter was often an issue for Hanley. It was common for the sound engineer to struggle to receive what was owed to him for his work. When he gained new clients, he did what he thought was best, and that was negotiating money later. Whether a good or bad decision, Hanley did what he had to do in order to get his sound system into the venue or performer's hands. The music business was still developing, and sound companies like Hanley's were often the last ones to receive their rightfully owed earnings.

During these decades Hanley had to work tirelessly to acquire clients because his ideas were so disruptive to existing practice. He worked hard to get paid because of the nature of his innovative work. Despite a slow financial start, eventually the sound engineer profited from the expanding music industry. A 1966 financial report pulled from the Hanley archives states that the company's volume was about $75,000, around $560,000 today, which for that time was fairly substantial. "Volume in 1966 is regarded by authorities to be close to $75,000 with the company in receipt of a modest profit."[1] In 1967 Hanley Sound had not yet experienced the financial boom of the festival years, a time when the company seemed to peak.

In the early days paying the contracted sound company was not a priority, and Bill Graham demonstrated that firmly. According to Bernstein: "We fought for every nickel that Graham paid us. We sent out invoices and we begged for our money from him. The place was a gold mine; it was full all the time. However, he did not want to pay the sound guy!"[2] There were times where the sound engineer worked and did not get paid at all.

It was a constant battle for Hanley to justify that his methods in sound were the answer, especially for music festivals. Promoters of the day were taking big chances on putting on outdoor events. Weather, permits, and licenses were all variables when organizing a festival or concert, and at the end of this list was Hanley.

It was fairly common for Hanley Sound engineers to record an artist's performances, per their request, from the soundboard. Over the years the sound company is credited for recording many albums including the Woodstock Festival soundtrack, *The Cowsills in Concert*, the Blues Project's *Live at Town Hall*, Herbie Mann's *Live at Newport*, and two successful James Brown live albums.

In January 1967 Hanley was hired to record and provide sound reinforcement for James Brown at the Latin Casino Theater in Cherry Hill, New Jersey, and at the Apollo Theater in Harlem. Brown is known as the "Godfather of Soul." The result of Hanley's work was Brown's 1967 *Live at the Garden* and his 1968 double album *Live at the Apollo Volume II*, released on King Records where the company is credited—"Audio by Hanley Studio." Hanley remembers: "We first recorded James Brown in New Jersey, but we were not credited on the album. I thought it was funny that it said "Live at the Garden" which made people think Madison Square Garden!"[3] In a 1967 *Billboard* article, Hanley explained, "Though I am not in the recording business, I have tried to master the business of live recording."[4]

It was on June 24–25, 1967, that Hanley's sound company was gearing up for Brown's appearances at the Apollo Theater in Harlem. This is where engineer Harold Cohen claims he got his "feet wet" with live recording: "Ray Fournier taught me location recording in the basement of the Apollo Theater while doing James Brown, which ended up being his second live album."[5]

When Hanley agreed to work with Brown in 1967, he charged the singer $1,400 to record his scheduled Apollo show(s). Because of the multiday recording effort, Hanley and his crew had to spend a couple of extra days in the city, which was costly. Following the performances Brown failed to deliver a payment. Eventually the performer's check was received back in Medford, but it bounced. Hanley was pissed.

Not all was lost and Hanley asked sound engineer Ray Fournier to step in and negotiate using the Apollo tapes as leverage for payment. Fournier had previous recording industry experience so he knew exactly what to do. Brown was livid that the sound company did not release the recordings. According to Fournier:

> I told Hanley that I would get him his money. I knew many guys in the musicians union because I lived in NYC. The guy in charge of recording contracts was a good friend of mine and I told him that James Brown didn't turn in a contract in order to do his live album. That got the ball rolling and soon after I got a call from James and he says, "Man you got the tapes?" I said yeah, I got the tapes! He wanted me to send them to him. I said you bounced a check for $2,000 and you owe us that money! He says, "What do you mean?" I said we waited around for two days. He told me he did the same set in Africa but

he wanted to use our tapes. I would not relent. He really wanted those tapes. You see I knew the union was pressing him. So we hung up and I didn't hear from him for another month or two. Then James called the office screaming "I NEED THE TAPES, WE ARE IN TROUBLE!" I said okay, we will send them to you C.O.D.—cash, no checks. We got our money.[6]

By the early 1970s the sound company had great difficulty making payroll. Not getting paid from a promoter was becoming the norm. At one show lead engineer Sam Boroda remembers an example of nonpayment that was particularly harrowing. Judi Bernstein gave direct orders to Boroda not to turn the sound system on until payment was received in full. Boroda recalls:

I remember walking into this hall and seeing the place half empty. This was a warning sign to me. I knew then that we were going to have a difficult time getting our money. I told the promoter that we needed to get paid up front and he dragged me into the ticket booth and paid me in one-dollar bills. He then grabbed me by the collar and said he was going to tear my face off! As he lifted my feet off of the ground the floors were thumping. The audience stomped angrily because the show did not start yet. I would not turn it on until I counted all of the money. I remember sitting on the plane to go home counting all of this money I stuffed inside a briefcase. I felt like a drug dealer![7]

THE ARTIST WANTS MORE MONEY

Bill Graham's influence proved instrumental in making bands like the Grateful Dead, Santana, and Jefferson Airplane famous. He gave them a venue to express themselves musically and because of that their popularity grew. Their new pop-level status resulted in a demand for more money from Graham. It seemed that the personalities the rock producer nurtured were outgrowing the comfortable nests he had built at both of the Fillmores. As the bands became increasingly popular, so did their appetite for more cash.

No longer able to keep up, the Fillmore East closed its doors on June 27, 1971. Graham called it quits. According to *New York Times* journalist Mike Jahn, "At the time Bill Graham blamed the greediness of some of the top rock musicians who said they would rather play a 20,000-seat hall like at Madison Square Garden (one hour's work $50,000) than the 2,600-seat Fillmore East (about four hours' work, roughly $20,000)."[8]

Within the pages of his autobiography, a frustrated Graham reflects explicitly on the situation: "The Fillmore East became what I had hoped it would be but the price was unbearable. At the end, the biggest acts would no longer support the place. The Doors were playing Madison Square Garden. The acts were saying, 'Our careers are more important than the scene.' They said to me, 'If you don't take us to the stadiums, we'll go with someone else.'"[9] Bill Graham was compelled to relinquish power to the artist. And it became clear that a shift was occurring within the music business.

During the Fillmore's demise, Madison Square Garden was hosting rock and roll shows through an antiquated public address system, audibly unsuitable for this influx of loud rock music. According to *Village Voice* writer Annie Fisher: "The [MSG] sound system is abominable. The Doors and the promoters got lots of money, and money is really what all these monster events, indoor and out, are all about. The music? Who knows?"[10] The shift to larger venues was a transformative occurrence that inevitably affected the business of concert sound reinforcement. When the Doors and other bands performed in arena settings, sound quality for rock music became a problem.

The artists' desire to make more money meant that Hanley needed to adapt, advance, and assist wherever and whenever sound quality suffered. Bands performing for huge audiences, indoors or outdoors, required sufficient power. With outdoor festivals drawing hundreds of thousands of fans, like Wood-stock and others, outdoor sound became a unique problem to solve. Calling on someone like Hanley, who happened to have enough equipment, was the obvious choice. Even though the technology had not yet fully emerged, and the market was still small, the sound engineer had been preparing his systems for these situations all along. Eventually the growing festival marketplace advanced Hanley's career to a new status in his field.

Hanley would see all of his Fillmore East friends again, including Graham, at the 1969 Woodstock Music and Art Fair. After learning of Hanley's success in sound application at Woodstock, those in charge at Madison Square Garden called on Hanley for his assistance. The venue's public address system could not withstand the rock acts they were hosting. This is where the engineer provided sound reinforcement for some of the most preeminent concert performances in rock music history, like folk singer Joan Baez, Donovan, and the Rolling Stones.

With memories of the Fillmore East fading fast, Graham was forced to pass the torch. To this day, his impact on the business of music and concert production is unassailable and too great to measure. Hanley, like Graham, fell victim to a similar sort of disruption within the emerging festival marketplace. However, one benefit of musicians becoming progressively more popular was a growing desire and need for better quality and reliable sound.

Chapter 29

THE FESTIVAL EXPERIENCE

The festivals that emerged in the 1960s gave musicians a platform for musically driven, politically charged responses to the injustices in the United States and abroad. To the counterculture that participated in these mass gatherings, the music was a force of social and politically driven interactivity, fueled by identifiable lyrics. The rallying cries of rock musicians encouraged disaffected youth, alienated from the "silent majority" of mainstream America, to join the growing festival movement. Although the festivals were often a celebration of utopian ideals, they also had downsides that included drugs like LSD, speed, marijuana, and alcohol.

At that time, the country was preoccupied with issues surrounding civil rights, the Vietnam War, the draft, nuclear weapons, the environment, women's liberation, and the sexual revolution. In an article "We Can All Join In: How Rock Festivals Helped Change America," journalist William Mankin observes: "As different as some of these transformative upsurges were from each other, most of them had a lot in common. By and large, whether politically shrewd, stubbornly idealistic or naively anarchic, they were based on the expectation that the world could—and should—be a better place."[1] Unlike some traditional sound engineers, Hanley wanted to improve the festival experience for those in attendance. He truly felt that he could make the world a better place by delivering quality sound.

The counterculture emerged and united at pop festivals during a span of just six years from 1967 to 1973. Hundreds of them were held in the United States, ranging in attendance from 20,000 to over 600,000. This movement was a cultural phenomenon. Some festivals lasted for days on end, but usually were held over a summer weekend on farms, fields, and even raceways. Although the 1969 Woodstock Festival is considered as setting the stage for the festival experience, there were many more that occurred before and after it.

In a relatively short span of time, Hanley provided sound reinforcement for at least half of the largest festivals held in the country. Eventually he became known as the "Father of Festival Sound" for his pioneering work at these mass gatherings. Over time the pop festival developed into a topic of contempt, facing severe cynicism. Due to this, by the early 1970s the sound engineer's business suffered greatly.

CELEBRATING THE PAST

The festivals of the 1960s counterculture were playgrounds of free progressive thinking, innovation, and extreme creativity. The emotional and physical exchanges that occur between thousands of people at a festival are ritualistic, celebratory connections shared through common experience. These musical unions offered the space, sound, and smells that foster acts of fantasy somehow connecting us to our ancestors who did the same.

In an article, "In Praise of Festivity," author and theologian Harvey Cox claims: "Man is *homo festivus* and *homo fantasia*. No other creature we know relives the legends of his forefathers, blows out candles on a birthday cake, or dresses up and pretends he is someone else."[2] When fully immersed in a festival environment we are not just acting in solidarity, we are innately tapping into to a cerebral fantasy hard-wired within us. We are viscerally rejecting societal norms and everyday reality by way of music, bucolic rolling green fields, makeshift communities, and perhaps extreme heat, mud, and soaking rain.

Over centuries communal gatherings have been essential to the human experience. The actions of men and women congregating in a communally celebratory manner remains connected to the past as far back as the ancient festival gatherings of Athens, Greece. Cox relates, "The festival, the special time when ordinary chores are set aside while man celebrates some event, affirms the sheer goodness of what is, or observes the memory of a god or hero is a distinctly human activity."[3] By 1969 a new age of the music festival surfaced in the United States, symbolic of modern humanity, without any ecclesiastical or sacrificial offering to a god or dead relative.

The Latin derivative of the word *festival* (or celebration) is rooted in the words *festum* or "feast." This is exactly what the music festival of the Woodstock era had evolved into: a feast of peace, love, unity—and for the promoter, money. For a time the youth culture ritualistically offered its innocence, long hair, nakedness, drug use, and undeniable dedication to the social and political issues of the day, all for the great spirit of music.

It was possible to attend a festival and not have heard or seen anyone perform. Wide-open spaces meant the music could drift and diffuse. And at best, accommodations for those who attended were minimal. In *Aquarius Rising: The Rock Festival Years*, Robert Santelli writes: "These people would have fared better catching acts in concert halls where good sound systems, decent visibility of the stage, and comfortable settings were more or less insured. The large out-door rock festival could offer none of these."[4] But eventually music promoters, bands, and managers began to see the enormous financial potential the pop and rock festival offered.

COMMODITY VS. COMMUNE

The Moondog Coronation Ball is an early example of organized multi-act youth-driven concerts. Held in the Cleveland Arena on March 21, 1952, this large indoor rock and roll event was organized by Cleveland radio deejay Allan Freed.

Freed had been convinced by Leo Mintz, the owner of a local Cleveland re-cord store called the Record Rendezvous, to change his radio broadcast format to the more popular R&B. Freed did just that. And soon after, a late-night show called *The Moondog Rock & Roll House Party* aired with Mintz as a sponsor. The *Cleveland Plain Dealer* reported: "By 1951, Freed started playing the music on his radio show WJW-AM, 'The Moondog Rock & Roll House Party.' Mintz not only sponsored the show, he also sat alongside Freed, feeding him slabs of wax to play."[5] Writer Jude Sheerin, in a BBC article, "How the World's First Rock Concert Ended in Chaos," explains:

> Mintz helped Freed, then a humble sportscaster, secure a new show on the city's WJW radio in 1951, devoted to playing this underground music. Freed would coin the term rock and roll—an old blues euphemism for sex—to describe the tracks. Using the on-air alias King of the Moondoggers, he would ring a cowbell, drink beer and howl in tribute as he played the records, while pounding out the beat with his fist on a phone book. The flamboyant Freed's late-night show caused a sensation with black and white listeners alike. Mintz and Freed's logical next step was to stage a live concert featuring the edgy new acts.[6]

Soon Mintz, Freed, and local promoter Lew Platt came up with the idea for a live concert at the Cleveland Arena. Headlining the event was the ev-er-popular Paul Williams and his Hucklebuckers, Tiny Grimes and his Rockin' Highlanders, the Dominoes, Varetta Dillard, and Danny Cobb. The tickets for the show were $1.50. Money was to be made, yet more tickets were printed than the 10,000-seat venue could handle. Thousands of angry ticket holders were outraged. The event was quickly shut down. Sheerin continues:

> Less well known is the reason why the Moondog Coronation Ball ended in disaster: a minor printing error. The mistake was caused by someone

forgetting to add the date to tickets issued for a follow-up ball, which Mintz
had set about organizing immediately after the initial one sold out. As a result, an estimated 20,000 people showed up on the same night for the first
concert—at a venue which could hold half that number.[7]

Sheerin adds: "The rock'n'roll genie was well and truly out of the bottle. The
Moondog Coronation Ball laid the foundations for every rock gig that followed,
from Woodstock to Glastonbury."[8] Although that event proved there was money
to be made in this format, any comparable rock and roll mass gathering would
not surface for some time to come.

By 1955, when Hanley was eighteen years of age, a young Elvis Presley was
making a big impression on teenage society. On weekends, at various high
schools and junior colleges the young engineer was operating sound at record
hops for many of the popular local disc jockeys. Record hops offered teens the
opportunity to freely dance to the music of the day. It was also a time when
parents and public were in an uproar over African American–inspired rock
and roll that captivated teenagers. The music itself, reinterpreted by white performers like Elvis, and later by English acts like the Beatles and Rolling Stones,
created a whole new genre of rock and roll.

Devices like the record player and transistor radio were giving the 1950s
teen a sense of new independence. This generation of American teenagers was
splitting from the routine practices of their parents and immersing themselves
in the popular music of the time. A new middle class emerged in the 1950s.
The children of this era became more of a consumer class. They were unlike
their parents, who were shaped by the impact of two world wars and the Great
Depression of the 1930s.

As a rebellious response to the seemingly frugal conservatism of their parents, the teenagers of this generation became more liberal and free. Saving less
and spending more money meant that, as quickly as the 1950s teen earned it,
they uninhibitedly spent it on the latest listening technology. At the end of the
1950s a lot had changed since the innocent record hops Hanley once knew. The
1950s generation grew up and found a voice in folk music.

In 1963 Hanley was in the center of a popular movement called the "folk
revival" at the Newport Folk Festival. The music then emanating from Newport and coffeehouses across the nation encompassed new ideals of purity and
honesty. Inspired by the music of Lead Belly and Woody Guthrie, this new wave
of folk music was a direct response to the civil rights movement in the United
States. It was a rejection of the gumball-laden sounds of popular radio-friendly
music. To some, topical folk dealt with the truisms of storied humanity.

Hanley, working out of his mobile studio at Newport's Festival Field in
1964, realized the potential impact these large-scale gatherings could have on

his sound company. In a 1964 *Boston Evening Globe* article written by Robert Glynn, Hanley speaks briefly about the influence college students had on his emerging business: "I credit student participants of the festivals for enabling me to go after the big stuff."[9]

The Newport Jazz and Folk Festivals not only launched a new direction for Hanley's career, but also prepared the sound engineer for a diverse series of larger and louder events later in the decade. Hanley's knowledge of festival production and infrastructure—sound, stage, and lights—was by then far more advanced because of what he had learned at Newport. An early leader of festival production, technical director Bob Jones, comments on the influence Newport had on the emerging industry:

> The Newport Festivals were more involved in the structure of a musical performances than you would find at other outside events. There were other outside events being done, like symphonies being played outside in New York; there was Tanglewood. But most of these entities were one artist or one group playing long sets or two halves of a concert. The festival concept became a lot more difficult for sound companies and the artist to contend with. This was because there were multiple setups, which had to be changed. The sound of one group, for example a traditional New Orleans band, could be followed by a big band, or a trio with a piano player, that are all very different. Sound now became quite an important element of each artist's presentation. The sound had to be right.[10]

Although Hanley's techniques and abilities in sound reinforcement expanded as the festival scene grew, his business sense did not. Any monetary potential these gatherings had were almost an afterthought for the sound engineer—as his *modus operandi* was to deliver quality sound and quality sound only.

During Hanley's career, rock and roll experienced several transformations. Ignited by the folk revival, most notable at Newport, the songs and musical performances around the civil rights movement catapulted the careers of Bob Dylan and Joan Baez. Bob Dylan's groundbreaking electric performance at the 1965 Newport Folk Festival single-handedly picked the bones of rock and roll out of folk's weathered skin, starting a new phase of folk rock.

But rock music didn't disappear. It remained hidden in the background, resurfacing full-on by the late 1960s and 1970s and dominating the music business once again. As a result, in a span of just ten years, 1964–74, there is evidence of distinct changes within festival audience culture, production development, and festival infrastructure.

The growth of the festivals revealed that the attitudes and expressiveness of this post–World War II baby boomer coming-of-age youth culture had transformed. Author Morris Dickstein writes in *Gates of Eden: American Culture in the Sixties* about these changes: "The spirit of the fifties was neoclassical, formal; the sixties were expressive, romantic, free-form. Rock was the organized religion of the sixties—the nexus not only of music and language but of dance, sex, and dope, which all came together into a single ritual of self-expression and spiritual tripping."[11]

The music festivals and emerging youth culture propelled Hanley's business toward its imminent success. It was a perfect storm of political, social, and cultural happenings driving the festival circuit with the sound engineer in tow. However, a new and louder style of rock music evolved as a response to the multilayered spectrum of social injustice, inspired by the turmoil in Vietnam and the Southern United States. To these factors, the sound engineer adjusted.

The eyes of music promoters widened as they realized the monetary potential these large events could offer, and having intelligible sound was a valuable asset. Quality sound meant that you could improve the audience and performer experiences, resulting in greater ticket sales. Organizers liked that a lot. It was a way for promoters to corral thousands of bodies into one place and to sell T-shirts, programs, and any other objects of their desire. Hanley was in such demand that his company provided sound reinforcement for at least three quarters of the jazz, blues, rock, folk, and classical music festivals at this time.

The new pop festival business marketplace was realized out of the Monterey (1967) and Woodstock (1969) festivals. It was especially the Monterey International Pop Music Festival (Monterey Pop) that displayed the counterculture as a marketable resource; even though it was a nonprofit event, it set an example of how the youth culture was willing to spend its time and money. At these events, promoters, managers, and bands figured out that they could make a bigger impact by appearing in one place, performing in front of thousands rather than hundreds.

After Monterey Pop in 1967, most promoters caught the festival bug. East Coast promoter Sid Bernstein saw the potential. In a 1967 *Village Voice* article, a writer claims the promoter was thinking about hosting a New York Pop Festival similar to Monterey that was supposed to take place in Central Park at the Sheep Meadow: "Every rock and roll promo wheeler dealer in New York seems to be bumping into each other's electric egos while trying to pull together something that resembled last June's Monterey Pop Festival. They all talk about what a groovy time everyone had out there in Califlowerland but they all get that same green glint in their eyes."[12]

The 1968 and 1970 concert films that followed the Monterey Pop and Woodstock Festivals helped elevate the careers and paychecks of its performers. After the box-office success of the film *Woodstock*, for example, entrepreneurial minded promoters, record companies, and film crews were foaming at the mouth to capture and capitalize on what had happened in Bethel, New York. Rock festivals made that same artist who previously played in front of smaller audiences very wealthy. Only a year after Woodstock, performance fees tripled for artists who had played the event. Following that, ticket prices doubled.[13]

Although the Woodstock Festival famously became a free event, it was not intended to be one. The festival's four producers, who formed Woodstock Ventures, had every intention of making money. According to producer Michael Lang: "After Woodstock the music business became an industry. Suddenly people realized what the marketplace was like, and what people would pay. Quality sound reinforcement was a big part of that. The industry grew from that point on. The bands started to demand it, and the promoters started to comply."[14]

Woodstock technical director Chris Langhart explains: "Woodstock was the point that vendors figured out that kids had money to spend. It wasn't clear that this was the case before. Youth marketing was born at Woodstock and suddenly kids have money, and kids spend money."[15] Not long after the Woodstock Festival, a flurry of mimics followed and for a very good reason—money. A lot of what we see within the contemporary festival infrastructure, its politics and its culture, is a derivative of what happened in Bethel. Figuring out how, when, where, and on what festivalgoers spent their money became a science.

In addition to paying exorbitant performance fees, promoters faced financial losses from things like gate-crashing, a trend born out of the dismantled fences at Woodstock. Instances like this turned into huge losses for festival producers, and were becoming the norm by the early 1970s. Reveling in a "free experience" was an ideal that seemed to fit nicely into the counterculture's anti-corporate utopian lifestyle. The notion that music should be free made it into a song "Music Is Love" by David Crosby, on his 1971 solo album *If I Could Only Remember My Name*.

> Put on your colors and run come see
> Everybody's sayin' that music's for free[16]

However, not all festivals were susceptible. Gate-crashing became such an issue that promoters for the Goose Lake International Festival, at the 390-acre Goose Lake Park near Jackson, Michigan, took matters in to their own hands. The festival occurred almost one year after Woodstock on August 7–9, 1970, and over 200,000 attended. To combat the free-loving and free-festival hippie onslaught, an "eight foot cyclone fence topped by six strands of electrified barbed wire" was put in place along the contained park. It didn't stop there. Local promoters "uncle" Russ Gibb and Richard Songer also hired "Security patrols . . . on horseback outside the fence; in jeeps inside the fence; and in boats along the lake."[17]

Some claim that Goose Lake was a success both in spirit and in organization, while others say it was wound too tight. This time many of the popular acts and production crews were treated to a working rotating turntable stage, while relaxed audiences enjoyed adequate sanitation and sprawling camping areas. Sound at this festival was not by Hanley.

National government and public opposition against festivals, coupled with rampant drugs and fatalities, proved to be very costly for promoters and Hanley. The frenzied mass media coverage often hyping festivals drove both public fear and curiosity. Despite this, for the most part the festivals were peaceful events, although a "peaceful festival" did not make exciting news. Just as these forces were taking place, a 1971 *New York Times* article reflected on the changing times within the festival marketplace:

> There have been signs that rock music can be expensive too, not only in terms of tickets but also in the toll of deaths, destruction of property, drug abuse, and general misery that seems to accompany today's rock festivals. Somewhere between Woodstock and Powder Ridge the idea became established in many minds that one need not buy a ticket to a rock festival. It was love, it was free, and you could crash the gate. But the idea came face to face with the skyrocketing fees of rock stars.[18]

Looking back, it's clear that the business of the contemporary rock festival was shaping itself as an industry leading up to the early 1970s. The streamlined facilitation, packaging, and artistry of things like sound, lighting, light shows, video, promotion, and marketing were all born out of this time. Over the past forty years the festival has gained a broader global identity, especially in Europe, where the festival market is currently bigger and more popular than ever.

Chapter 30

THE FESTIVAL PHENOMENON BEGINS

THE TRIPS FESTIVAL

Four significant West Coast musical gatherings can be identified that paved the way for the Woodstock Music and Art Fair. On the weekend of January 21–23, 1966, the Trips Festival held at Longshoreman's Hall in San Francisco launched the psychedelic West Coast festival experience. Author Ken Kesey, leader of the communal group the Merry Pranksters, American writer Stewart Brand, and visual artist and composer Ramon Sender produced the event. Kesey is best known for writing the novel *One Flew Over the Cuckoo's Nest* and Brand for his work as editor of the *Whole Earth Catalog*. The three-day indoor gathering hosted bands like Big Brother and the Holding Company and the Grateful Dead. The event drew 6,000 or more attendees.

Famed Grateful Dead sound engineer Owsley Stanley supplied an ample amount of LSD to those in the packed room. Friend Don Buchla—a pioneer of electronic music and co-inventor of the voltage-controlled transistorized modular synthesizer—provided the sound. Attendee and Merry Prankster Ken Babbs explains:

> We had Don Buchla build us a soundboard.... He lived in San Francisco and he built us this thing called the Buchla Box. I think he worked on the Moog synthesizer. This guy was unbelievable.... he had ten speakers set up ... in the balcony. He had this board in which he could run the sound around in circles ... [and] ... would isolate one, and have sound wheeling around the room. He had this thing like a piano that was just flat and you ran your fingers across it and it would play the notes. Made it himself, absolutely fantastic. He made up this box for us that was essentially a mixer and a mike amp and a speaker box and an earphone box."[1]

According to Tom Wolfe, in his book *The Electric Kool-Aid Acid Test*: "The music suddenly submerges the room from a million speakers . . . a soprano tornado of it . . . all electric, plus the Buchla electronic music machine screaming like a logical lunatic."[2]

Later the Buchla Box was installed on the Pranksters' converted school bus called *Further*. *Fact* magazine reported, "Here the psychedelic troupe broadcast their LSD-fueled ramblings around the bus and to the bewildered passers-by of Middle America."[3] Buchla eventually worked with the Grateful Dead building part of their sound system, mixing shows, and incorporating sounds from his Buchla Box during live performances.

The Trips PA and liquid light show established a format that caught on and continued at some of the area music venues. Wolfe adds: "For one thing the Trips Festival grossed $12,500 in three days, with almost no overhead, and a new night-club and dance-hall genre was born. Two weeks later Bill Graham was in business at the Fillmore West Auditorium . . . and his first weekend was advertised as the 'sights and sounds of the Trips Festival.'"[4]

THE HUMAN BE-IN

A year later, on January 14, 1967, a one-day counterculture-driven event of around 20,000 was held at the polo fields in Golden Gate Park in San Francisco. It was called the Human Be-In. Less of a festival and more of a gathering, it's most often referred to as the "Gathering of the Tribes." Inspired by a California law banning the use of LSD, the Be-In kicked off the Summer of Love counterculture movement. Poet Allen Ginsburg chanted mantras, while speakers Timothy Leary, Jerry Rubin, and others took to the stage. The gathering also showcased music by familiar Fillmore West and Family Dog bands like Jefferson Airplane, Big Brother and the Holding Company, Quicksilver Messenger Service, and the Grateful Dead. Dead engineer Owsley Stanley most likely provided the sound. Aside from turning on and tuning the sound system, Stanley also helped the Be-In crowd "turn on, tune in, and drop out" with substantial amounts of his custom-made white lightning LSD. It has been said that the Hells Angels were invited by the event organizers to provide security for the equipment.

THE FANTASY FAYRE AND MAGIC MOUNTAIN MUSIC FESTIVAL

Within the continuum of developing festivals on the West Coast, the KFRC Fantasy Fair (Fayre) and Magic Mountain Music Festival was held on June 10–11, 1967. Rather unknown and often referred to as the "Mount Tam" Festival, the event was staged atop Mount Tamalpais in Marin County, California at the Sidney B. Cushing Memorial Amphitheater. Mel Lawrence, promoter and producer of the Second Annual Miami Pop Festival and eventual Woodstock veteran, was involved with this first ever multi-day rock and pop festival. Producer Tom Rounds assisted Lawrence.

The event was produced for San Francisco radio station KFRC, where Rounds was the program director and Lawrence the promotions director; the festival was held as a charity benefit in support of a local organization called the Hunters Point Child Care Center. Lawrence recalls, "We came up with the idea to do this festival on top of a mountain, yet not many people knew about it."[5]

Around 15,000 people or more showed up over the course of the weekend. The festival offered no camping but getting to the top of the mountain proved easy since an outfit called "Trans-Love Bus Lines" provided transportation. Known for their involvement with previous West Coast festivals, the infamous motorcycle club the Hells Angels assisted with shuttling folks throughout most of the weekend. Lawrence reflects: "The Hells Angels were there to perform security. This was before Altamont. It was the Summer of Love. I negotiated with the California Highway Patrol and the Hells Angels to allow them to be the only ones who can bring their motorcycles up to the top of the hill. Everyone else had to park at the bottom and we bused up over 15,000 people over two days."[6]

The festival acts occurred on two stages; one at the main amphitheater, and a smaller one for additional performances. Lawrence recollects, "It had an incredible lineup including all of the San Francisco groups who were big at that time."[7] The event promised an eclectic performance roster including Jefferson Airplane, the 5th Dimension, Tim Hardin, the Byrds, Captain Beefheart, Dionne Warwick, the Grass Roots, Marvin Gaye, and the Doors. All bands donated their time and fees to the Hunters Point Child Care Center.

The event provided the audience with vendors, refreshments, and arts and crafts booths that surrounded the mountaintop amphitheater. A light and sound show took place inside a large geodesic dome light chamber made of a pipe structure covered in plastic sheeting. Local company McCune Sound parked their panel van at stage right next to a speaker cluster on top of some short wooden scaffolding. On the side of the sound truck it read "Harry McCune Sound Service, San Francisco."

Just a week later, on June 16–18 the Monterey International Pop Music Festival was held. Festival promoters Lou Adler and John Phillips hired Lawrence to help with operations, leaving little time for him to prepare.[8] According to Lawrence, Adler and Phillips came up to the Mt. Tamalpais Festival site before Monterey Pop out of curiosity: "They had a lot of questions on how we put things together, how we handled traffic, how we did the staging."[9] Considered a high-profile event, Monterey Pop quickly overshadowed the KRFC Fantasy

Fair and Magic Mountain Music Festival. Lawrence reflects: "Monterey often gets acknowledged as the first festival. Monterey had more money for publicity than we did but ours was truly the first pop festival."[10]

THE MONTEREY INTERNATIONAL POP MUSIC FESTIVAL

By mid-June 1967, the United States had spent billions of dollars on the escalating war in Vietnam. The 40,000 who received draft notices every month faced the possibility of joining the 15,000 American lives already lost in Southeast Asia. There seemed to be no end in sight. President Johnson's actions were met by the antiwar sentiments of the American people. From New York to San Francisco the country erupted in protests, some ranging over 100,000 in attendance. At these demonstrations draft cards were burned. Further, the impact of the war exploded within the evolving music scene. Evidence of such responses can be seen and heard at the first rock festival of its kind called the Monterey International Pop Music Festival (Monterey Pop).

New York Times writer Kevin D. Greene claims Monterey Pop ushered in a new age for American music: "But just as important, it was a signal moment in a cultural and political upheaval that had been incubating in the Bay Area for nearly a decade. The Beats and their bohemianism of the late '50s and early '60s built a crucible from which a new anti-establishment worldview would emerge."[11]

Monterey Pop was a three-day event, held June 16–18, 1967, at the Monterey Fairgrounds in Monterey. The festival glittered with memorable performances and elevated artists like Jimi Hendrix and Janis Joplin to pop-star status. Monterey Pop saw a decent-sized crowd that weekend but there have been many speculations regarding the exact number that attended. According to producer Lou Adler, approximately "100,000 people attended the event over its three days."[12] Most were spread out around the twenty-one acre fairgrounds, with several thousand contained within the actual performance arena. Journalist Steven K. Peeples claims, "The vast numbers attending Monterey validated this music's importance as rock and roll came of age."[13]

Some say that there was nothing but "good vibes" with "no trouble" to be had at Monterey Pop, demonstrating to the world that an event of this caliber could be done. It showed organizers that putting on a festival of this magnitude was possible. With the infrastructure in place and the festival successfully executed, an example had been set. Adler claims that Monterey Pop was one the first festivals of its kind that created a template for how festival facilitation for rock and roll was to be done. Like George Wein's East Coast Newport Festivals, Monterey Pop changed the standards of the performance arena: "After that weekend, artists would demand and get preferential treatment. That's one of the side benefits of

Monterey Pop—it taught the acts how to perform under the right conditions, and how to get the right conditions in the first place. Our idea for Monterey was to provide the best of everything—sound equipment, sleeping and eating accommodations, transportation—services that had never been provided for the artist before Monterey."[14]

Lou Adler, partner John Phillips of the Mamas and the Papas, and the Monterey Pop production team had moved in just ten days prior to the festival start. Adler remembers: "There was nothing there, not even a proper stage to house the kind of amplification that was coming in. We had to build the speaker systems right on the site." In addition, a communications center was set up, and those who were on staff communicated through walkie-talkies. Adler continues: "The transportation crew we organized included not only cars and drivers for all the acts, but scooters, motorcycles, bicycles, whatever else it took to get around. We had cleanup crews, and an arts committee to oversee the booths and displays."[15] A good deal of what impresario producer George Wein had done at Newport was mimicked at Monterey. "Organizers sought out the best musicians, sound and lighting systems and food in order to lift the level of what rock and roll should be."[16]

The Monterey Pop production staff was made up of a talented team of engineers, technicians, and musicians that included Hanley's Newport colleague Chip Monck, who was in charge of lighting. It also included Mel Lawrence, who was brought over by Adler and Phillips for his work at the Magic Mountain Festival the week before. "I was in charge of operations for Monterey Pop. I brought my crew from Magic Mountain and took care of all the fencing, campgrounds, and that kind of thing."[17] McCune Sound was hired. Sound engineers Jim Gamble and Abe Jacob mixed, while Wally Heider, Bay Area engineer, recorded the event.

In the 1968 D. A. Pennebaker concert film *Monterey Pop*, you can hear artist David Crosby at sound check proclaiming, "Groovy, a decent sound system at last!" His voice echoes into the empty fairgrounds in naïve excitement. And rightly so, since until then West Coast groups had never experienced a large outdoor PA like they had at Monterey Pop. According to Abe Jacob: "My involvement with Monterey was that I was already doing sound for the Mamas and the Papas while at McCune. John Phillips and Lou Adler were the two major producers of the festival. So they asked me to put together a sound system for them since I was working for McCune and the band. It was a perfect fit."[18]

For the festival, McCune Sound used a combination of Altec 612 cabinets (on top, ten-cell horns with 280 drivers), 825 A7 Voice of the Theater cabinets (on top, two-cell 203 long-throw horns with 288 or 291 drivers), and Altec 9844 speakers. Most of the vocal mikes were Shure's Unidyne III SM56 microphones, equipped with A2WS windscreens with built in shock mounts. Last, McCune used Altec 80-watt rack-mount mono amplifiers. Jacob explains: "It was all Altec equipment. Basically it was a bunch of A7s with separate sectoral horns and a mixing console we had just built which had sixteen channels and two outputs with bass and treble equalization."[19]

Ken Lopez, attendee, sound engineer, and former vice president of JBL, claims that, for the time, the Monterey Pop system was typical:

> It was a little pile of Altec theater boxes and some side fill monitors. I would think that by today's standards this would be considered more of a club-sized system, unsuitable for 6,500 people. But in fact it was adequate, and more than met the needs of what we were expecting to hear. It was not loud, but it was musical in that the system disappeared into the performance. This was textbook sound reinforcement in the purest sense.[20]

Blues Project band member Steve Katz reflects on his group's sound as "absolutely horrendous and the reason was because our keyboard player's vocal and keyboard mike wasn't working! I don't remember stage monitors at Monterey Pop; we must have had them, it was a big event."[21] On the other hand, drummer Roy Blumenfeld claims, "In the evening during the Otis Redding performance I was backstage and it sounded very clean and clear."[22] According to Lopez, stage monitors were present: "I had access to the sound checks and the arena between performances and took special note of the monitors and their construction. A number of them were 'appropriated' by some of the bands, including the Dead, due to their uniqueness."[23]

In a *Crawdaddy!* article, music journalist Paul Williams affirmatively stated: "Certainly, the festivals will continue."[24] And they most certainly did. Pennebaker's *Monterey Pop* documentary, released in December 1968, paved the way for future concert films like *Woodstock* as a means to showcase up-and-coming talent. A 1969 *Chicago Tribune* article by Robb Baker, "Music and Art Fair of Woodstock Is Another in a Batch of Pop Festivals," reflects on the rush of festivals taking over the nation inspired by Monterey: "The success of Donn Pennebaker's film documentary of the 1967 festival in Monterey, California, may in part account for the festival explosion. At any rate, major gatherings of clans already have been held in Miami, Los Angeles, Denver, Atlanta, and Atlantic City."[25]

Many of the acts that performed at Monterey Pop were also to play at Woodstock a couple of years later. Monterey Pop brought a huge amount of information to the table, giving future organizers a blueprint to go by. Tom Law, who attended both festivals, reports: "Monterey didn't have the problems Woodstock had, the weather and the impact of all those people. Monterey was the beginning of those concepts that we took to Woodstock: To treat people responsibly, with the expectation that they would act responsibly in return."[26]

Haines Square in the early 1900s. Medford, Massachusetts. Thomas Convery collection.

Bill Hanley in 1939 at around two years old. Hanley archive.

The Bal-A-Roue in 1952. Medford, Massachusetts. John Kane collection.

Bill Hanley looking at the Boston Garden sound system in the 1950s. Hanley archive.

Bill and Terry Hanley working on their Rebel speakers in 1957. Hanley archive.

Bill Hanley in 1960 working on the speaker set-up at the Metropolitan Arts Center. Hanley archive.

Hanley Sound Equipment Company in 1961 setting up for the Kingston Trio at Castle Hill in Ipswich, Massachusetts. Notice block and tackle hoist of Voice of the Theater speaker. Hanley archive.

Bill and Terry having a discussion outside of the mobile trailer in 1964. David Roberts collection.

Bill Hanley working in his mobile studio in 1964. Hanley archive.

Hanley Sound mobile studio at Newport early to mid 1960s. Hanley archive.

RCA mixers set-up for the Newport Folk Festival in July 1964. Hanley archive.

Newport Jazz Festival speaker set-up. Festival Field. Late 1960s. Hanley archive.

Bill Hanley, twenty-eight years old in Washington DC for the Johnson inauguration at the Capitol Building. Setting up for the Mormon Tabernacle Choir in 1965. Hanley archive.

Lyndon B. Johnson Inauguration feature in 1965 Shure newsletter, Sound Scope. John Kane collection.

Lyndon B. Johnson speaker set-up in front of Capitol building. Hanley archive.

The Beatles concert at Shea Stadium in 1966. Hanley Speaker set-up (lower right). Tony Griffin collection.

The Remains on tour with the Beatles. Ed Freeman Photography Los Angeles, CA

The Beach Boys on tour in 1966. Notice Hanley's custom plywood stage monitor just below Mike Love. From the book Surfboards, Stratocasters, Striped Shirts by Bill Yerkes

Bill Hanley on stage at Miami Pop Festival. December 1968. Henry Diltz collection.

Hanley Sound Kenworth tractor trailer at the Miami Pop Festival. December 1968. Henry Diltz collection.

Janis Joplin live video screen projection at the Fillmore East. 1969. Joshua White collection.

Bill Hanley at the recording console in the basement of the Fillmore East. Late 1960s. Joshua White collection.

Bill Hanley. Fillmore East. Late 1960s. Joshua White collection.

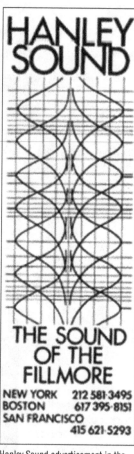

Hanley Sound advertisement in the Fillmore East program. 1969. John Kane collection.

Hanley Sound manager Judi Bernstein 1960s. Hanley archive.

Chapter 31

ALL ROADS LEAD TO WOODSTOCK

MIAMI POP II

The Monterey International Pop Music Festival was an inspiration for the Second Annual Miami Pop Festival (Miami Pop II), held on December 28–30, 1968. This event saw over 100,000 in attendance at Gulfstream Park in Hallandale, Florida. Hanley was hired for what is considered to be the first official rock festival on the East Coast. Miami Pop II festival producer Mel Lawrence brought the sound engineer on board after he was referred by industry professional Stan Goldstein, although Lawrence was considering other vendors for sound. According to Lawrence, "I met Hanley, and we liked him."[1] Chip Monck was a natural choice for lighting and production, and ended up working alongside Hanley once again.

The production notes of producers Mel Lawrence and Tom Rounds report:

Bulletin #8, 10/16/68—73 days to go

Sound—Neither Swanson or Criteria has enough manpower or equipment to really handle the job, so it now appears that Swanson and Criteria, together, will make a joint proposal for sound services.

Bulletin #17, 10/31/68—57 days to go

Hanley Sound called today from Boston and is interested in making a bid on both sound and lighting.

Bulletin #21, 11/6/68—52 days to go

Chip Monck and Bill Hanley . . . are meeting with Mel this weekend in Miami and may come up with a new total bid. . . . Meanwhile, Swanson Sound,

Criteria ... Peter Polofian and 18 subcontractors in Miami are being stalled until we get a reading from Monck and Hanley....

Bulletin #24, 11/11/68—47 days to go

Production: The Hanley Sound bid will come in around ... $9,000.

Bulletin #27, 11/14/68—44 days to go

Monck's proposal includes everything ... and is combined with a sound bid based on Hanley.

Bulletin #29, 11/18/68—40 days to go

Production deal is being solidified with Chip Monck et. al. Hadler and Festival Group has been finally turned down; Hanley Sound is in and notified as such by Monck.

Mel Lawrence's post-festival report

... The sound system ... operation and set up was flawless ... Hanley's crew are pro's ...

Mitch Fisher's post-festival review

... Monck and Bill Hanley did an excellent job ...[2]

Stan Goldstein worked as chief mixing engineer at Criteria Recording Studios in Miami. Goldstein had previously witnessed Hanley's work at the Fillmore East and was impressed. Goldstein observes: "I knew of Bill Hanley from the Fillmore at the time and attended events he did the sound for, but up until then, I never met him. He was the guy who was doing large outdoor and PA gigs."[3]

A week prior to the festival launch, Hanley and his crew began sound reinforcement setup for thirty-five diverse acts at the Gulfstream racetrack. Hanley deployed sound systems for both the Flower and Flying stages (overseen by Chip Monck). These were two separate main stages located several acres apart. Both operated simultaneously, which allowed musical acts and audiences a unique experience. With art exhibits and concessions in between, the Flower Stage was located in front of the Gulfstream racetrack grandstand; the Flying stage was in a tree-lined meadow.

According to Lawrence: "There was a stage in front of the grandstands and another one further away near a parking lot. This was the first time a festival ever had two stages."[4] *New York Times* journalist Ellen Sander, in an article called "The Miami Festival: An Inspired Bag of Pop," claimed that music from Hanley's sound system (on both stages) could be heard throughout the grounds of the festival: "A constant magic and movement filled the air as crowds wandered comfortably from area to area."[5]

The scheduled performances included Joni Mitchell, the Grateful Dead, the Paul Butterfield Blues Band, the Charles Lloyd Jazz Quintet, country music by Flatt and Scruggs, and Steppenwolf, to name just a few. Days before, while in Los Angeles, Steppenwolf lead singer John Kay, along with his band, bought costumes to wear for their Miami performance. Each band member was to represent a figure of the "establishment." Hanley Sound's Harold Cohen recalls Kay's actions while he and Terry Hanley were present working the festival's second stage:

I will never forget this because Steppenwolf came in on a golf cart as we were changing over stands in between sets. We didn't know it was Steppenwolf at first because the lead singer John Kay came on stage with a police outfit and helmet on, one other member was dressed as Uncle Sam! Kay demanded the microphone, said something, took off the outfit and started singing! We weren't ready! We had no idea that he was coming in dressed as a cop. The whole band was in costume. It was a surprise to everybody![6]

As the spotlight hit Kay's face, donned in a helmet and signature aviator shades, he shouted into Hanley's microphone at the confused audience. Many thought the event was getting shut down until the lead singer took off the helmet revealing his identity. Cohen explains: "The stage was only two feet high. It was very low. However, there weren't people right there. They were in the grandstands and isolated. So there wasn't any notion of a problem occurring."[7] According to Cohen the Flower Stage setup was as follows:

We had two RCA and one Altec 15167A mixers stacked up that were gaffer taped together. One RCA was on the bottom, the Altec in the middle, and the other RCA on top. The RCA mixers were four input devices with no tone controls, a master volume control and a meter. They were very good. They had been used mainly for remote broadcast applications. These were metal encased and had a metal cover with a collapsible handle. In total this gave us around fourteen inputs. We were using Rebel monitors on stage and Altec Voice of the Theater speakers with 311 horns. I believe Bill was using 410s for the main stage. We also used McIntosh 275s as well.[8]

Hanley and his team had done a lot of work for Steppenwolf in the early days. According to Hanley, he and his crew often referred to the band as "Step-in-shit" because of the difficulties they faced working with them, especially John Kay. Lead sound engineer Sam Boroda recalls a run-in with the singer on a tour in 1969: "I was sitting on the plane in first class and John Kay was sitting across the aisle from me. He looks at me through his dark mirrored glasses and says "Who the fuck are you?' I said, 'I am the fucking guy who has been doing sound for you for the past three weeks!' I thought, if you don't know who I am after seeing me at all of those concerts then something is wrong!"[9]

Miami Pop II demonstrated a unique way of facilitating a rock festival for enthusiastic audiences in a suitable setting and easily accessible viewing areas. Journalist Ellen Sander, took notice of such standards in the *New York Times*: "The event was a resounding success in both organization and programming," adding that the "Concerts were staggered in sets of forty five minutes each with a fifteen minute overlap, making it possible to see everything or stay in one area for those portions of the program."[10] Miami Pop II saw minimal violence with only a few drug busts.

Cohen remembers: "After Miami Pop myself and truck driver John O'Toole went over to the Miami Beach Playboy Club. John was young and good looking and picked up girls like he was picking up a cup of coffee. I noticed Bill didn't show up. I saw him the next morning and said 'Where were you last night?' He said he was at the Fontainebleau Hotel playing pool with Tiny Tim in the Boom Boom Room all night! I just laughed! Tiny and Hanley were good friends. He came out of Steve Paul's The Scene nightclub along with Johnny Winter. We had a system on rental there for many years. This is how Bill got to know him."[11]

Most people don't know that there was another Miami Pop Festival earlier in 1968. It's often confused with the larger and better publicized Miami Pop II. Miami Pop I is sometimes referred to as Miami's "Underground Festival" or the "1968 Pop Festival." It was held on May 18–19 with around 50,000 in attendance. Both events were held in the same location at the Gulfstream racetrack in Hallandale, Florida.

Miami Pop I was a milestone production for one of its promoters, eventual Woodstock Festival producer Michael Lang. According to Lang, Miami Pop I was where the "seeds to Woodstock were sown."[12] Lang recalls having to scramble for sound and staging for the event. One of his ideas, for example, was to line up a series of six separate stages, using flatbed trucks to make the process of setting up and breaking down more efficient. This innovative system allowed one band to quickly follow the next. Lang explains: "We couldn't figure out how to build stages quickly that would hold everybody securely. We called around and found six flatbed trucks and rolled them onto the track. We set them up two by two so that we could structure the show on three rotating stages."[13]

Miami Pop I was a two-day festival with no camping. There were two shows a day, in the afternoon and the evening. The Jimi Hendrix Experience headlined the festival, with many other notable acts highlighting the lineup. For sound reinforcement Lang used Stan Goldstein of Criteria Recording Studios. According to Goldstein: "The audio community was minimal back then. Number one, the number of people working in the field was very small. Number two, the amount of people working with large sound systems at that time was even smaller."[14]

FROM MIAMI TO BETHEL

Individuals from the Fillmore East and Miami Pop crews formed the impetus that eventually became the core of the Woodstock Festival production staff. According to Lawrence: "We were the first people contacted to work on Woodstock. A lot of the original crew for Woodstock were Miami Pop people."[15]

Michael Lang recalls, that around September 1968 he and partner Artie Kornfeld began looking for a site for a new venture. By January 1969 they met future Woodstock co-producers and financiers John Roberts and Joel Rosenman. Combined, the four individuals formed Woodstock Ventures. Lang brought Stan Goldstein in to work as a headhunter for technical and production support for the intended event. When the time came to make a decision about sound reinforcement, Hanley's name came up. The sound engineer's quality work at Miami Pop II and the Fillmore East was the catalyst that established Hanley as Woodstock's chief sound engineer.

As soon as Michael Lang firmed up his agreements with Woodstock backers Rosenman and Roberts, he drew up organization charts to allocate key roles at the festival with Goldstein.[16] Goldstein's primary task as technical operations manager for Woodstock was to effectively staff the festival. This required finding prime vendors for lighting, sound, stage, and production. Hanley was the most logical and suitable choice for the promoters, unless Goldstein wanted to seek out a West Coast company. According to Goldstein, "There was really no one else for the job and capable to handle this size of event."[17]

Hanley and his crews had been facilitating sound reinforcement for festivals with ease leading up to the Woodstock Festival. But Goldstein still needed convincing that his sound was going to be sufficient enough. During the early stages of organization, Goldstein continued to feel the engineer out about his methods and philosophies on large-scale sound reinforcement facilitation. Goldstein reflects: "It's not that I thought I was some sort of super techno geek with knowledge surpassing his. But there were different ideas on how to do a job and I wanted to make certain that the job Bill was prepared to do was the job I wanted him to do."[18] Hanley and Goldstein spent a number of weeks discussing this before Hanley's position as Woodstock soundman was finally secured.

Early in 1969 Hanley got a call from Goldstein at his Medford office. Worried, Goldstein gently interrogated the sound engineer about how he would actually support a festival of this size. Hanley was apprehensive, actually becoming somewhat defensive about giving up his ideas so easily. The sound engineer had seen a number of new sound companies who had witnessed his work borrowing from his innovations. Goldstein was simply trying to emphasize that this was going to be a large-scale event, and if Hanley were awarded the gig, how would he address the sound? What sort of equipment would he use and how would he execute and apply his skills to the event?

According to the wary sound engineer, he felt that "talk was cheap" and that he should not reveal all of his secrets: "Stan was picking my brain and I didn't like that idea. How do you pay the bills if you give up all of your ideas? I felt I started this business long before some of these other companies came along. I needed to pay my bills, even though I didn't do it for the money."[19]

Both men remained leery of each other's intent throughout these negotiations. This has been documented in a 1979 book by Bob Spitz, *Barefoot in Babylon: The Creation of the Woodstock Music Festival*. According to the imaginative author, the "conversation" between Goldstein and Hanley went something like this,

> "You say you don't know what you want?" Hanley inquired in a mocking tone of voice. "I don't have time to give you a Dick-and-Jane primer in outdoor sound right now, Goldstein. You see, I'm a busy person. Everybody wants me to bail them out of trouble, but nobody knows what they want. I'll talk to you when you get your scene together. Just decide what it is you need, call me back, and I'll come down there and plug in my equipment one-two-three," mercifully leaving out "'cause you've got no place else to turn."

> [Goldstein] "No, man, you don't seem to understand what we're building and what it's gonna entail from . . ."

> "Hey, Goldstein—you're a fine person to tell me I don't understand. Get it together pal. . . . I'll be hearing from you."[20]

Spitz claims Hanley had an "unflagging, boy wonder, egotistical confidence" regarding his abilities. The author's assessment of Hanley's attitude might be a fairly accurate one, given reports of the engineer's interaction with many of his crew and clients. The successes fostered by Hanley were often celebrated, but it was not always smooth sailing for those who worked for and around him.

Some claim that his unique personality and methods of executing things often created intense frustration.

Those who worked for Hanley quickly realized his transactional style, referring to him as a "mad scientist," "odd," or even "arrogant," and at times "difficult." Known as the "Johnny Appleseed of Sound," Hanley was the "eccentric Boston guy" to all of the hippies who worked around him. Older than the longhairs, the engineer most often wore a white shirt buttoned to the collar, always in deep thought and occasionally talking to himself. Lee Mackler-Blumer, assistant to the director of Woodstock security, Wes Pomeroy, claims: "Bill was really the mad scientist. The only thing he didn't have was a white coat flying behind him. Truly he had these visions, and he had this knowledge. He obviously had spent years in a room with a pencil, figuring all this stuff out."[21]

Regardless, Hanley always seemed to get the job done even if he pissed off a few along the way. He always kept calm in the most tense of moments and Woodstock turned out to be the truest test of this character trait. Crew member David Marks—whose job for two nights was to sit on one of the Woodstock sound towers to keep photographers and others off of them—relates: "It was Hanley's coolness at Woodstock that dissipated any angst towards any problems, and there were lots of problems! The rain caused all kinds of hassles; amps kept blowing, cables kept shorting. It was really a dangerous situation."[22]

Expecting a larger than normal turnout, Lang was unsure if a suitable sound system even existed. The young promoter had been considering Grateful Dead sound engineer Owsley Stanley to provide sound for Woodstock, but he was unsure if the "Guru of LSD" could do the job. All along, Lang admired Hanley's intrepidness, a personality trait that made Lang feel confident. This paid off for Hanley, since it was Lang's decision in the end to hire him. Through it all Hanley himself remained confident that he could do the job, and Lang "liked that."[23]

Lang made a wise and conservative choice with the Irish Catholic kid from Medford. He reflects on why he chose Hanley, claiming that what Hanley did was "incredible" in terms of "strength" and "clarity":[24]

> We came to a situation where I wanted great sound. Bill said to me that great sound for an event like yours doesn't exist, but I can build it, and he sold me. He had the credentials and whatever quality sound existed in our industry at the time (which was not an industry then) was a product of Bill's talent. I am sort of a vibe person, and after meeting Bill, I believed he could do it and quite frankly I never thought about it again. Maybe this is naïve of me but I had no doubt that he would deliver it. One of the greatest moments at Woodstock was when Richie Havens took the stage and the sound system worked. Without that we would have been lost and it worked beautifully; we never looked back.[25]

A majority of the Woodstock staff approved of Hanley's appointment. Lawrence recalls: "He was a little quirky, but with Bill there was an air of confidence. I thought that this part of the festival is covered and this guy knows what he is doing. It was a brilliant move to hire him not only because of his focus on clarity but because he made sure the audience was covered."[26] Artie Kornfeld remembers: "Lang had his own offices and hired some of the best people money could buy. Hanley had done the sound on basically every major tour and for every major artist that was hired."[27] According to Joel Rosenman: "Hiring Bill wasn't really my role. However, when Hanley's name came up I just assumed we were getting a top professional."[28]

Michael Lang's assistant Ticia Bernuth-Agri recalls a feeling of assurance when Hanley was brought on board: "Hanley never got his due. He wasn't a person who steps forward. He just did the job. He was a behind-the-scene type of person. When Hanley came on we were relieved because we thought, he's got it. We don't have to worry about that anymore."[29]

As the months went on, Hanley and Bernstein firmed up negotiations with Lang. While on one of Hanley's many visits to the Fillmore East, he and Goldstein discussed logistics with other essential team players. But Bernstein had some concerns. She recalls:

> Bill and I were there for the original meetings. Months before Woodstock we were in NYC with everyone from the Fillmore East at their office. We ended up in a conversation with the entire production staff like Chip Monck, and Stan Goldstein, pretty much everyone. They laid the plans out on the table. However, it seemed like nothing was solid and I became worried. We constantly had trucks on the road. How were we going to pull off something so big? How were we going to get paid?[30]

Additional meetings occurred between Goldstein, Hanley, Lang, and Lawrence leading up to the festival prior to it changing locations from Wallkill, New York, to Bethel. Lawrence recalls: "When Woodstock was getting put together and we all met in New York, we talked about how Woodstock should be staffed. Through Stan Goldstein we got the crews together from Miami Pop to work on Woodstock."[31]

When the Woodstock lineup became known in the spring of 1969, Hanley Sound had worked with most, if not all, of the performers except for Sha Na Na. According to Hanley: "By that time we had virtually worked with all of the acts that were on the Woodstock bill, so everyone knew us. We also worked with some of the same people from the second Miami Pop Festival at Gulfstream racetrack in 1968."[32]

Hanley knew Alan "Blind Owl" Wilson of Canned Heat who performed at Woodstock. The Wilson family lived in Arlington, Massachusetts, not far from Medford. When Canned Heat hit the Woodstock stage on Saturday night, the sound engineer had no idea that Alan was the shy kid he had met a few years prior. Through mutual acquaintances Hanley had become friendly with Alan's father Jack Wilson, who was a well-known ham radio buff. Occasionally the sound engineer visited Jack's hilltop home in Arlington Heights and exchanged ideas about audio. Alan's sister Jayne Wilson observes:

> Our house had quite a few antennas attached to it. There were many affixed to the front and back that extended pretty high up. There was actually one that was a dish and that rotated! My mother didn't like this at all. She was quiet and sweet. But my father would come home and go into his ham radio shack in the yard and fool around with this stuff. He was always winning contests for this hobby. He became well known in the area for this. He had his ham radio commercial operator license and his call sign was W1QXX and everyone knew it. My father knew everyone, so this is probably how Hanley and my father became friends.[33]

As Hanley recalls, Jack's wife Barbara would greet him at the door. Upon entering their home he met the young Alan sitting alone in his room perfecting his unique tenor vocal style and blues guitar playing. Jayne remembers, "My mother used to tell me that Alan was too embarrassed to sing in front of people, so he used to hide in his room."[34]

According to some, Jack drove a truck from Arlington to Bethel filled with some of Canned Heat's equipment for their weekend scheduled performance. Jayne recalls, "My father and one of his coworkers drove an old truck up to Woodstock. He was on stage when the band was playing. Alan said he was up on the actual staging on a catwalk and singer Bob 'The Bear' Hite was screaming at him to 'Get down!'"[35] Hanley remembers: "I never really put two and two together until much later when I found out it was Alan playing with Canned Heat at Woodstock! I spoke to Jack over the phone and he told me he was there! I had no idea! Alan was a strange kid from what I recall."[36]

Back in Medford it was business as usual. In the spring of 1969 Bernstein booked the Woodstock Festival, wrote it into the Hanley Sound calendar, and didn't think much about it thereafter. By the summer of 1969 that calendar was bursting at the seams with events. Various employees were coming in and out of the office, picking up and dropping off sound gear.

For those involved in these early days of sound reinforcement execution, most recall it being a "fly by the seat of your pants" operation with no guidebook.

In 1969 sound reinforcement equipment was not something you could just go out purchase at a store. Either you made it, went without, or adapted. This year—especially—Hanley Sound crews were left with very little time to prep for each gig. And then as the Woodstock date came closer, the office staff was catching wind that there were some issues with the site at Wallkill.

Chapter 32

THE GOLDEN AGE OF THE ROCK FESTIVAL

1969

The phenomenon of the 1960s pop festival in Western industrialized societies occurred during a period of uncertainty and renaissance. In the lore of sixties culture, the festival experience has been naively misconstrued and glorified as a rain-soaked drenching of peace and love. It was a time of intense passion, a time of instability and extreme turmoil. It was a time in history when individuals came together to spontaneously celebrate. Festivalgoing flower children imagined that everything would be all right, even if for only three days. But often this was not the case. This compelled Hanley to provide innovative sound and also to balance it with peacekeeping, sometimes under difficult circumstances.

By the beginning of the 1970s, there was a rising tide of gate-crashing, tear gas, and police arrests. But despite the inevitable negative media attention attached to these events, it was a magical time of hope and unity. In an article titled "We Can All Join In: How Rock Festivals Helped Change America," William Mankin wrote: "For most of adult America, all these unsettling cultural awakenings suddenly became things to be questioned, to be skeptical of, scoffed at, rejected, feared, condemned, opposed, suppressed, or prohibited by law. But the counterculture would be propelled by the young, so it was their views that mattered most."[1]

During 1968–69 Hanley Sound provided sound reinforcement for many notable music festivals throughout the United States. A partial list of Hanley's festival bookings, and the number of people who attended, is shown below.

By 1970 the company had become widely known for its outstanding work in the large-event market. You were lucky if you could catch Hanley even for a moment back at the Medford office. *Boston Globe* reporter Nathan Cobb in 1970 claimed: "To know Bill Hanley however is not necessarily to be able to find him. He is usually crisscrossing the country with one of his two 40 ft. trailers or his eighteen wheel truck, each filled with enough equipment to amplify sound to delight 20,000 people."[2]

HANLEY FESTIVALS OF 1968–69			
Date	Event	Location	Attendance
1968 December 28–30	Second Annual Miami Pop Festival	Gulfstream Park, Hallandale, Florida	100,000
1969 June 21–22	Toronto Pop Festival	Varsity Stadium, Toronto, Ontario	60,000
1969 June 27–28	Denver Pop Festival	Mile High Stadium, Denver, Colorado	50,000
1969 July 4–5	Atlanta International Pop Festival	Atlanta International Raceway, Hampton, Georgia	140,000
1969 July 11–12	Laurel Pop Festival	Laurel Race Track, Laurel, Maryland	15,000
1969 August 15–17	Woodstock Music and Art Fair	Max Yasgur's farm, Bethel, New York	400,000

With so many festivals occurring all across the country, Hanley Sound crews had to be extremely focused. What they encountered during this time of musical gatherings is like no other within the history of sound reinforcement. Hanley's influence and the hard work of his team is the common thread woven through the many festivals of this period.

DENVER POP

Barry Fey, "Denver's most active and successful rock concert promoter,"[3] hired Hanley for his Denver Pop Festival at Mile High Stadium in Colorado on June 27–28, 1969. It was a month or so before Woodstock, and only weeks before the first Atlanta Pop Festival. Denver Pop promotion boasted a stellar lineup of sixteen top-notch bands. It was a huge undertaking and over 10,000 people were expected to attend each day.

Fey was worried since it had been a rainy June. Getting rained out was always a risk for promoters when events were held in a field or open-air stadium; a single rainy weekend could be disastrous. However, Denver Pop had fairly dry weather. According to Fey, he was nervous about holding an event in an outdoor arena: "The only place we could put on rock concerts was at the Stadium. There was no way I'd be able to sell the 11,000 plus seats, plus the acoustics were awful. It was all concrete and sound bounced all over the place."[4]

Unlike other festival environments, this venue offered an infrastructure of bathroom facilities, water, fences, electricity, and seats. Fey hired Hanley's colleague Chip Monck as stage manager and emcee. Under pressure to finish on time, Monck recalls Hanley running behind schedule: "Bill had found a 110 volt power outlet and brings his soldering iron and a dozen rolls of two conductor shielded zip cord. He is soldering away most of the day then rolls in lots of trailers of very large speaker boxes. He was a clever person; it was always a concern though if he would be finished on time."[5]

The event, which totaled 50,000 in attendance, saw violence. Members of the Students for a Democratic Society and American Liberation Front were mixed into the crowd charging the air with anti–Vietnam War sentiment that soon spilled into the streets. On the first night, tension built outside the stadium as some festivalgoers, believing they should not pay to hear the music, attempted to climb over the fences. It took the local police force to keep them at bay. Members of both groups passed out antiwar literature as Frank Zappa, Iron Butterfly, and Three Dog Night performed. Fey remembers: "The festival was right in the middle of anti-establishmentarianism, and because I was a businessman, I was viewed by the young people as part of the establishment."[6]

On Saturday evening the problems reached a climax as the angry crowd grew in size with even more gate-crashing. The episode caused a big stir with local law enforcement. As people scaled the fences, police in full riot gear were summoned in an attempt to restore order. When the first wave of officers arrived, hundreds of individuals greeted them with rocks, bottles, and beer cans. Hanley, who was located inside at the south end of the stadium, witnessed the successful fence climbers throwing objects down at the police from the top of the grandstands. "At Mile High Stadium we were trying to keep the trouble under control. The kids were dropping bags of water on top of the cops from the grandstands. The cops began to use tear gas; it was a real mess."[7]

Officers were going down, while political obscenities were heard throughout the stadium. Chaos ensued and any sense of order diminished. No sooner than tear gas was launched into the crowds, it was thrown back at the officers. According to Hanley, Monck and Fey were "taking turns going up on stage trying to calm people down."[8] The chaos continued as performer Johnny Winter played on. Chip Monck tried to calm those near the stage who were in fear of their lives.[9] He could be heard pleading with the nervous crowds in a charmingly calm voice "to cover your faces with whatever piece of clothing you have."[10] Eventually the thick clouds of tear gas drifted in and settled on paying customers in the grandstands. It forced hundreds to scramble out of the venue crying and choking.

Fey increasingly became dissatisfied with how the Denver police handled the matter. Hanley shared his thoughts with the promoter on how to ease the situation. The sound engineer thought that by letting some of the outside people in for free, it could have eased the situation. Fey was reluctant, since a decision like this could cost him thousands. Eventually he relented under Hanley's encouragement, but it came a bit too late. Hanley recalls, "I told him, if you just let those kids in for free, it would make things a lot easier."[11]

The local government didn't like the embarrassment of having their asses kicked by a bunch of hippies, so police efforts were escalated. By the last day of the festival (Sunday), riot gear–ready platoons and police dogs surrounded the venue. A pepper fog machine that could generate thick clouds of tear gas was ready to be used if necessary.[12] Around 7:00 p.m., all hell broke loose and once again it was festivalgoer against officer.

The lineup for Sunday evening included Joe Cocker and Jimi Hendrix, promising an even larger turnout. Helping mix sound was Terry Hanley, who had turned twenty-three years old over the weekend. Adding to the crowd size were the curious who were there to witness the insanity. Fey remembers: "When it was Joe Cocker's time to go on, I found him hiding in a restroom in the north stands. He was terrified. 'So, this is America?' he asked."[13] Some claim that during Jimi Hendrix's set, thousands of gate-crashers filled the festival grounds to catch the performer. "Hendrix ditched the event halfway through . . . escaping the stadium in a panel truck."[14]

Amid the chaos, Hanley realized his role wasn't just as sound engineer but as a peacekeeper. He urgently pleaded with the police to back off without success, remembering "Talking to senior members of the police department asking them to stop and they wouldn't. They wanted to fight."[15]

After Fey was convinced to go outside the venue and distribute a few thousand free tickets, he was taken aback by the response: "We planned to hand them out to the gate crashers and that would take care of that. I went up to one guy and handed him a ticket. He brushed it away and said, 'I don't want any fuckin' ticket. We don't care about the show, we came to fight the police and cause trouble.'"[16]

According to Hanley: "I urged Barry to let the kids in and not have a fight over it. The cops wanted to have a fight, which was typical of police operations of that time. I was telling Barry to give tickets to the troublemakers to get them inside and out of the way. This way we could stop the ruckus."[17]

Even though Fey eventually declared Denver Pop as a free event and let the remaining agitators in, nothing seemed to work. That weekend Barry Fey only made around $50,000 in profit, an event the promoter considered one big mess. The turmoil that occurred over the weekend at Mile High Stadium cursed any chance of future festivals ever happening in Denver again.

The following month Hanley was hired to support full production in Atlanta for the First International Atlanta Pop Festival, a gathering with an impressive lineup that fortunately was rather peaceful. With the Denver event over, local and federal governments fought hard to put an end to other mass gatherings. Not all festivals ended like Denver Pop, but some did.

THE 1969 NEWPORT JAZZ FESTIVAL

Nineteen sixty-eight had been a turbulent year full of protests, race riots, rising death tolls in Vietnam, and the assassinations of Robert F. Kennedy and Martin Luther King Jr. The 1969 soundtrack of the maturing baby boom generation was quickly becoming rock and roll. For some but not all, folk and jazz music were now passé. As a result, the Newport Jazz Festivals, booming in previous years, were not as successful. According to NJF producer George Wein, "I was coming to the unfortunate realization that the youth of America were no longer interested in jazz."[18]

In an effort to broaden interest in his upcoming July 4–6, 1969, Jazz Festival, Wein made an interesting move. By arranging a lineup of jazz and rock groups he felt that it might attract a new (and diversified) audience to his faltering festival. Wein explains, "I was letting the Newport Jazz Festival capitulate to the pressures of a rock-and-roll generation."[19]

Festival turbulence and gate-crashing seemed to be on everybody's mind in 1969—especially for promoters. Up until now Wein had done an excellent job of keeping his festivals fairly orderly. By moving away from a traditional jazz format, Wein was hoping to connect with a new generation of festivalgoers. This decision ended up being a huge mistake for the producer. This was the second time Hanley's old boss experienced a violent backlash from audiences. It had been in 1960 that Wein was met by angry crowds at his NJF.

Hanley's last year at Newport was in 1967. In 1969 local engineer Bob Bennett was providing sound. The unique experiment—the alternation of rock and jazz groups at the festival—caused a lot of tension. An amalgamation of groups created the eclectic lineup: Jeff Beck, Miles Davis, Frank Zappa, James Brown, Sly and the Family Stone, B.B. King, Johnny Winter, John Mayall, Led Zeppelin, Ten Years After, Jethro Tull, Blood Sweat and Tears, Herbie Hancock, Buddy Rich, Dave Brubeck, Jimmy Smith, and George Benson.

The four-day festival brought 80,000 rock and jazz fans into Newport that July weekend. The location, Festival Field, was only suitable for around 20,000 and it quickly became un-accommodating to the crowds attracted by the fusion lineup. As fast as the tickets sold out, a mass of frustrated festivalgoers congregated outside the festival site.

Conscious of disastrous outcomes, Wein tried to avoid the violence that had occurred at the Denver Pop Festival less than two weeks earlier. According to Wein, the '69 Newport Jazz situation was "Ripe for a gate-crashing extravaganza," adding "Twenty four thousand festivalgoers, mostly rock oriented, squeezed into every available section of the festival site, more than ten thousand people gathered with no prospect of acquiring tickets. They could do nothing except hang around and listen to whatever strains of music escaped the amphitheater."[20]

Desperate, Wein approached the Newport public address system, and pleaded with the unruly audience to keep calm and remain orderly. Nothing seemed to work that weekend, as spikes of violence, fires, gate-crashing, and audience overcrowding ensued. This event demonstrated that experiments like this would jeopardize jazz festivals from ever happening in Newport again. The *New York Times* reported: "George Wein, producer of the festival expressed disappointment at the consequences of his attempt to add rock groups to the sixteen year old Jazz Festival."[21]

Hanley Sound crew member David Marks attended Newport that year. Marks, from South Africa, shares a unique perspective on the festival turbulence in the United States: "Maybe Americans don't want to admit this but festivals in America were usually violent events. There was always a fight or a shooting or some bullshit. Ironically in Africa in all of the years I have been doing sound there was never any fighting or shooting in all of the black audiences. At white festivals there were always guns and always knives."[22]

This same weekend, in Hampton, Georgia, southern festival promoter Alex Cooley was preparing for a festival of even greater magnitude, the First Atlanta International Pop Festival. Hanley had been hired, his contract signed. A week or so prior to the festival, the engineer was preparing his crew and equipment for a potentially massive crowd, blazing heat, and humidity.

ALEX COOLEY AND THE FIRST ATLANTA INTERNATIONAL POP FESTIVAL

Many Atlantans had attended both Miami Pop Festivals in 1968. Among them was renowned southern festival promoter Alex Cooley, who became inspired to work in the festival business after attending Miami Pop II. *New York Times* writer Ellen Sander observed, "The Miami Pop Festival was a monument to pop, an excellent model for future events of this kind."[23]

While on his way to Key Largo in 1968, Cooley and some friends heard a promotion for the Miami Pop Festival on a Florida radio station. After some soul searching, Cooley decided to go to the three-day event, leaving his friends behind. According to Cooley: "I had the most fun I've ever had. At the time I was looking around for something to do, trying to come up with a direction for my life."[24]

Miami Pop inspired Cooley to take on a career in music promotion and the moniker "Alex Cooley Presents" has stuck ever since. Ultimately Cooley managed to corral seventeen associates to support a "Miami Pop inspired festival" in his home territory of Atlanta. Cooley eventually hired Hanley for his Atlanta Pop Festival.

When Cooley began producing festivals he made Hanley his go-to solution: "I had heard of Bill purely by reputation, before I even did my first show. I hired him because he did it the best. But at that time he was about the only one that was doing it. If there were other people as capable and had the equipment he had, I didn't know them. Bill was reasonably priced and had a strong work ethic."[25] Hanley remembers: "I provided full production for his festivals. Alex was really great to work for, a really nice guy. I remember these festivals in Atlanta to be very hot, but going over fairly well."[26]

Cooley was finally able to bring the music of his generation to Atlanta. By introducing the music of the counterculture into this area of the South, the promoter truly felt he could dissolve the political and cultural divide Atlanta was going through at the time: "You have to understand the context. It was the height of the Vietnam War and Lester Maddox was Governor. I wanted to do something that made people where I lived understand that we could change."[27] Kathy Masterson, Cooley's twenty-two-year-old assistant, claims: "That generation of promoters was like the gunslingers of the wild west. But Alex was the best people person I had ever met. He was able to draw people in and he was a good salesmen who could talk to managers and other promoters."[28]

On a scorching hot Fourth of July weekend in 1969, over 150,000 people attended the First Atlanta International Pop Festival to see a stellar lineup of acts, including Joe Cocker, Led Zeppelin, Johnny Winter, Canned Heat, Grand Funk Railroad, Chuck Berry, Al Kooper, Chicago Transit Authority, Creedence Clearwater Revival, Dave Brubeck, Delaney and Bonnie and Friends, Pacific Gas and Electric, Paul Butterfield Blues Band, Johnny Rivers, Spirit, Sweetwater, Ten Wheel Drive, and the star attractions of Janis Joplin and Blood, Sweat and Tears. Referred to as Atlanta Pop, the event was held at the Atlanta International Raceway in Hampton. The *Atlanta Journal-Constitution* headline claimed, "Music Fans Stay Orderly Despite Heat Wine and Drugs,"[29] and it was true. By comparison, Atlanta Pop was nothing like the gate-crashing riots that occurred in Newport that same weekend.

Despite such a memorable roster of performers, and over 100,000 in attendance, Atlanta Pop received less press coverage than was anticipated. Some say that local networks didn't even know Atlanta Pop was happening until that Sunday, catching everyone by surprise. In the history of large-scale festivals, Woodstock, only a month later, seems to absorb most of the spotlight, overshadowing any significance Atlanta offered. Cooley, who worked hard to put the event on, found this disappointing: "The Atlanta Festival was before Woodstock and sometimes it's a little aggravating that people know that festival so well. They did a festival in the media capital of the world and that's why they got the spotlight. We never got the publicity that Woodstock got and ours was a great festival, but there were many other great festivals in this country at the time."[30] Cooley and Co. saw only $15,000 in festival profits that year. The unconventional promoter recalls feeling uncomfortable about making any profit at all as it clashed with his counterculture ideals. He claimed, "We felt like we needed to give it back."[31]

An eager seventeen-year-old high school student named Alun Vontillius was helping crews at Miami Pop II when he met Hanley. The sound engineer invited Vontillius to assist with stage construction and sound system deployment at the upcoming Atlanta Pop Festival that following year. Vontillius recalls: "Hanley was an incredibly inviting guy. He and other stagehands showed me the ropes and after that I was led into the industry."[32] Wanting an adventure, the affable Vontillius took Hanley up on his offer: "It was like joining the circus. In 1969 I drove from Miami to Atlanta with my hippie friends in my hippie van and arrived at the Atlanta Pop Festival gate with no credentials and no badges. I saw Alex Cooley at the box office and he told me where Hanley was. I remember it being so hot."[33]

The young crew member recalls the backbreaking work setting up Hanley's equipment in the southern heat with no cranes or lifts used to deploy the heavy equipment:

> Hanley brought in lots of scaffolding in order to create these huge towers, which took an incredible amount of manpower to erect. Once erected things were handed up platform by platform. You literally had to have a deck that was six feet off of the ground and a cabinet was passed up to that person, and then that person passed it up the next person who was six feet up to the next. After walking the stuff up, then we put the multi-celled horns on top. It was all hand over hand, it went up this way and it came down this way. After the festival was over we packed it the way it was set up. We packed the bass bins at the nose of the truck and they were stacked two high with the horns on top.[34]

Vontillius recalls the stage-to-sound communication throughout the weekend being very "primitive." With Hanley about fifty to sixty feet into the crowd and no radio or intercom communication from the stage, it was difficult. "Hanley devised a laundry clothesline so he could attach notes to it. Crews would literally clip stage plots to them and fly them back and forth to each other. For future festivals we used Terry Hanley's intercom system 'Terry-Com,'" recalls Vontillius.[35]

The first Atlanta Pop Festival went over fairly well, although there were some typical activities such as public nudity, illegally parked cars, and rumors of LSD-spiked water supplies.[36] According to the *Atlanta Journal-Constitution*, Atlanta Pop was devoid of any real issues, proclaiming: "The crowd was orderly, attentive, and responsive. No major outbreaks, no rioting, and no mass arrests."[37]

According to crew member Nick Burns, the LSD-spiked water report was true: "The electrical crew brought out these bright orange igloo water containers filled with water because it was so hot, and someone laced them all with acid. There were people down there who probably had nothing stronger than Southern Comfort or bourbon and now they are all tripping. They had to take away a few of them in ambulances."[38]

Regarding the quality of Hanley's sound system, according to journalist Paul Beeman, the blaring sounds were clear and audible, even for those who were in the rear, too far from the stage to see:[39] "The music, for those who appreciate modern pop music, was exceptional. The sound was good and the performers gave their best."[40] As spectators crowded into the Atlanta International Raceway to see and hear their favorite acts that weekend, the vibe was calm and "audiences were reasonably well behaved and quiet," a feeling that lasted until the festival ended.[41]

Notable was the boisterously loud Grand Funk Railroad, whose impactful sound Hanley faced many times in his career. It was in 1969 at the Atlanta Festivals when Grand Funk first gained acclaim for their thunderously loud performances. The group had not yet been signed to a record label. Mark Farner, the young singer and guitarist of the band, recalls his memorable festival debut:

> Atlanta Pop was the first time we had experienced anything to that level of sound. I said "holy shit man look at these speakers! Look at these racks of amps!" We were freaking out! When we started singing I was like "holy crap!" I didn't know what playing in front of 180,000 people meant, then I got twenty feet over their heads and out and said wow!!! It was an ocean of people man, and I had to piss like a racehorse. But the sound just boomed out there![42]

Cooley reflects: "After I got to know Bill at the first Atlanta Pop and saw what he was capable of, I hired him again for more of my festivals."[43] The relationship between the sound engineer and promoter was long lasting. Cooley hired Hanley Sound for many festivals, among which were his Texas International Pop Festival (1969), the Second Annual Atlanta International Pop Festival (1970), and the Mar y Sol Festival in Puerto Rico (1972).

Chapter 33

WOODSTOCK

THE CATSKILLS

Cradled in the Eastern Catskill Mountains of New York State, the bohemian village of Woodstock and surrounding areas had hosted numerous festival and artist gatherings for many years. It's certainly no coincidence that the Wood-stock Festival occurred in this location. The utopian ideals expressed in 1969 had been passed on here for generations.

Over the years the area saw many creative transformations. By 1903 a group of wealthy utopian Englishmen opened an arts and crafts colony named Byrd-cliffe to "pursue the ideals of the arts and craft movement."[1] In 1912, sculptors and painters from the New York Art Students League established a summer program in the area. In the 1920s, festivals offered "bacchanalian fetes, with eccentric celebrants wearing handmade costumes for all night revelry."[2]

Hervey White (1866–1944), one of the founders of Byrdcliffe, eventually created the Maverick Art Colony (MAC). White was a "writer, socialist, so-cial reformer, and above all idealist" who had a "singular vision of a utopian community where creative freedom could thrive in an environment of rustic simplicity."[3] Following his vision, White and a partner purchased a large farm just outside of Woodstock and formed the MAC.

Out of White's newly established community came the Maverick Festival (1915–31). This helped fund the upkeep of the MAC. The Maverick should be noted as one of the primary inspirations of the Woodstock Music Festival. According to the history of these annual summer events—which at one point had over 6,000 in attendance—the Maverick was a "bohemian carnival filled with communal spirit."[4]

During the 1920s the festival was held on the grounds of the colony in the afternoon and evenings of the August full moon. "The culmination of the eve-ning was a theatrical spectacle that began after dark with riotous performances by locals, followed by a costume ball that lasted to the morning."[5] Author and attendee Alf Evers, who was ninety-nine years old when interviewed by the *New York Times* in 2003, recalls: "There was a lot of drinking, mostly local cider, and men dressing up as women and women dressing up as men. One year I even saw a goat floating about with its horns painted blue."[6] The raucous behavior of the growing festival and the onset of the Great Depression negatively impacted the festival and colony. It never rebounded.

Although White passed away in 1944, his visionary spirit lives on. The Maverick Concerts Chamber Music Festival he founded in 1916 is still active to this day. It is considered one of the oldest summer chamber music series in the country. The legacy of the arts in the area also continues to be preserved and fostered by the Woodstock Artists Association, founded in 1919.

In the 1960s the music scene began to pick up in the town of Woodstock when the famed music manager Albert Grossman settled here. Many of the artists he managed followed. Performers like Bob Dylan and members of the Band fell in love with the bucolic artist community. According to Alf Evers's book *Woodstock: History of an American Town*, Woodstock in the early sixties was a home and playground for rock musicians: "Bob Dylan bought the big former Stoehr place in the East Riding area of Byrdcliffe. Joan Baez, Peter Paul and Mary and other folk singers became part of Woodstock life after 1962, singing at the Espresso, the former gift shop soda fountain known since the 1920s as The Nook."[7]

Between 1967 and 1970, several summer concerts took place at the Peter Pan Farm located near the town of Woodstock. Pan Copeland (Pansy Drake Copeland) a local who owned the property and operated a delicatessen, organized the concerts. These mini-festivals were referred to as "Sound-Outs." Held across the town line in Saugerties on Copeland's farm, the Sound-Outs were organized in a "relaxed non-commercial way."[8]

The 1967 Sound-Outs were well received. They drew audiences as large as 1,000 to their weekends of music, marijuana smoking, and a pleasant sense of togetherness within a natural setting. Many local and national acts performed for this bohemian community as they relaxed in their tents, sleeping bags, and parked campers. According to those who were in attendance, the gatherings included "simple sanitary facilities as well as a single food concession." These were constructed from "re-furbished chicken coops."[9] For just fifteen dollars, you got into the show for the entire weekend. The ticket price included parking, water, and firewood.

Over the years different producers partnered with Copeland, and by 1969 a musician from Seneca Falls, New York, named Cyril Caster was tapped to head the festival production team. At that time the Sound-Outs were renamed the Woodstock-Saugerties Sound Festival, or simply the Woodstock Festival. According to a 1969 *New York Times* article, this peaceful mini-festival had grown to 2,000 in attendance with full production: "Many hippies, most of whom came from NYC, earned ... the price of admission each day by performing such tasks as building a stage, fixing the lighting and sound equipment, clearing the field and organizing traffic."[10] That season the production team planned eight concerts, signing headliners like Van Morrison, Paul Butterfield, Cyril and his band, Tim Hardin, Chrysalis, and Children of God. But due to inclement weather very few of the concerts were staged that year.[11]

Some Hanley Sound employees and even Hanley himself recall supporting the Sound-Out concerts prior to the larger Woodstock Festival. According to Cyril Caster: "There were various situations with as many as twenty-one concerts per summer. I had two Shure vocal masters linked together at some shows. I'm sure Bill did come up with something better than that at some point."[12]

Inspired by the goings-on at the Peter Pan Farm, producer Michael Lang claims he modeled the Woodstock Festival after the Sound-Outs: "They had a joyous, healing feeling to them—a result of that bucolic setting—with little kids running around, people sharing joints and lazing around on blankets as the sun set. Three or four artists would perform on a makeshift stage, about 6-inches off the ground, in what had been a cornfield."[13]

The Sound-Outs made a lasting imprint on Lang. He felt strongly that this hippie revelry could be facilitated on a larger and more elaborate scale. Apparently this is one reason Lang and his partners called their event the Woodstock Music and Arts Fair.

Not far from the Copeland farm, Michael Lang's dream came to fruition after intense organization. On August 15, 1969, around 5:00 p.m. Woodstock Festival opener Richie Havens performed nervously from the monumental Woodstock stage to a densely populated audience. Hanley sat just seventy-five feet stage left out in an ocean of people. Elevated above the audience, as if it were moored to its makeshift scaffolding, the engineer's modest mixing (banquet) table seemed to float like a raft atop a vast ocean of hippie heads.

As the intensity of Havens's performance peaked, it set the stage for the rest of the weekend, igniting the positive vibes associated with the event. While the folk singer wailed his anthemic counterculture assertions over the rolling landscape, Hanley sat like a captain at the helm, attempting to control the sound waves from escaping Yasgur's expansive alfalfa fields.

While Havens stomped in time to the rhythm of his beat, Hanley twisted and turned the knobs of his mixers to lock in the right fidelity. Screaming with melodic conviction, "Freedom, Freedom, Freedom, Freedom!!" the Woodstock storm had begun as Havens and Hanley continued to find their bearings. The rest is history . . .

WHAT WAS WOODSTOCK?

The Woodstock Music and Art Fair of August 15–17, 1969, was and will always be the epochal event of the 1960s counterculture generation. However, it was more than that. It has transcended its time and place, and by now its effect on culture, politics, and society can be felt on a global scale. Time has revealed the events that occurred on that mid-August weekend are some of the most significant in the history of popular music.

The music emanating from one of the largest sound reinforcement systems of that decade blossomed in the pastoral rolling fields of Bethel, New York, in Sullivan County. The energy coming out of Hanley's speakers touched the hearts and minds of over 400,000 people scattered throughout the 600-acre Catskill dairy farm.

The Woodstock Music Festival was to be held in Woodstock, NY, but the owner of that location informed Lang he had changed his mind. Exploring parcel after parcel of land thereafter, Lang and partners (with a blank check in hand) found a suitable site in Wallkill, NY. Subsequently the Wallkill site failed to materialize and Lang's search continued.

Lang, "The baby faced producer with a Shirley temple hairdo,"[14] eventually met dairy farmer Max Yasgur and agreed to lease some of his 600-acre farm. Once the deal was done, it set the whole production into fast forward, with crews working around the clock up until opening day.

Artist Arnold Skolnik designed the event's poster displaying a white catbird perched on the neck of a guitar. Next to it is "Woodstock Music and Art Fair presents An Aquarian Exposition: 3 Days of Peace & Music in White Lake, NY." The well-known advertisement boasted thirty-two varied acts that eventually performed in front of one of the largest audiences ever to gather at a musical event in the United States.

Woodstock's lineup included artists like Richie Havens, Sly and the Family Stone, Canned Heat, Ten Years After, Jefferson Airplane, Janis Joplin, Joe Cocker, the Band, the Grateful Dead, Crosby Stills Nash and Young, the Who, and others. Lasting nearly four days, the festival ended on August 18 with one of the most memorable performances ever of "The Star Spangled Banner" by Jimi Hendrix.

"Hippies Mired in Mud" and "Traffic Uptight at Hippiefest" were the dubious headlines that characterized coverage of the Woodstock Festival by national media outlets. This made parents and government officials extremely uneasy. Eventually authorities called the producers threatening to shut the event down during the height of the festival. All the same, people were actually getting along, and for this, Woodstock will be remembered as peaceful.

For those few days Max Yasgur's farmland represented a city in motion with only two deaths, a birth (or two), several overdoses, and many others guilty of having too much fun amid the rain and mud. Aside from an obvious opportunity to see a few of your favorite bands, this mass gathering was a direct and blatant response to tumultuous times. War, assassinations, civil rights issues—people were fed up. The Woodstock audience broadcast a message to the world that a community of a half a million people could live in social harmony.

Woodstock was intended to be a money-making venture. The four partners forming Woodstock Ventures Inc., wanted to fund the creation of a recording studio from the profits of the festival. But the event famously lost a huge sum of money, about $1.6 million. The last-minute move from Wallkill was not only costly but left minimal time to facilitate many of the important production details earmarked for the festival. Insufficient at best, when crowds swelled, the fences came down, and with not enough ticket gates in place, Woodstock became free.

Partner John Roberts eventually had to take out loans against his personal wealth to pay the accumulated debt from Woodstock. According to *Rolling Stone*, Roberts's wealth is from a "family fortune built on Polident and other products of the Block Drug Co. It is what is keeping everyone out of debtor's prison and in smiles."[15]

John Morris, managing director of the fair's production, says Roberts has taken loans against his own personal fortune to the tune of $1.3 million to pay debts incurred. These included unexpected expenses for power, emergency food and medical supplies, helicopters, limousines, telephones and moving everything from the original site in Wallkill. (The cost of the performers is estimated to be close to $300,000, with each act collecting from $10,000 to $15,000.) To pay off the debts, Woodstock Ventures is counting on the profits of a film to be released by Warner Bros., hopefully at Christmas. Approximately twenty five percent of the film's net profits are earmarked for the company.[16]

In an effort to recoup losses, festival producers Michael Lang, Artie Kornfeld, Joel Rosenman, and John Roberts ultimately caved in and agreed to a one-million-dollar flat fee in exchange for the rights to the film, plus a small percentage of the movies profit. Warner Brothers had already negotiated distribution rights to Michael Wadleigh's documentary film that he shot for around $100,000. By 2009 *Woodstock* had brought in over fifty million dollars worldwide through box office earnings and even more through rentals.[17] Although a major loss to those who invested time and money, the legacy of Woodstock now exists as a time capsule in the form of an Academy Award–winning film and stellar soundtrack. This legacy also lives on within the countless stories told by attendees, each and every one having a seemingly unique Woodstock experience.

Even though Woodstock was marred with production failures, the music never stopped on that weekend. Yes, there was an occasional lull in music or a technical mishap; however, the beauty of Woodstock lies within these imperfections. In between the music there were numerous stage announcements meant to serve the large village that gathered in and around the area. Information conveyed through Hanley's PA allowed thousands to better communicate. Ongoing messages from the stage helped folks meet a friend in a specified area, find something to eat, or know when an impending rainstorm was rolling in. Hanley's sound system was truly instrumental in keeping this large mass of people feeling as if they were not alone.

A *New York Times* journalist wrote, "As the announcer on stage praised the crowd, and as the bands registered their excitement at playing such a gathering, the crowd felt an increasing sense of good in itself."[18] Yet with all the tension and confusion between the outside world and the world contained on that farm, it all seemed to come together in the end. With the help of Hanley's reliable sound system, which continued to work efficiently throughout the weekend, Woodstock was a success.

In "A Joyful Confirmation That Good Things Can Happen Here," *New York Times* journalist Patrick Lydon wrote: "Out of the mud came dancers, out of the electrical failure came music, out of hunger came generosity. What began as a symbolic protest against American society ended as a joyful confirmation that good things can happen here, that Army men can raise a "V" sign, that country people can welcome city hippies."[19]

Later, in September 1969, Hanley found himself in a Lewisville, Texas, hotel room reading an article about what had occurred in Bethel. The sound engineer was there in Texas to provide sound reinforcement for the Texas International Pop Festival. While scanning the paper he noticed his name mentioned, although it had nothing to do with sound. According to Hanley, what he read was the only negative thing he ever heard mentioned about Woodstock, and it involved his truck:

> At Woodstock we left the truck constantly running for loading and unloading equipment. One truck may have been on its way out to another gig? So we had the lift gate down and there was a car trying to get around us. So he couldn't wait and got hung up on the post attached to the lift gate on the tractor-trailer. We had to pick up the car and move it over. There was a big skirmish over this and it ended up in the paper! It was funny to read that this was listed as one of the problems occurring at Woodstock![20]

At the end of December 1969 Hanley was slated to provide sound for yet another Miami Pop Festival in Hallendale, Florida. Word was out that a Woodstock-sized occurrence was being planned. As a result, the permits for this festival were quickly halted. *Rolling Stone* reported: "If officials in New York were relatively unconcerned by what went down at Bethel, the City Council of Hallandale, Florida, wasn't. The Tuesday after Woodstock, it revoked the license it had issued the month before for this year's Miami Pop Festival, scheduled for December 27–29."[21]

Chapter 34

BY THE TIME HANLEY GETS TO WOODSTOCK

The festivals leading up to Woodstock varied in size and production, but cannot compete with the magical happening of peace and love on that August weekend in 1969. Out of all of the festivals Hanley supported, Woodstock is what he is most known for. It was, for all intents and purposes, a true test of his expertise as a sound engineer. Woodstock demonstrated to the world that Hanley's acoustical philosophies and technical application in sound were unmatched. For the festival, the thirty-two-year-old sound engineer not only provided sound reinforcement but also facilitated its recording. Additionally, he had a hand in some of its production design, like the Woodstock security wall. The techniques and innovations Hanley learned throughout his career proved to be critical factors in how the historic event evolved, keeping music as the central focal point.

Very little information about Woodstock sound reinforcement exists. How was the equipment prepared and transported? How was it constructed and deployed? What type of equipment was used? How does one go about projecting sound for over a half a million people? These are important questions, since during this time most of the sound companies available were small, unsuitable, and insufficient. There were a few on the West Coast that probably could have pulled it off, but overall most were not capable of handling such a complex job. Sound reinforcement was still at a primitive state of development.

WHO FOUND THE WOODSTOCK SITE?

In Woodstock lore there are many scenarios as to who discovered the festival site first. Almost fifty years have gone by and memories are fuzzy. According to Michael Lang's assistant Ticia Bernuth-Agri, everyone that went to Yasgur's bowl for the first time thinks that they were the one to discover the location: "After Michael and I found the site and told everyone, a lot of people went up there to look at Yasgur's land."[1]

Hanley's influence is unknown on Lang's final decision to settle Woodstock at Yasgur's farm. The sound engineer's review of the natural bowl did occur sometime between the Wallkill fallout and Lang's negotiations with Yasgur, most likely Sunday morning July 20. Whomever it was that first laid claim to the magic spot is irrelevant. What is most important is that Max Yasgur approved it. If not for this honest, friendly, and sympathetic farmer, there would have been no festival at all.

According to Hanley, upon arriving at the bowl it felt right from the get-go: "When I walked the field I realized that it was ideal. Even though no one mentioned to me that this area was previously chosen, I innately felt that it was ideal for a large festival site."[2] What follows describes the events leading up to the discovery of Yasgur's farm.

MID-JULY 1969

Hanley and some of his crew arrived at the 300-acre festival site, the Mills property (Howard Mills Industrial Park) in Wallkill, New York, in mid-July 1969. They were there to begin preparations. On Tuesday, July 15, no sooner than they got there, he and his crew were informed that the local Wallkill zoning board of appeals had banned the event. Hanley recalls: "I never got to design the Wallkill site after I looked at it. The trees were cut down and the field was prepped. Then it was canceled."[3] Fortunately, finding the Yasgur site happened fairly quickly thereafter, and the festival ultimately took place in Bethel, New York.

Agri recalls what happened the day after they lost Wallkill. White Lake resident Elliot Tiber called her claiming he had permits and available land to use. Tiber heard the town of Wallkill had placed an injunction against the festival and wanted to help. According to Agri: "Elliot said 'Hey our town wants your festival! I got the permits and I got the land! Come to White Lake!'"[4]

Agri called Lang at his lawyer Miles Lourie's office in New York to tell him the good news. When Lang answered the phone he had been in a meeting with his Woodstock partners Rosenman, Kornfeld, and Roberts. Agri remembers, "Michael said he would be right over to pick me up."[5] From Manhattan, Agri and Lang drove his white Porsche 911 upstate to White Lake, a hamlet in Bethel. Agri arranged to meet Tiber at the El Monaco, a dilapidated eighty-room motel that his parents owned. According to Lang in *The Road to Woodstock*, it took about "ninety minutes" to reach Bethel in his rental car: "Ticia and I zipped up the New York State Thruway to Route 17, followed it to Route 17B and Country Rd. 52."[6]

Before they left Manhattan, Lang called Lawrence and Goldstein, who were at the Wallkill site thirty-five minutes away from the El Monaco. The festival producer explained the situation and asked to meet them at the motel. Law-rence explains: "It had become known in the area that we were no longer going to be at Wallkill. I recall people calling us saying 'I have this field, and I have that field.' Others on staff were in helicopters looking for sites. John and Joel were driving around looking for sites. We were frantic!"[7]

When they arrived on the motel grounds (about fifteen acres) they quickly found out that what Tiber was offering was nothing more than a swamp located behind the business. Lang observes, "We started descending straight down into a large swamp filled with nubby growth and amputated saplings."[8] Lawrence and Goldstein had not arrived yet, so instead of waiting around, Agri and Lang took a drive to survey the beautiful countryside. As they cruised along the winding farmland roads, Lang reflected on the memories he had as a boy while on family vacations in the area.

Along 17B Agri glimpsed a street sign that said HAPPY AVENUE, "I said, Michael we got to turn around! Just because of that name! I figured it was an omen! I mean we just left WALL-KILL! So we went up the road and then eventually turned onto Hurd Rd. and there was Yasgur's farm! That's when we said 'Oh my God that's a perfect site!!'"[9] Lang remembers: "We took a right turn off 17B onto Hurd Rd. About a quarter mile up, we broached the top of the hill and there it was. It was the field of my dreams—what I had hoped for from the first."[10] Agri remembers: "I felt like it was meant to be. Like I was meant to be there at that very instant to tell Michael to turn up that road!"[11]

Lang and Agri were elated by their discovery and couldn't wait to get back to the motel to tell everyone. Lawrence and Goldstein had arrived and were waiting. The two had just seen Tiber's swamp and didn't look thrilled. The group took another glance at the marshy landscape while Agri waited back at the motel office. Lawrence recalls being upset when he saw the less-than-suitable area: "After Tiber called, Stan and I drove up there. When we arrived we met Elliot and he showed us this swamp out in the back of his motel. We got really pissed off."[12]

The group congregated at a table located off of the motel office to discuss things. By then Goldstein had left. Lang asked Tiber if he knew of anyone that could connect him to the owner of the land that he and Agri had stumbled upon earlier. Embarrassed and happy to help, Tiber said he knew someone in real estate and connected the festival producer with a local realtor named Morris Abraham. Lang recalls Abraham as "sleazy looking."

Abraham arrived in his Buick. Leaving Tiber and Agri behind, Lang, Law-rence, and the realtor sped off in the Buick. Lang was anxious to show Abraham the natural bowl he had found earlier, plus a few other possible locations. Law-rence recalls: "Elliot introduced us to this real estate guy who took us around to various sites. He said he knew a guy who has a dairy farm and that's how it happened."[13]

After surveying the desired spot in the mist and fog, Abraham connected Lang with dairy farmer Max Yasgur, the owner of the property. At the time Yasgur owned several farms and over 2,000 acres. His dairy farm was one of the largest in Sullivan County.

Yasgur was finally reached by telephone and invited the party to his house, only a few minutes away. While at his home the farmer and Lang discussed the logistics of the event and the land needed to support the festival. Yasgur was sympathetic as he heard about the Wallkill fallout and felt bad. After the discussion at Yasgur's home, Abraham, Lang, Lawrence, and Yasgur left for another look at the proposed area. When the group arrived, Lang and Yasgur walked into the depth of the field for a more intimate talk. Agri remembers: "It was amazing it all happened in one day. Michael and Max shook hands in the field after that."[14] According to Lawrence: "Michael and I drove out there with Max to the site. We arrived at the top of the hill and looked down and there was a lake, a bowl, and some flat areas. We looked at each other and said, 'This is it!' Michael did the deal right there."[15]

Along with the natural bowl that offered the perfect concert location, Lang explained to Yasgur that he needed more land surrounding the area, approximately "600 acres with additional acreage for parking and camping."[16] As both men stood at the crest of the hill, the dairy farmer figured out his price and the deal was struck. Yasgur would lease his land to Woodstock Ventures for $50,000 plus another $75,000 to be held in escrow to cover any damages.[17]

Lang notified his staff immediately and by late that afternoon trucks were being loaded at the now former Wallkill site. The following day, Wednesday, July 16, Lang, Roberts, and a few others, including lawyers, met with Yasgur and his son to firm up further details. Yasgur had a set of standards for what the festival producers were and were not allowed to do on his land. Lang recalls, "We signed papers at 10 p.m. that night."[18] According to Roberts: "Max and I shared a carton of chocolate milk while the attorneys retired to the den to draw up papers. At 10:30 p.m. the deal was closed. Hands were shaken all around and the Woodstock contingent stumbled out into the night."[19]

By Friday, July 18, word was out that a major festival was coming to Bethel. Permits were still needed, but according to Lang, "By Saturday evening (July 19th) everything was moving its way to White Lake."[20]

With very little prep time available, literally less than four weeks, this change in plans sent the Woodstock Ventures crew scrambling—including Hanley. Once the official word was out, most everyone migrated to the El Monaco to set up shop. Lang rented Tiber's parents' entire motel. The run-down business was also used for offices and a ticket outlet. According to Lang, this saved the motel from going into foreclosure: "We established festival headquarters in the old New York Telephone Building in neighboring Kauneonga Lake."[21] Woodstock

Ventures rented another run-down Borsch Belt motel called the Diamond Horseshoe. It housed even more workers and was closer to Yasgur's farm.

After Wallkill was canceled, Hanley waited patiently to find out what his next move should be. On Saturday afternoon July 19, Lang informed the sound engineer that he should evaluate the new site. Hanley then made a call from the El Monaco back to Medford to inform his staff.

On Sunday, July 20, Hanley was called in to have his first meeting with Yasgur and Lang at the farmer's home. The men greeted each other, shook hands, and without hesitation jumped into Lang's stretch limousine. Speeding out of Yasgur's long dirt driveway, they took a left turn onto 17B, another left turn, then a final right turn that led them into the thickness of Yasgur's alfalfa fields. Out the right-hand side of the dusty limousine window, the sound engineer noticed that the farmer had smaller fields mostly covered in trees. Eventually the vehicle came to the intersection of Hurd and West Shore Road(s) at a large, treeless field. Hanley recalls asking to stop the car. Stepping out, he gazed in awe over the soon to be hallowed Woodstock Festival ground. Hanley recalls being excited: "I said stop the car! That's it! I knew it when I saw it! It was the perfect spot for a festival!"[22]

Hanley saw that the topography of Yasgur's farmland formed a natural bowl. It sloped down to Filippini Pond located on the north side of the field. Once Hanley examined the area he knew immediately that its amphitheater shape was suitable for sound. It was then that the engineer envisioned a layout and construct of the stage and speaker towers at the bottom of the hill. They would be positioned on a slight rise with the picturesque pond behind them that was eventually used for skinny dipping. According to Hanley the area had similar characteristics to Festival Field at Newport: "I walked the site, designed it, and laid it all out in my head. The field's natural sloping wave like tiers went down and up and down again. It was the best that we were going to find at such a late date."[23]

LATE JULY 1969

By July 21 and 22 the site layout was finished, with trucks arriving throughout the two days.[24] With less than thirty days to get the festival up and running, most if not all on the Woodstock staff felt overwhelmed, including Hanley. Leading up to the event the sound engineer continued to examine the chosen site. Intermittently traveling back to his Medford office, Hanley gave his team and office manager Judi Bernstein updates and finalized preparations.

Hanley Sound crew member Brian Rosen recalls the buzz in the office after Hanley filled them in on the festival: "We had been doing rock festivals all along and they had been getting bigger. Woodstock looked great! However, predictions didn't seem as big as previous festivals but everyone wanted to be part of it because of the acts that were lined up."[25]

Hanley Sound crew member Bill Robar recalls: "Nobody thought it was going to be big. It was just another gig. We did lots of big concerts because we were the only ones on the East Coast doing sound for big shows. Hanley was the guy and we were doing all the festivals."[26]

Dick Butler handled a good chunk of the driving responsibilities for Hanley Sound. Butler recalls a "murmur" floating about the office regarding the event: "I remember thinking, 'What is this going to be?' This could be something, and it was. But we had heard that before with other things we had done."[27] According to sound engineer Sam Boroda, "We had no concept on how big Woodstock would be."[28]

Chapter 35

THE WOODSTOCK CREW

Hanley's Woodstock sound crew consisted of about twelve employees, all at various levels of experience and skill. Boroda recalls: "I remember when all of the hoopla was going around leading up to Woodstock. Hanley had a lot of guys that he picked up from other festivals. Some were good in carpentry, others in electronics, and so on."[1] Even so, Hanley had hardly enough manpower to get by. Those employees who were not at Woodstock were working other gigs and this required additional staff to be brought in at the last minute. According to Harold Cohen: "We weren't heavily staffed with people at Woodstock. It's not like you see today. I laugh at what you see during concerts and festivals today with regard to the number of people that are involved."[2]

Both Hanley and Bernstein were onsite at Woodstock. Hanley delegated Boroda to oversee most of the Woodstock setup and facilitation. He began mixing sound on Saturday night. According to Cohen, Boroda did most of "the mixing and the original sound checks at Woodstock."[3] Boroda recalls having no intention of joining the crew at Woodstock until the very last minute: "I wasn't even supposed to go. I wasn't on the crew at all until Judi said to go ahead, we need you on the sidelines!"[4]

Tractor-trailer drivers John Brodie and Dick Butler hauled most of the equipment to the festival site. Brodie was known to be extraordinarily vocal and extremely boisterous. Boroda recalls: "Every word that came out of John Brodie's mouth was the F-word! F-this and F-that! He was a teacher in Medford and I think it was because he could not talk that way in school so he let it out at Hanley Sound! He pissed and moaned from the moment he came in, to the moment he left but everyone loved him!"[5] Aside from his driving duties, Brodie acted as an occasional stagehand. At Woodstock he captured many photos, but according to Hanley they were lost.

Shop foreman Nick Burns, engineer Harold Cohen, sound engineer Brian Rosen, technicians Scott Holden, Ken Sommers, David Marks, and occasional helper Bill Robar were also on hand. Engineers Lee Osborne and Eddie Kramer (not a Hanley employee) managed the sound recording for Woodstock. Hanley colleague Tom Field (Tom Field Associates) supplied lighting equipment and other personnel.

Marks had come to the United States on holiday from South Africa in 1969. That June he arrived in Nashville, Tennessee. A year earlier "Master Jack," a song that Marks wrote, famously sung by the South African folk rock group, Four Jacks and a Jill, hit the American charts. According to Marks, "The song did very well, it peaked the American charts and came in at eighteen on *Billboard* Hot 100."[6]

Not finding what he needed in the songwriting capital, Marks hitchhiked from the South to the seaport town of Newport, Rhode Island, looking for work. It just so happened to be the weekend of July 6 when the Newport Folk Festival was occurring. While in attendance, Marks began to chat with Dick "Schlitz" Kopanski, a friend of Hanley's. Marks explains: "Dick gave me Bill's number at Hanley Sound in Medford. I reached them from a call box and Judi answered; she called me back and we spoke for a long time. I think she liked my accent. Years later I realized that she was interviewing me!"[7]

After arriving at Newport, Marks found his way to Greenwich Village. This is where he settled while in the United States. In New York he occasionally worked for Hanley Sound alongside Harold Cohen. When reporting to Medford over the phone, he could sense that Bernstein was impressed with his work ethic: "She sounded pleased. She told me to report to a show at Madison Square Garden in late July."[8] Proving himself during each gig, his responsibilities eventually grew. Yet, he never imagined that he would end up at Woodstock.

Marks is a significant character within the lengthy and multilayered family tree of Hanley Sound employees. Marks recalls: "I was just a young sound roadie doing his "thang" and enjoying this wonderful newly found and expressive freedom in a pale blue safari suit top. I was hearing and seeing the USA for free."[9] Eventually, after being with Hanley Sound for a little less than a year, Marks moved back to his home country.

Billy Pratt and Nelson Niskanen were Hanley Sound crew members assigned to the New York metropolitan area. Both became part of the Woodstock crew. According to Pratt: "When Hanley did Woodstock he was looking for help. I asked Nelson to come along because we needed warm bodies."[10] Niskanen was not with the company long, joining Hanley Sound in June 1969. He had been out of college and was in need of a job when Pratt introduced him to the sound company. Niskanen recalls: "Billy Pratt asked me if I needed work. He was at the Fillmore East and also worked for Hanley. When I learned of what he was doing I said sure! I ended up working for him during the summers of '69 and '70."[11]

The two arrived at the site in Pratt's Volkswagen bus a day or two early. As soon as they parked behind the stage area they began to assist with the setup.

Pratt helped with various duties, acting more or less as a go-between from the sound tower to front of the stage. Nelson and Pratt hoisted equipment up by ropes from the ground to the stage, "almost two stories," as Pratt recalls.

Toward the rear of the stage near the road, a makeshift freight elevator was installed so equipment could be loaded expediently. This device, which many on Hanley's crew refer to as an "equipment lift," was very helpful. The setup used an outer frame made of scaffolding and a pulley system. At the top, one end of a cable ran across a series of pulleys connected to a heavy-duty winch. The bottom end attached to the platform that was raised up and down. An elevator operator was on duty around the clock lifting crew and equipment to the Woodstock stage.

By 1969 Bill's brother Terry Hanley was working independently and no longer part of Hanley Sound. Nevertheless, Hanley had asked him to help with an upcoming Blind Faith tour that coincided with Woodstock. Terry recalls the rock and roll music business exploding in 1969, claiming, "That year Hanley Sound did several other major US tours and festivals."[12] A big fan of the supergroup, Terry obliged. "I really was a huge Cream fan, and Blind Faith came together for a tour. I was a fan of Clapton, Winwood, and Baker. That tour came onto the agenda before Woodstock. It was a really big deal! However, I did drop some equipment off at the Woodstock site, but left after that."[13] On August 16, after wrapping up a Blind Faith date at the Earl Warren Showgrounds Arena in Santa Barbara, California, Terry caught word of the conditions crew members were dealing with 3,000 miles away in Bethel.

David Freese, short-term business manager to Hanley and Bernstein, did not attend the Woodstock Festival. Hanley had assigned Freese to wait in New York with a monitor system for the post-Woodstock Jefferson Airplane performance on Dick Cavett's ABC television show. Additionally, Hanley Sound technician Rick Slattery was there assisting with the performance setup. Cohen recalls that Freese was a great person but didn't last at Hanley Sound very long. Cohen claims that some of his business sense was not very practical: "David bought an expensive calculator much to Bill's dismay. This was a sore point with Bill. The footprint of this thing was around the size of a cable TV box and it had big buttons on it. David bought it on credit, and Bill was pissed because he was paying for it well after he left. Every time the subject of payment came up, you could hear Bill grumbling!"[14]

Tufts engineering student Rick Slattery had joined Hanley Sound in early 1969. His old college roommate, who lived in an apartment behind the sound company, noticed Hanley's young crew members packing trucks, and one day he got curious. Having an interest in music himself, Slattery's friend inquired about the company. Slattery recalls:

My friend called me after he had met Bill, who was preparing for a Chambers Brothers tour. Hanley needed wheels put on these road cases. So my friend Nick called me because he knew I was a good carpenter and referred me to Bill. I needed the money and it sounded like a great job! In the beginning I was working at Hanley Sound at night, building various things. I hadn't met Bill yet since he was on the road a lot. It was just Judi and David Freese at the office. As time went on I just kept doing bigger and bigger jobs, and I eventually met Bill.[15]

For the event, Woodstock Ventures recruited volunteers, as did Hanley. Deploying a sound system for a festival of this caliber was an epic undertaking, requiring extensive manual labor. Like at most festivals, early on, the sound engineer grabbed volunteers from the crowds to help set up his heavy equipment. Sound reinforcement was relatively new and exciting, and this seemed to allure the young and interested festivalgoers. Woodstock technical director Chris Langhart recalls having to enlist help from the crowds:

Every morning I would look out of the window of the trailer. There was a whole bunch of people who wanted to work. I just looked at them in the face and you decided which ones looked like they had seen some work somewhere. I asked a few interview questions and you took your crew from yesterday and we went off and did plumbing or wiring into the forest so the Christmas tree lights could be out there. So Bill never really had to hunt for people because he was doing the grooviest thing out there and people came in droves. They all wanted to do it, and why not? If you were looking to have a good time in sound, he was it. He had the clients, it was interesting, and he was into what made it go right.[16]

Chapter 36

GOING UP THE COUNTRY

By August 1, 1969, Hanley's preparations for the festival were in full swing. While tractor-trailers were being loaded, speakers and other essential equipment were being built. Hanley's custom-designed Hanley Sound Inc. (HSI) 410 lower-level and 810 upper-level speaker cabinets measured 7 ft. x 3 ft. x 4 ft., each roughly weighing in at a whopping half-ton. The front panes of glass of his shop windows had to be removed when the speakers left Medford. These speakers, known as the "Woodstock Bins," are now as iconic as the festival itself; they were based on Hanley's specifications. According to Boroda, "We were building the speaker cabinets at Hanley Sound, one by one."[1]

Around a week or so before the festival, Hanley traveled from Medford to the site via a four-seat commuter plane. He chartered the aircraft from his pilot cousin Mike Goulian, who owned a small shuttle service. This proved to be an amazing convenience and time-saving necessity. Having already been at the site, the sound engineer was curious about how things were coming along when viewed from the sky. While in Medford, Hanley somehow convinced Judi to accompany him for the aerial overview. The office manager was afraid of heights, and reluctant. Nevertheless, she agreed and departed for Bethel with Hanley from Hanscom Field, a public airport in Bedford, Massachusetts.

Once over the Woodstock site, Bernstein looked down onto Yasgur's farm in despair. It was obvious to the office manager that the last-minute obstacles of securing this location left little time for the Woodstock staff to efficiently set up. She remembers: "I was so afraid of heights. When I looked down I said to Bill, either no one will be able to get in or out, or there is going to be no festival at all. I looked down again and realized it had been raining, and the stage was nowhere near finished."[2] After the survey, Hanley's cousin flew them to the local Monticello Airport. On the runway, Hanley yelled over the humming propellers, emphatically detailing what arrangements Bernstein should make. The plane took off with Judi, leaving the sound engineer behind.

What Hanley told her, and what she saw, remained fresh in her mind. The anxious office manager arrived in Medford with detailed notes on additional

Woodstock preparations. Once at the office, she called for a meeting with the Hanley Sound crew. They needed to organize and prioritize what was to get done in the coming week.

Hanley's orders included the rental of a Dodge camper that would act as a home base at the festival. Bernstein was to load the motor home with essential provisions. Shop foreman Burns was in charge of delivering the vehicle to the site. Hanley also requested that Bernstein fly to Philadelphia and meet friend and colleague Bob Siegel, co-founder of the Philadelphia Folk Festival. She was to accompany Siegel to Woodstock, being sure to arrive on time. Bernstein sensed that Hanley somehow knew Woodstock was going to be bigger than expected.

THE WEEKEND OF AUGUST 8–10

On the weekend of August 8, 1969, Hanley was getting things in order at the festival site. His pilot cousin had flown back to Massachusetts only days earlier. By now, almost everyone was in panic mode. Getting equipment in and out and communicating with Medford about additional resources needed was becoming difficult. *New York Post* writer Alfred Aronowitz was reporting from one of the Woodstock office trailers and recalls Hanley's mood: "Bill Hanley was shouting at the top of his lungs demanding his own private airplane. Hanley argued, 'If you don't get me a plane, I'm going to put it on my invoice anyway for lost time.'"[3]

As Hanley stood outside the entry door of his trailer-office located behind the stage across West Shore Road, he watched the madness of the festival's ongoing production. Through the thick late summer haze he saw his fifty-foot speaker towers being erected beyond the massive scaffolding structures supporting the Woodstock stage deck. The sound engineer's vision for the sloping alfalfa field was coming to fruition. Aronowitz wrote: "Inside the trailer, Bill Hanley kept arguing for a private plane. Only a couple of years ago he used to wear suits and ties. Now he had long hair and a goatee and he was one of the country's leading specialists in the public amplification of rock and roll performances."[4]

The festival location, which no one could have ever imagined, sat within the heart of the Catskills, in the shadow of the Borscht Belt, where many Jewish New Yorkers vacationed for years. Hundreds of people were getting production under way over the sprawling acres of Yasgur's farmland. New paths in the forest called "Groovy Way" and "Gentle Path" were carved out. Other roads were built so crews could effectively remove trash and waste from the Port-O-Sans. Performer and audience accommodations were all being constructed, including toilets, wells, plumbing, staging, scaffolding, lighting, fencing, camping, parking, and other essential infrastructure. For sufficient power, festival leaders had the electric company bring in 220-volt lines and cable to support the lights and stage. According to Lang, "The telephone and power guys got behind us and made an all-out effort to get the job done in time for the show."[5]

By Saturday August 9, Bernstein had rented the Winnebago-style camper from a rental facility in Rhode Island. Crew members Burns and Cohen began to stock and load the vehicle with food, pillows, blankets, and equipment. The mobile home, which slept six, served as shelter for various members of the crew. Burns remembers: "I loaded it with packages of Oscar Mayer hot dogs, cans of tuna fish, Franco-American spaghetti and coffee . . . lots of coffee! I figured these were the easiest things to cook. I also packed soap, clothes, and shaving stuff . . . so these kids are out in the back pond and I am taking a nice warm shower!"[6] Burns claims additional recording equipment was stored in the mobile home last minute: "We loaded the camper with two Scully and two Revox decks. Somebody rolled in cases of ten-inch reel quarter-inch Scotch 203 tape as well."[7]

THE WEEK OF AUGUST 11

By the week of August 11, leading up to the festival opening day, those at the site were hard at work. Back at Hanley Sound it was no different. According to Woodstock artist coordinator Bill Belmont: "A week before the show, things went into an all-night phase. The stage kind-of got finished. Three days before the show Hanley finally started getting the sound system hooked up."[8]

In Medford, Cohen and technician Scott Holden were preparing additional equipment that Hanley needed. Holden, who was local to Boston, began his job at Hanley Sound at age nineteen, mostly assisting with the physical setup and stage organization of events. Holden remembers: "I had been making boxes at the shop when someone on the staff said 'Kid we have a show in upstate New York, do you want to go?? We need help!!' I of course obliged."[9] Woodstock was Holden's first official gig with Hanley Sound.

On Monday August 11, Holden and driver John Brodie packed Hanley's Peterbilt, a white tractor-trailer called the "White Dove." The semi was loaded with speakers, cables, harnesses, and other equipment. On board was Holden's 650 BSA motorcycle. This turned out to be a smart move, since it served as an essential means for efficiently getting around the farmland and through the Woodstock traffic congestion. Both individuals left Medford the following morning.

By early Tuesday afternoon, August 12, Brodie and Holden arrived at the Woodstock site. Cohen remained at the office to work on a specially designed sound console for the festival. According to Holden: "The site was a madhouse when we arrived, with everyone trying to frantically set up the equipment. Carpenters were still preparing the stage, and we had a lot of heavy equipment that needed to be lifted on to the sound towers."[10]

Hanley truck driver Dick Butler was already on his way to the site. He was driving the gargantuan Kenworth eighteen-wheel air-ride tractor-trailer known as the "Yellow Bird." The distinctive yellow semi sported a "Hanley Sound Inc." sign on the driver side portion of its container. Butler was the principal driver of this vehicle.

Butler recalls bringing a number of truckloads to and from the Woodstock site: "From what I remember I left around the same time as Scott and John, around four days before. There were about four or five loads that we brought out to the site. We hauled equipment out there, unloaded it, and then went back to Medford for some more. It was nothing more than a cow pasture when I arrived."[11]

As Butler pulled up, Hanley and his crew helped guide him and the Yellow Bird into the service road behind the stage area. The jumbo rig was parked there for the entire weekend. When Butler made his final delivery to Yasgur's farm, he stayed at the site and watched the show. Like many others on Hanley's crew, he figured out how to stay fed, dry, and warm for the entire weekend.

The camper that Bernstein rented was bursting at the seams with Woodstock survival provisions. By Tuesday morning its designated driver Nick Burns was on the road. Alone in the camper without the convenience of GPS, cell phone, or any other twenty-first-century navigational device, Burns set out from Medford with a full tank of gas and an old road atlas with the Woodstock route penciled in. Oddly, someone forgot to let Burns know the festival site was moved. He recalls the confusion: "That's all I knew, there was no communication i.e. cell phones etc. So I am going to Woodstock, New York and go right by Bethel. When I stopped for gas someone told me the festival had been moved. I was in panic mode trying to double back and get to Bethel. I made a collect call to Hanley Sound and could not get a hold of anyone."[12]

Finally on track from being lost, Burns took a side road into the farm. It was stop and go the last three miles to his destination, taking him an hour and a half.

Crowds of people and traffic converged as word spread about the event. As the shop foreman inched his way along, a couple of hitchhikers made repeated attempts for a ride in the mobile home. Burns was reluctant to do so. With expensive recording equipment on board he could not take any chances. "I pulled my forty-five pistol out and set it on the dashboard and politely told them, 'I have been hired to get this thing there and nothing is going to stop me.'"[13]

By late Tuesday afternoon, Burns reached the crew at the festival site. "Why I didn't go to jail for driving that thing the way I did—I don't know!"[14] He pulled into the backstage area with the recording equipment and provisions that he and Cohen had loaded into the trailer days before. Hanley and others were overjoyed to see the much-needed amenities. Burns noticed that they were visibly exhausted.

David Marks, now living in Greenwich Village, began to catch wind of the festival brewing in Bethel. "When I was on the phone with Judi, she mentioned that in a couple of weeks there's going to be this festival in upstate New York. I told her I already saw all of the posters in the Village where I was staying at the time. I had no idea that when she told me to catch a bus to Bethel, that it was the one-in-the-same festival!"[15]

Per Bernstein's request, the musician left his part-time job in the Village. On Tuesday morning August 12, he boarded a bus headed to Bethel. "I worked so many gigs with some of the biggest names in folk and rock history in 1969. It was at Woodstock where I finally met Judi in person! Someone eventually picked me up in a VW Bug at the station."[16]

Before the weekend of Woodstock, David Freese gave Cohen some cameras for Marks. According to Marks, having these in his possession was additional incentive to leave his other "regular" job in the Village. As he was on his way toward upstate New York he hoped to capture some images to bring back to his homeland. It had been less than a month since he had joined Hanley Sound on a fluke, and in a rather short time he was shooting photos, working sound equipment, and absorbing what American counterculture was like in 1969.

Shooting with a Pentax Spotmatic 35mm camera, Marks recorded the most detailed visual data of any Hanley Sound employee in the company's history. "It had no flash, a 400 ASA, and a telephoto lens. This is how they were back then."[17] Marks also kept a detailed journal detailing the goings-on at Hanley Sound.

Later on, Hanley donated and shipped parts of his Woodstock Sound System to Marks in South Africa. Because of Hanley's generosity, that sound system helped fight apartheid through a series of "Free People's Concerts" that were held in South Africa during the early 1970s.

Scheduled to work another event on Woodstock weekend in Bronxville, New York, Hanley Sound engineer Brian Rosen was on his way to New York in a Chevy station wagon packed to the gills with equipment. On Tuesday somewhere along the way to his gig, he made contact with the Medford office. His plans quickly changed. Someone informed him that Hanley needed the equipment he was carrying.

Rosen was elated with the request. He wanted to work Woodstock from the get-go. "We all wanted to work it, and I wanted to go from the beginning."[18] Excitedly, the engineer turned the unprepossessing company car around and headed two hours northwest to the site. As he got closer he noticed traffic for the event was starting to build. While navigating to Exit 16 on the New York State Thruway, and then the Quickway to Exit 104, and finally onto 17B, the traffic became even worse. It took some time for Rosen to finally turn onto Happy Avenue.

As Rosen drove up the Thruway it had been a sunny cloudless day. But at the site, it had been raining off and on since production had begun three and a half weeks prior. According to Lang, "From the first day, foul weather plagued our progress on the Bethel site."[19] When Rosen turned in to Yasgur's muddy fields, he remembers: "It was clear to me that it was going to be amazing. But I still had a gig to do in NYC so Bill wanted to send me back. However the traffic was so thick a state trooper told me I couldn't leave. Even though Hanley talked to the cop, I got to stay . . . I didn't mind."[20] By Friday, the trek from New York to Bethel would take almost five hours.

On Wednesday August 13, crew member Bill Robar drove to the festival with his wife in his Bahama yellow Porsche. Arriving early in the morning, he parked the vehicle next to the performer bridge for the entire weekend. If you look closely, you can see his sports car in various archival photographs. Robar recalls what he saw when he got there: "It was beautiful green grass. I began helping Sam with sound checks and helped set stuff up. The scaffolding was up but they were still working on the stage that I don't think ever got finished. But all of Hanley's gear was under the stage and we were running cable to the mixing console."[21]

Holden, who had arrived the day before, offered Robar a ride on the motor-cycle stored in the back of the White Dove. According to Robar, the feeling of whizzing through Yasgur's field was liberating: "I remember one of my fondest memories was driving Scott's BSA to the top of the hill to check the sound when the area was still empty. It was an absolutely gorgeous green field and the sound was crystal clear; it was simply amazing."[22]

Attendees started to arrive early, and by Wednesday night there were 50,000 gathered at the site. According to John Roberts, "Several hundred frantic long-hairs swarmed over the stage and applied some last minute finishing touches such as lights and sound equipment and a roof."[23] With no ticket booths in place, and gaps in the fencing, Roberts quickly realized Woodstock would be a bust. According to him, once non-paying attendees were inside the performance area, there was no reason for them to leave. "Knowing that there was a lot more audience where they just came from, they wanted to get good seats—and that meant they weren't intending to leave."[24]

Boroda arrived at the Woodstock site on Wednesday morning in his new Mustang Shelby GT350. After he carefully parked the vehicle backstage, he located Hanley and assisted him in wiring the sound system:

When I got there I basically took over after I consulted with Hanley. He visibly had his hands full and was overwhelmed with so many things like the recording etc. So I took over and assisted. Many of the amplifiers had problems and needed attention. When the amplifiers came in they needed to be checked and double checked and repaired. I selected the area under the stage as a sufficient location for the amplifiers and I also wired the towers. At the site I created wiring schematics, which Hanley approved of.[25]

By early Wednesday afternoon Holden was helping Boroda with some of the preliminary Woodstock sound checks. These lasted until 3 or 4 a.m., Thursday morning. By then almost 60,000 had gathered at the site. According to Boroda: "On every single event I did with Hanley I used to tell him to go and leave me in peace. I would say go do whatever it is you need to do, and in the morning it will all be done. I would work overnight so I could work at peace and I preferred it that way."[26]

As the Woodstock stage was being constructed, most everyone on the crew was clambering and hammering right up until the last minute on Thursday August 14. One could not help but notice that Yasgur's land was a beautiful bucolic setting for a music festival. Beyond the flimsy chain-link fence surrounding the site was a picturesque cascading hillside, spotted with groves of trees. The hamlet of White Lake sat in the distance. Amid this natural grandeur arose Woodstock's eighty-foot stage.

Boroda entertained those who arrived early with music. This helped pass the time as attendees were anxious for the event to begin. Holden recalls: "The crowds were growing and Sam was playing 'Spinning Wheel' by Blood Sweat and Tears to check the sound. I did a lot of grunt work so I didn't see Bill very much. But I remember the sound being clear as a bell, which hit me later as Joan Baez sang 'Swing Low Sweet Chariot' a cappella on Friday night."[27] According to the 1979 book *Barefoot in Babylon: The Creation of the Woodstock Music Festival* by Bob Spitz, "Echoing through the less-than-half-filled-fields of Max Yasgur's farm was Steppenwolf's newly released *Easy Rider* anthem "Born to be Wild.""[28] For Boroda it was important to appropriately set the mood for the festivalgoers.

Boroda explains that playing music through a sound system at a festival was unheard-of at the time: "I was the one who turned on the system and tested it. I am one of the firsts when I was doing sound with Bill to bring in a cassette and a cassette deck. Before that people used to say 'testing 1, 2—testing 1, 2.' The Blood Sweat and Tears tape and song that I used was hand selected because it had an incredible amount of drum, bass, and horns."[29]

Like Hanley, Boroda walked the hillside of the nearly packed alfalfa field for multiple listens. At dawn, Boroda opened up the sound system, pushing the sounds of his chosen repertoire of music. As he surveyed the site, Woodstock documentarian Michael Wadleigh shot the engineer's every move. Boroda observes, "Somewhere in those film reels I can be seen setting up the sound system, and walking the field."[30]

Hanley thought it was a good idea that Boroda was piping pre-recorded music to the gathering masses. Hanley believes the "in between" sounds at Woodstock (and other festivals) are almost as important as the live performance: "When you are at an event the worst thing to do is not have some sort of music playing through the house system in between acts. At Woodstock 1999, for example, I noticed the hired sound company did not have any music playing during performance down time. This was a big negative, you have to have music playing at all times."[31]

Music during the festival lulls might have influenced reluctant Woodstock performers like Crosby Stills Nash and Young (CSNY). Initially the band did not want to play the festival. It seemed that the size of the crowd, and being a newly formed band, made them hesitant. In David Crosby's autobiography *Long Time Gone* he relates: "The other notable thing about Woodstock was that we were scared, as Stephen said in the film. We were the new kid on the block, it was our second public appearance, nobody had ever seen us, everybody had heard the record, everybody wondered, what in the hell are they about?"[32]

That Thursday afternoon a phone rang in one of the production offices. Bill Belmont, artist coordinator and manager of Country Joe McDonald, was handed the phone. It was musician Stephen Stills on the other end of the line. Stills conveyed to him that he was unsure if they could pull off the performance. Belmont recalls:

Right around this time, for some weird reason Hanley was testing the sound system. And Suite: Judy Blue Eyes is blaring over the sound system. This was a very popular song from CSN's first album, which everybody was playing that summer. So I stuck the phone out the window and told Stephen ok they are playing the song, everybody knows it, and you should come. So he yells at Crosby on the other end, HEY CROSBY THEY ARE PLAYING SUITE JUDY BLUE EYES! So he said we will be there![33]

Without Hanley's sound system, CSNY might not have taken the Woodstock stage at 3:00 a.m. on Sunday, August 17. A performance to remember, it was about a minute or so into their opening number "Suite: Judy Blue Eyes" when it seemed Stills's guitar was out of tune. The singer asked the sound engineer over the Hanley Sound system, "A little less bottom end on the guitar, please!" Billy Pratt, who was assisting at the sound console then, recalls the situation: "I was fortunate enough to mix certain groups. This was an infamous moment, which can be heard on the recording. Much to my embarrassment it was me who he was talking to when he asked for more bottom end. We heard him beautifully out in front but it must have been the side fill monitors throwing him off. I am still unclear about this. However, it was still an embarrassing moment."[34]

Hanley colleague, lighting professional Tom Field, arrived earlier that week. He recalls the uniqueness of one of the first Woodstock sound checks: "As the festival opening grew nearer I heard Sam over the PA. He was this French sound engineer that worked for Hanley. When they do sound checks they say 'testing, testing, 1 2 3 4' etc. The first sound out of the sound system I heard was "*un, deux*, and *trois*." So that means the first words said at Woodstock were in French and here we were in the middle of the Vietnam War."[35] Hanley can be heard at the beginning of the Woodstock soundtrack. Speaking through his sound system, he says, "testing—1 2 3 4 5—5 4 3 2 1—testing."

Fillmore East engineers Chester and Goddard arrived at Woodstock late on Thursday August 14. They had been manning the Fillmore East Hanley Sound System on Tuesday August 12, for the Who, Jefferson Airplane, and B.B. King. The concert was held in Lenox, Massachusetts, for Bill Graham's "Fillmore Night at Tanglewood." This performance was part of the 1969 Berkshire Music Series. According to Chester: "The Fillmore did very few concerts over the summer, some in July but most of August we didn't do shows. So we took the system out of the Fillmore and a bunch of other stuff from Hanley and put it all together."[36] During the sweltering late summer months, the Fillmore East Hanley Sound System was broken down, moved to outside gigs, and reassembled on the spot.

After the Tanglewood show, they packed everything up and trucked it back to New York, putting it all back in storage. They were curious to see how Hanley and their other Fillmore East colleagues were doing at the Woodstock site. Chester recalls: "At this point the Fillmore East crew (and Hanley) had been at Woodstock for two weeks or so, madly trying to put it together at the new location. When we arrived, they looked at us and said, 'Fresh bodies!!! Had some sleep? We haven't! You are on staff!!'"[37]

When the two unlucky individuals walked through the Woodstock trailer door, they jokingly asked John Morris for a backstage pass. According to Goddard it didn't work out that way: "John and I showed up at Woodstock and we were nothing more than hangers-on at that point. Instead, I think we ended up fixing a gas hot water heater and a few other things."[38] Later, Goddard worked on several major tours for Hanley Sound.

Throughout Thursday, and the subsequent weekend, the use of marijuana became visible, as pillow-like clouds of white smoke hung over various areas in the audience. The buzz in this large-scale and growing commune indicated that something very special was about to happen. Hanley sensed this. And it was becoming evident that more than 100,000 people would be showing up.

As the day of hard work closed and evening came, the Woodstock production staff noticed an army of people had congregated on the field. Once night fell, glowing faces were lit by thousands of little campfires around the natural bowl. As Hanley and his team made the final touches from their sound tower, they could hear the faint and intermittent hum of people chattering and music playing from portable radios.

By late Thursday evening and into the next morning, Route 17B leading to the festival looked more like a parking lot. People abandoned their cars, choosing to hitchhike. The connecting roads from the New York State Thruway were getting jammed too. Back at the site, production people were attempting to get the show ready in time, amid the occasional rain and resulting mud.

When the Woodstock Festival opened on Friday afternoon August 15, 1969, the Medford office lay dark and empty. Unlike most days, it sat in silence, without the typical sounds of doors slamming, phones ringing, the buzzing and banging of its technicians hard at work.

Hanley's crew was in place by Friday, with just a few trickling in, like Judi Bernstein. The Hanley office manager arrived in Bethel from Philadelphia on Friday afternoon August 15. She and Bob Siegel drove his compact car to the festival from his home without incident until they reached the Thruway. As they got closer they hit a snag on route 17B. Traffic was at a standstill. Cars were strewn about the long stretch of road. "I said to Bob, 'this is not good.'"[39]

An opportunist, Bernstein noticed a police officer close by and asked for assistance. "When I got the policeman's attention I told him who I was and that we are providing the sound for the entire festival. I said if you don't get me through there would be no sound. Nobody will hear anything!"[40] The officer responded by immediately escorting the two individuals through the crowded road. With Siegel toggling his clutch and gas pedal the whole way, they finally made it to the site later that evening.

When Bernstein arrived, Hanley and his crew, who had been up for days, were wiped out. Wearing a black Woodstock windbreaker spotted with various burn holes from his soldering iron, the engineer was scrambling to fix a sound console that Cohen had flown in earlier that day. Bernstein remembers, "When I got there Bill was running around with his soldering iron, fixing amps and microphones."[41]

Woodstock security, stagehands, and production crew members were issued T-shirts and windbreakers with the Woodstock logo on them. Specific colors identified what roles they performed: black for stage, blue for staff, green for operations, red for security and medical, etc. You can still see the burn holes on Hanley's Woodstock windbreaker to this day.

Chapter 37

NO CONSOLE, NO FOOD

THE SOUND CONSOLE THAT NEVER WAS

Harold Cohen was last to leave the Medford shop the week of Woodstock. Even the most essential Hanley crew members traveling to the festival found that it was difficult to get in and out of the area. Most used different routes, traveling by car, camper, tractor-trailer, helicopter, plane, bus, and motorcycle. Because of the extreme traffic conditions, Hanley arranged for some of his Medford team to be flown in—Cohen being one of them.

For the entire week leading up to Woodstock, Hanley was awaiting a custom sound console that he had designed for the festival. When the sound engineer learned he was hired, it became his wish to create something new and innovative. With little time to spare, he began his design on a new console only weeks before the event. "We didn't have anything but the recording console at the time and we wanted to designate our best equipment for that. So we needed to design something new and advanced."[1] With almost no time to prepare, the engineer still thought he should build something better to support so many acts.

On Monday August 11 Hanley learned that Cohen was still frantically building the unit back at the shop. Hanley was worried. Others at the site witnessed the building tension. Tom Field arrived on Tuesday and recalls the vibe he felt at the site: "The traffic was just amazing but I was able to get through. By Wednesday when I was backstage, I noticed Hanley's sound console wasn't there and people were getting extremely concerned."[2] Field had previously collaborated with Hanley Sound at various festivals and events. Hanley was anxious and called Cohen in Medford and ordered the console to be flown in ASAP regardless of its state of development.

By early Friday morning Cohen, with console and other essentials in hand, made his way to Hanley's pilot cousin. They flew to the Monticello airfield. According to Cohen: "I was literally working on this until I got on the airplane. A

265

good portion of it was working, however by the same token, it wasn't finished."[3]
Cohen landed in Monticello at 5:00 a.m. Friday morning on August 15.

Because of the traffic, Cohen thought he would need to be transported to the
site via helicopter. "The new airport wasn't even open yet and to my recollection
had no established control tower. The runways were completed only two weeks
prior. Bill tried to fly me in, but all of the helicopters at the festival were being
used for the performers."[4] Cohen had to stay put until Hanley found a way to
get the console to the festival site safely.

With only hours left until showtime, a decision needed to be made on how
he could retrieve the unit from Cohen in time. Luckily, Holden had packed his
motorcycle in one of Hanley's tractor-trailers. That motorcycle was now at the
Woodstock site. With no time to spare, Hanley asked Holden to meet Cohen
at the airport and pick up the sound console. Holden agreed. Niskanen, nearby,
jumped on the back of the bike clutching on.

Taking off in a dash, the two didn't get very far. They were forced to thread
their way at a snail's pace through the thick traffic and hordes of hitchhik-
ing hippies. When they finally reached the engineer around 10:00 a.m. Friday
morning, he was visibly distraught. Cohen remembers: "Hanley had sent Scott
and Nelson by motorcycle to grab the console and a LA 2A limiter from me.
They left me behind with some nonessentials that were requested by Bill."[5]
According to Niskanen, the unit was awkward and difficult to hold on to while
on the bike.

Hanley claims the console was around 2 ft. x ½ ft. x 1½ ft. Balancing an
awkwardly shaped device on a motorcycle in Woodstock traffic was extremely
precarious. With the console and a limiter in their possession, Holden and
Niskanen managed to race back, navigating the BSA motorcycle through the
dense crowd. Holden explains: "It was bad. I was sitting in traffic, the BSA was
without an oil cap, and my leg was getting sprayed with hot oil! I was trying
to balance everything through the traffic while using my clutch. My hand just
about petrified because I was holding the clutch so long. We arrived chafed,
sweaty, and thirsty, just in time. The whole situation was intense."[6] According
to Nelson they thankfully made it back unscathed, and by the grace of God.
Niskanen remembers:

> Hanley or Judi asked me to accompany Scott to go meet Harold and get this
> mixing console up at the airport. The roads were just snarled. The kids came
> up and couldn't go any further so they just parked their cars and walked the
> rest of the way. It was like this all the way to down to where Route 17 joins
> the Thruway. Scott had his 650 BSA and I was asked to sit on the back and
> hold it. When we met Harold I noticed this was a pretty sizable and heavy
> unit! On the way back I basically held on with my knees while I had my arms

around this thing. It was so big that it sat in front of my face and I couldn't see anything! Scott is frantically weaving his way through traffic and pedestrians. I could see out of the corner of my eyes how close we were to hitting some of these objects; it was probably good that I couldn't see anything.[7]

Once near the site, Holden opened the throttle, flying down the dirt road to meet Hanley who was waiting expectantly. At the head of the road the BSA was escorted by Hells Angels, clearing the way in order to meet the show opening. Holden remembers: "As we got closer we fantasized that we would make a dramatic sliding stop right up to the mixing platform, plug everything in, and Chip Monck would announce 'And now Richie Havens!' That never happened."[8]

The console that Holden and Niskanen were sent to retrieve never functioned properly. According to Hanley: "It never worked. I tried. It was having parasitic oscillation problems."[9] Although delivered only hours before show time, the custom unit never saw the light of day. It was a unique and advanced piece of equipment for its time. If working properly, it would have allowed for additional inputs and also had monitoring capabilities. All of which were needed for a large event with multiple performers like Woodstock.

Switching gears, Hanley had to find another solution. With no time to build something new, he ended up using a combination of Shure, RCA, and Altec mixers. The only other console Hanley had on hand was designated for recording purposes.

So the question remains: Why didn't Hanley's Woodstock console work? Cohen suspects that Brian Rosen, the nineteen-year-old Hanley Sound engineer, questioned its performance and altered it at the site. Before Cohen left the office for Woodstock he called Rosen on the telephone and explained to him what needed to be done to get the unit running correctly. Cohen recalls: "In all of his naïve wisdom, Brian thought I didn't do it correctly and tore out a good portion of it, making it unusable. He thought my grounding was incorrect. Being a student Brian had a lot of theory in his head, but he just didn't have the practical experience."[10]

But according to Rosen: "I had been doing some experiments, and Hanley had an oscilloscope, which I knew how to use because of my studies at Carnegie Mellon. I started watching what was going on with the system before the event, and saw we were getting an overload on the board. Using a stack of Shure mixers allowed us to avoid this."[11]

According to Hanley, the console design did not require any pads and had at least twenty audio inputs, three mix buses, and was switchable between three individual sets of microphones. The unit had a direct transformer input that had low impedance in, and high impedance out, allowing for voltage gain. Cohen explains there just wasn't enough lead time to complete the unit:

"Potentially it was capable of handling the swing. All of this went on a mix bus that had a line amp in it. It was not possible to overdrive it because there were no electronics on the front end. It had to be well shielded and built into a large electrical conduit box."[12]

Back at the airport, Cohen was starving and thirsty as he continued to wait patiently for a lift to the festival site. Fortunately, he saw a friend he knew from the White House staff boarding a helicopter. Cohen pleaded with him for a lift: "He was used to seeing me in a suit and tie and here I was in a T-shirt and jeans. He offered me a ride via helicopter if I promised him a live sound feed off of my board. So we agreed knowing he would have given me a ride otherwise."[13]

Early Saturday morning August 16 Cohen finally flew into the Woodstock site by way of a twin-engine Bell military helicopter. Through its clear plastic window, he could see a mass of sleeping festivalgoers. Cohen was exhausted and hungry when he landed. He quickly looked for something to eat and a place to crash.

EATING AND SLEEPING

Although food eventually found its way to the people, staying fed was an issue for many that weekend. Those who didn't want to lose their prime spot near or around the dense stage area suffered the most. Locating something to eat during Woodstock wasn't just a difficulty for festivalgoers; it also affected production staff.

Prior to the festival happening in Bethel, finding a food vendor to handle the job was extremely difficult. Lang remembers: "As it turned out, the large food-vending companies like Restaurant Associates—that handled ball parks and arenas—didn't want to take on Woodstock. No one had ever handled food services for an event this size."[14] At one point Nathan's Famous Hot Dogs were considered but dropped out. Finally the producers settled on a concessions company called Food for Love. However, underestimating crowd size, Food for Love became overwhelmed, running out of supplies fairly quickly.

In the beginning things weren't so bad for those on Hanley's staff. Before arriving at Woodstock, Siegel and Bernstein stopped at a market and purchased a cartful of steaks. As Siegel grilled them backstage and fed the crew he watched the crowd thicken. Nervously leaning over and joking with Hanley, he asked, "And what's going to keep all those starving hippies from moving down here and eating these steaks?" Hanley replied, "My sound system will." Eventually Siegel's steaks ran out. Hanley was right.

As the weekend wore on, hunger set in for the Woodstock staff and audience. Soon New York Governor Nelson Rockefeller declared the site a "disaster area." Eventually word was out that Woodstock was dealing with a food shortage resulting in thousands of donations being airlifted in including water, sandwiches, canned goods, and fruit. For those few days, Hanley's sound system became essential in letting over 400,000 hungry participants know when and where they could locate something to eat.

Grabbing one of Hanley's microphones, Hog Farm commune leader Hugh Romney announced to the hungry crowd what their options were if they wanted to fill their bellies. Romney proceeded to let them know that the Hog Farm was serving food and for those near the stage, cups of granola were being passed around. "What we have in mind is breakfast in bed for 400,000! Now it's gonna be good food and we're going to get it to you. We're all feedin' each other."

Typically, festival food was not very good, especially for Hanley's crew. According to Boroda, "It was generally horrible." While working at Hanley Sound he recalls that there was very little attention to comforts like sleeping and eating. For Boroda, it was all about getting the job done. Rest and nourishment came later. Boroda remembers:

> It was a misery. I could not eat peanut butter and jelly or fast food. At Woodstock Judi was a yenta; she provided us with some food. I would rest here and there in the trailer. But any sort of comfort at Hanley Sound was a major problem. Later on I requested no fast food in my contract. Bill was the kind of person who was on the go so much. He had plane tickets in his pocket all the time. It was not a worry to him.[15]

Boroda's description of Bernstein at Woodstock as a "yenta" is accurate. *Yenta* is a Yiddish word meaning a noble or gentlewoman. When trying to feed the hungry Hanley Sound crew during the festival, the office manager felt she was reenacting the biblical story of the "fishes and loaves" from the Gospel of John. As the cans of tuna fish packed in the trailer dwindled down to a few, she stretched them out, similar to the five small barley loaves and two fish. Like Jesus feeding the multitudes, many sandwiches (and potato chips) were made with limited means that weekend. "It was like an act of God. I made a very little bit go a long way!"[16] adds Bernstein.

According to Rosen: "Judi organized everything from food to getting paid. It was an all-male crew and she was the only female we dealt with; I liked to call her Mama Cass. I remember we ate food that the organizers provided: it was Horn and Hardart boxed lunches as well as the Hog Farm's rice and beans."[17]

Pratt and Niskanen recall finding food wherever they could. Pratt explains: "We got there a day or two early. There was a trailer we could crash in that had some cots in it, and Judi had some food. The platform that was in front of the stage was kind of my raceway. I re-plugged many of the mikes that were switched out throughout the weekend."[18] With so much constant work being done, Niskanen remembers he and Pratt were hungry a lot: "People from town were feeding folks and we got some of that. Billy and I bought bread and peanut butter at the store. Occasionally someone would have some food and share it around."[19]

Sleeping at the festival site was another problem. Even though Woodstock Ventures assigned the crew rooms at a local hotel, it was more than a mile away, and that didn't make sense if you needed to remain onsite. So Hanley Sound people did not use these accommodations. According to Hanley, "When I had to, I either slept under the tractor trailer or in the camper."[20]

The rain, traffic, and mud made resting difficult. Some crew people slept on trailer floors, on scaffolding, under the stage, and other shelters. Rosen recalls, "We set up cots in some of the empty trailers we brought."[21]

Holden remembers spending a good amount of his time under the stage, where it was warm and dry: "I was under the stage where all of our amplifiers were in a row, in one long line. I had no idea what I was doing at all. It was crazy! I didn't sleep for three days! But I was warm and dry!"[22]

Truck driver Dick Butler recalls losing two mattresses and one seat out of his truck: "I still don't know where they went!"[23] According to Niskanen, he and Pratt were some of the lucky ones: "At some point we slept at a cottage someone rented. We had it made! We were able to go back for a couple of hours, rest and get a shower! It was wonderful!"[24]

Nick Burns spent a lot of his time perched high up on one of the four Woodstock light towers. In total they supported twelve Super Trouper carbon arc lamp follow spotlights. The towers that had four spotlights on them supported almost a thousand pounds of equipment and were difficult to climb up through the mud and rain. Burns controlled a tower that held two. Throughout the weekend, intrepid festivalgoers attempted to climb them, compromising the structures. Although the towers were guyed, the heavy spotlights did not have safety chains on them.

Burns recalls the challenges of survival he encountered, because sleeping eighty feet in the air for long periods of time was both precarious and dangerous. Often unable to leave, he ate by cooking hot dogs over one of the extremely hot carbon arc lamps. If nature called the crew hoisted up bottles of water, and urine-filled bottles got sent down:

> From Tuesday night until Monday morning it was flat-out spending nights up on the light tower. If you could grab a half an hour catnap you were lucky, once it got dusk you were up there for the night. It was so muddy at one point; I got in between the two Super Troupers so I wouldn't fall off. When you are climbing up a tower and you sense movement or motion it gets pretty dangerous. There were a couple of times there where it got a little hairy![25]

Chapter 38

STAGE ANNOUNCEMENTS

HOLY SHIT!

What some recall as the first words spoken through the Woodstock sound system varies. This perhaps is because there were multiple sound checks leading up to Richie Havens's opening performance. According to John Morris, production coordinator and emcee, the first words were "HOLY SHIT!" When Morris walked up onto the Woodstock stage he didn't realize the sound system was being tested: "I was standing near a mike. I could see that people had filled up the area during the night. I went HOLY SHIT! About a half a million people laughed at me! That's when we all knew Woodstock was on!"[1]

Despite the many semi-calamities that occurred during that weekend—stage failures, the weather, and the mud—the sound was constant and never failed. This meant that the Hanley system was not just a vehicle for music, but likewise, a vehicle for verbal communication. Morris reflects on Hanley's work: "A lot of things failed at Woodstock but the sound did not. You could pick up a mike in the middle of a storm as we did and calm everybody down. You could announce acts and it all happened. It was just there."[2]

Bernstein claims Hanley Sound was instrumental in keeping Woodstock together: "We were the main source of communication!"[3] Throughout the weekend, desperate attendees wrote on whatever scraps they could find to relay messages to emcee Chip Monck, so he could broadcast them to the audience. Monck recalls the experience as terrifying: "I was petrified. But I got to the point where I was practically paternal in some periods of dire emergency, for example, 'Sit the f!@# down!' or 'Do you have any idea what you're doing?'"[4]

According to emcees Morris and Monck, managing a crowd of this magnitude was a major responsibility. Although sharing duties, they were often inundated and overwhelmed with requests for announcements from all directions. Morris recalls: "We were able to control the crowd fairly well, which always amazed me. It was Hanley's system that allowed us to be the source of communication at Woodstock."[5]

When you hear stage announcements like "Holly your insulin is in the medical tent," "Please stay off of the sound towers," or even "Don't take the brown acid," imagine what might have occurred if Hanley's sound system was not working properly? Because of his work, we are left with many memorable quips that came off the Woodstock stage.

Fifty years later, many of these soundbites have been etched into the psyche of popular culture. What if Hugh Romney's (known as Wavy Gravy) reassuring morale-boosting pep talk to the crowd did not come through with clarity? We would have never heard the comforting and poignant, "We must be in heaven, man!" Also, think of all those who would have missed out on that free granola.

At some point at the beginning of the festival, Monck asked the people in the back of the crowd if they could "hear well?" The majority raised their hands in affirmation. This proved the sound system served as a source point of communication for those camped out in the back of Yasgur's fields. Even folk musician Richie Havens asked the audience if they could "hear okay?" At the beginning of his impromptu song "Freedom" he asked for more guitar in his monitor from Hanley, who was at the controls just then. Clearly, having good sound was on the minds of both the audience and musicians throughout that incredible weekend. Joel Rosenman recalls the significance of Hanley's sound system:

> At Woodstock there was a core sense of community, decency, and love that half a million kids had at the festival. Our infrastructure wasn't designed for such a large crowd. The weekend would not have held together as beautifully as it did except that the audience re-created another infrastructure. A more natural one maybe, but one that people shared with each other, where there was little or no violence. I think that one of the things that contributed greatly to that was Chip Monck and John Morris. Those guys didn't sound like each other but they had this kind of reassuring and supportive (anything but panicked) sound to their voices. The clarity and immediacy of their voices thanks to Hanley felt like they were speaking directly to you. Because of Hanley's sound system we could remind people that there was no need to panic, that the world is watching and you are all doing fine. Being able to communicate that without hum or feedback helped us reinforce that message.[6]

The lack of security at Woodstock made Hanley's sound system even more essential. How to effectively maintain order of a large crowd was on the mind of producer John Roberts. The *Louisville* (Kentucky) *Courier-Journal* reported: "Roberts and others worried constantly about the potential for a riot. Some of the production people on the stage looked in awe at the crowd as a sleeping monster. 'Those kids are going to snap,' someone said, 'They're going to realize that it's not really fun to slap, slap, slap, through the mud. Then they are going to rise up and kill us all.'"[7]

As elementary as this concept is, if there had not been a successful application of sound reinforcement, Woodstock might have ended in chaos. Joshua White claims: "Woodstock should have been a riot, yet was glued together by the fact that you could always hear the announcement from the stage."[8] Woodstock artist Country Joe McDonald of Country Joe and the Fish reflects on the performance of Hanley's Woodstock system:

> It would have been completely different if the audience was unhappy. . . .
> A large audience that is unhappy with the sound is an unruly and difficult
> audience. The way Woodstock happened, like the size of the crowd coming
> way beyond anyone's expectations, the gates coming down, the bad weather,
> without good sound it would have been like Altamont. Not only would it
> have been a complete disaster for the audience and bad vibes, but also we
> wouldn't have been able to communicate.[9]

Elliot Tiber recalls that Hanley's sound system could be heard for miles outside the perimeter of the natural bowl. In his book *Taking Woodstock*, he romantically describes the situation: "And then I could hear it, as clear and illuminating as sunlight. Richie Havens was singing 'Freedom.' Richie's unmistakable voice was rolling down Route 17B like thunder. It bounced over the hills, valleys, and lakes that joined Yasgur's farm and the El Monaco. I wasn't able to go to the concert, so the concert came to me."[10]

On Saturday night Niskanen borrowed Holden's motorcycle for a listen from the top of the hill. After weaving his way around the natural bowl, he parked the bike and ventured into the hemlock- and fern-laden forest. Stumbling upon an area known as "Bindy Bazaar," it was a peaceful change in scenery from the densely packed audience area. Here, he saw hippie vendors in multicolored kiosks selling clothing, jewelry, posters, leather goods, and drugs. Trails marked Gentle Path, High Way, and Groovy Way networked throughout the woods. The trails had been cleared in early August by a team from the Hog Farm. As Niskanen walked along, Hanley's sound system could be heard echoing in the background:

> As I got up there I was amazed. The sound was as clear up on the top of
> the hill as if I were on the stage. When I went into the woods the sound
> was still also really strong. They had little booths up there and there were
> lights strung everywhere. I couldn't figure out where they were tapping the

electricity? They were selling leatherwear, beads, and I remember one booth had the words DRUGS on top of it. This guy had weed set up and pills and other things on this little counter. It was like something out of a fantasyland. It was beautiful.[11]

WOODSTOCK GARBAGE

A big issue at the festival was the mounting level of trash. Hanley's sound system helped provide a solution to the problem. By Saturday morning there were thousands of kids waking up in the mud and garbage that was accumulating fast. It had been raining and most of the staff was exhausted. By now everyone behind the scenes had been working for weeks in an effort to get stage, sound, and electricity running efficiently. In shifts, most of the crew congregated under the stage so they could stay dry and warm next to the hot amplifiers. Holden remained under the stage protecting the expensive equipment throughout most of the festival.

Being awake for days on end took its toll on Hanley. Back at the trailer the sleep-deprived sound engineer crashed on a camper bed. Bernstein recalls: "Bill comes in the trailer and collapses, spread-eagle on this double bed in the back. He was out like a light. I couldn't move him."[12] As soon as the engineer had fallen into a deep sleep there was a knock on the door. It was raining torrentially. When Bernstein answered, it was rain-soaked operations manager Mel Lawrence in a yellow hat and slicker. He needed to turn the sound system on as soon as possible and required Hanley's permission. According to Bernstein: "I could not wake Bill up for the life me! With every try he fell dead into the comfort of his pillow. Finally, I woke Bill and told him that Mel needed permission to operate the sound. Bill answered, 'I grant him permission' and clunk he fell back to sleep!"[13] People were sitting in trash. Lawrence was concerned. The only way to deal with it was to turn on the sound system.

It took Lawrence and Bernstein around fifteen minutes to find someone from the sound crew awake and authorized to turn on the power for the microphones. Bernstein explains: "I found one of our guys in one of the other trailers. He was sleeping so I kicked him a few times, and yelled at him to turn this thing on!"[14] After an engineer was located, the system was turned on in the rain.

Lawrence and Bernstein sped off to the stage to ask thousands of sleeping audience members to move back far enough in order to collect the accumulating trash. Convincing a crowd of this size to push back hundreds of feet was no easy task. If not properly choreographed it could have been a disaster. According to Bernstein, this was more difficult than it sounds: "I said to Mel,

'Good luck! If anyone rushes the stage, just shut off the sound system!' And then I ran back to the trailer."[15]

Lawrence's stomach turned as he grabbed the microphone, unsure about how to wake the drowsy, rain-soaked audience. What he intended to say was, "Good morning, everybody, rise and shine! We're going at it again today! Let's clean up our areas!"[16] But what he actually blurted out was: "GOOOOOD MOOOR-NINGGG!" . . . as three hundred thousand heads vaulted up in unison.[17]

The volume of the sound system was too loud, so Lawrence asked the engineer to turn it down some. He then kindly asked a favor of the Woodstock crowd who lay huddled in different positions around the stage: If he were to pass out garbage bags would each person throw away the trash around them? The crowd would then pass along the filled-up bags until they reached either side of the stage, where they were to be collected. "We've got to keep this place livable so we can prove to the rest of the world that we can make it together in peace and in comfort. And we're gonna do it, too."[18]

Chapter 39

THE NUTS, BOLTS, AND OCCASIONAL ZAP!

BZZZT . . .

Wet weather compromised Woodstock production. In *Guitar: An American Life* by Tim Brookes, Grateful Dead singer and guitarist Bob Weir relates that on late Saturday evening August 16 he felt and saw a giant blue spark. It was powerful enough to push him back several feet: "I got a shock. The stage was wet and electricity was coming through me. I was conducting! Touching the microphone and the guitar was nearly fatal."[1]

The situation was so bad that bass player Phil Lesh claims he heard helicopter pilot communications coming through his speaker. Power surges also knocked out the motors on eight of film director Michael Wadleigh's cameras. It was the beginning of a terrible grounding issue that had to be dealt with. According to Dead sound engineer Owsley Stanley: "The radio was heard as leakage at a considerable volume in the PA and in the . . . amps. No one from the venue seemed to know how, or care to fix it."[2]

Mid-afternoon on Sunday during Joe Cocker's performance, a thunderstorm set in. A soaking rain and forty mph winds compromised the sound and light towers, scaffolding, and other steel structures on stage and off. Most of Hanley's crew was frantically covering the mixing console, amplifiers, drivers, and microphones with Visqueen, a durable polyethylene sheeting. Hanley often had this on hand in large rolls. The crew on stage and production staff did the same.

According to the *New York Times*, "Amplifiers and other electronic devices were covered to avoid damage and recorded music was played for the crowd."[3] Hanley recalls: "This storm was a close call when it came in. I never felt anything live though. I remember being under the stage frantically covering everything in plastic. But I never got zapped. I believe it was a grounding issue."[4] Nick Burns remembers it differently, claiming that most people didn't realize how harrowing of a situation this really was:

This was a close call and it could have become a complete and total disaster! Everything was live out there because it was wet. Every time you touched a connector you got zapped! And some zaps were bigger than other zaps. There were people running around underneath the stage in their hip boots trying to keep all of the amplifiers going. That was one of Scott Holden's jobs. We were running around trying to keep technical stuff functioning.[5]

According to Jefferson Airplane bass player Jack Casady: "Woodstock was a test for everyone. We were all trying to figure out how not to get electrocuted. The grounding wasn't as thought out as it is today. I am sure it was absolutely nutty chaos for Hanley and others."[6]

Through Hanley's still functioning sound system emcee John Morris could be heard (through a hot mike) pleading with people who remained perched on the steel structures to come down. As the fifty-foot towers swayed in the wind, jeopardizing the safety of thousands below, Morris urged, "Please get down off of those towers." After this essential message was spoken, the power was cut. According to Morris, in Woodstock: The Oral History: "We cleared the stage of electrical equipment. It was raining like crazy, the sound system was on, the mike was shorting into my hand and it was like zap, zap, zap. I knew I had to try to get people off of these towers and try to keep everyone calm and cool."[7]

Some say that the microphone lines were the source of the zapping. Initially buried in the ground, the cables now lay exposed, poking out of the wet mud. According to Rosenman, it was a nightmare. Problems were surfacing almost continuously. In the production trailer he received a phone call from the chief electrician, who was located backstage: "He sounded pretty nervous." In a 1989 NPR interview, Rosenman explains the exchange he had with the electrician:

[He] said, with the rain and all of those hundreds of thousands of feet scuffling over the performance area, the main feeder cable supplying electricity to the stage—the musicians, the amplifiers, whatever has been—unearthed. And with additional abrasion from these sneakers and whatever, sandals, it may wear away the insulation on these cables. I'm worried with all those wet bodies packed together that we may have something approximating a—and he paused for a moment and I couldn't believe that he was searching for the words that he came up with—But he came up with mass electrocution.[8]

The thought of shutting down the power was not an option and neither was a mass electrocution. Luckily the electrician thought of a solution in the nick of time. He told Rosenman that by working a shunt from the power source to the stage he could then bypass the main feeder cables. Rosenman and Roberts waited twenty minutes for the best or worst of scenarios to arise. "John and I sat there looking at each other. I guess we were waiting like in the movies for the lights to dim a little bit, the way they do when they throw the switch in the electric chair chamber. And I think it probably took a hundred years for those twenty minutes to pass. The phone rang and it was the chief electrician again. He said, 'I did it, I did it, everything is fine!'"[9] Once the cables were identified, audience members were cleared for safety concerns. The wires were then buried, and plywood boards were placed on top.

The drenching rain washed over everyone and when it settled it left the audience equalized and cleansed. Steam rose off their bare bodies and into the air, the mood of the gathering began to shift. Monck claims that this rain shower "saved" folks by bringing audience and production people back to "normality" in a rather unorthodox way. He reflects on the challenges of being exposed to the elements: "You were all drowned rats sitting in mud whether you were the performer or audience. So here we were all exactly the same. Everybody is miserable, everyone is wet, everyone is cold, and everyone is covered with mud. Nobody can find his or her shoes and the sleeping bag has been long trampled into the muddy ground."[10]

Other production mishaps occurred like the roof structure not being completed, and the last-minute collapse of a three-half-moon-shaped stage turntable system. Designed in advance, the sixty-foot turntable was supposed to be set in and rotated so that the acts could change over quickly. The design was made up of three half sections, two actually attached together to rotate. The third half section was to be preloaded with equipment when the others were ready to go. At all times Hanley expected to have three sets of mikes ready on set for each half. "The cheap eight-inch casters eventually tore out of the rolling stage platform since someone did not use washers with the screws, they eventually ripped right out. This made us scramble last minute in miking the groups, we did the best we could. We intended on rotating them around when the acts changed over, but they were never used after that."[11] Monck recalls these production problems in some detail:

These half turntables that we had made were supposed to connect with each other and spin. It was supposed to make our turn around something like three minutes. They were made of wood and since everybody stood on them the casters broke off of the support so that was the end of that. This made us go back to the half hour turn around between acts. The blessing was that if we didn't have continuous music for three days what are these kids going to do for the eight hours they have off? Big question. So here we are with the broken turntables. Everything seems to be working fine with Hanley which is good but I have no roof to hang my lights on and the Joshua Light Show has nothing to hang their screen on. For all intents and purposes it is an absolute wreck.[12]

Even though many claim the system didn't work at all, one theory was that the turntables rolled until the Grateful Dead performed at 10:30 p.m. Saturday evening August 16. When the band finished around midnight, production staff realized that the half-moons were inoperable. To make things worse, the band's performance was lackluster and shock-filled due to a bad electrical ground. It seemed the band's extraordinarily heavy gear was the culprit and as a result the rolling system laid dormant for the rest of the weekend. According to Owsley Stanley, the band's sound engineer:

> The rotating "cookies" were built on furniture casters, not commercial units. . . . I took one look and begged that we be allowed to set up on the stage floor, because we knew our gear was so heavy it would collapse. . . . We were refused this option, and sure enough the setup had rotated only about a foot when the casters simply folded over. . . . this caused a lot of strife, and somehow we were blamed for it.[13]

A little-known fact is that the Joshua Light Show was intended to be part of the Woodstock experience. According to White: "We couldn't get the light show going. We did it one night and it was great. Nobody gave a shit. It's one thing to do a light show in front of 3,000 people; in front of 400,000 you are small. You might as well not be there. It was all about the sound, which didn't fail."[14] At one point over the weekend the show's large screen acted like a sail as heavy winds came in. Avoiding a catastrophe, those close by wielded their knives to puncture holes in the screen so wind could pass through.

Even though there were technical calamities over the weekend, Hanley's sound system was not one of them. Bob Pridden, sound engineer for the English group The Who, reflects on his Woodstock experience and the impression Hanley's sound system had on him at the time: "I thought Woodstock was a bit of a mess but the sound system was great and he had enough of it. The Who used the Hanley System at Woodstock and I think for the time it was an absolute excellent job to cover that many people. The system was one of the pioneering systems. He did invent the PA as it stood at that time."[15]

According to Jack Casady: "Hanley was out there trying to wire stuff together to make it work. He really had his job cut out for him. What he was able to do under these circumstances was a real test. Kind of like being out at sea, are we going to capsize?"[16]

Chapter 40

WOODSTOCK, HOW SWEET THE SOUND ... SYSTEM

A HEAD-IN-THE-CLOUDS GENIUS

According to one highly observant Woodstock attendee—a twenty-two-year-old Boston College student—you could hear and feel the detail of an artist's performance through Hanley's Woodstock sound system: "You can hear every sound, every click of the guitar pick. When you get a really heavy group you can feel the music actually hitting you."[1] These are accolades that Hanley and his crew can be proud of.

Over the years, other attendees like Nick and Bobbi Ercoline have praised Hanley's Woodstock system. Before they were married, Nick and Bobbi were living near Bethel when they heard the festival was arriving in their backyard. The couple decided to go, and good thing they did. For fifty years now they have decorated the cover of the Woodstock soundtrack album. It was a by-chance photo taken by Burk Uzzell. The two can be seen embracing, wrapped in a muddied blanket while listening to music through Hanley's system.

By the time Nick, Bobbi, and their friend Jim "Corky" Corcoran drove to the site in a borrowed station wagon, Route 17B was too congested to travel. As they turned down various country side roads, they could hear Hanley's sound system in the distance. They eventually had to leave their car and walk in, following the masses to the site. Bobbi remembers: "We weren't even near the site yet and we could hear it. As you got closer the background began to fill with music. The closer you got, the sounds and smells became more intense. The sound was always there. It was the backdrop. You could hear the announcements and music. This was a constant. Looking back, what Hanley did was really amazing considering what we have now."[2] Nick recalls:

From where we were we couldn't see the stage, but we could see the speaker towers. Even so, you could hear the music and announcements. It wasn't overwhelming at all. It was a clear sound. That natural bowl made the perfect amphitheater and it allowed the sound to fill. There were so many people there that you could easily lose touch with whomever you were with. This is when the stage announcements became important. Hanley's sound system was a lifeline for all of us.[3]

The Woodstock sound reinforcement system was likely the most expensive PA assembled at any festival and concert site up until that time. But the details of its deployment and the specific equipment used, up until now, have been incomplete and incidental. Hanley's team worked tirelessly in facilitating the setting up and taking down of the innovative PA. The equipment was compiled of various levels of technologies and was adapted and systemized to Hanley's specifications. According to Harold Cohen, it was powerful for its time: "The Woodstock sound system was approximately 10,000-watts RMS in audio into 8-ohms. It would be the equivalent of a sound company today going out and doing a 125,000 to 150,000-watts."[4]

To protect his innovative designs from independent, smaller sound companies, Hanley famously constructed his main speaker systems in black boxes, otherwise known as "bins." These custom enclosures were designed exclusively by the sound engineer. Cohen explains that the black boxes were a way to maintain Hanley's innovations as trade secrets: "There weren't ten companies in the country like there are now that would be capable of being able to amass that kind of power. We did all of this using equipment that was owned by Hanley Sound. We didn't pool it with five other companies."[5]

At the festival, Hanley Sound also installed a small sound system for the Hog Farm free stage. It hosted a puppet theater and performances by the Grateful Dead as well as Joan Baez, who reportedly played there for two hours. Cohen recalls, "We not only had the main stage covered but we had small systems available for the free stage at Woodstock."[6]

John Morris, who was head of production at Woodstock, reflects: "Bill always had a soldering iron. He literally made the pieces and put the pieces together to make the sound work, and it got bigger at Woodstock. Bill made it possible for half a million people to hear."[7] Essential elements like crowd control, staging, miking, power amplification, recording, and large speaker systems all relied on the "eccentric, head-in-the-clouds genius."[8]

According to Country Joe: "I think it was remarkable that the audience heard good music at Woodstock. No one had ever built towers like that before. Although Hanley prepared sound for fifty thousand people, he made it work for half a million. I don't know how Hanley did this or what enabled him to do it. He was way ahead of his time, and his time came at Woodstock."[9]

Stan Goldstein claims that Hanley's operation made a distinct impression on him. In Joel Makower's book *Woodstock: The Oral History*, Goldstein reflects on Hanley's work: "Bill in many ways was very innovative, very creative, and very sloppy. His equipment didn't get the maintenance it required. But Bill was extraordinary. He put together what was not a bad PA system of relatively enormous size, and he hauled the shit all over the country."[10] This is a slight criticism that Hanley accepts. However, Hanley observes, more money could have gone into his equipment if promoters had invested in his ideas: "You couldn't do this without having enough money. Most promoters were not willing to pay for good sound back then."[11] Goldstein adds: "PA was really in its infancy and Hanley was capable of putting together a large system. Bill was the original tractor-trailer man. He created touring sound of magnitude."[12]

MONITORS

By Thursday morning August 14, Hanley's Woodstock PA was up, tested, and fully functional. According to author Bob Spitz, "Bill's JBL monitors were set up on the stage and the main current was turned on for the first time to test the microphone levels and it worked!"[13] For monitors Hanley used two side-fill dual JBL C55 (or JBL 4530 LF) bass cabinets powered by Crown DC300 amplifiers, each containing Altec 311-60 radial horns and Altec 290E drivers (for the highs). According to Hanley, "We had done a lot of tests, and we got into 290s because they stayed together longer."[14] The low-end drivers were assigned around 50-watts. The 290s, for example, could handle the midrange at around 100-watts each. "The 290s (with 311-60 radial horns) sat on top of base cabinet stacks."[15]

Various audio processing equipment was located at the side of the stage, such as controls for side-fill stage monitor levels, stage intercom system, and some lighting controls. The sound engineer compressed the audio using a Teletronix LA-2A pushing the limits of his amplifiers, so he could get more energy out his system.[16] The stage monitors were in mono while the audience sound was in stereo—all were on the house mix.

SPEAKERS

By Friday morning August 15, more than two hundred thousand people had arrived at the Woodstock site. In the midst of the gathering masses, Hanley and his crew finalized the sound system right up until the last minute. Hanley delegated various duties to his team including lead engineer Sam Boroda, who was instrumental in wiring the speaker system.

Cranes were needed in order to get the heavy speakers and horns up into the air and firmly placed onto the sound towers. Eight cabinets were arrayed on each tower, four on top and four below. Gently placed onto the scaffolding, each one weighed over 450 pounds. Additional guy-lines stabilized the speaker scaffolding in place. According to Hanley, "We taped strands of bulbs to the guy-lines on all the towers for decoration and safety."[17]

Hanley's design planned two elevations for speaker placement: The lower elevation at around twenty feet for the front of the audience and a higher elevation at seventy feet for those further away. Expecting only 50,000 to 150,000 people or so, Hanley designed this initial layout to cover everyone in what he refers to as the "funnel."

The engineer's speaker placement at Woodstock was directional, meaning that he aimed the speakers where the people were. Having multiple levels gave him options. At the time, Hanley perched the speakers high up because he could not afford to buy costly delay systems. Hanley claims that ideally he could have used two delay units with one for backup. However, at a whopping $16,000 each, this was more than he was to be paid for his services. Even without a delay unit, the sound engineer realized that he could still achieve great fidelity with minimal distortion.

For the upper speakers (placed at seventy feet high), it meant that the far-away audiences in the back would be able to hear with clarity. "These speaker stacks, four on each side, were custom-built HSI 810 cabinets, with four Altec model 1003B, 5x2 ten-cell, multicellular horns on top (300Hz. min freq)."[18] Each speaker contained eight direct radiator JBL D-130F fifteen-inch drivers in them, that when combined, made a grand total of sixty-four woofers for the upper level system. According to Hanley: "I reversed the 288 diaphragm (with the back covers off) of the Altec horns that sat on the top of the towers to handle the highs. This was a tri-amped sound system with crossovers at 500Hz and 8kHz. The sightline was to the plateau of the hill. That's why the sound towers were so high."[19]

Hanley and his crew would often use empty soft-drink cans propped underneath the drivers on the back of the horns. The height of the can(s) would support the desired position depending on where they wanted to aim them. At times taping two together, or even crushing a can would give them an exact position.

Hanley claimed that he designed the special upper speaker cluster in this way to "compress the sound,"[20] adding: "The longer the column, the narrower the beam of sound, horizontally. I contained all that energy and aimed it at the back of the audience."[21]

For the lower speakers placed twenty feet up, the sound engineer was able to avoid blowing out the audience closest to the stage. The lower stacks contained four custom-built HSI 410 "horn loaded" cabinets per side, with four fifteen-inch drivers in them, which combined made a grand total of thirty-two woofers for the lower-level system. Each lower stack had four ten-cell, multi-cellular horns placed on top with 290 drivers for the highs.

Once all of the speakers were tuned and positioned, Hanley and his crew patched them together and plugged them into a 400-amp, three-phase AC power distribution system (located backstage) wired from nearby White Lake. A total of "220 volts was on tap,"[22] recalls Hanley. Again, this was a design the engineer claims was supposed to be for 100,000 people maximum.

MIXING

Mixing and recording a festival like Woodstock was a complicated and cal-culated feat. The mixing console platform was constructed of scaffolding and plywood boards, and was positioned about seventy-five to one hundred feet stage left into the crowd. Boroda was instrumental in choosing the location of the platform: "I never wanted to be in the center of the stage. I only wanted to be on the left side or the right side. I opted for the right side because of easy access to the equipment. The distance from the stage allowed me to hear any problems before they reached the audience. This was a comfort zone for me."[23]

Hanley's equipment, minimal at best, sat on a folding banquet table on the mixing platform. When it rained he protected his equipment with Visqueen clear plastic. Cohen recalls, "We had no canopy or top cover like you see to-day." According to Monck, this was typical of what Hanley brought to festivals: "When you did a Hanley show you usually got an inordinate amount of scaf-folding, huge amounts of boxes, and lots of mixers that were state of the art at that time."[24]

Monck explains you could not help but notice the sound engineer's setup during the festival: "There is a riser somewhere out in the field, with a banquet table out on top of it. There is Hanley with no cover over him, and it's raining. There are a bunch of mixers on top of the table with splitters going back to Lee and Eddie who are recording in the trailer, the others are split to the PA—it was all very simple."[25]

Even though the intended mixing console had to be abandoned, the sound engineer was prepared with a backup. The actual console used at Woodstock had twenty-four inputs, along with four output busses, and was constructed from API Automated Processes Modules. Hanley explains: "I used four mod-ified Shure M67s with input pads, two Shure M63 Audio Masters for EQ, an Altec 1567A mixer, four Teletronix LA 2A limiters (leveling amplifier) between the mixers, and the power amplifiers."[26]

Hanley often used the Teletronix LA 2A model to protect amplifiers and speakers from being overloaded by the varying levels of concert sound. The sound engineer considered using the newly designed Altec Acousta-Voicing equalizer at Woodstock but did not see a need for it then.

At any given time there were over twenty Shure Unidyne SM545 microphones being used during the event, some having special transformers in them. Hanley remembers: "On the stage, we used custom microphones I had built from Shure factory parts. They closely resembled what Shure Brothers came to sell as the popular Shure SM58, the most noticeable difference being that mine had a brushed-chrome finish like the older Shure 'bird cage' mikes."[27]

The microphones were fed to two custom-built Belden 19 pair snakes, which the Hanley brothers constructed before the festival. These were fed into a custom switchover box that allowed for a quicker changeover at the console. Another one of Hanley's innovations—the microphone snake—was highly uncommon for the time. This device was developed out of a need to organize, simplify, and systemize complicated miking from a long distance. Last, there were no custom mike arrangements for specific acts at Woodstock. According to Hanley, "They weren't into that then, this was jam city."[28]

Various engineers on staff mixed sound during the Woodstock weekend. Busy with keeping the whole sound system working properly, Hanley was only at the platform intermittently. Boroda usually mixed sound at night, claiming: "Initially at Woodstock I was not part of the mixing. Then at the end of the day, on day two, I began to mix on the platform. This is when everything changed. Judi could actually tell when I was mixing. I used to do all-nighters at the board."[29]

As soon as the sun set over Yasgur's farm, Bernstein pointed at the French sound engineer and said, "Go mix!" He obliged. At 3:30 a.m. on Sunday, Boroda mixed Sly and the Family Stone. After their set ended at around 4:20 a.m., Boroda heard through the makeshift intercom system that the band had praised his work during their performance. They were so impressed that at future gigs, the band requested the engineer to mix for them. "After Woodstock I became more in demand as a mix engineer."[30]

RECORDING

Prior to Woodstock, Nick Burns heard rumors of a big festival. Fairly soon, Hanley was hiring extra help from local college kids to build and prepare cabinets. Burns remembers: "Someone called and asked Bill if he could record the festival, Bill Hanley being Bill Hanley said sure! Next thing we know we are putting together all this stuff to record this thing. Then Judi started working with Tom Field and all of the producers to get stuff done."[31]

According to Hanley, the recording board was constructed from Langevin equipment. It supported a 16-channel parallel-feed microphone snake, split to the recording trailer located 110 feet offstage. The recording trailer housed two Scully 280 eight track (reel-to-reel) tape recorders responsible for the Woodstock film music soundtrack (and LP) and also a Nagra mono tape recorder that recorded the house mix for film sync.

Hanley's recording console arrangement had inputs in groups of ten, and multiple tiers that positioned the mix engineer to see and feel every knob. Without moving his torso, this unique design allowed chief mixer and master recording engineer Lee Osborne to mix effortlessly while the Warner Brothers representative (and engineer) Eddie Kramer assisted.

The work Hanley did on the Woodstock recording console inspired him to begin an even more elaborate one after the event—a console in the round: A semi-circular shaped unit, made up of three tiers, with two two-band equalizers per channel including presets. According to Hanley: "Speed is important in a sound console. Having everything in the engineer's field of vision as well as at your fingertips was and is very important."[32] Because of the post–Woodstock Festival financial fallout at Hanley Sound, the console in the round remained only partially built.

On the console that Osborne and Kramer used, each channel contained Geiling Potentiometers (pots), EQ, and amplification. Hanley explains: "It had over twenty inputs and twelve outs. It performed at only 3 dB under what the best do today. The console's EQ covered 50 Hz, 1 kHz, 2.5 kHz, 5 kHz, and 10 kHz in 2 dB steps with slider controls."[33] Individual vocal miking, and onsite submixes were fed to the recording engineers with Hanley overseeing. According to Niskanen: "Hanley was all over the place at Woodstock. He was a busy man. I remember him backstage working in the recording trailer diligently."[34]

Being so far away from the stage, the recording engineers had minimal communication from the mixing console. As a result, mixing and recording was all done blindly. However, there were efforts throughout the weekend to rectify the situation. To be better informed of a band's particular microphone positioning while mixing, engineers sent notes via a clothesline or hand delivery. According to Boroda, "If you were good at it you could come up with a mix in about thirty seconds."[35] Hanley recalls: "When someone finally got out to us through the thick audience with—let's say a microphone list for example—the bands had started playing already. It would have been fine if the turntables on the stage didn't collapse on us. We really did the best we could."[36]

The failed clothesline system, although innovative, was Cohen's idea, claims Boroda: "Sadly this makeshift device did not work for us. Harold suggested that we write down the stage set-up and wheel it over to them. I said to Harold, 'It's pouring rain!' It worked for a short time, then after that we had a delivery boy going back and forth. After that, we had headphones."[37] With trial and error, they got the intended stage intercom system up and running making it easier for the crew to communicate. "Before the note system, we were trying to get our intercoms to work, eventually they worked, and then they didn't. There was very little in the market for intercoms then,"[38] recalls Boroda.

Engineer Eddie Kramer arrived around 6:00 a.m. on Friday, August 15. He had been sent by Warner Brothers to assist with the recording. According to Kramer in *Acoustic Sound News*: "Hanley supplied the sound for the Fillmore East and they also had this eight track Scully sitting in the basement, which is where I used to record a lot of bands from underneath the stage of the Fillmore East. So they figured 'Oh, we'll get Kramer.' We've got the gear. He does Hendrix. And we'll send him on up there."[39] Hanley notes: "In comes Eddie Kramer dressed in a purple-blue crushed velvet blazer with his expensive boots covered in mud. Judi told me she heard him say that he was getting frustrated because he was getting dirty!"[40]

Hanley gives specific accolades to engineer Osborne, whom he hired to do the master recording of Woodstock. The sound engineer acknowledges Osborne's tolerance when dealing with challenges associated with the event: "Lee Osborne was a fine mixer. He was great under pressure and was able to handle most of our problems because of his vast knowledge and experiences in the live sound field."[41]

The recording of Woodstock stretched over three and a half days in three continuous eighteen-hour sessions. The result was sixty-four reels total on eight-track tape. According to Kramer it was an arduous task: "The stage crews were the ones who we were trying to communicate with, which was rather difficult. We had vitamin B12 shots to keep us all going [and] slept on the floor of the truck for a few hours. [We] rolled tape all the time, whereas the poor film guys they were struggling with five cameras trying to keep it all going."[42] Eric Blackstead, producer of the multiplatinum album *Woodstock*, wrote on the back of the jacket: "The recording of the music at Woodstock was a challenge of unprecedented scope and complexity requiring a level of endurance from both man and machine previously unheard of in location recording."[43]

Hanley was to be paid $15,000 for his sound reinforcement and recording services. This was an agreement he had made previously with Woodstock Ventures. After the festival the Woodstock recordings stayed with the sound engineer. But if Hanley wanted to be compensated, he was required to turn over the recordings to Warner Brothers immediately. Nevertheless, the sound engineer held on to the tapes until it was confirmed that Osborne would be in charge of the mixdowns for the film and soundtrack. A mixdown is one of the final steps a multi-track recording goes through before mastering. Since Osborne oversaw most of the onsite recording task, Hanley felt this appointment made sense.

In the meantime, Warner Brothers was getting anxious. They required the original master tapes in order to produce the film and soundtrack. On October

9, 1969, back in Medford, Bernstein received a letter from Sidney Kiwitt, vice president of Warner Brothers-Seven Arts, Inc. It demanded the recordings:

Dear Judi:

This will acknowledge receipt of your invoice to Woodstock Ventures, Inc. in the amount of $15,000 for services performed by Hanley at the Woodstock Music & Art Fair from August 15 thru August 17, 1969, in connection with your eight (8) track recording of the Fair.

You warrant and represent that the aforesaid invoice and the monies due you represent a proper charge by you to Woodstock Ventures, Inc. for services performed by Hanley.

In reliance of the aforesaid warranty and representation and in consideration for your assignment to us of your mechanic's lien on the aforesaid eight (8) track original tapes and further for your delivering such tapes to us, we agree on execution of this letter by you in the space provided below to pay you the sum of the $15,000.

We have advised you, and shown you the appropriate part of our agreement with Woodstock, re the aforesaid tapes, wherein the tapes are to be delivered to us and will hold you harmless from any claims by Woodstock against you because of your delivery of these tapes to us.[44]

Hanley would not give up the recordings until his requested arrangement was mutually settled. He waited until Warner Brothers and Woodstock film director Michael Wadleigh of "Wadleigh and Maurice Productions" agreed. Wadleigh was in charge of editing the film in Los Angeles. Hanley remembers: "I wouldn't give up those tapes until I knew that Lee was in charge of doing the mixdown. He was the best man for the job, it made sense, he was familiar with it. I felt that someone else doing it would not have worked. When they agreed, I gave them the tapes. Lee did a fabulous job!"[45]

Warner Brothers rented Hanley's equipment for around seven weeks for the mixdowns of the film. During this time, Osborne worked tirelessly cataloging and mixing the music for the movie (and album) for Warner.[46] The rental included eight- and four-track Scully tape recorders, one Teletronix Limiter, one graphic equalizer, one McIntosh amplifier, two 604 monitor speakers, and a custom Hanley console. But Warner Brothers never replied to Hanley's invoice of $13,727.00 for these services. "They never paid me for the rental of my equipment. I never heard back from them,"[47] adds Hanley.

The Woodstock soundtrack was released on LP in the United States on May 11, 1970, on a subsidiary label of Atlantic Records—Cotillion. It sold over one million copies in the United States, and earned gold record status. Hanley Sound is credited on the album.

The documentary film *Woodstock* was theatrically released on March 26, 1970, garnering acclaim. According to its Academy Award–winning director Michael Wadleigh, Bill Hanley and the work of his sound company was exemplary. Wadleigh recalls:

> On rare occasion when I had a chance to go out in the audience, I became massively impressed with the kind of sound coming out of Hanley's system. I always had this impression of Bill that he had great respect for the music. He saw the importance of the lyrics and message from many of the groups at the time. You needed intelligibility. I think that Artie and Mike strove to get the best and that's why I think Hanley got the gig.[48]

Because of these efforts, we can thank Hanley and his crew for one of the greatest concert film soundtracks of the twentieth century. According to a *Boston Globe* article, the Woodstock concert recording was "Technically the finest live recording ever made. Thanks go to Medford's Bill Hanley for that."[49]

McINTOSH AMPLIFIERS

Throughout most of Hanley's career the sound engineer exclusively used the McIntosh brand. Most of the McIntosh equipment came out of the Hanley Sound rental inventory. At one point, he was even interested in franchising the product. However, McIntosh vice president Gordon Gow would not have it. According to Hanley: "We never could quite figure it out? I guess it was the image he was trying to create. I suppose he was trying to stay with the high fidelity image."[50] Gow was a traditionalist when it came to music and technology, and simply didn't understand the need for his product within the developing field of sound reinforcement.

This was yet another innovative technological tipping point for Hanley. By using the McIntosh brand often, he created its inevitable need at live events. Hanley observes: "Electricity came along and freed the musician. Gordon was still of the belief that if you hang this single microphone in the right location, then you get the right mix, and you reproduce that in the living room of the listeners."[51]

In a jokingly heated conversation between the two gentlemen, Hanley reminded Gow that Hanley Sound's extensive use of the McIntosh product was pioneering. Hanley claims Gow felt that pioneers didn't make money, recalling what Gow said: "Pioneer Shmioneer! The Indians shot your ass with an arrow!"[52]

Case enclosures for McIntosh amplifiers were nonexistent at the time, and had to be custom built by Hanley Sound. According to Cohen, "We incorporated a number of modifications to the MC3500s, made the equipment roadworthy, including a road box with standard rack fittings, doors front and back for access."[53]

Just days before Woodstock, Hanley was looking for additional backup support for the product. The sound engineer put in a call to McIntosh Labs headquarters in Binghamton, New York. Unfortunately, McIntosh service manager Al Hyle could not leave.

Under the stage Hanley had a mix of tube and transistorized McIntosh amps, claiming, "Transistors were new and a bit risky, but we needed all the power we could get."[54] It was Holden who often ended up under the stage keeping watch of the more than seventeen amplifiers. If not carefully tended, they could get very hot and compromise the system. Helper Niskanen occasionally relieved Holden of his duties when he needed a break. "We were watching the VU meters on Hanley's amplifiers to make sure they weren't overdriving. Back then if they started to drive too hard they could pop speakers."[55]

To prevent overheating in extreme outdoor temperatures, Hanley kept his amplifiers cool by using bagged ice and blowing fans. "Below the stage we had 10,000-watts of tube power amplifiers for 11 kw of quiescent power dissipation. We used a few Crown DC 300s and Phase Linears as well. We used McIntosh MC3500 series (and 275's) 350-watt RMS high fidelity tube amplifiers that switched from 1-ohm to 62-ohm impedances. It was very difficult keeping them cool!"[56]

THE WOODSTOCK SECURITY WALL AND ALTAMONT

Often overlooked in Woodstock history is that Hanley was a key player in designing the security wall. It was an essential component for audience crowd control and accessibility. Unique for the time, the wall was intended to create an imposed distance from the audience to the stage.

Hanley's philosophy behind the security wall was a design that can be seen more clearly from an aerial view. According to Hanley, when seen this way, on either stage left or right: "The construction of the twelve-foot-high fences show how the crowd was kept orderly and funneled in. I designed and laid these out. I made the fences this tall on the sides for better crowd control."[57] After a rain, wet clothes, sleeping bags, and other items were randomly hung to dry along these outer fences.

The wall stood around ten feet high where it met the stage area. The massive stage deck was approximately seventy-six feet across and eleven and a half feet tall. A six-foot-high chain-link barrier fence was constructed and placed about eight feet in front of the stage. Beyond that was the ten-foot security wall. This area between the wall and stage contained space for camera platforms, made from plywood and sawhorses.

With a wall this high, if the crowd pushed forward they would no longer be able to see the stage; everything would be out of their sightline. Hanley's design allowed for better egress in front of—and on the side of—the stage, something he learned from watching crowds at other festivals: "If you look at the shots from the plane—the rest of the fences in the funnel—you'll see that there's these great big green areas. That was my idea to keep everybody within range of the sound system."[58]

By comparison, just three months later, the Altamont free concert in Tracy, California—held on December 6, 1969—is a good example of what not to do when preparing for a huge crowd. Woodstock technical director Chris Langhart, who attended Altamont, claims: "The West Coast mentality concerning space was much different than the East Coast. It was just stupidity, the stage was too low and the Hells Angels as security was a bad idea."[59]

The allure of the Rolling Stones drew over 300,000 attendees to Altamont. From the Maysles Brothers 1970 film *Gimme Shelter* the stage can be seen, a mere four feet off the ground. Merged with a crowd of pushing and shoving festivalgoers, heavy drugs, and a security force led by the Hells Angels, with no real sense of crowd control, it's no wonder this event ended in tragedy.

According to Robert Santelli's book *Aquarius Rising: The Rock Festival Years*: "By 11 am there wasn't a free square foot of space within seventy-five yards of the stage. The first aid tents were already handling cases of overdoses, and those who dropped impure acid were coming in regularly."[60]

The Stones hired Monck for Altamont as stage manager because of his excellent reputation. But the festival site moved and this posed a major and now familiar problem for all involved. This last-minute change resulted in minimal time to adapt the original design for its new location at the Altamont Speedway.

The original stage was designed to be on top of a rise. But now it was to be placed at the bottom of a slope. According to Monck in an interview for the *Canberra Times*: "The stage was one meter high—39 inches for us—and (originally at Sears Point) it was on the top of a hill, so there was no audience pressure on it. Because of the short notice in the change of location, the stage couldn't be modified."[61]

To Monck's dismay, the Altamont stage was one that could easily be climbed by determined and overzealous fans. The minimum height of the stage should have been twelve feet. Monck knew all along that a lower stage could pose real problems, which it did. "Building the low stage was an egregious mistake and a major reason for the violence at Altamont."[62]

This Stones tour had seen its share of chaos, and the Altamont concert was no exception. In the end, for Monck, this resulted in a hostile confrontation with the Hells Angels. "Monck confronted a member of the Angels stealing a large custom carpet that was part of the Rolling Stones stage set and lost teeth after being hit in the mouth with a pool cue. He later tracked down the person and managed to trade the carpet for a case of brandy."[63]

If the Woodstock stage had been the same height as the stage at Altamont, the scene at Yasgur's farm would have been very different. The problems that the Altamont concert faced are an example of what could have gone wrong at Woodstock if the sound and stage accommodations were insufficient. Hanley's design of the Woodstock security wall provided not only security but protection for the stage, and also corralled the audience into positions where they could hear best. According to Hanley: "Altamont was a disaster. If I had done the sound there I would have suggested a similar measure of security around the stage like we had at Woodstock."[64]

Chapter 41

I'M GOING HOME

Just before another thunderstorm, thousands of festivalgoers began to leave the Woodstock site on Sunday morning August 17. The evening's campfires still flickered around the hillside fed by those who remained, while a steady stream of others attempted to leave. By the late afternoon the great migration home revealed the 600-acre farm muddied and littered with mounds of trash. As far as the eye could see, parked cars lay abandoned along most of the access roads. The audience that braved the elements over the weekend was as tired and beaten down as those who put the event on.

Harold Cohen recalls mixing for New York doo wop group Sha Na Na at 7:30 a.m. on Monday morning August 18. The engineer was caught by surprise as emcee Chip Monck yelled, "HAROLD TURN UP THE MIKE!" during the group's performance. A huge fan of doo wop, the usually meticulous and attentive engineer was so awestruck he hadn't been paying attention. This slight goof-up at the mixing board has been forever immortalized in the Woodstock film. Cohen remembers: "I had mixed for Hendrix before and really enjoyed his music, but I was really into Sha Na Na's set! I didn't realize that their mike wasn't at a sufficient level. Because of this I was slightly chastised by Chip for good reason. It was one of the only times I had ever missed a cue!"[1] According to Sha Na Na singer and musician Jocko Marcellino: "We always knew it was Chip who said it and we assumed it was the sound engineer he was talking to. To this day during our performances we still say 'HAROLD TURN UP THE MIKE!' The audience gets a real kick out of it!"[2]

Burns was in the immediate area and recalls the audio fumble: "I mixed stuff at Woodstock too. If we were doing the sound for a show, almost anybody could mix and Harold just happened to be up there. I remember being out there setting levels and going through stuff and hearing Chip yelling at Harold."[3] According to Boroda, "Harold was technically smart, but back then he had no balls with regard to the pressures of mixing and was often scrambling at the board. In my opinion he was not cut out to be a live engineer."[4]

Though as fate had it, Boroda was one of the last engineers to mix sound at Woodstock beginning at about 2:00 a.m. until early Monday afternoon on August 18, with Cohen assisting. "I typically mixed at night,"[5] adds Boroda. Cohen recalls he wasn't really all that excited about Woodstock to begin with, claiming, "I really didn't want to go," adding he "would have rather stayed at home," and that he "had enough of festivals, by then."[6] To this day, Woodstock is one the most memorable moments of Cohen's career, even though at the time he considered it to be just another gig: "I was sick of the mud and rain."[7] With about 40,000 left in the audience, the engineers mixed Jimi Hendrix's Gypsy Suns and Rainbows performance, beginning around 9:00 a.m., lasting until a little after 11:00 a.m. and closing out the festival.

The entire Hanley crew was completely wiped out by late Monday afternoon. Cohen went home to New Jersey briefly for some needed rest. "I finished up with Hendrix. I was the only one awake aside from Judi who was talking to me through the headset to keep me awake. Someone drove my Firebird up, which was waiting for me after the festival. I got a couple hours of sleep and went home to my parents' house."[8]

After much needed rest, and with Woodstock mud still caked to his shoes, Cohen rushed back to Boston to meet his colleagues. On Tuesday August 19 he was to assist Boroda, Marks, and Holden for a sold-out Joan Baez concert at Harvard Stadium in Cambridge. According to Cohen, this show was done with the speaker stacks pulled from one of the Woodstock towers. Hanley Sound was also set to provide sound reinforcement for a series of upcoming concerts in Boston called Summerthing.

After Boroda mixed Hendrix he was flat-out exhausted. The sound engineer rarely participated in the breakdown of festivals. "I told Bill I would set up the festivals, no problem. But I would not break them down. I had my fill."[9] The burned-out and muddied Boroda grabbed his belongings and got in his Mustang, which was parked backstage. Although he tried, Boroda could not get out of the area quickly: "I drove as far I could and parked on the side of the road somewhere. I woke up twenty-four hours later and drove to Fall River to see my girlfriend. I met the crew at Harvard Stadium the next day."[10]

Among members of the Hog Farm and hundreds of others who stayed behind to clean up the mess, the Hanley Sound crew began the breakdown of the sound system right after Jimi's set. Holden remembers: "I was on stage when Hendrix was playing the National Anthem as the last act on Monday morning. When he stopped we struck the show and packed everything up. There were a lot of people on stage at that point."[11]

According to Dick Butler it was business as usual, just another show; however, navigating his rig through the Woodstock muck and mire was a challenge: "After the show during breakdown, that's all I was thinking, don't get hurt and don't get stuck in the mud. We were young and dumb, who's going to get hurt?"[12]

As soon as the breakdown was in process Bernstein in her authoritative style turned to Burns and ordered the shop foremen to grab some equipment for Harvard Stadium. Burns recalls, "After Jimi's set, which still gives me goosebumps when I hear it, Judi says to me 'Nick grab microphones and monitors and other essentials, we have to be in Boston by the next afternoon for a Joan Baez concert.'"[13]

Late Monday evening into early Tuesday morning, Nick and Judi packed the rented Dodge camper that was due back at the rental facility. They loaded it up with his tools and the two of them headed back to Boston. In Springfield, Massachusetts, the camper engine gave them trouble. "We needed to get back quickly so I went under the hood and disconnected the governor and boogied back pretty fast. After the engine finally gave up, we rented a car packed it up and proceeded to drive to Harvard Stadium."[14] Burned out and cross-eyed when they arrived at the venue, they were told to go home and get some rest. "And that was the end of Woodstock for me,"[15] adds Burns.

Tuesday morning after the final Woodstock breakdown, Holden and Brodie loaded the tractor-trailer and drove straight to Cambridge. According to Holden: "I can remember lying on the grass for that show and seeing Sam pulling up in his Mustang to mix the show at Harvard. I was so exhausted."[16]

The day after Woodstock the taping of *The Dick Cavett Show* on ABC, episode three, season thirty-seven happened using Hanley's equipment. Although recorded on Monday August 18 it did not air until the next day. The band had requested the use of Hanley's monitor system. David Freese, former Jefferson Airplane sound engineer assisted with the setup. Rick Slattery, who had recently joined the crew, was working in New York and was at the studio to assist. "The Jefferson Airplane needed a sound system, it might have just been the monitors I don't recall? But whatever it was, it was pretty small, because the audience was small."[17] For years to come, this episode has been a highlight for many "Woodstock" hungry fans.

During the taping of the show, Stephen Stills raved about the quality of Hanley's sound system over the Woodstock weekend. While sitting in the round, Cavett asked the group of musicians that included David Crosby, Joni Mitchell, and members of the Airplane two very interesting questions, beginning with, "Was the sound okay?" followed by, "Could everybody hear?"[18]

Stills responded immediately to Cavett's question with a hand gestured peace sign, emphatically proclaiming "Bill Hanley Sound!"[19] Here the musician offered the sound company a much-deserved plug. Slattery suggests that the show host might have asked this question because he saw the equipment being brought in: "Cavett probably needed some talking points to keep the show going, I am not sure why he asked such a question?"[20] But what followed was a reflection-based exchange on sound quality, monitor systems, and the ability to hear well while on stage. Crosby contributed a good description of what it's like being on stage and not being able to hear.

A brief, yet related conversation then ensued when Cavett asked Crosby, "What is it that drives you most up the wall about bad sound equipment?" Crosby responded: "Bad sound and especially bad monitor sound, of course that's how we sing. If you noticed earlier when Grace (Slick) was putting one finger to her ear. It just closes one circuit, and then you can hear by conduction and then you can sing. Even if you can't hear, and when you see someone doing that, that means there is no monitor sound."[21]

The accolades by Stills make perfect sense. Even before Woodstock Hanley had supported members of Crosby Stills Nash and Young on various occasions. In 1967 he had helped Stills's other band, the Buffalo Springfield, when they were having issues hearing each other while performing live.

Hanley Sound Inc. was a well-oiled machine by the time they arrived at Woodstock. From Hanley's beginnings at the Newport Jazz Festival in the early 1950s to the presidential inauguration of Lyndon B. Johnson in 1965, and several festivals thereafter, the growth of his company was evident. While Woodstock proved that people could join together peacefully in what was a disaster area, it likewise demonstrated to the rather small community of sound reinforcement companies what could be achieved technically. There was only one sound company that could have given Woodstock its voice, and that was Hanley's, a company led by a savvy sound engineer who was able to execute (with intricate detail) the applied science of sound reinforcement for one of the largest festivals to ever happen. Woodstock skyrocketed Hanley to fame, resulting in an influx of calls requesting his service from all over the country.

Nevertheless, after Woodstock Hanley experienced major financial losses. The bubble was about to burst on what was projected to be a lucrative market. The sound engineer estimates he lost over $400,000 when later scheduled festivals fell apart. Hanley reflects: "There were some, but not as many. A lot of idealism is what got me involved with the festival market in the first place. I wanted to make the world a better place to live by making it easier to communicate. This is what Woodstock was all about for me."[22]

THE WOODSTOCK BINS

Hanley and David Marks shared a mutual passion for human rights and peace through music. In the nine months Marks was with the company, the two forged an extremely strong bond that grew after Bethel. Sharing similar philosophies,

Hanley shipped a portion of his Woodstock equipment to South Africa in an effort to help assist Marks's fight against apartheid. According to Marks: "After I left, Judi and Bill sent my baggage by sea from Boston. It was too much to fly with in those days. I had two metal trunks full of records, posters, programs etc. As a goodbye gift, Hanley Sound gave me a state-of-the-art radio cassette speaker combo."[23]

In 1970 Hanley made a humanitarian contribution in fighting social injustice outside of the United States when he sent Marks's organization, 3rd Ear Sound (est. 1969), a sound system. Throughout the 1970s, equipment labeled "Hanley Sound" surfaced at many free, open-air, folk festival music events held in the South African townships. Marks recalls, "For many years throughout Southern Africa this system was collectively referred to as the Woodstock Bins."[24]

Hanley was very anti-apartheid and expressed to Marks that he didn't want his system played to segregated audiences. According to Marks: "This was the genuine article from Hanley Sound of Medford, Massachusetts, USA. Bill sent an impressive amount of equipment to help us with our cause."[25] The equipment Hanley graciously sent to South Africa included the following:

> Altec designed 4–15 marine ply-cabinets that weighed in at half a ton each, stood 6 ft. straight up, almost 4 ft. deep and a yard wide. Each loudspeaker carried four 15-inch JBL D140 woofers. The tweeters consisted of 4x2 cell and 2x10 cell Altec Horns. At the mixing end of the system were 3x4 channel ro- tation-pot Shure ME mixers, daisy-chained to give us the awesome capability of mixing twelve microphones all at once. We had one 10-band home-made graphic equalizer for the entire system and a Teletronix Limiter/Compres- sor—our single most impressive and revolutionary bit of gear; all driven to distortion by a bunch of Crown D300 power amplifiers.[26]

When Marks received Hanley's Woodstock Bins, he was able to launch the first Free People's Concert on a beach in Durban, South Africa. The following year it moved to Wits. Marks recalls: "My goal was to organize a free concert and offer our great musicians a platform. Wits University was one of the very few venues in the country where we could present mixed bands and audiences; it was a place where township and suburb could meet."[27]

According to Marks his event used a loophole in South African law: "If the event was a private function then the entrance was free. If the musicians played for free, then there could be no restriction on who attended."[28] This system worked for a number of years until permits were eventually required.

> The focus was unreserved freedom with an anti-apartheid undertone. South Africa was at war; the townships were in flames and many South Africans,

including Wits' students and academics, were in jail or underground or in exile. Anyone who opposed the status quo was shot at and dragged into police vans, even if they were doing nothing more dangerous than dancing. The Free People's Concert offered respite. It opened its doors to people from every race and sector of South African society and explored the nation's shattered psyche and unclaimed future through the music of the times.[29]

By 1971, Marks had received the first of his royalties from his hit song "Master Jack." Soon he invested his money into presenting, promoting, and recording South African musicians. Marks reflects, "What struck me was how lucky I'd been with 'Master Jack' when there were so many musicians in South Africa who weren't being recognized."[30]

By 1985 the Free People's Concert grew into a nationally known event seeing audiences over 30,000. Still intact and slightly worn, Marks sent the Woodstock Bins back to Medford that same year. Marks is known as a self-appointed preservationist of South African musical heritage and continues to work tirelessly for the preservation of his nation's music repressed under apartheid, music that Marks claims has "been forgotten." He still performs music and produces shows through his organization 3rd Ear Sound. As of 2019 Marks is working to preserve his history and involvement with Hanley Sound.

The South African sound reinforcement industry owes a lot to Hanley's contributions, but also to the work of Marks, who had foresight and vision back in 1970. This is the first time we see Hanley's influence emerging in another country. According to Marks:

As the years wore on and the dog-eat-dog newly born sound industry of South Africa started to stretch in the late 1970s, many root-stories would grow into urban legends; war stories by rival sound companies and their crews began to surface. Without exception all those early music sound companies evolved out of 3rd Ear Sound and Bill Hanley's Woodstock Bins, no matter how much money they have made. But the industry has matured and become extremely hi-tech and internationally comparative and innovative; pity then that the campfire or barroom war stories that one hears from music and sound-people have become so boring by comparison to what was happening in the USA and UK. The reason? Just as with music, you can't reinvent history. The world doesn't work like that.[31]

Marks and Hanley did not meet again until 1997 at an annual Woodstock Festival reunion on Yasgur's farm. "At the reunion, I was embarrassed to say to Bill that a new breed of South African sound-people had no idea of their recent roots and the Woodstock connection."[32] And in this way the distinctly powerful global influence of Woodstock has been felt for fifty years.

POST-WOODSTOCK

With word spreading fast, Hanley's expertise in sound was in high demand. On September 6, 1969, a one-day festival called the First Annual Midwest Mini Pop Festival occurred at the Cincinnati Zoo Amphitheater, with around 5,000 in attendance. It was produced by concert promoter Jim Tarbell, who first met Hanley at the Fillmore East in July 1969. When the crew arrived to set up the sound system, Tarbell claimed it was still "caked in Woodstock mud."

Tarbell also contracted Hanley Sound for an installation at a club he was opening in October called the Ludlow Garage. When he closed the popular Cincinnati venue in February 1971 its marquee had included "Sound by Hanley." According to Tarbell, Hanley was very easy to work with:

> When I decided to open my club I went to the Fillmore in July to shadow Bill Graham to see what he was doing. I remember Hanley being there and I told him about my pop festival and that it was kind of a teaser promoting the opening of Ludlow Garage. I thought if Hanley was good enough for Bill Graham he was good enough for me. So I asked him if he would do my festival and a permanent installation at the Garage. He told me after he did Woodstock in August that he would leave some of the equipment there for my festival and later the Ludlow Garage. At the time there wasn't much choice for sound and I was a perfectionist of sorts. I had heard of Hanley before I went to New York; it was pure coincidence he was there.[33]

Another major call after Woodstock was from one of New York's premier venues, Madison Square Garden (MSG). The arena had been plagued with sound problems. A relatively new venue, those behind some of its original design were not anticipating the extreme loudness of rock and roll.

This rise of big arena rock did not please Hanley's former customer, Fillmore East rock promoter and producer Bill Graham. As soon as MSG sponsored rock events, word spread about its poor sound and Graham reveled in it. The producer claimed it was a monstrosity unworthy of even playing an "acoustic guitar in" calling it the "biggest fucking rip off in the city" more suited for "chariot races, roller derby, and boxing."[34]

Clearly the producer was worried that audiences were getting too comfortable seeing bands in big arena settings. One after another, acts dropped off the smaller theater and club circuit, playing larger spaces, and this negatively impacted his theater. According to Hanley, "MSG was the next venue after the Fillmore East where the acts made more money, it was bigger and louder."[35]

Graham was not supportive of the many pop festivals surfacing around him, although he eventually came to accept them. Other promoters like Sid Bernstein saw the monetary potential early on. Bernstein was actually interested in bringing acts into bigger venues, willing to embrace this shift into large arena rock. In his book *It's Sid Bernstein Calling*, he writes: "What was coming across loud and clear was that the gamble I had taken by putting the Beatles in Shea had set a precedent. Rock and roll had proven that it was capable of drawing huge crowds to very large venues."[36]

At the same time, in the mix of Hanley's growing clientele were the massive anti–Vietnam War demonstrations popping up all over the country. No other sound company seemed to want these dangerous jobs except for Hanley. His passion for making the world a better place was almost as great as his passion for sound. Organizers who learned of Hanley and his sound company's reputation tracked him down, placing the sound engineer and his crew in front of hundreds of thousands of protesters.

Chapter 42

MADISON SQUARE GARDEN AND THE UNIONS

MADISON SQUARE GARDEN

New York's new Madison Square Garden (MSG), commonly known as the Garden, was fully constructed by 1968. It was an improvement over its previous location at 8th Avenue and 50th Street; the new MSG offered "deeper" and more "comfortable" seats arranged for improved accessibility. Along with wider and better organized aisles, the lighting and sound were designed to be the finest at the time. In 1966, a *New York Times* article had extolled, "The acoustics, always a problem in the present building, will be the best possible, with a series of electronically controlled speakers synchronized to serve each area individually."[1]

The 20,000-seat, five-story-high MSG was constructed over Pennsylvania Station, with its main entrance located between 7th and 8th Avenues. An ambitious architectural venture, the indoor arena was built in several stages. Before the main arena was complete, the smaller 5,000 seat Felt Forum, located four levels below the Garden (at street level) opened in November 1967. It presented boxing matches and other events. Finally on February 11, 1968, MSG opened its doors with great anticipation. The *New York Times* claimed, the modern arena could offer ". . . basketball, hockey, track meets, boxing matches in quick succession. Additionally, it could host circuses, horse shows, concerts, various ice shows, and many other exotic events."[2]

During the final construction of the main arena, bad reviews surfaced about the Felt Forum's acoustics during live events. As the venue hosted more and more performances its inadequate public address system was a noticeable problem. In a 1968 *New York Times* article, "Music: Echoless Caverns; Felt Forum, Meant for Sports, Lacks Bass," writer Harold Schonberg reviewed an American Symphony Orchestra performance at the venue: "The sound is very bad. Even the old studio 8H of Radio City was better than this."[3]

Sound systems for arenas like the Forum were designed with a focus on the delivery and quality of the spoken word rather than musical sounds. Schonberg acknowledges this: "It is basically a sports palace and was not designed with concerts in mind. In the rear, the orchestra had weak treble and virtually no bass, and it sounded like a 1930s recording. Closer to the raised stage, the sound improves, naturally, but again the lack of reverberation makes chords come out with a dull thud."[4]

Other sports complexes across the country faced similar acoustic problems with live music, especially rock and roll. When this style of music was introduced to the larger MSG "state of the art" sound system, it proved no different. According to Gary Keller, then assistant superintendent of the facility: "MSG was brand new and it just wasn't designed to handle the loud sound that rock concerts were putting out. Our system didn't have huge bass speakers or enough distribution capabilities for rock and roll sound, it just couldn't handle it. We needed to hire somebody to install these outside systems."[5]

Before Hanley arrived, MSG managed sound through two public address systems. The first was an end arena system on a track suspended from the ceiling that had the capacity to move in forward and downward motions. The second was a system in the round that measured about twelve feet in diameter. Permanently located in the center of the arena, it could change elevation. According to Hanley, racks of 50-watt Langevin amplifiers powered these systems: "They were located stage right under the grandstands in head electrician Norman Leonard's office."[6]

MORE BAD REVIEWS

Word spread fast that the Garden PA lacked quality. This resulted in many rock groups bringing in their own amplification systems. Even so, nothing proved sufficient. As more acts were booked, key decision makers at the arena became worried. MSG president Irving Mitchell Felt, who was known to "closely follow the trends in rock music,"[7] was among them.

According to David Hughes, assistant to Gary Keller, bands started demanding more than what the Garden could offer: "It was a system from the early sixties, a center speaker cluster with a bunch of horns and some amplifiers. Bands like Cream and Blind Faith would come in and bring in their own speakers and sit them on stage and that didn't work out well."[8] New Jersey concert promoter and producer John Scher claims many bands weren't used to playing venues this size. This included Cream, who performed at MSG during their farewell tour on November 2, 1968. Scher remembers:

The old MSG on 50th St. never had a rock concert in it. Mostly because during the 1950s and even throughout the 60s bands rarely played arenas.

Usually acts would play the Capitol or Orpheum in Boston as well as various other theaters across the country. I believe one of the first rock bands to play the new Garden was Cream. I used to manage Jack Bruce and he told me quite candidly, "We didn't know what the fuck we were doing!" Nobody had ever played a place that big, other than outdoors. Cream played in the round on a rotating stage so everyone could see them. Everyone was experimenting back then.[9]

On January 24, 1969, West Coast group the Doors performed to a sold-out crowd at the arena. The lackluster sound quality at this event was an imminent sign of things to come. Writer and reviewer Mike Jahn claims: "Rock concerts in the Garden and places of similar size are always a dubious enterprise and this was no exception. The microphone system made the group sound like the music was being played through a broken transistor radio. It was hard to hear the lyrics, and a large measure of The Doors' value is based on those lyrics."[10]

By 1970, the Doors were traveling with a more advanced sound system. According to one of their crew members, touring with a PA was difficult: "It sometimes gets a bit hectic going from town to town on a tour. The group came in by plane from Cleveland the night before. We had a couple of people bring the equipment by bus. This made it with no problems, but sometimes it isn't that easy." By then, being able to hear was top priority for the West Coast band. According to *Billboard*: "The Doors use their own PA system, thus insuring themselves of being heard in a large auditorium. A member of The Doors crew said this is done since some promoters do not furnish a PA system and if they do, it may not be adequate for the group's needs."[11] Dick Gassen, a promoter who booked the band many times, said, "Good sound is a sense of pride to the group."[12]

In July 1969 supergroup Blind Faith performed on a rotating stage in the center of the Garden to a packed audience. It was the band's first US performance. Critical of the English rock group's positioning within the arena, *New York Times* writer Robert Christgau claimed the stage (which made a complete rotation every four minutes) "revolved like a "Lazy Susan," causing "acute acoustic and visual problems," adding, that the "sound fluctuated drastically as the amplifiers turned."[13]

For the same show, *Village Voice* reviewer Lucian Truscott recalls the group providing their own amplification instead of using the house PA, claiming that the Garden sound system was only good for "vocal announcements of fights, and tennis matches, not for amplification and singing."[14] Comparing MSG's acoustics to that of a "high school gym," Mike Jahn (who most likely didn't realize the band wasn't using the house PA) observed: "As usual, the Garden sound system was bad, and the breaks between songs were punctuated by indignant shouts to that effect. Considering the acoustics and the size of the house, Blind Faith did rather well."[15] Truscott continues that opening act Delaney & Bonnie trudged through the Garden's terrible acoustics: "The position of their mikes had them picking up sounds of the amps as well as their excellent singing. The result was a buzzing mess that made all but the backbeat of the vocals virtually inaudible."[16]

According to Hanley: "The rotating stage was not an ideal position for a sound system and it took some time to get it right. I used my system for Joan Baez in early August, but the Garden didn't want to spend the money on a new system. I couldn't convince them until I was finally called in after Woodstock."[17] By late summer of 1969 Hanley and his crew had created a unique sound system design for the rotating stage.

HANLEY ARRIVES AT THE GARDEN

By August 1969 Mitchell Felt ordered Garden executive vice president Alvin Cooperman and superintendent Dick Donopria to bring in Hanley for help. They finally realized that a new sound system was needed. Felt, Cooperman, and Donopria knew the experienced engineer was the right man for the job because of his recent work at Woodstock. Keller relates: "We had hired Bill Hanley because he had done the Woodstock Festival. He came in as a consultant and to install sound systems for several concerts that were coming up."[18] Hanley remembers: "They were getting panned in the papers and right after Woodstock I got a call from Mr. Cooperman. They really needed my help. They wanted me to solve their problems and were worried about losing the venue as a theatrical, musical showplace. MSG didn't have much of a sound system."[19]

When Hanley arrived at the venue, Donopria had the engineer work closely with Keller and Hughes. Both men ran physical plant at the arena and were anxious about improving the sound. According to Hughes: "Hanley was getting some sort of fee to help us out when rock and roll was developing. He had some fame under his belt, and started helping us hang sound. We were struggling because we had a very inadequate sound system at the Garden that wasn't designed for rock music."[20]

Hanley's new strategically placed sound system was first used on October 17, 1969, for folk singer Donovan. A newspaper advertisement for the performance included "Sound by Hanley" touting "A superb new sound system in a magnificent new arena. A perfect setting for the world's greatest audiences, to hear the world's greatest music."[21] For that show, Donovan sat on a raised platform amidst a stage covered with flowers. Garbed in a white flowing shirt and acoustic guitar, the Scottish folk singer performed in front of a sold-out crowd.[22]

In the coming months at MSG and the Felt Forum, the sound engineer provided sound for the biggest acts in concert history like Raphael, the Isley Brothers, Herb Alpert & the Tijuana Brass, the Rolling Stones, Johnny Cash, the Everly Brothers, Janis Joplin, and Sly and the Family Stone. Hanley's office, where he stored most of his equipment, was located on the first level of the Garden in the old boxing department on the 31st Street side.

Hanley was at MSG for about a year and a half. During this time he encountered difficulties working with the unions. Hanley's job was time sensitive, and the union crews slowed down his process. Once Hanley arrived for a show, he and his team were forced to hand over their equipment to the waiting union members. This caused a lot of friction. Such roadblocks cost promoters and Hanley thousands of dollars. According to Keller: "Bill Hanley built the framework for hanging all these speakers at the Garden. He installed and removed them. Hanley also had to deal with the various unions that were there at the time. I don't think we paid Bill Hanley to do the sound, the promoter did."[23]

UNIONS

When Hanley arrived at the 20,000-seat arena for a show, the big question was, who was going to actually install his sound system? Would it be him and his crew, the Teamsters, the IBEW (International Brotherhood of Electrical Workers), or I.A.T.S.E (International Alliance of Theatrical Stage Employees)? After heated negotiations between the parties, the final agreement was that the IBEW would install Hanley's systems not I.A.T.S.E. Hanley explains: "The IBEW was in charge of the in-house sound system. However, in the beginning there was a fight between the two unions on who would handle all of my equipment."[24]

According to Keller, "The decision was made that when the outside sound systems came in from Bill Hanley, the electricians, who were permanent employees, ultimately became responsible for the Garden sound system as well as the operation and maintenance of Hanley's system."[25] This was always an ongoing "dilemma and a challenge" at MSG, Keller adds, because the stagehand union (I.A.T.S.E.) had been involved with the installation and removal of equipment for ice shows and concerts in the past.

It became clear to Hanley that the IBEW had little or no concept of what quality sound reinforcement application and implementation was. During his time at MSG he fought hard so that the IBEW would not get their hands on his mixing console, something they had no knowledge of. Hanley often felt that the union workers failed to grasp the show business side of what the performer and management required to facilitate quality sound on such a large scale.

Hanley recalls: "The IBEW had a lot of electricians whom they ordered from union hall, but many of them knew nothing about sound. The IBEW wanted to be in charge and in control of the entire sound system. They just weren't familiar with this type of work. They screwed in light bulbs and wired buildings."[26] According to communications scholar Dr. Donna Halper, it was not always the fault of the union: "Back then, there were lots of turf wars, and many of the IBEW and I.A.T.S.E. folks were older and more conservative. Not necessarily politically, but in terms of what they were used to. They knew how to set up big band concerts, but some of the intricacies of rock were a mystery to them. But unions themselves were not the villain here."[27]

With the sound engineer's hands tied, he had to step aside and watch union workers set up and tear down his equipment, sometimes at a snail's pace. According to John Scher, working alongside unions took some getting used to: "The union guys were smart and held their ground. And for someone like Bill Hanley who wanted to do things on the fly, the union rules conflicted with that. It would drive him crazy. But if you were going to promote and work at the Garden you learned to live with it."[28]

During the unloading of Hanley's tractor-trailer, the engineer was usually on stage directing unions where they should place his equipment. According to Hughes: "The teamsters unloaded the sound system off of the trucks, the riggers would rig the ceiling for the trusses, and the IBEW would mount the speakers on the trusses and fly the speakers and run wires. I remember watching Bill telling the union people where to put the stuff."[29] According to Nick Burns, the protocol after the Hanley Sound crew arrived at MSG was somewhat annoying:

> You would pull the truck up in front of MSG. You would get out and go sign in. They would have a driver that would back the truck in to the loading dock. We would open up the back doors and pull the lift gate down. The crew would come in and move the cabinets, then the rolling cabinets to the lift gate, then to the loading platform. Then another crew would come in and move them to back stage. Then another crew would come in and move them to out in front of the stage to where we would set up. Then, if we were lucky, we were allowed to hook up our own equipment. They were bastards.[30]

Truck driver Dick Butler recalls the Peterbilt 356 Cabover Air-Ride tractor-trailer was fairly new at the time and equipped with a lift gate that helped with the efficiency of loading in and loading out: "The Peterbilt had a Maxon Lift Gate, which was unheard of at the time. It really helped us unload equipment at some of these farm shows and arenas back then."[31]

As Hanley's systems got bigger and more intricate, promoters grew increasingly frustrated because of the extraordinarily long setup times. Hanley explains: "It took a long time to make sure there was a midrange horn pointed at every ear in the house. They didn't have a great foreman, and I wasn't a great foreman either, so that was part of the problem. They just didn't understand the business."[32] Promoters were paying thousands in labor for electricians, while the sound engineer was making a mere $800 to $1,000 a night.

Hanley Sound was an independent contractor and was most often engineering on the fly. If the union brought in more men, that meant additional labor costs, deflating the promoter's profit. After all the trucking, fuel, and labor expenses, the sound engineer often came out in the red. According to Scher, putting on a show in a union run house could be costly:

> The MSG union situation is very simple to understand. A number of different unions had a contract with the Garden. NYC is a tough union town. These union contracts continued at the new arena and they were reasonably onerous. They worked hard but it was the work rules that cost a promoter like me a fortune. In those days they would only let a teamster drive the truck up the ramp and there weren't that many teamster drivers in the rock and roll business in those days. So concert crews would have to stop on the street and wait for a crew of teamsters (four or six) to greet them at the ramp. Then they would unload the tractor-trailers where they were parked. Then you had to pay them for four hours even though it only took them an hour. Then the stagehands came in and took the equipment up on the stage. Then once it was on stage the electricians' union came in. They are the ones who plugged everything in. So you had three unions doing the job that really just the stagehands could have done. These were the work rules, and the minimums that you have to pay. It was all in the contract that you signed. No matter how much you could discuss and negotiate with them they have a minimum of hours per job and that's what they are going to get.
>
> You could play a show at MSG and then go to the next city and your production bill would be half. When they opened up the Meadowlands they negotiated a deal where the stagehands did everything except for one electrician. So production manpower was reduced tremendously. MSG didn't ever try to do much about this to my knowledge. But it didn't matter much, because it is an iconic venue and acts wanted to play there. I am a pretty big union believer, except for when they take advantage. However, to this day MSG still sounds great and the union guys are all hard working and know what they are doing. If you treat them well they will work with you.[33]

According to Hanley, head IBEW house electrician Norman Leonard carried the electrical license for the building, giving him the power to allow a show to go on or not. The sound engineer claims Leonard was more interested in being paid rather than doing quality sound work. Always keeping a constant and watchful eye over his domain, the headstrong electrician was constantly barking at Hanley and his crew to not touch "this wire" or that "that wire."

Hanley Sound engineer Brian Rosen recalls crew members having to learn the rules quickly: "We had to find out the rules of what you could do and what you couldn't. Then you had to work within these rules. Could we plug this into the wall? Could we put this down somewhere?"[34]

Most in the business back then used a two-wire system. However, the circumspect and experienced electrician turned ornery if things weren't grounded right. According to Hanley, Leonard was infamous for his concerns over grounding issues: "I remember him being worried that everything be properly grounded. On one occasion I even watched him drop one of the trusses and put his meter on it just to see if it had current running through it. If there was, it was a nightmare because you knew you would have to take everything apart."[35]

According to Fillmore East alum Bob See (See Factor Industry, Inc.), he and others like Hanley butted heads with the electrician: "When we first started grounding, we took a #2 conductor and ran it directly to the truss from a ground. And he allowed that for a short period of time. I'd have battles with him. He demanded that every piece of cable be three-wire."[36]

With Leonard's influence, a three-wire system was eventually developed, making system-grounding connectors a standard in the touring industry. Anytime a show arrived at the Garden, Leonard ordered it to be rewired, and if it wasn't to his specifications he made sure it would be. After several tour hang-ups with Leonard, word spread that if you had a date at MSG it better be Garden-Proof. Leonard stopped anyone, even a three-ring circus, if they weren't "three-wire" compliant. According to Richard Cadena's book, See claims that when Leonard introduced system grounding conductors to live entertainment touring, he "single handedly changed the industry."[37] At many festivals systems were left outside in the elements and this worried Leonard. According to Hanley and See, if there was dampness of any kind Leonard became extra concerned.

Many innovations in the evolving concert and touring business started in the late sixties and early seventies. Concert production technologies like portable sound, portable lighting, and power distribution were ideas rooted in the theater, motion picture, and television industries. These developments involved a dedicated group of innovators in order to meet the flow of challenges and risks. But production safety was often left by the wayside. This resulted in some individuals risking their lives in order to build the infrastructure of concert production.

Nick Burns, who had a hand in concert lighting, recalls some of the dangers while working at MSG. On a number of occasions he modified and used a mountain climbing harness so he could dangle hundreds of feet off the arena floor. The MSG cable suspended ceiling structure was an impressive 425 ft. in diameter. Burns remembers:

> There was no OSHA back in those days. If you fell they would grab a dustpan and sweep you up and keep going. The unions were eating donuts and coffee while we were hanging stuff, sweating our asses off in mountain climbing gear. We were trying not to kill ourselves hanging high up from the sub ceiling. We ended up buying carabiners, mountain climbing rope, and harnesses at Eastern Mountain Sports and made our own safety lines.[38]

JOAN BAEZ AT THE GARDEN

During a sold out "Sid Bernstein Presents" Joan Baez performance on August 8, 1969, Hanley found the unions particularly difficult. Earlier that day his eighteen-wheeled Yellow Bird tractor-trailer arrived at MSG. Dick Butler, known for his ability to navigate the Peterbilt through the narrow streets of Manhattan, precariously parked the behemoth at the Garden's loading dock: "I always made sure I was on time. They would say, 'You can't fit that rig in here!' and I would say 'Oh yeah? Watch this!' and I would have fruit carts ducking me! You had to make yourself a bull. Most of the drivers in that day would say 'you never say you can't do it until it won't fit!'"[39]

The Baez show was David Marks's first as a Hanley Sound crew member. Accompanying him was Hanley Sound veteran Rick Slattery, who warned the newbie of the constant tension with the unions. It was hard for Marks to realize the complexity of the issues; he was ordered to wait for union workers to arrive, instead of unloading the Yellow Bird. Once the gear was off the trailer, Slattery and Marks were not allowed to touch the equipment. As the day wore on, Marks observed the painstakingly sloth-like production process.

Knowing which union was responsible for what, and where equipment should go was a big problem for Hanley and his crew. From riggers, to stagehands, to electricians, all of the jobs were partitioned. Butler recalls: "They would look at us like, 'Who are these longhaired hippy freaks coming in here?' They were used to theaters and acts like Frank Sinatra and Guy Lombardo. You had to talk their language in order to get things done. Back then they didn't want anyone touching anything unless they did it."[40]

However, according to Boroda, playing a union hall was not always so bad. He recalls:

With the festivals we would often have volunteers helping us set up all of the sound towers for example. However, the tear down wasn't like this. I would find myself alone quite often when things needed to come down. I would call the office in tears because I had a forty-five-foot truck to load and no one around to help. Only when we played union halls was there guaranteed help to set up and reload.[41]

Per Hanley's direction, Marks was to make sure the union stayed away from his soundboard. Needing something to do, he was asked to clean the scuffed up and dirtied floor monitors during the folk singer's scheduled sound check. As Marks approached the rotating stage it was being decorated with flowers. Halfway into his task, Marks recalled what Slattery had said about not touching any equipment and stopped. This caught the eyes of union workers and caused a stir.

Although frustrated, Marks was empathetic. It was only two years prior in 1967 that he was working at the Free State Gold Mines in South Africa. Recalling the abysmally backbreaking labor and horrid working conditions he experienced while working underground, he could relate to the effort the union workers were putting into their jobs: "I thought of those poor buggers that I had shared my time with underground for five years, earning thirty five cents or less a day—almost a dime in American money; digging out the gold that helped make the USA the most powerful nation on earth. They had no union and no way to bargain."[42]

That afternoon after Hanley's lunch, which always consisted of chicken, his favorite food, he calmly negotiated with the angered union workers. Marks was a South African and the sound engineer suspected they might not have liked that. Hanley was known to do well in adverse situations, and according to Marks he always worked through problems in the calmest fashion. The crew member gives a descriptive account in his journal of what it was like working with union crews at MSG. At this particular show, time was of the essence:

Bill, ever the New England Boston gentleman, would patiently point and guide the shaking union hand toward the correct connector—no not in that one, in that one, male into female, the traditional way around. That cable, not there, there we are . . . that's it. Little more to the left, as the bifocals steamed up like a granny squinting one-eyed with a thread and needle up to the light, aiming and hoping for a hit. The cables slowly got laid and plugged in—the crew struggled to keep their hands off the connectors and the union men's throats. It wasn't easy—it was just incredible—like out of a movie even Andy Warhol couldn't cook up in a drug stew. Wanting to get our little hands on to move this and that to the right spot, into the amplifier, across the crossover back to the mixer, over to the speakers and so on. What a painful union operation.

Then the clock struck home, time and the changing of the guard. Three hours to show time and we weren't near to connected yet. Bill sprung into negotiation mode. Remember that the next shift of stage-hands and riggers were now on over-time; so no matter how long it took not to get the job finished, so that it didn't have to get done in time, you could earn a few extra union hours if you just sat back and let the sound crew do their job. It worked—and in an hour the gear was set up and sounding. There was a little more flexibility up where the lighting consoles were, because the electrical union's overtime man took one look at the mess of faders and knobs and his eyes glazed over. 6 o'clock being well past his brandy and bedtime, he fell asleep. We were not so lucky down at the sound desk in the auditorium.[43]

For the Baez performance, Hanley and fellow engineers Boroda and Cohen designed a special sound rig for the Garden's revolving stage in the middle of the arena. The stage turned slowly, giving the audience a unique 360-degree view of the folk singer as she performed. According to the *New York Times*, "Joan Baez, standing barefooted on a circular platform strewn with ferns and flowers, sang last night to an audience of 20,000 that filled what she referred to as 'this cozy living room.'"[44]

According to Marks, the uniqueness of this stage and previous sound problems inspired the engineer to develop the—Hanley Hula Hoop—a horn-loaded ring placed above the center of the revolving platform. Using block and tackle (pulleys and straps) crews hoisted the Hula Hoop and his surrounding heavy HSI 410 bins up into the Garden's ceiling structure. Marks recalls: "The high-end sectoral and multi-cell horns were strapped around a huge metal ring that was set and pulled up over pulleys from the stage. All the cabling was skillfully twisted so that when the stage turned, the cables would untwist and not get tangled up with the artists and stage monitors far below."[45]

Hanley explains: "We built this thirty-five-foot in diameter circular pipe to be suspended above the rotating stage. It was constructed in four sections and bolted together with eight speaker horns (tweeters/squawkers) clamped to it. Then we suspended (by cable) eight custom HSI 410 woofers/speakers dead center of the arena completing the sound system for the theater in the round."[46] This speaker configuration allowed the audience to be able to hear completely.

That evening the folk singer held the audience's attention "like a master show-woman," crediting the Hanley Sound system with outstanding performance. According to the *New York Times*, the sound for this show was "Great, and for the record, since the Garden's sound system has been castigated so frequently, the amplification was excellent except for a slight, occasional background hum."[47] The sound engineer recalls the familiar "hum" coming from the Garden's noisy air conditioning system.

Brian Rosen remembers the concert as something special, claiming it was "The most amazing single performer concert I had ever done while with Hanley Sound. Especially her solo of Amazing Grace."[48] Although a success, there had been commotion at the soundboard. According to Rosen, who was at the mixing console, Hanley wasn't allowed to turn a knob without asking permission:

> I remember very distinctly the rule was that their guy had to run the board but Bill could tell him what to do. However, the guy wasn't very good and it was loud and hard to hear. This guy was a fifty-year-old electrician and he didn't know much about sound. I recall there was a union steward who was nearby and I remember him telling Hanley "You tell Jim what to do and he will do it." Bill was trying and trying to get him to do the right thing, but he was getting increasingly frustrated. So Bill just reached over and twisted the knob to the right position, pissing off the union steward.[49]

THE BAND AT THE FELT FORUM

Hanley remembers the difficulty he had at the 5,000-seat Felt Forum supporting the Canadian roots rock group the Band, claiming the "sound issues" stemmed from conflicts he had with the IBEW.

In those days, members of the Band were perfectionists about their sound. Although they had the option to perform at the larger Madison Square Garden, they were more interested in the "intimate" environment the Forum offered. Uninterested in screaming fans, guitarist Robbie Robertson said, the Band "just wanted to be heard." Yet at this event nothing was going right, and they took it out on Hanley.

At one of the four shows performed on December 26 or 27, 1969 (two performances per night), writer Alfred Aronowitz claimed: "You could hear Bill Hanley grumbling about what perfectionists they are. Hanley is one of this country's great mad wizards of sound amplification and is grumbled about often as a perfectionist himself. But his speakers were garbling some of The Band's harmonies and the vocal mic's weren't being mixed right."[50]

Brian Rosen remembers the group actually getting physical with Hanley's equipment at the show: "If they didn't like the sound they would take it out on us. The Band would kick Hanley's monitors!"[51] In true Boston fashion, Hanley later said that the Band were "pains in the asses" to work for.

Chapter 43

THE GREATEST ROCK AND ROLL BAND
IN THE WORLD

THE ROLLING STONES TOUR OF 1969

Facilitation of production in a timely fashion while working with the IBEW was difficult for Hanley and his crew, but it in no way disrupted his focus. The sound company's roster of concerts and festivals in 1969 was impressive, ending the year with the historic Rolling Stones tour of 1969. Of special note were their performances on November 27–28 at Madison Square Garden. Shots of these legendary Garden performances can be seen in the Maysles brothers' documentary film *Gimme Shelter*. According to some, this tour was like no other. Rock critic Robert Christgau referred to it as "history's first mythic rock and roll tour."[1] Some of the performances can be heard on the Rolling Stones album *Get Yer Ya Ya's Out!*

Following the tour, and three days after the Stones' famed West Coast Altamont concert, on December 9, 1969, Judi Bernstein issued an invoice to Stone Promotions Limited for $85,000 for sound services rendered:

> Sound system design & installation for Rolling Stones Tour of U.S.A. for the following dates now confirmed: Detroit—11/24/69; Spectrum/Philadelphia—11/25; Baltimore—11/26; Madison Square Garden—11/27 & 11/28; Boston—11/29; West Palm Beach—11/30/69[2]

Even though these performances were considered some of the band's finest work, portions of the tour were marred with sound issues and other mishaps. This was a tipping point not just for Hanley but rock concerts in general.

According to author Robert Greenfield, this was one of the most noted tours in rock and roll history. In his book, he describes the 1969 tour as "A mad, chaotic adventure run by show business hustlers and out-and-out grifters, with

313

planes leaving empty in the middle of the night and landing at strange airports, and concerts getting underway five and six hours late."[3]

Regardless, it was a healthy money-making venture for the English rock group, with gross estimated earnings for twenty performances at around $2 million dollars. For three MSG shows they grossed $286,542; with the Stones net earnings at $160,000, this was not bad for 1969. *Rolling Stone* claimed, "Assuming this proportion held true nation-wide, the Stones' share from the whole tour comes to about $1,120,000, before taxes and before paying a small battalion of sound men, recording engineers, stagehands and so on."[4] It was clear that this new and bustling industry was now evolving into a seemingly cohesive business.

On September 14, 1969, the Rolling Stones announced they would tour the United States. Behind the scenes they had been putting the finishing touches on their upcoming album *Let It Bleed*. By late October they were rehearsing for the tour in Los Angeles. During an October 27 press conference at the Beverly Wilshire Hotel in Los Angeles, questions were raised regarding the high ticket prices, something, that Mick Jagger defended: "We aren't doing this tour for money, but because we want to play America and have a lot of fun. We're really not into that sort of economic scene. I mean, either you're gonna sing and all that crap or you're gonna be a fucking economist. We're sorry people can't afford to come. We don't know that this tour is more expensive. You'll have to tell us."[5]

The Rolling Stones had not toured since 1966, and a lot had changed in the music business. Hanley had not seen the band since he last worked with them at their chaotic Lynn Manning Bowl performance in Massachusetts on June 24, 1966. After the band's three-year touring sabbatical in 1969, Mick Jagger described these changes regarding sound quality: "The sound system improved, and we got better accustomed to performing again. It's really a matter of confidence. It takes a while to get that up."[6] Drummer Charlie Watts added: "People didn't scream anymore. The music was taken seriously. In '69 you had proper amplification. Suddenly you could hear everybody. Nobody had heard DRUMS before. We must have sounded a joke before. But in '69 you really had to be on top of it to play."[7]

A concert at Fort Collins, Colorado, on November 7 marked the group's tour opener. On November 8–10 they performed shows in LA, Oakland, and San Diego. At the LA Forum (November 8) the assigned sound company was having issues, problems that Hanley learned of and wanted to avoid.

LA Times writer Robert Hilburn reported that the Forum show was badly compromised with delays and sound issues: "The Stones sound people just overestimated their abilities. They only had eight men. It was too big a job for them. They could have used three times that many. There were delays of twelve and twenty minutes while sound equipment was moved and retested between acts."[8]

At this performance, promoter Jim Rissmiller attempted to get the sound people to "hang the speakers from the ceiling." Like most promoters, Rissmiller wanted to sell every seat in the house, and getting the speakers up high enough "avoided blocking anyone's view." Such modifications needed to take place in advance, avoiding delays. Consequently, Rissmiller was expected to "refund concertgoers whose views were blocked, $4,000–5,000." This was a major setback for anyone trying to make a dollar in the concert promotion business at that time.[9]

Hanley had been reading newspaper reviews about problems the Stones were having in LA. Learning from others' mistakes, the sound engineer devised a plan. He thought that by lifting his heavy speakers up into the air and tilting them outward, away from the stage, he could solve logistic and acoustic problems. By doing this, he could free up otherwise valuable obstructed seats, which promoters liked. In addition, he felt that this plan helped project sound more efficiently over a broader range.

MSG promoter Howard Stein felt that a "good rock show should be a well-produced form of theater." The former Broadway theater producer ordered a proscenium stage to be built that blocked off 4,000 seats per show that were not be sold. According to *Billboard*: "A backdrop is being built and extensive stage and lighting, including scaffolds in front of the stage, is being installed. Chip Monck is handling the lighting, with Hanley Sound installing an extensive system. . . . even the spotlights are being brought in, rather than using existing Garden facilities."[10]

On November 11–16 the Rolling Stones performed concerts in Phoenix; Dallas; Auburn, Alabama; and Chicago. On November 24–25 Hanley and his crews traveled to meet the band for shows in Detroit and Philadelphia, followed by a concert at Baltimore's Civic Center on the 26.

On November 27–28 the group had their first performances at MSG, followed by two concerts at the Boston Garden. On November 28 the sound engineer prepared the Rolling Stones' press conference at Rockefeller Center's Rainbow Room in New York. Here the Stones "arrayed themselves behind a tangle of microphones" while a mob of journalists "pressed forward and started shouting questions."[11]

The madness of this benchmark tour continued and the MSG performances were no exception. On Thanksgiving Day, November 27, 1969, an audience of over 16,000 fans waited three hours for the group to take the arena stage. Accompanying the band were three other acts: Terry Reid, B.B. King, and Ike and Tina Turner.

On the first night, Janis Joplin jumped up on stage after an invite from the Turners. According to the *New York Times*, the place was elevated with anticipation: "A sellout crowd filled Madison Square Garden last night to absorb the rock sound of the Rolling Stones. After three hours of waiting it finally came, roaring through the arena like an electrified blast on the Thanksgiving horn of plenty."[12]

As the band performed, Hanley and crew sat and mixed the shows on a raised platform located at the 32nd Street side of the Garden, near the stage right exit area. The position was not entirely front of house, since promoters wanted to sell as many seats as possible. Hanley remembers, "They made me put a piece of plywood on top of the entrance way and this was where I put my amplifiers and mixers."[13]

Hanley recalls union labor costs at around $8,000 to bring in and take down his MSG Rolling Stones sound system. This raised a red flag for the venue's conservative leadership, who were concerned about incautious expenses. Exorbitant expenditures struck a chord with superintendent Dick Donopria, known for his extreme frugality. When it came to spending money on the venue, including Hanley's equipment, Donopria was the first to know.

According to Hanley, the Garden was a costly structure to maintain: "They were trying to keep up with the overhead of the building and wanted to make MSG a profitable venue. They were doing everything they could to get acts in, so as to pay the mortgage. The Garden was on a unique piece of property resting on an active running train station."[14]

WINCHES AND BASKETS

When Hanley arrived at the arena, he had a blank canvas to work with. At first, his sound system was manually hoisted by a block-and-tackle pulley system. In order to cut down on labor costs, save the promoter money, and provide the best sound possible; he designed special hoists to raise the steel rigged hanging frames that his speakers rested on and were affixed to. These were referred to as "baskets." Such devices allowed Hanley to hang three or four cabinets at once on either side of the stage. Each basket weighed in at over 2,000 pounds. A third separate basket structure was centered above the audience that held high-frequency horns.

Deployed in this way, the "line array" system of powerful midrange horns and high-frequency speakers became an integral part of the foundation and future application of sound design. When properly splayed, Hanley's system hung about fifty feet in the arena air. Many recall that this design was unlike anything they had ever seen; for others, it was frightening. Even so, Hanley's main stage Garden speaker arrangement was the beginning of yet another sound engineering innovation.

According to assistant superintendents Keller and Hughes, this dangling spectacle was a potential liability because the heavy speakers hung tilted over the stage and slightly over the audience. They had concerns because they were used to seeing speakers stacked on scaffolding on either side of the stage. This was the first time they had ever witnessed this type of application. It took convincing by Hanley to assure them that this arrangement was safe, even though it looked dangerous.

Keller and Hughes did not want to take the risk and called in consultants to confirm what sort of loads the Garden's roof structure could sustain. According to Keller: "We were concerned with the design and how it was to be hung. They were huge speakers and they were stacked on top of one another on this framework Hanley built. They were then hoisted up towards the ceiling! We were concerned about the safety of the people that were underneath it!"[15]

A hoisting system that could lift such weight was crucial for elevating each of the huge 400-pound, custom-designed HSI 410 (Hanley Sound Inc.) woofer speakers. In order to successfully lift the baskets into the air, Hanley devised a plan, recalling what he had seen a decade and a half earlier at the Boston Garden near his Medford home. At a traveling Ice Capades show, Hanley witnessed a hoisting device that lifted heavy curtains and aluminum lighting trusses into the cold arena air. Gary Keller recalls that he too had seen similar installations: "They would sometimes hang lighting grids because the lighting was inadequate for ice shows. They would lift the framework using chain hoists to elevate the light grids up, but they were relatively light. These were just aluminum trusses with light fixtures on them."[16]

Realizing it was cheaper to electrically hoist his speaker/basket configurations rather than have union crews manually elevate them, Hanley purchased (out of pocket) four one-ton CM industrial winches with eighty-foot chains on them. The sound engineer figured that if you could reverse the winch mechanism, the speakers could be successfully hoisted off the floor. Hanley explains: "We reversed the contact relays on the hoists so they could climb upside down. The hoist is normally in a fixed position. So we affixed the chain onto a joint in the ceiling, and made the hoist climb up the chain pulling the baskets."[17]

Keller remembers he had never seen a winch used this way in a concert setting: "So Bill made these hoists climb this chain, which they were not designed to do. The motor actually climbed up the chain along with the weight that was fastened to the hoist."[18] According to Hanley: "When I ordered four of these from CM they asked what I was going to use them for. I explained, and not long after I received a letter letting me know that they relieve themselves from any liability."[19] By solving this problem Hanley established an innovative development in concert sound reinforcement deployment.

Fillmore East and Woodstock staff attended some of the Rolling Stones performances. Among them were photographer Amalie Rothschild, Chris Langhart, Bob Goddard, and Stanley Goldstein, who was hired by documentarian Albert Maysles to record audio for his film *Gimme Shelter*. Earlier on Thanksgiving Day, Langhart ventured over to see what the sound engineer was up to at the arena. He was awed by Hanley's device:

> Bill was the first to fly anything! I remember Dick Donopria who was the superintendent of the Garden standing there next to me and saying, "I don't know about that. You got something up there that's doing the work for you and you can't get at it." From a scientific standpoint this is where the speakers needed to be, and Hanley was one the first to put them up there and they still put them up there today. These concepts were being sorted out then. Yes, his winch would labor and the house wiring was questionable, but the speakers were up there. CM should put up a page honoring Hanley for this. After this innovation, many of these winches were sold. He started it and everybody copied. They all thought he must have bought this one because it's the best one! You couldn't sell any other brand of winch to the rock and roll trade because CM was the only one anyone ever wanted and that's because of Bill Hanley.[20]

Engineer Bob Goddard recalls working on some of these devices for Hanley:

> Bill realized that it was more effective to have the chain affixed to the ceiling, and the winch climb up the chain. This is not how CM thought the chains would be used. The winches that Bill got when you turned them upside down didn't work right. There was a contactor in there that expected to open by gravity. So we had to open them up, take the relay out, turn it the other way around, drill new holes and bolt them back in. People think that these winches could always be functioned this way. But it's not true. CM eventually changed the design so they could be run in either direction. Bill was the one who early on figured this out.[21]

Some of these winches were even used in theater environments. In March 1972 Hanley was called upon to install a sound system for the premier of the *Concert for Bangladesh* documentary film when it opened in New York. "I remember climbing up those big A1's that were hoisted with Hanley's winch,"[22] adds Goddard.

After Hanley left MSG, his baskets remained. As other shows rolled in, those in charge of sound continued to borrow his ideas. According to Hughes: "People were coming in after Hanley with their groups and equipment and used the baskets he designed. We used that system for quite a while."[23] Soon, a flurry of sound companies that witnessed Hanley's applications mimicked the technique. Keller adds: "After I left MSG I went to the Astrodome and everybody was doing it there. It seemed like every time we had a show come in they were bringing in chain hoists and running them up backwards, just like Hanley did."[24]

Prior to the Rolling Stones 1969 performances at MSG, band manager Sam Cutler had never seen speakers suspended in the air like this. Cutler recalls that the Garden performances were the first time the Rolling Stones ever experienced "really great sound": "Given that this was at the time period the Stones were playing through house PAs, the 1969 tour was the first tour where the band had consistently good sound and Hanley was a big part of that. It had a significant effect on them because it was the first time they could hear each other. The music definitely went up a notch because of these shows thanks to guys like Hanley."[25]

By 12:30 a.m. the band exited the Garden stage. They would meet Hanley and his crew again on November 30, 1969, at the West Palm Beach Pop Festival in Florida. This was the event that marked the end of this memorable tour. Soon thereafter, the Rolling Stones put the final touches on their formative album *Let It Bleed* at the famed Alabama recording studio Muscle Shoals Sound. The album was released on December 5, 1969.

On December 6 the group headlined a free concert in Livermore, California, at the Altamont Speedway in front of 300,000 people. The disastrous chain of events that occurred at this concert displayed a distinct and dark change that closed out the 1960s. Even with the promise of a new decade on the horizon, 1970 represented an unpredictable future. Fortunately Hanley Sound was not contracted for Altamont, although it almost was.

TRANSITIONS

Toward the end of Hanley's tenure at MSG he recalls losing money. Providing sound at the Garden was not profitable. "I was getting paid peanuts. With the Stones I could only afford to buy four hoists and MSG did not want to buy any of it. They scoffed at the cost, and the reason it was so expensive was that these were big structures and they had to be reassembled each time we did a job at the 20,000-seat arena."[26]

It was a hand-to-mouth business for the sound engineer. It seemed that he was competing with his own innovations and developments. As the concert business grew, Hanley had difficulty keeping up with the demands of quality sound and efficient service. He often spent his own money to appease arena management and promoters.

By the time Hanley was ready to leave MSG for good, he was storing a large amount of his equipment at the facility and Donopria didn't like this. According to crew member Rick Slattery: "Bill had a tendency to move into places with his stuff and used MSG as a relay station. We would go to MSG during all hours of the night to get equipment that we needed. We kept blowing gigs. All of the sudden we got a reputation for being over budget, overtime, or not at all."[27]

At the same time, other competing sound companies were itching to do a better job, and they did. Keller remembers: "Bill was there and then another sound company came in after him like the Clair Brothers, who had a whole different concept about how things should be done."[28] Soon companies like the Clair Brothers abandoned Hanley's basket system, designing sophisticated trusses to hang their speakers. According to Hughes: "A lot of people learned from Hanley. He could have been the sound master of the whole country if he had pursued all of the stuff he knew. He was a crazy genius of a guy and a sound guru."[29]

After Hanley, New York sound engineer Jack Weisberg of Weisberg Sound came into MSG. Hanley observes, "Jack was offering a cheaper price and less sophisticated systems. The Weisberg System was easier and cheaper to install."[30] According to Weisberg: "Hanley had a lot of equipment. Big equipment. He was, as I heard, late for many of the gigs, and slow to get started. I suspect this was probably because of the unions and or that Hanley was spreading himself too thin."[31]

Weisberg Sound, founded in 1968, was known for several innovations in the field of sound reinforcement. Hanley Sound employee Fritz Postlethwaite, who worked for the outfit in 1973 after his year and a half stint with Hanley Sound, recalls: "Weisberg was an interesting guy. He designed his own horns like Hanley but they were all based out of the old Altec and Klipsch theater systems. Jack was inventing his own weird shapes. In his basement he had this mad scientist laboratory with these mock up horn shapes. Some even looked like fish and gigantic blobs from outer space!"[32]

Many of Hanley's innovations in sound reinforcement were concepts he learned throughout his career. He often gained knowledge by trial and error, and the Rolling Stones tour was no exception. These shows put every bit of Hanley's raw innovativeness to the test, proving his tenure at the Garden a success. Concepts like the Hanley Hula Hoop and his CM winch and basket system were all ideas, developments, and innovations he deployed at the Garden.

After MSG, Hanley continued with the festival and political gathering movements occurring in the United States. The anti–Vietnam War demonstrations and mass gatherings of the 1960s and '70s required good sound. These were often viewed by the White House as a form of antigovernment activity, and as such were fearless endeavors for the sound engineer. Both Hanley and protesters were closely watched by the Nixon administration.

Thousands of people marched in the streets as demonstrations occurred almost every other month. A tidal wave of protesters flooded Washington and made headlines. Hanley Sound was one of the only companies in the country to engage in this new territory of demonstration sound. Afraid of a collective counterculture mutiny, the government often quelled the demonstrations during the early stages of planning and deployment, including sound reinforcement.

Chapter 44

DEMONSTRATION SOUND

THE RIGHT MAN FOR THE JOB

Bill Hanley is known mostly for his work in jazz and rock music, but his expertise was also needed during the civil rights and anti–Vietnam War demonstrations of the 1960s and '70s. Hanley and his team were providing full production for some of the largest protest demonstrations in US history, many of which were in Washington, DC. Whatever the message, the antiwar movement needed a consistent and reliable sound system so audiences could hear. The role of a sound engineer at these rallies proved essential.

Hanley Sound was among the few companies that took on such daring and monumental tasks. The sound engineer often subcontracted concert lighting from the local Boston lighting company, Tom Field Associates. Owner Tom Field speaks highly of Hanley's unique ability to work under intense and occasionally dangerous political pressures: "Hanley could talk to government people, and they would listen to him. I think most technical people would not be comfortable doing that. This I believe is what made him so effective at these demonstrations."[1]

As with the festivals, music played an essential role during the demonstrations. These events sparked and inspired the music of a generation. This transformative cultural happening was a melting pot of folk and other classifications of music. It empowered the youth culture with an oppositional voice. The identity of the antiwar movement was realized through popular music even more than through organizations or large demonstrations.[2]

During the early to mid-1960s, the folk music occurring at the Newport Festivals became a backbeat to the impending protest movements. It was here that the identity of the antiwar movement formed its voice. Protest songs would soon infiltrate the radio dial, like Jimi Hendrix's rendition of Bob Dylan's "All Along the Watchtower," Barry McGuire's "Eve of Destruction," Pete Seeger's "Waist Deep in the Big Muddy," and Country Joe's "I-Feel-Like-I'm-Fixin'-to-Die Rag."

Crosby Stills Nash and Young's "Ohio" was released as a response less than two months after the 1970 Kent State shooting. Eventually messages like these seeped into mainstream music, forming the classic counterculture soundtrack as we know it.

Bob Dylan's "Blowin' in the Wind" is a protest song about the civil rights movement that has become part of the popular music ethos. However, the folk singer denounced any implied political ambiguities in the lyrics of the song, and took little credit for any political influence in his songwriting. In author Peter Doggett's book, *There's a Riot Going On: Revolutionaries, Rock Stars, and the Rise and Fall of the '60s*, he writes, "No one symbolized the ambiguous relationship between music and revolution more than Bob Dylan."[3] Peter Paul and Mary gained success with their version of Dylan's song that spent five weeks on the top of easy listening charts, and peaked at #2 on *Billboard*.

For political folk singers like Joan Baez, the path from the Newport Folk Festivals to Woodstock and then on to Washington protests was easily navigated. While Baez and others became the face of many of these political events, she also grew in commercial popularity, leaving some of her purist followers slightly disappointed.

By the end of the 1960s it was obvious who was rising to the top of the record business game and who was not. In *Music and Social Movements: Mobilizing Traditions in the Twentieth Century*, Ron Eyerman and Andrew Jamison claim, "By the end of the 1960s with the commercialization, fragmentation, and de-politicization, the 'movement phase,' of that development had largely come to an end."[4] Hanley was front and center while many of these stars were rising in popular music. Without good sound their ascent to fame would have been difficult.

Political anthems expressed in the 1960s folk revival reemerged in early 1970s rock and roll. This resulted in a new version of music that projected the social commentary of the day. Social and political issues were changing rapidly by the end of the sixties. As the impact of folk music dwindled, new forms of music arose out of the old. The aggressive sounds of rock and roll gave a voice not only to the frustrated public who fought against US policy but also to the thousands of soldiers risking their lives in Vietnam.

The message of the antiwar movement was becoming audibly clear at mass gatherings, festivals, demonstrations, and especially on progressive "freeform" FM rock radio. In Jeff Kisseloff's book *Generation on Fire*, he interviews folk rock singer-songwriter Barry "The Fish" Melton. Melton reflects on his shift out of folk and into a new genre of electric music, claiming that his "musical ideas had changed." At the beginning of his career, as a folk purist Melton was opposed to electric music, perceiving it as music of the "establishment." Melton heard the Paul Butterfield Blues Band in Berkeley, California, shortly before he and his band (Country Joe and the Fish) made their influential psychedelic album *Electric Music for the Mind and Body*. Melton reflects on the Butterfield band: "They were the first white guys who played electric music that were any good, or so I thought. I said to myself, 'Man if I could do something like that, I'd play electric, and I did on our second record.'"[5]

Even though many of the musical events Hanley supported had some level of political message, he didn't officially provide sound at demonstrations until 1969. But things were moving quickly for the sound engineer. In just a few short years he had shifted from the folk crowds at Newport into the maelstrom of the festival movement. Now it was the heated streets of Washington, DC.

When Hanley became involved, demonstration and festival were merging into one political and musical driven occurrence. Hanley's willingness to do these jobs came at a great cost. Because of his connections with the antiwar movement, several of his politically related accounts were suddenly dropped. Hanley recalls: "The issues and content of the civil rights movement were still being represented, just merging onto one stage. After my involvement with mass demonstrations, I was never called back for formal political work—this was fine with me."[6]

Hanley provided sound reinforcement for over a dozen demonstrations throughout his career. Friend, pacifist, and peace activist Bradford Lyttle organized most of these large demonstrations with a coalition of antiwar activists. Lyttle is known for his work with the Committee for Non-Violent Action and the National Mobilization Committee to End the War in Vietnam (formed in 1967). The latter was eventually renamed the New Mobilization Committee to End the War in Vietnam, sometimes called the "Mobe" or "New Mobe."

According to Lyttle, Hanley's involvement was significant to the movement. Their work together on major anti–Vietnam War, Washington, DC, demonstrations included the November 15, 1969, March Against Death/Moratorium to End the War in Vietnam; the May 9, 1970, Moratorium Day/Student Strike; and the May 1–3, 1971, May Day Demonstration. Hanley assisted with other smaller rallies in New York, for example, the Women's Strike for Equality on August 26, 1970, and a Central Park Rally for the National Peace Action Coalition on November 5, 1971. Lyttle, now in his nineties, resides in Chicago.

Before the upcoming November demonstrations, in September 1969 Lyttle and members of his team, activist leaders Fred Halstead, and co-chairperson Cora Weiss began their search for a competent sound company. It was crucial that organizers acquire a sound system that could reach large audiences. This was a difficult situation. Reliable, qualified, and road-tested public address systems for events that had over 200,000 in attendance were not readily available like they are today. Lyttle recalls that most of the sound systems used for rallies in Washington had been insufficient.[7]

Before contacting Hanley, Lyttle actively sought out a sound system that he thought might be adequate for the job. He and his team had realized that the systems they were using had limitations. For the upcoming November demonstrations, an average turnout of 200,000 or more was expected. Lyttle explains that he and his team could "Not take any chances with losing sound for such a large rally."[8] Volunteers proved to be an unreliable source, another reason why Lyttle needed a professional firm like Hanley Sound. He reflects on the bidding process with various sound companies in a self-published report written in late 1969:

> We then consulted local sound system firms. The company that had installed the systems for the October 21, 1967 Lincoln Memorial and Pentagon rallies was friendly but didn't want the job. They said that their equipment would be tied up in another contract. Another Washington firm, that handled the Poor Peoples' Campaign March on Washington, was interested, discussed our public address requirements with us in detail, and submitted a bid that fell between $12,000 and $20,000. This was not an unreasonable bid. They would have provided not only the sound for the main rally, but also systems for the Arlington Cemetery starting point of the March against Death, and the mass march assembly area on the Mall, west of the Capitol.[9]

According to Lyttle, before Hanley, professional political sound reinforcement companies sympathetic to the cause were hard to come by. The organizer recalls that these firms lacked the power needed for a crowd of 500,000: "In order to deliver the important political message, you needed the best sound system possible."[10]

Lyttle was always concerned that the CIA and other political agencies could infiltrate his operation by pulling the plug on the sound. For this reason a sound engineer he could trust was essential. He and his team toiled over this decision until they finally agreed to use Hanley because of his political reliability. An unknown protester once said that "True peace, like true love, cannot be made: immediately, unilaterally or without sacrifice" and in Hanley's case this same set of values holds true. The sound engineer recalls his passion for the movement and his focus on getting the job done:

> I was never into war. I felt that they were bad and shouldn't go on. I also felt that good audio was needed so that people could hear and understand what was being said during these demonstrations. Nixon was trying to kill these events. However, we had a security system and a devoted group of people who made sure trouble didn't break out at the staging area. We were afraid of the F.B.I. or somebody of that ilk trying to break in or break up these demonstrations.[11]

Bill Hanley was introduced to these events sometime around mid- to late 1969 through friend and folk musician Peter Yarrow of Peter, Paul and Mary. Hanley had befriended Yarrow at Newport. A passionate organizer himself, Yarrow and his group performed at a number of peace demonstrations during the 1960s.

In early 1969 while in New York, Yarrow called movement organizers Lyttle and Weiss. According to Lyttle, the folk singer highly recommended the sound company for upcoming demonstrations scheduled for that year: "From NYC, intriguing phone calls from Peter Yarrow kept us up-to-date on his search for and negotiations with Bill Hanley."[12]

The organizer recalls that Yarrow was enthusiastic not only about Hanley but about singing at the upcoming November 15 rally. Yarrow was interested in managing the entire entertainment roster for this event and other scheduled demonstrations. Based on these recommendations, lead organizers and other members of the New Mobe staff decided to go with Hanley Sound.

Getting in touch with the busy sound engineer during this time was a challenge for the organizers. Lyttle remembers, "Hanley was scurrying back and forth across the country with two semi-tractor trailers filled with sound equipment; this was before mobile technology and email."[13] Most often if you wanted to get in touch with Hanley, you needed to contact the main office in Medford and leave a message with Judi. With any luck Hanley might get back to you within a reasonable amount of time. Lyttle recalls: "Hanley however, remained a mythological character to us. He had done the sound for Woodstock. He traveled around the country with his equipment in two semi-trailers. His was the only system that could handle rock music. He was elusive."[14] In Lyttle's 1969 report, he reflects on Hanley's personality and presence within the industry:

> By the time I met Bill in Washington, his empire had expanded to impressive proportions. He had offices in NYC and Medford, Massachusetts. Seven crews toured with rock bands and other musical groups. His major system roamed the country in the two 40 ft. semi-trailer moving vans. No cold-blooded technician, Bill fit into the hippy and rock music scene. His main crew of twelve that worked with the two vans shared his outlook. They combined technical competence with the new life style.[15]

Most of the demonstrations Hanley was involved in were very large, yet by now he was qualified for jobs this size. It was three months after Woodstock, and the sound engineer had the largest cache of sound equipment in the country. Hanley explains: "I was mobile enough, and so well equipped that I could have provided sound reinforcement for all, or most of the festivals and demonstrations going on in the country back then."[16] It was during these gatherings that organizer Lyttle came to know Hanley on a personal level and on a few occasions both were arrested for their involvement.

Outrage against the Vietnam War was in full swing, and the public's contempt toward President Richard Nixon for not pulling troops out of South East Asia was climaxing. In Henry Kissinger's book *The White House Years*, he reflects on these tumultuous times in US history: "So it was. As the months went by in 1969, we were confronted by public protests, demonstrations, and the quickening demands in the media and the Congress for unilateral concessions in negotiations."[17] Kissinger snidely recalls that the weekly demonstrations at the Pentagon "included such charming gestures as pouring blood on its steps."[18]

Moratoriums and May Day protests continued, with emotions escalating. People were frustrated with the government and crowds at demonstrations were getting bigger. Hanley and his crew were in the throes of it once again. According to Lucy Barber's book *Marching on Washington: The Forging of an American Political Tradition*, "In 1965, the first substantial anti-war march had been relatively small, attracting some 25,000 people; in 1967, the march against the Pentagon was joined by about 100,000 people; in 1969 the Mobilization against the Vietnam War surrounded the Washington Monument with at least half a million people, making it the largest on record."[19] At the height of these protests it was becoming more and more dangerous for Hanley and his team to conduct their work.

Chapter 45

THE MARCH AGAINST DEATH

NOVEMBER 15, 1969

Hanley provided sound reinforcement for an anti–Vietnam War moratorium on Boston Common on October 15, 1969. The protest marked the beginning of a nationwide series of teach-ins, demonstrations, marches, and rallies against the war. According to the *Harvard Crimson*, "More than 100,000 demonstrators demanding an immediate end to the war in Vietnam massed on the Boston Common yesterday in the largest anti-war demonstration in New England history."[1] As part of the event, local revolutionary radio station WBCN sponsored a skywriter to create a giant peace sign in the air above the common. Although this was a fairly large turnout of protesters, one month later an even larger demonstration took place in Washington, DC, on November 15, 1969.

On Sunday, November 9, Lyttle and his team finally met Hanley in Washington. Straight from a concert in Virginia, the sound engineer was visibly exhausted and needed rest. By Monday the 10th most of his crew and equipment began to arrive at the Washington site from locations across the nation. It was Lyttle's responsibility to connect all the variables, like scouting locations for the stage, booking hotel rooms, and connecting Hanley with his crew. It was a huge undertaking.

Hanley's portable stage arrived Tuesday around midnight. According to the organizer, "The first van that could be converted into the heart of the stage, arrived."[2] By Wednesday morning the construction of the stage area began. However at this point, Lyttle recalls having difficulty solidifying a permit for the rally. He claims this obstacle was yet another example of "unusual cooperation" by the government.

While at the Washington Monument, in typical fashion Hanley began to analyze the proposed area for his sound and stage construction. Realizing it was not an ideal location for sound reinforcement, he decided that his system would work best closest to the southwest corner of 17th Street and Independence Avenue.

Organizers Lyttle, Eric Small, and John Gage were all in agreement with the last-minute change. According to Lyttle this alteration made sense: "This location enabled Hanley to focus his sound over a 90 degree arc."[3] Hanley and his crew began to deploy the complex stage and sound system into the cool crisp fall air. The impending Thursday deadline was on their mind.

There was a peripheral buzz around the engineer as organizers worked to establish the new demonstration site. Hanley recalls: "The organizers took care of the permits and other things. I took care of all the technical responsibilities and the security wall system."[4] Fred Halstead reflects on some of the event's production and facilitation in his book, *Out Now! A Participant's Account of the American Movement against the Vietnam War*: "John Gage, and a crew of volunteers were working on the stage, the press tent, and a set of bleachers for the speakers, entertainers, and their guests. Lyttle was checking out the assembly area on the Mall west of the Capitol where the March was to be begin."[5] It was a dedicated team of volunteers, committee members, and various groups responsible for the demonstration, all of whom strove for peace.

The March against Death began at 6:00 p.m. on Thursday November 13, 1969. According to the *New York Times*, "The 40-hour demonstration that the protestors called a 'March against Death,' in which 40,000 filed past the White House bearing the names of the United States dead in Vietnam, ended at 7:30 am [November 15]."[6]

Organized by the New Mobe, the March against Death ended on the 15th on the Capitol steps with a mock memorial service. The *New York Times* reported: "In the lead were three drummers, followed by youths carrying aloft 11 wooden coffins that contained placards bearing the names of the dead. The placards had been paraded past the White House. The coffin bearers were surrounded by a cordon of young who were joining hands."[7]

Folk singer Peter Yarrow recalls the profound effect this gathering had on the country then: "This was called the March against Death where the names of the war dead had been placed in coffins all through the night in candlelight procession. It was intended to be more than simply a protest against the war. It was intended to be an affirmation of a way of life that the demonstrators wanted to see put in place."[8]

After the March against Death, a moratorium march and rally was slated for Saturday, November 15. The event attracted over 500,000 demonstrators, including many activists and performers that Hanley knew. The march extended from the foot of the Capitol and sprawled over blocks. Around midday, over 250,000 were slated to walk many blocks up Pennsylvania Avenue to the Treasury Building, and then continue four blocks down 15th Street and on to the Washington Monument.[9] Leading up to the Saturday event there had been a cold north wind blowing as the tired sound engineer and crew worked frantically under clear blue skies. It was imperative that they finish before the scheduled performances and speeches.

According to the *Boston Globe*, the November 15 demonstration crowd was large enough to showcase "many causes" representing a diversified list of radical groups: "Its members were for freedom for the Chicago Eight and the Panther 21, for peace through prayer, opponents of 'the struggle of the Greek People for Freedom,' for independence for Puerto Rico, advocates of larger wages for General Electric's striking unionists, and for the removal of Spiro Agnew from office."[10] Hanley's hometown paper acknowledged the Medford sound company, too: "Audio equipment for Saturday's rock show and speakers was leased from Hanley Sound Inc. of Medford."[11]

A 1969 *Billboard* article titled "Galaxy of Stars to Make Capital An M (Music & Moratorium) Day," reported: "The list of performers was longer and more impressively 'now' than any other performer group ever gathered in the political city for any other event, including presidential inaugurals."[12] Yarrow was responsible for booking all the performers, including Arlo Guthrie, Richie Havens, John Denver, the cast of *Hair*, the Cleveland Symphony Orchestra string quartet, Earl Scruggs, Peter, Paul and Mary, and Pete Seeger.

Through Hanley's sound system, activist and folk singer Pete Seeger performed John Lennon's antiwar anthem "Give Peace a Chance" to an attentive crowd. Seeger recalled the "sea of humanity that gathered on the National Mall that day."[13] Seeger, an old friend of Hanley's from Newport, spoke in high admiration of the engineer's "performance and expertise" in delivering clear, audible sound—sound that gave voice to the movement.

According to *Billboard*: "At a giant post-march rally, on three stages built on the monument grounds, the Capitol was set to see an unprecedented performance by a parade of top recording stars never before gathered together under the sky. The stages and sound equipment for the rally were donated free by Bill Hanley, who produced the sound for the awe-inspiring Woodstock Festival."[14]

Providing a service like this at no charge reveals that Hanley shared ideals that were at the very foundation of the antiwar movement. He sacrificed his time and finances because he yearned for a world that was fair and just. Hanley simply wanted a world that was free of racism, war, and one that was not driven by monetary pursuits. This made him more than the ordinary sound engineer.

According to Lyttle, Hanley was not a political activist, but he did have "sympathy towards the movement."[15] The sound engineer worked well with the organizers, and adapted to most of the movement's radical elements. It was because of these values, politically driven or otherwise, that demonstration organizers went with Hanley for sound. Hanley and his crew carried out a nearly impossible task during the most intense demonstrations in American history. Amid the tear gas thrown by police, the threat of being jailed, and the backbreaking work, he never flinched.

The sound engineer often sought assistance from volunteers that he pulled from the crowd. Some were even familiar to him, like Woodstock alum and Hog Farm members Campbell Hair and David "Bucko" Butkovich. Hanley had met the two at Woodstock and other festivals where the Hog Farm's services were instrumental. Known for their work at the festival acting as the "Please Force," they quelled problems before they escalated, using non-intrusive tactics.

Organizer Halstead played an integral role in facilitating and organizing some of the largest peace demonstrations on American soil. He describes the significance the Hog Farm had on the November 15 demonstration:

> There are also a number of specialized groups involved, including the Hog Farm community from New Mexico, who were experts at gently cooling problems in the countercultural milieu at rock concerts; and a large squad of trade unionists who protected the speakers and prominent guests. This latter group was not Quaker trained, but they weren't armed either. The Hog Farmers also staffed the kitchen at the Marshal Center and kept everyone filled with hot soup.[16]

After Woodstock, Hair went back to New Mexico. From there he and others drove the "Road Hog" bus to Washington for the upcoming November protests. It was here that he saw Hanley putting together his portable hydraulic "Magic Stage" and sound system: "The second time I met Bill I really got to know him. Myself, and Hog Farmer David Butkovich just started working for him. I remember Bill setting up his portable Wenger Wagon Magic Stage. It was a hydraulic stage that Bruce DeForest helped him build. We didn't get paid, but we built the stage and the sound system for Hanley."[17]

At the 6:00 p.m. curfew Hanley was still dismantling the PA and packing up his gear. Just then members of the Hanley crew began to smell the distinctly strong stench of tear gas. With very little time left before the rally permit expired (ten minutes or so), a line of police forcefully charged over a grassy knoll with the Washington Monument as a backdrop. Many attendees were waiting for buses to leave the area. Soon tear gas cartridges began to explode around those still milling about, keeping warm by bonfires. Instinctively crew members hid in one of Hanley's semi tractor-trailers until the thick looming clouds wore off. Nixon could not have been any clearer in his message to exit the area.

According to some of the crew, while Hanley was out in front of the stage, he tried to explain to the police that he had a permit: "I was attempting to strike all of my heavy equipment. However they were unconvinced, and would not listen to me. So myself and others went off to jail until the following day."[18]

On November 17, 1969, Hanley's arrest was documented in the *New York Post*: "The cops crashed through the snow fences around the stage arresting everyone in their path, including Bill Hanley, the Boston sound specialist who had rushed out of his trailer-truck to defend his delicate electronic equipment."[19] Spectator Rick Wurpel attempted to sneak a hookup from Hanley's board so he could record the show. Wurpel saw most of the event unfold:

> Hanley caught me trying to sneak a feed, but was nice about it. I was using one of the very first Advent Dolby Cassette decks. When he realized I didn't know what I was doing he brought me into the (semi) trailer, hooked me up to one of the McIntosh amp racks, and shared his granola! The next day there were 500 cops in full riot gear surrounding us. Fortunately we found shelter in Hanley's trailer for helping him with the load-in.[20]

According to Hair, who witnessed the occurrence, the stage area was chaotic, teeming with people and musicians. When the show ended and the audience cleared most of the field, the police moved in. Hair recalls: "Then we saw a whole army of masked police coming over this ridge and heading towards the stage. It was a long line across a whole field at the Washington Monument. It was all cops and riot gear. We hid in the truck when Bill was getting arrested."[21]

Others on Hanley's crew weren't as lucky as Hair. According to the *New York Post*, "Not only did they arrest Hanley, but they arrested members of his crew who worked around the clock for two days to assemble the stage and the sound towers and who now were busy disassembling them."[22] Butkovich recalls the police were "arresting everybody, near or around the stage area, no matter who you were." But for Hanley and those who did not have a chance to hide, they were either tear-gassed, arrested, or both. Butkovich reflects:

> There had been speakers and musicians on the stage before all of this went down. We all bought jumpsuits of the same color so that we all looked like we were working at the place, you know on the payroll? We hoped this would keep us immune to any problems, but the National Guard got a hold of us and put us in jail. When they first came down from the tall part of the amphitheater they were throwing big tanks of tear gas on us. When you are twenty-one years old and you are living that sort of life nothing much scares you. You are pretty much ready for anything and it doesn't matter. I am sure Hanley had everything insured. We had gone into battle before. We knew what we were doing and there was only one way to do it. I had never been held at gunpoint. I had been arrested and had problems before; but when

you got an eighteen-year-old guardsman wearing full battle dress (with a gas mask on) pointing a gun at your head, while he has you down on the ground—that's a little scary! Campbell was with me at the time. When they got me I was right out in front of the stage, at stage right. They just marched us off and that was it, in our jumpsuits.[23]

The next morning, Washington, DC, mayor Walter Washington ordered the release of Hanley from jail. According to Butkovich: "I remember the next day after Hanley was let go he walked along inside the cells where they had put everybody. They didn't even look at our I.D.'s . . . nothing. So Hanley walked along and said 'that one'—'that one'—'that one' as he pointed to us. After that, they took us out of the cells and turned us loose."[24] Back at the rally site, the sound engineer and his crew moved quickly to pack up the stage, and in Hanley's words—get the hell out of Washington!

In the end, those who were falsely arrested received $10,000 each since they were targeted before the permits actually expired. Hanley remembers: "I got out early because the mayor sent some lawyers down to release me. Unfortunately, I missed out on the ten thousand dollars!"[25] Woodstock colleague and Hog Farm leader Hugh Romney, known as Wavy Gravy, hails Bill Hanley for his loyalty to the cause: "When the police were starting to pop people, Bill was ready to go down with his system into the slammer. I was very intrigued by his dedication."[26]

Hanley's good judgment, coupled with his helpfulness and ability to improvise, made him an invaluable component to demonstration organizers. Lyttle remembers, "Hanley did not buckle under constraint."[27] Hanley's passion for clarity and audibility became essential to the messages being conveyed.

According to Yarrow, "Hanley was devoted not just to sound, but to music as a form of telling the cultural story of who we were and where we wanted to go."[28] November 15 demonstration attendee and protester Al Braden recalls: "We were there to gather together, to let our voices be heard, and to try to get the Vietnam War stopped. I remember the sound was clear, we were able to hear, and because of this we were able to participate and know what was going on. We felt very much a part of it."[29] Braden's testimonial serves as a good example of the impact Hanley had on the movement. It explains how good sound is often overlooked. According to Hair: "Nobody else at that time could have done an event like this. Someone else could have done it, but it would have been ticky-tacky. What Hanley did was all very professionally done."[30]

Lyttle observes that the facilitator for political events and rallies like these needs to be prepared for government opposition: "There's always the potential risk of provocateurs or saboteurs plotting around you."[31] He admits that even though these tasks were dangerous, the sound engineer carried them out with courageous fervor: "He would simply pack up, wipe the tear gas out of his eyes, and move on."[32]

During the 1971 May Day demonstrations in Washington, DC, Hanley and his crew deployed several big sound systems for a multiday event. At the gathering, he constructed a system near Potomac Park. This area was tear-gassed heavily. According to Hanley, "The threat of being tear-gassed and physically hurt was imminent always."[33]

This was the era of the Chicago riots and Kent State shootings, when a violent response lurked around every corner. Yarrow recalls Hanley's ability to focus under such duress: "Anybody who stepped into that, as a person who had to deal with these opposing forces, needed to have a remarkable capacity to remain calm in the midst of chaos. Whereas other people would freak out, Bill would become more focused and think very methodically. His mind was on the task."[34]

Soon Hanley and his crew would begin preparations for a demonstration the following spring on May 9, 1970, when over 100,000 college students attended a protest gathering in Washington. Here the message at this demonstration was similar to the last—the withdrawal of the US military from Vietnam and Southeast Asia.

During these times Hanley was not driven by money, or even the music. He felt it was his responsibility to the American people that every word said be heard. Yarrow observes, "He who has control of the microphone, the quality of musical sound, and can facilitate its communication, is playing an essential role that will make a huge difference in the way we go as a society."[35]

Chapter 46

STUDENT STRIKE

MAY 9, 1970

On Thursday April 30, 1970, Nixon ordered the invasion of Cambodia. This sent a grim signal to many Americans that a potential end to the war was nowhere in sight. The situation was escalating while many in the United States became increasingly intolerant of hearing the same old government rhetoric. As a result, student strikes prevailed.

According to Halstead, the angry mass reactions began before Nixon could finish his televised speech for the Cambodian action: "The biggest of these early spontaneous outbursts, occurred at Princeton, New Jersey, where some 2,500 students and faculty (out of a university community of 6,000) met immediately and voted to strike the college. By morning the strike was virtually solid."[1]

The nationwide strike included thousands of high schools, 500 colleges and universities, and involved over four million participants. On Monday, May 4, 1970, the Ohio National Guard shot and killed four unarmed student protesters at Kent State University. It was a critical moment that became the face of the movement, forever memorialized in Neil Young's protest song "Ohio."[2]

The May 9, 1970, anti–Vietnam War demonstration in Washington was organized soon after the Cambodian invasion. Speakers and performers at the event ranged from Dr. Benjamin Spock, Coretta Scott King, and Allen Ginsberg to Rennie Davis (of the Chicago Seven), Abbie Hoffman, Jane Fonda, as well as folk singers Judy Collins and Phil Ochs.

Once again, Lyttle and his team negotiated with the government in using the seven-acre Lafayette Park for the event, just north of the White House at H Street. Lyttle's initial plans were to gather over 100,000 people there. Most were young students. However, plans changed and the location got pushed south of the White House, just outside of the fifty-two-acre park known as the Ellipse.

On Monday May 4 Lyttle phoned Hanley and asked if he could handle sound reinforcement for May 9. Even though he was unable to give the sound engineer location details or crowd size information, Hanley readily accepted the offer. This is yet another example of both Lyttle and Hanley's commitment to the movement during this turbulent time of mobilized protest.

By Thursday May 7 Lyttle phoned Hanley again with specifics on where the rally was to take place. Lyttle recalls: "I talked with him again Thursday night and said as far as we knew he would be on H Street north of Lafayette Square and the crowd would be 10,000 to 30,000. We might not have a permit for a sound system and he might be gassed. His reply was, 'Well, gas masks will have to be standard equipment for sound engineers.'"[3]

Nixon and his aides had a long history of refusing to grant permits to the New Mobe. In *Marching on Washington: The Forging of an American Political Tradition*, Lucy G. Barber writes: "The proliferation of protests between 1963 and 1971 significantly affected the reactions of federal and district officials. The increasing number of protests and participants, required officials to refine their responses."[4] It was commonplace for organizers to switch plans because of Nixon's intense monitoring of these rallies. "Officials had to negotiate with organizers over how protesters would use public spaces."[5]

According to Lyttle and Halstead, "switching gears" was all too common when negotiating with the government. Halstead explains:

> At first we expected a crowd of ten thousand. This would fit into Lafayette Park, or if that were denied, into H Street on the north with plenty of room to evacuate the crowd in the event of a police charge or gas attack. Later it was clear we would have a much bigger crowd. We decided to ask for The Ellipse, a field just south of the White House, where there was room for 100,000 or more. The government adamantly refused both The Ellipse and Lafayette Park, saying it would keep us out of both areas, but not out of H Street, just north of the park, or the Washington Monument area, south of The Ellipse. The monument area was too far from the White House, so we proceeded with arrangements on the H Street plan, though Brad and I were not pleased with it. A crowd of 100,000 would jam not only the H Street area adjacent to the park but the streets leading into it for several blocks back.[6]

Nixon had a disdain for protests. However, around 4:00 a.m. on Saturday, May 9, 1970, during this particular demonstration the 37th president of the United States did something rather unusual. Driven by immense national scrutiny, without Secret Service protection, Nixon paid an impromptu visit to the Lincoln Memorial to chat with the protesters. Later, that same morning, Hanley arrived with two tractor-trailers expecting to set up at H Street for a proposed crowd of 30,000. Later that number grew to 150,000. Even so, Hanley easily adapted, and was fully prepared.

Plans changed again when it looked like the crowd size was going to be bigger than anticipated. This forced organizers to push the demonstration toward The Ellipse. Lyttle explains: "Changing plans on a dime did not faze Bill at all. He could have opted out for such a huge change in venue for this demonstration but he did not."[7] As soon as they moved the equipment to the new location, Lyttle realized the "cherry picker" crane they ordered did not arrive. The picker was essential for Hanley to place the giant HSI 410 lower-level and 810 upper-level speakers thirty to forty feet high on the towering scaffolding. Hanley needed to quickly devise a plan to get his speakers above the crowds.

With no heavy lifting device in sight, Hanley proceeded in a way that was as unique as it was dangerous. Lyttle remembers: "Bill analyzed the situation, and quickly built a staircase out of this scaffolding. With one tower being the highest, he then built steps leading up to the highest point."[8] With help of volunteers from the crowd, Hanley and his crew walked the 700-lb. bass speaker cabinets, midrange horns, and other heavy equipment up a manmade stairway to the top of the towers in ninety-degree heat. Hanley recalls: "At the May 1970 demonstration in Washington, DC we had no cherry picker to lift these heavy speakers up the scaffolding. So I had no other choice but to roll them over and walked them up."[9] Fortunately no one was injured during the arduous effort.

Lyttle has great admiration for Hanley's ability in getting the job done under the most difficult of circumstances. Like an old war comrade, the peace activist reflects on the sound engineer's dedication to the cause. The organizer recalls that although Hanley was unpaid, he still faced the threat of pepper spray and gas bombs. In Lyttle's journal published shortly after the demonstration he writes of Hanley's good work: "You could hear the speakers clearly at the base of the Washington Monument. There isn't a sound company in the country other than Hanley Sound of Medford, Massachusetts that would've committed itself to a job like this."[10]

Once again Hanley drew on his courage and applied his skill in sound reinforcement, and in effect mobilized the nation. His dedication and commitment allowed performers and speakers to use their voices as tools of empowerment. According to Yarrow: "I know of nobody more important in that role than Bill Hanley who was always there to bring his heart and soul to make it possible. Bill was the single most important figure in this arena and he should be remembered for that."[11]

Chapter 47

MAY DAY

OUT OF VIETNAM NOW!

By April 1971 Hanley and Lyttle shared a work ethic. They were a team when it came to planning large mobilizations. On April 10, members of the National Peace Action Coalition (NPAC), an umbrella organization made up of around a hundred activist groups, were frantically trying to get things in order for a scheduled protest and rally on April 24 on the Mall in Washington. Lyttle and Halstead were among its organizers. According to Halstead, NPAC was an "authentic united front of the masses" with only one message—Out of Vietnam Now!

Working from what the *New York Times* called a "shabby office" on the eighth floor of a building in downtown Washington, Lyttle called Hanley once again. The *New York Times* reported: "Harried young people shuffle papers, answer constantly ringing telephones, type press releases and occasionally stop to eat. This office of the National Peace Action Coalition and the workers, mostly volunteers, are trying to get half a million people to come to Washington on April 24th to protest the Vietnam War."[1] If successful, this was to be the largest nonviolent civil disobedience demonstration in US history, with Hanley providing sound reinforcement.

Lyttle informed the sound engineer that he was having difficulty securing permits for the event. Nevertheless, Hanley and his crew began to prepare. The *New York Times* reported on April 11: "Permits for the demonstration are still being negotiated with Government officials . . . Neither has permission been obtained for the use of the Washington Monument grounds behind the White House, the planned assembly point for the march, or for the use of the Capitol grounds, where a massive rally is scheduled."[2]

Eventually, with permits obtained, on April 24 over 200,000 people of various groups marched up Pennsylvania Avenue to the steps of the Capitol, where they rallied. This peaceful mobilization included a group called Vietnam

Veterans Against the War, who had already held a five-day protest on April 19–23. On April 22 one of its members, John Kerry, gave a statement in front of the Senate Foreign Relations Committee.

John Denver and Peter, Paul and Mary were among the artists who performed at the April 24 event. Protesters heard new songs with tougher antiwar lyrics than previous years, along with some old familiar ones. From the stage Kerry and Indiana senator Rupert Vance Hartke spoke. On this gusty spring day, Hanley's sound reinforcement system could be heard clearly throughout the Capitol grounds all the way back down Pennsylvania Avenue.

MAY 1–3, 1971

After nearly two weeks of extreme antiwar protests, a series of demonstrations referred to as "May Day" were held in Washington. Demonstration and event preparations for May Day took place over a ten-day period. The objective was to shut down the government for a day. By now, Lyttle was working swiftly and could call Hanley at any time for help. Together they planned for separate placements of sound reinforcement at different locations. According to Lyttle, "The May 9th 1970 protest had demonstrated to us that the anti-war movement had reached the point where mass demonstrations, involving more than 100,000 people could be called, on a few days' notice."[3]

On Saturday May 1, 1971, during some rather warm spring weather, an estimated 35,000 to 70,000 demonstrators flooded into West Potomac Park (close to the Washington Monument) for a concert event preceding the demonstration. By this time organizers had secured a camping permit for the area. This marked the beginning of the scheduled May Day activities. The park, adjacent to the Lincoln Memorial, lies on a fairly narrow, grassy "peninsula" in between the Tidal Basin and Potomac River. It was here, that the youth culture encamped. They parked their cars, vans, and their "ramshackle tents" gradually covered the park grounds over the course of the day.[4]

While folks funneled in, Hanley and his crew were busy constructing the stage and sound system that was positioned west of the Capitol steps. Lyttle recalls that prominent performers like the Beach Boys, Charles Mingus, Elephant's Memory, NRBQ, Mother Earth, and others were in attendance.

Throughout the growing encampment, into the early night before the show, the smell of marijuana permeated the grounds as a white low-lying pillow of smoke became visible. Throughout the park multilayered sounds of radios played the songs of the day, placards danced, and antiwar chants echoed in the distance. According to the Boston College newspaper the *Heights*, the "May Day tribe demonstrators" were not only there to protest, but also for the music: "Saturday's events focused on a marathon rock-soul-folk concert at West Potomac Park. Phil Ochs, the Beach Boys, and others played, but reports of appearances by John Lennon and the Jefferson Airplane proved false."[5]

As the music pulsed through Hanley's towering speakers, it reverberated across the Mall. For those in tents, sleeping bags, or simply lying on a blanket under the night sky, they could thoroughly enjoy the music. However, the behavior of the protesters agitated the police, who were waiting anxiously in buses on the outskirts of the park. They were ready to move in.

Lucy Barber wrote: "Many people stripped off excess clothing. Into the evening, they listened to a diverse set of performers. Speeches condemning American imperialism came right after announcements about lost children and bad acid. In the relative quiet before dawn, the police officers were probably edgy."[6] The May Day demonstration was evolving into a highly tense situation.

To many, this part of the May Day event seemed more of a concert and less of a protest. According to those present, it lacked the intensity of previous demonstrations. One *Washington Post* writer even referred to it as a "gigantic pajama party." Initially about 60,000 or more attended, but by nightfall only 35,000 to 40,000 remained in the park.

Lyttle and Hanley were emphatically against the ever-present heavy drug use and extreme radicalism that surfaced at the event. Lyttle recalls, "The hippie lifestyle often involved contempt for the nuts and bolts of organization."[7] From an organizer's perspective this changing behavior was a hindrance to the facilitation and representation of the movement. Lyttle adds: "Hippies collect money? How materialistic and crass. Revolutionaries help erect sound systems? That was manual labor."[8]

As bands and announcers concluded that evening's performances and speeches, the grounds grew quiet. All was calm for the time being as smoke arose from the trash-fueled fires that were left smoldering. Unknown to those who remained, at 6:00 p.m. that evening, the Justice Department had decided to revoke the organizer's permit and shut down the entire demonstration. "Police perceived the crowd as a threat to general public safety, and listed use of drugs and setting of campfires as other reasons for moving the demonstrators out."[9]

Days before, Hanley and his crew had set up a smaller sound system for a rock concert at the open-air Sylvan Theater slated for Sunday May 2. According to a 1971 article in the *Toledo Blade*, "The May Day tribe hopes to hold demonstrators in Washington over the May 1–2 weekend with some rock concerts, speeches and 'religious activity' at the Sylvan Theater on the Washington Monument grounds."[10] The concert never happened, because at dawn on Sunday an entire convoy of buses packed with law enforcement emptied out near Potomac Park, just south of the Mall. Hundreds of officers waited in silence, itching to remove the thousands who had come to protest and had enjoyed Saturday evening's rock concert. Mounted police were called in, keeping their horses calm until they were given the green light to go ahead.[11]

The White House attempt to dismantle the May Day protest had begun. While attendees exhausted from the previous day's activities slept, helicopters began to swarm above. The demonstrators were already savvy about low-flying helicopters, having launched helium filled balloons connected to cables that could potentially get caught in the rotors of the aircraft. But not much could be done when over 2,000 DC police, in full riot gear and equipped with tear gas, surrounded the area. They sealed off exits and began ordering the sleepy-headed crowd to leave.

Officers charged through the early morning light ordering everyone out of the park. Blaring announcements could be heard from the police loudspeakers: "This is the Metropolitan DC Police Department! Your permit for this encampment has been revoked! Anyone remaining will be arrested!"[12] This was enough to wake most everyone in the area, leaving them startled and scrambling, including Hanley's crew. With sleep still in their eyes, protesters leaped out from their encampments, grabbing whatever they could, and leaving all of their trash behind. By noon, two sweeps of the park had been conducted. A line of over "300 helmeted and riot-equipped" officers emptied the area entirely. About one hundred arrests were made.[13]

Hanley's then girlfriend (now wife) Rhoda Rosenberg recalls the ominous words "Your permits have been cancelled! You must vacate the area by noon!"[14] echoing throughout the park. The two had been fast asleep in Hanley's car. According to Rosenberg: "I remember a lot of activity at night; Phil Ochs didn't go on until very late. They didn't have any speakers so it was really much more of a concert."[15]

Unable to get a hotel room, the couple had slept in Hanley's Buick station wagon, handed down to the engineer by his father. A rather large vehicle, it was perfectly suitable for sleeping and hauling gear. Rosenberg remembers that, although it was a harrowing experience for all involved, she was witness to Hanley's expertise in dealing with law enforcement at mass gatherings. She recalls:

I remember going to sleep earlier than Bill. He must have crashed after I did. I got up around 6:00 a.m. the following morning, and since I arrived that evening, I did not know about the extent of the inhabitants of the campers and tents. I remember teepee tents made out of canvas all over the grounds. When I got up I walked around and saw other campers waking up, cooking their brown rice, etc. As I walked back towards our car I then noticed a line of police in full attack gear with masks on their faces. All I could think of was, what are they going to do to these poor kids? All they are doing is sleeping and have their campfires going? This is not good! Later Bill told me that there were hundreds of police. At this point none of the organizers were at the site. So I proceeded to the car to wake Bill up.

I said: "Bill wake up!" (I insisted he get up) "They are out there!"

He said: (In a groggy voice) "Who's out there?"

I said: "The police!"

He said: "What do you want me to do?"

I said: "You have to do something! You just can't let this happen! Go talk to them!"

He replied: "What am I going to say?"

I convinced him to do it, which he did. Bill then got out of the car to go talk to the police, as they surrounded the camp. I was hanging on the snow fence and they were massively photographing us. I am sure we are in some files somewhere. Later, Bill told me they said there was no permit and of course there was one. We couldn't be there without any permit, as Bill told them. Bill knew this because he worked with Brad Lyttle, and although Brad was courageous he was not stupid. As a facilitator of these events Lyttle dotted all of his i's, and crossed his t's when it came to permits. As I stood there and watched Bill talk to the police at this particular location of the camp, some of them got on their buses and left. If I hadn't woken up Bill who knows what would have happened.[16]

On Monday morning May 3, a Day of Civil Disobedience was planned. Lyttle recalls, "Collectives, largely of college students, tried most of the morning to close down Washington by sitting in streets and blocking traffic."[17] According to one May Day spokesman, "Demonstrators are scheduled to take over all the principal bridges across the Potomac and many of the traffic circles between downtown and the Maryland line, in an attempt to stop federal employees from coming to work."[18]

Near the east side of the Washington Monument, Lyttle waited for thousands of people to gather for the walk to the Pentagon. By this time, demonstrators were to sit down in the roads. "In the early morning of May 3, nearly 20,000 protesters gradually spread out around the district. Rather than occupy the center of Washington, the protestors wanted to block it off."[19] Lyttle was arrested during the chaos and describes the brutality of the situation:

We did not get far since the police blocked the demonstrators near or around 14th and Jefferson Drive. When we did not disperse as ordered, officers pounced on me, and placed me under arrest. When I went limp, one began to hit me in the face with a club. After several blows, he succeeded in cutting my lip, and drawing blood. Seeing the blood, he stopped hitting me. Officers then took my arms and began to drag me to the paddy wagon.[20]

The organizer was charged with resisting arrest. Later, while in federal court he was charged with three felony counts that included assault on a police officer with a dangerous weapon. But the only weapon the pacifist had with him that day was a bullhorn, which he had discarded just before the arrest. Lyttle was faced with ten years of jail time if convicted. Fortunately he was not.

Around a quarter of a million people were involved in the large-scale demonstrations that occurred in April and May. On May 3 approximately 15,000 to 20,000 people blocked the streets of Washington, DC. Smaller protests continued, and an estimated 13,000 were arrested. Most were held at the Washington Coliseum, resulting in the largest mass arrest in US history. The May Day demonstration diminished several days thereafter.

Hanley has maintained his relationship with Lyttle over the years. After the event, the sound engineer assisted the organizer with other demonstrations including a 20,000 to 30,000 person event at Battery Park in New York. Lyttle observes, "Bill's ability to meet the challenges and requirements regarding the diversity of the events was remarkable."[21]

Without Hanley and Lyttle's political, physical, and financial sacrifices, the public would have been audibly removed from the important messages at these marches and rallies. Without sufficient sound, the music and message that was intended to inspire and empower demonstrators might not have been achieved. Lyttle explains: "I watched all of these political developments with considerable naiveté, dismay, and ambivalence. It hadn't occurred to me that the anti–Vietnam War movement might become a revolutionary vehicle. I saw the movement as a way to end the war, and perhaps, promote nonviolent resistance."[22] Both men altruistically made considerable contributions to society during this time.

By the late 1960s and early 1970s the music scene had achieved its own voice. Even though most of the antiwar activity took place in Washington, DC, it also occurred at college campuses where Hanley and his crew continued working. Hanley was one of the only sound engineers at the time that took such risks, and did so because of the cause and not for cash. The protest movement needed dedicated people like Hanley, and like so many of his colleagues have said, "He made it happen." Even when he didn't have the right equipment or was faced with a court injunction, Hanley simply—made—it—happen.

Chapter 48

A SYMPOWOWSIUM AND THE
FATE OF THE FESTIVAL

After Woodstock, Hanley and his crew delivered sound reinforcement for many other notable tours, festivals, and anti–Vietnam War demonstrations. Moderate to large in size, some reached over 200,000 (or more) in attendance. A partial list of Hanley's festival bookings from 1969–72, and number of people who attended, is shown below.

HANLEY FESTIVALS OF 1969–72			
Date	Event	Location	Attendance
1969 August 30–September 1	Texas International Pop Festival	Dallas International Motor Speedway, Lewisville, Texas	120,000
1969 September 13	Toronto Rock and Roll Revival	Varsity Stadium, University of Toronto, Ontario	20,000
1969 November 28–30	West Palm Beach Pop Festival	Palm Beach International Raceway, Jupiter, Florida	50,000
1969 December 27–29	Miami Rock Festival	Miami-Hollywood Speedway Park, Hollywood, Florida	Unknown
1970 March 27–29	Winters End Festival	Econ Dude Ranch, Bithlo, Florida	40,000
1970 June 24–July 5	Transcontinental Pop Festival (Festival Express)	Toronto, CNE Stadium; Winnipeg, Winnipeg Stadium; Calgary, McMahon Stadium	37,000, 4,600, 20,000
1970 July 3–5	Second Atlanta International Pop Festival	Middle Georgia Raceway, Byron, Georgia	200,000
1970 July 17–19	Love Valley Rock Festival	Love Valley Ranch, Love Valley, North Carolina	75,000
1970 July 17–19	New York Pop	Randall's Island, New York	40,000
1970 July 30–August 1	Powder Ridge Rock Festival	Powder Ridge Ski Resort, Middlefield, Connecticut	50,000
1972 April 1–3	Mar y Sol Festival	Manatí, Puerto Rico	30,000

A familiar face at many of these events was activist Wavy Gravy. According to the Hog Farm leader, the festival experience was something to remember: "By this time the festival scene had gone on so long, that people had devised this positive and creative anarchy which I just loved!"[1] Eventually Hanley and Wavy Gravy's friendship developed, lasting until this day. Although their roles differed, they shared a similar sentiment regarding the festival phenomenon occurring at the time. Wavy remembers:

> We created these instant cities that rock and roll was the soundtrack for. I have a line in the Woodstock film where I say "there is a little bit of heaven in every disaster area" and I maintain that to be true. Even until this day if we are filling sandbags in Mississippi, its rich people, poor people, black people, white people, yellow people, red people, and everybody chipping in. At Woodstock we had a better soundtrack thanks to Hanley's dedication, which I really came into full "bleshing" of at the West Palm Beach Pop Festival with the Rolling Stones.[2]

In October and November 1969, crew members Marks and Butler went on a five-week tour with the Turtles, Three Dog Night, and Hoyt Axton. From Medford they traveled in the Yellow Bird to North Carolina, through South Carolina, Tennessee, Georgia, Oklahoma, Florida, Louisiana, and New Mexico. Eventually they connected with Steppenwolf in El Paso, Texas, to assist the band. Marks recalls, "We off-loaded a few tons of loudspeaker bins and a mixer for Steppenwolf and Al Kooper at the University of Albuquerque."[3]

Between tour dates, Marks and Butler slept in the semi's sleeper cabin and occasionally a seedy hotel. According to Marks, as the truck rolled through the night the hum of its treads on the pavement and the engine roar lulled him to sleep: "My share of sleep would usually be in the cab behind the fired up ... anti-hippie driver, rocking 'n rolling between gigs up or down Route 66 in the twenty two ton Peterbilt Horse 'n' Trailer; dozing off to the country music radio shows at two or three am I'd always wake up before sunrise and shoot the breeze with Dick as we pulled into another State."[4] Marks recalls the quality of motels in the US at the time:

> Each U.S. rock and jazz tour was designed to fit into a particular chain of motels; they didn't only all look alike, they also smelt and felt the same. From the franchised green and gold of the Holiday Inn to the sweet 'n stale bedroom sweat of the orange and blue Howard Johnson; franchised air fresheners, menus and bar-room lounge muzac as well. Often, after a late night gig, you'd wake up and wouldn't know where you were or what state you were in.[5]

On tour Butler was primarily in charge of driving, while Marks unloaded and set up. Marks remembers, "Dick never cared much for rock and roll music," so Butler usually slept during the gigs while Marks was working. The next day they were on the road again. Marks adds, "You could tell we were crossing a border by the changing sounds of the music on the truck radio and the accents at the truck stops."[6]

Only able to use one of his arms, Butler became known for his crafty driving skills. Precariously manipulating such a large vehicle in and out of antiquated music venues and tight theater locations was extremely difficult. Marks explains:

> Dick genuinely had a gammy right arm—the one he somehow used to artfully manipulate all thirteen gears with—without you even noticing. The arm wasn't broken for not joining the Teamsters, he proudly told me—he was born that way. He was also short, chewed a match at all times, wore dark sunglasses, good looking and born to impress. And believe you me, behind that Peterbilt's wheel, when he wasn't flying one-handed down a highway, he surely could move those twenty-two tons sixty feet backwards and forwards around a tickey. We needed him—not too many of the older southern sports arenas were built to care or cater for rock and roll rigs and gigs, that's for sure.[7]

THE SYMPOWOWSIUM REPORT: SAVING THE FESTIVAL FROM GOING SOUTH

A gathering of music business minds occurred over Halloween weekend in October 1969. The event was called the Sympowowsium and was held at the Hummingbird Music Camp deep in the Jemez Mountains of New Mexico. A lot had been learned and celebrated about Woodstock. But the complications and subsequent rise in popularity of this event and other copycat festivals seemed to require this problem-solving-festival-think-tank.

The Sympowowsium was organized to further understand and sort out what the future of rock festival mass gatherings should be like. The dominant theme and question during the weekend long gathering was: What comes after Woodstock? Activist and counterculture figure Lisa Law, who was in attendance, recalls: "The Sympowowsium was a gathering of people who were interested in putting on concerts. It was right after Woodstock in Jemez, New Mexico put on by the Jook Savages, Tom Law, and Reno Kleen."[8] Bill Hanley was also there.

On Friday October 31, 1969, Hanley landed at the single runway WW II–era Los Alamos Airport located on the Pajarito Plateau within the Jemez Mountains.

After grabbing a cold Coke to quench his thirst, he called the office from a pay phone to check in. After an hour or so, Merry Prankster leader Ken Kesey pulled up to the dusty airport. Hanley's ride had arrived. According to Hanley: "Kesey came in at screeching halt and whisked me away! He was clearly tripping his socks off at one hundred mph!"[9]

Children attended the Hummingbird Music Camp during the summer months. By October it was empty and perfect for accommodating the event. Equipped with small cabins and larger lodge-like structures it was ideal for the morning and evening planned meetings. When Hanley arrived he found his room and settled in. Later he greeted the familiar faces that were socializing over their meals and occasional joints. This elite roster of music business leaders included record producer Paul A. Rothchild, journalist and author Paul Krassner, Woodstock producer Michael Lang, Ken Kesey, various members of the Hog Farm, and more than sixty other individuals who gathered to discuss the festivals' preservation (see appendix).

According to Reno Kleen, one of its primary organizers, Hanley stood out from the counterculture crowd that weekend:

> Bill was a unique character, I must say. I think he was very hip, so to speak, as to what the scene was because he was dragged into it. He had the capacity to do large sound; he was knowledgeable, had a good head, and worked well with other people. But he had kind of a straight image. He wasn't like the freaks around him or wannabes or whatever they were.[10]

Kleen claims that he and friend Tom Law thought the idea of a Sympowowsium seemed logical since they knew so many promoters, musicians, and technicians. Getting them in one room all at once made sense to him: "We invited them all to New Mexico, and had a fairly good turnout!" The organizer explains how the unique name of the event emerged:

> I came up with the name "Sympowowsium," which was a cross between native American powwow, like an old Indian gathering, because we were up here in native territory in New Mexico; then symposium to have a dialogue over this whole thing about this phenomenon of these large festivals and this new outdoor arena environment and how to deal with it. The objective was to find some way that people understood that they were getting shortchanged—that the people coming to these concerts were getting ripped off.[11]

Forever optimistic, Hog Farm member and counterculture figure Tom Law felt that there was a real sense that something could be done to preserve what had happened at Woodstock:

After Woodstock we realized that this was happening, we saw what we did at Monterey and we saw what the Hog Farm could do. After Woodstock, we said to one another that this thing could go one of two ways: it could go the way of the power takeover and money, or it could be part of a conscious- ness—the way a festival should be. So we got a lot of people up into the Jemez Mountains, and Hanley was there. We talked the whole thing out for three days because we had a feeling that something bad could happen too, and sure enough, soon after, Altamont happened. We were trying to avoid this. The Sympowowsium was a highly further-left-than-Jesus-Christ-think-tank. All we were trying to do [was] get the people who were involved in festivals to come there and think about what they were doing and the effect it had on culture at large.[12]

According to Kleen, after Woodstock, he noticed promoters of the big festi- vals were giving people from the hippy commune, the Hog Farm, freedom to do their thing like setting up trip centers, food stands, camping provisions, etc.—an idea that Kleen supported, because festival organizers had no idea about how to deal with festival issues, adding, "We at the Hog Farm knew what to do." Kleen recalls: "They barely had crappers on the scene. So the promoters had to be told what to do. They had no established relationships with the police or anything. There had to be a change to do this. Outside of Woodstock, there were a lot of players involved with these large events. A rapport started to develop slowly; guys like Bill Hanley were being referred to do these large events."[13]

Throughout the weekend the group exchanged ideas about possible festival fixes associated with shifty promoters, violence, gate-crashing, and heavy drug use. The general consensus was that the festival, a force that gave strength to the counterculture, should be presented in a positive way. The Sympowowsium was an honest attempt to build on and preserve what happened in Bethel. It was in the hearts and minds of its attendees to try and avoid any, if not all, foreseeable festival disasters from occurring.

Although many interesting and progressive ideas were exchanged, nothing really came of it. Robert Santelli, in *Aquarius Rising*, writes: "The conference could have played a major role in the future presentation of rock festivals, but due to lack of leadership and an eventual parting of philosophical ways, the meeting turned out to be nothing more than a pleasant get together of old friends."[14]

The December after Woodstock, the ugly side of rock and roll mass gather- ings emerged at the Rolling Stones Altamont free concert. Leading up to the event, the rise of heavy drug use, four high-profile assassinations, the Vietnam War, and the civil rights movement created a threshold. Then at the end of 1969, the trial of the Chicago Eight materialized. What happened at Altamont caught the brunt of the culminating societal and political issues at the end of this tumultuous decade. As soon as the counterculture grasped its identity in free love, it lost its grip on Woodstock idealism.

Weeks after the Sympowowsium, a ten-page document called the *Sympowowsium Report* was sent to its attendees. In the report authors Kleen and Tom Law write about the utopian views expressed over the weekend: "We speak of festivals as spiritual events as massive gatherings of people in New Age proportions. To speak the truth they will be designed to exercise minds with awareness, promote consciousness, foster mutual respect, collect and center creative energy, and learn discipline not by law or club but by experience and example."[15]

The report served as a detailed and structured plan about how a festival could be run more ethically and efficiently. Kleen and Tom Law itemize and emphasize crucial festival details like the facilitation and deployment of security, communications, food services and preparation areas, types of food served, sanitation, showers, drinking water, transportation, mother nursery, spiritual facilities, trip centers, drug advice, and camping gear. Lisa Law recalls that the report was designed to "educate people on how to put together a concert" without ripping them off, while creating an environment where they could enjoy themselves rather than suffering.

Kleen recalls: "I did the drawings and Tom Law and I wrote the text of this workbook that we created. We sent this to all that attended. In those days there wasn't much stuff around to work with. We probably produced that book over a day or two, it was one of those mimeograph things, it was the sixties."[16]

The interior of the Hanley
Sound Office. Medford,
Massachusetts. Late 1960s.
David Marks collection.

Bill Hanley in deep thought
on the Woodstock Festival
stage. August 1969. Henry
Diltz collection.

Hanley Sound right hand
speaker stack at Woodstock.
Jim Shelly collection.

Sam Boroda and Nick Burns at the Woodstock mixing console platform. August 1969. David Marks collection.

Bill Hanley at the Woodstock mixing console. August 1969. David Marks collection.

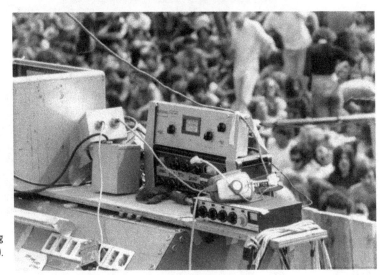

Woodstock side-stage, monitor audio processing equipment. August 1969. David Marks collection.

Scott Holden underneath Woodstock stage falling asleep watching over McIntosh amp racks. August 1969. David Marks collection.

Hanley Sound Woodstock trailer office. Backstage. August 1969. David Marks collection.

Woodstock speaker stacks. Monday morning. August 1969. David Marks collection.

Bill Hanley backstage at Woodstock.
August 1969. David Marks collection.

David Marks somewhere in Texas standing outside of the Hanley Sound "Yellow Bird."
1969. David Marks collection.

Mixer and Woodstock Bins in Soweto, South Africa in 1971. David Marks collection.

Speaker set-up at the Second Atlanta Pop Festival in Byron Georgia 1970. Ric Carter collection.

Madison Square Garden Productions
and Howard Stein Proudly Present

The *Rolling*
Stones

2 Performances Only · Thursday and Friday
November 27-28 at 8:00 P.M.

Sound by Hanley
Prices: $8, 7, 6, 5, 3.50

Madison Square Garden Box Office
Opens Tomorrow (Thurs.)

For mail orders make check or money order payable to Madison Square Garden Center.
Enclose self-addressed stamped envelope and add 25¢ per order for handling. Never mail cash.
Tickets also available at over 100 Ticketron outlets. Call (212) 759-2734 for location nearest you.

Rolling Stones MSG advertisement. Notice "Sound by Hanley." October 1969. Hanley archive.

Bill Hanley at Madison Square Garden on rotating stage for Joan Baez holding speaker cable affixed to Hanley Hula Hoop. 1969. David Marks collection.

Bill Hanley at mixing position for the West Palm Beach Pop Festival. November 1969. Bob Davidoff collection.

Aerial view of Hanley Sound truck and speaker set-up at West Palm Beach Pop Festival. November 1969. Ken Davidoff collection.

Bill Hanley at console at Varsity Stadium for Toronto Rock and Rock Revival in 1969. David Marks collection.

Bill Hanley, Judi Bernstein, and Billy Pratt. Wall Street Anti-War Protest. 1969. David Marks collection.

Hanley Sound set-up at Anti Vietnam War Demonstration in Washington, DC. October 1969. Al Braden collection.

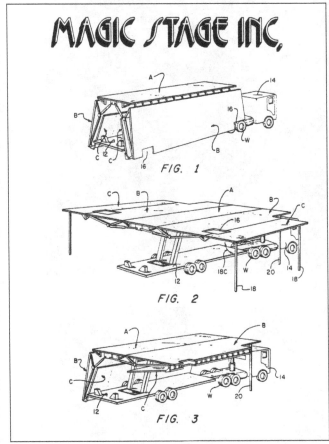

MAGIC STAGE INC,

FIG. 1

FIG. 2

FIG. 3

Bill Hanley featured in Esquire Magazine "Top 100 in Rock." 1970. Hanley archive.

Magic Stage illustration diagram. Hanley archive.

Bill Hanley and Rhoda Rosenberg somewhere on the road. 1970. Hanley archive.

Author John Kane with Bill Hanley and Stan Miller at Boston Garden in 2015. John Kane collection.

Bill and Terry Hanley posing in front of their historic marker at 430 Salem Street. Haines Square. Medford, Massachusetts. 2013. Hanley archive.

Chapter 49

WOODSTOCK SOUTH

PALM BEACH POP

In the *Sympowowsium Report,* Tom Law provided a brief analysis of what he witnessed at the 1st Annual Palm Beach International Music & Arts Festival. This festival has come to be known as the West Palm Beach Pop Festival or Palm Beach Pop. The event was held the weekend after Thanksgiving, November 28–30, 1969 in the town of Jupiter, Florida, location of the Palm Beach International Raceway.

Law was emphatic that what he saw represented everything wrong with the way festivals were being handled. A sleazy promoter, poor sanitation, coupled with bad weather, and Palm Beach Pop set a sour trend for festivals to come. This is what Law had to say about this poorly organized gathering:

The "First Annual Palm Beach International Music and Arts Festival" just happened and was the essence of what the "Sympowowsium" was trying to prevent from happening, maybe pose the alternative. The promoter, inspired by the gross smell of money, had raised some, punched the buttons to book the necessary talent and scheduled this music and arts festival at his drag strip in the everglades not far west of Palm Beach. I met him the day after the Sympowowsium when Hanley, Lang, Cowing, Pate and I flew to Florida to check it out. In the three hours I spent with Rupp I noticed he was vacant of respect for the performers he booked, the audience he hoped to attract, the police, local officials or himself. Not one word or vibe that smacked of respect. At festival time a month later at the request and experience of Hanley, Wavy Gravy, Reno Kleen, Richie Ruffin and I flew to the scene and did what we could to ease the pain or make it more fun and to somehow relate that this is not necessarily where festivals are at. It was an unsuccessful artificial imitation plastic "Woodstock" minus warm spirit and real good vibes. The promoter didn't order the first nights rain but by unaware and negligent planning he set it up for mud and garbage.[1]

Hanley called on Hog Farm members David Butkovich and Campbell Hair to help with the event. Butkovich had met Hanley at the Woodstock free stage, where he helped run sound. He was an original communal member of The Merry Pranksters, spending time driving around in Ken Kesey's psychedelic 1939 International Harvester school bus known as Further. Butkovich recalls: "Bill liked what I was doing at Woodstock so he asked me to go back to Boston with him. I got in the truck and started driving to Medford."[2]

After Bethel, Butkovich stayed on with Hanley Sound ending up in Washington for the anti–Vietnam War demonstration of November 15, 1969. Butkovich remembers: "That November I drove down to Washington, DC, from Medford and we all got arrested. The next time I saw Bill was when he came down to get us out of jail. Then Campbell and I went down to Florida for the West Palm Beach Pop Festival."[3]

Hair, a member of Wavy Gravy's Hog Farm, became an official Hanley Sound employee in Washington, when he and the sound engineer hit it off during the November demonstration. He remembers: "Bill asked me at the Washington demonstration if I wanted to go down to West Palm Beach to help with a festival that was happening there. So I jumped off of the Hog Farm bus 'the Road Hog' went back to Medford and joined the Hanley crew down to West Palm Beach."[4]

After the harrowing experience in Washington, Hair and Butkovich were the first to arrive at the Palm Beach Raceway on Sunday afternoon, November 23. Hair recalls that he had no idea what he was getting into when they finally arrived at the festival site: "I was so naïve. Just a few months ago I came out of the mountains in New Mexico and ended up at Woodstock. Then I met Hanley at the biggest peace march ever, and now I am driving down to Florida with Bucko!"[5]

In a steady rain, hauling a portable stage, the crew members pulled up and parked Hanley's White Dove tractor-trailer. Waiting out the bad weather, Hair and Butkovich surveyed the area. Hearing a shaking sound, they spied a pile of rattlesnakes on the other side of the racetrack fence. According to Hair, the site was "in the sticks," feeling remote and deserted. Butkovich adds: "When we got there, I noticed the wheels on the stage were flat and worn from traveling so much. We had to replace them ASAP."[6]

After unloading equipment, they got situated at a local hotel and waited for Burns, Pratt, Boroda, Bernstein, and Hanley to arrive. The following day they began the sound and stage setup for the opening date, November 28. However, there were some snags. David Rupp, festival promoter, used car salesman, and speedway owner, was having problems. Hair explains: "As soon as we checked in to the hotel, we called Judi and she filled us in on the issues the promoter was having. We could also sense it wasn't going to be easy."[7]

A court injunction had stalled the initial facilitation of the festival, but was miraculously overturned on November 20 and Rupp was able to walk away with a "declaratory decree" allowing him to hold the event. In a *New York Times* article, "Rock Festival a Show of Youthful Good Manners," journalist Ben Franklin writes: "Only two weeks ago after $20,000 worth of tickets had been sold at twenty dollars each for the three-day event, the county zoning board unanimously denied the special permit it had insisted was required. Ten days ago Mr. Rupp, a used car dealer and raceway owner, won a reversal of the zoning decision."[8]

Not long after the decision, the community responded by burning down Rupp's used car lot, further demonstrating the level of anger from residents.

Leading up to the festival, fierce opposition from locals, city and county authorities, and Palm Beach County sheriff William Heidtman continued. Woodstock had created a national stir. It was so great that the conservative, churchgoing West Palm officials were terrified about an impending flood of counterculture youths taking over their town. Sex, drugs, and rock and roll, which had become synonymous with events like this, frightened the townspeople. The happenings at Woodstock worried the southern town of Jupiter, Florida. Sanitation and traffic issues occupied everybody's mind. According to the *Palm Beach Daily News*, "Tales of nudity and narcotics at rock festivals near Seattle, Wash., Atlantic City, N.J., and Woodstock, New York, backed up by a movie film and still photos of scantily clad girls that one attorney said 'shocked my conscience' marked the representations of festival opponents."[9]

Like any promoter, Rupp was hoping to cash in, but because of low ticket sales the festival was headed for financial disaster. Hundreds of counterfeit tickets were sold and this didn't help revenue. To avoid the twenty-dollar ticket fees, fearless attendee wannabes risked their lives by swimming in the nearby gator-infested canals and lagoons to gain access to the festival grounds. Additional gate-crashing and food shortages sent local officials into a desperate frenzy. Adding to the chaos, the November weekend was fraught with extraordinary cold and rainy weather. The *New York Times* reported: "Soggy blankets paved the ground and mud clung to bell-bottom trousers and granny dresses. Festival employees warned the crowd, 'Keep cool, keep your drugs dry and watch your brother next to you. The narcs (narcotics agents) are here.'"[10]

During the festival, to Rupp's dismay, Sheriff Heidtman had positioned over 150 deputies across from the raceway. The paranoid sheriff set up surveillance cameras, and from all sides the promoter was feeling the mounting pressure of his creation. Some say that the thirty-one-year-old Rupp lost over $500,000 in the end. "It was all the cash I had and all that I'd borrowed."[11]

According to Butkovich, Rupp was not as organized as he should have been leading up to the event. He remembers that by Friday, the festival's opening day, the promoter disappeared from the scene because of the mounting pressure. With an absentee promoter, the event lacked key decision-making, and as a result, organization of the production suffered. On a fluke, and out of desperation, a short-straw contest was held back at Butkovich's hotel room. To his misfortune, he won, and was appointed acting head of festival production. Butkovich remembers:

> We all met in a hotel room and it came down to all of us drawing straws to see who was going to run the show, I got the shortest straw. West Palm Beach was a lot more chaotic than Woodstock. There wasn't that much adult supervision. The Florida National Guard was on standby and they were ready to move in at any moment. There was no one to run things. So for the next six or seven days I ran the show. I was up for days.[12]

Known for his work at Woodstock, Wavy Gravy was flown in by helicopter to help keep the peace. Hair recalls: "We got the promoter to bring Wavy Gravy down and Tom Law and many others from previous festivals. To me West Palm was more about the music business than at other festivals. Although, some of the acts were incredible, the whole thing was different because of the police presence and politics."[13] Wavy Gravy (Hugh Romney) reflects on the event in his book, *The Hog Farm and Friends*:

> The festival is set-up in the middle of a swamp. Chock full of gators and poisonous snakes. Coupled with fences, the Army, and cops, Mission Impossible could not get in without a ticket. The operations trailer had only one phone, which was constantly busy. The sound in the swamp was provided by Hanley. Bucko was havin' the time of his life answering the phone and talking to helicopters movin' amps and musicians up and down. Outside it was raining and raw. Tom Law takes me on a tour of the site. Mostly it is a mud dance with a fairly large crowd. And I stand in a puddle pissed off at the promoter. I wonder what kind of fiend could put this together? Promising Palm Beach in the press. Using the Rolling Stones for bait. Net all you can. Then land them in a speedway in the center of the Okefenokee.[14]

By Saturday November 29, Palm Beach Pop had drawn over 50,000 attendees. Even though the festival was faced with many adversities, its lineup was nothing short of stellar. Acts like the Rolling Stones, Janis Joplin, Jefferson Airplane, Johnny Winter, Grand Funk Railroad, King Crimson, the Byrds, Sly and the Family Stone, Steppenwolf, Iron Butterfly, and Vanilla Fudge might have made for a weekend to remember.

A twenty-seven-mile-long traffic jam caused performers like Sly and the Family Stone, Jefferson Airplane, and Janis Joplin to be flown in by helicopter from the nearby Colonnades Hotel on Singer Island, which was also the festival headquarters. Boroda claims that for him, Palm Beach Pop was one that really stood out. He recalls being invited to hang out with some of the acts back at the hotel:

> This girl who was working for a promotions company invited me back to the hotel where the bands were. So that afternoon I got on the helicopter located backstage and left. At the hotel, Sly from the Family Stone remembered my work at Woodstock, and raved about me. I had to get back because I had to mix that night. When Judi saw me get off the helicopter she was blown away! She asked "how the hell did I manage that?"[15]

Hanley Sound stagehand Alun Vontillius shared his bottle of Southern Comfort with Janis Joplin as well as Johnny and Edgar Winter. At the event, he was in charge of setting up the microphones. Vontillius remembers:

> There was definitely a smaller crowd at this festival and the weather was really bad. I pretty much helped with bringing the mikes out and miking the bands. On Saturday night of the festival I was the last guy to pack up all of the microphones in this briefcase and load them into the trailer. I fell asleep in the trailer and the crew member that was supposed to start the show saw that there were no microphones. That's because they were under my head in the trailer where I was sleeping! Everyone thought they were stolen![16]

Hanley brought in full production for Palm Beach Pop. He recalls: "We used the Wenger Wagon as a base of operations and that also kept things dry. We brought the sound, stage, and lights to West Palm Beach."[17] Hanley adds that the audience area at the festival site was much narrower than at Woodstock: "The speaker placement was set up to drive sound further back rather than off to the sides because of our position on the raceway. The area was typical of the Florida landscape carved out of the Okefenokee Swamp."[18]

For performance efficiency, the Palm Beach Pop production team used a carting system on stage. This was inspired by the failed half-moon turntable design attempted at the Woodstock Festival. But the engineer felt that a carting system could work here, adding: "The stage system design broke down at Woodstock and I learned from that. I was determined to not have that happen again."[19] A functioning stage cart system could cut down on wait times in between performances dramatically. Hanley observes: "The logic for such a device at both festivals was to get bands on and off the stage quicker. We built and designed this to get the efficiency up when we preloaded various groups' equipment. Chip played a big part in this."[20]

Hanley's crew and production team built fifty-foot-wide carts that could move stage left and stage right across the hundred-foot stage. It worked great if crews could preload the carts with performers' equipment. Hanley explains: "At Palm Beach we had angled track that was lying on top of the stage deck and the carts had v-grooved wheels that sat on the track. We were able to switch all the microphones at once so we didn't have to re-mike everything when the group got to center stage."[21]

Hair claims that the crew learned a lot from the mistakes made at Woodstock. He recalls everyone wanting to make things technically better at Palm Beach Pop: "Overall the event ran pretty good except for the weather. Fortunately, we were able to store the stage in Florida at a relative of Bill's. With Hanley's guidance we developed carts to roll the acts on and off for quick set changes."[22]

Because of a stipulation written into the Rolling Stones' contract, the fence in front of the stage was made twenty feet high. The structure was so high it caused problems for the audience sitting in front. This forced festivalgoers to stand up, making it difficult for those in back to see. By Saturday, the crowd was getting angry about the height of the fence. Fortunately, Woodstock veterans Wavy Gravy and Tom Law were well versed in festival facilitation; at Palm Beach Pop they were in charge of crowd control. Realizing the issue at hand, they scrambled backstage to give the production crew an update. They urged Hanley and others to address the problem immediately before a riot broke out. Wavy Gravy recalls the situation could have unraveled if Hanley didn't make a quick decision:

> So I run inside and tell Judi Bernstein the story. She thinks I'm on an ego trip. So I find Tom Law and the Savages and take them out to meet the people. The people are anxious to saw the fence. They give us an hour to get it together or they'll tear it down themselves. That would be the move the man was waiting for. The promoter would panic and the Guard would charge. As a last resort we lay on the floor of the operations trailer singing "We Shall Overcome." Everybody's got to walk over us to do business. Bill Hanley comes in and he can't believe it. I told him we have about four minutes left! Hanley is a good guy but he's in a difficult position. After all, "The Rolling Stones are pretty important." Tom Law says he's got two minutes and he cracks. Tells us to cut down the wall and I really respect him for that.[23]

Those within the Stones camp were catching wind of the fence being cut. They were wary about any staging or lighting problems developing at the festival and threatened not to go on. Worried, David Rupp assured the band's management over the telephone. According to the *Palm Beach Post*: "Rupp sees the issue as a professional feud between Chip Monck who handles lighting and staging for the Stones and Hanley Sound Co., holders of the festivals contract. 'Hanley's the best,' Rupp said. 'He's done more festivals by accident than Monck's done on purpose.' Rupp said, 'Lets not get caught in the middle of a feud. The Stones have been paid, they'll play.'"[24]

The tired, wet, and shivering audience waited for the very tardy Rolling Stones to arrive. They were scheduled to close out the festival on Sunday, November 30 at midnight, but didn't arrive until Monday morning. As the crowd waited, Wavy Gravy got permission to make a special announcement. He told the audience they could break up and burn the wooden bleachers for warmth. According to Hanley: "There was a heavy police presence, and it was freezing! The kids were making campfires out of the wooden bleachers and outhouses to stay warm!"[25] Vontillius remembers: "It was so cold that I was wearing my rubber boots. About four or five times I walked into the audience and stood on top of the embers of someone's fire until I could feel my feet again."[26] Crew member Nick Burns recalls the unusually cold weather:

> It was the end of the Stones tour and they decide to do a festival in West Palm Beach and everyone is jumping for joy; we all thought we're going to go down there and be warm, yeah right! The area was weird. It was some reclaimed marshland. Three nights in a fucking row and it was freezing! It was 33 degrees! Nobody except for a few idiots from New England had warm clothes. Backstage I saw Janis Joplin swigging Southern Comfort. I offered her my jacket to keep her warm ... she looked at me like, "are you kidding?!"[27]

Anticipating the Stones' arrival, Hanley worried that crowds might rush the stage. According to Wavy Gravy: "Tom Law and I had promised Hanley to guard the stage with our lives against an onslaught from the audience, but it really didn't matter at all. People were too wet, and too tired."[28]

The Stones had been delayed at the airport, coming straight from New York, where Hanley had provided sound for them at the Garden on November 27–28. They would not hit the Palm Beach Pop stage until 4:00 a.m. Monday morning, where they performed a short set to a thin crowd. Stones road manager Sam Cutler made the following announcement to the sleepy, wet audience before they went on:

> Uhm, we've eventually arrived in Florida, after eleven hours and fifteen minutes in the same airplane. We had amazing hassles getting here but we managed to get through and we are here and we're gonna come and play for you and we wanted everybody to know that if you had to wait and you're cold

that we are sorry, we got here as quickly as we could. We were hung up by things like the president landing at the airport where we were trying to take off and that made it all very difficult. Uhm, anyway when we come we'd be very grateful if everyone could wake up which will be really nice and if you wanna get warm then the best excuse is to sort of dance around and flip out if you want to over the music. We'll be here quickly, we're sorry to hang you up over it, right. As soon as we can get here we'll be here thanks.[29]

Mick Jagger was unaware of the adversities that the audience was going through. Before he took to the stage for the band's set, someone whispered the situation to him. Wavy Gravy remembers: "Mick Jagger really didn't understand. Stepping out of his heated helicopter, he started comin' on like an electric fop. Talkin' down to the folk. Then he realized what everybody had been through, started to empathize and beg for forgiveness."[30] The Rolling Stones commanded a whopping $100,000 dollars for their brief appearance at Palm Beach Pop.

By the end of 1969 a majority of Americans perceived the rock festival as a vehicle of moral collapse, highlighting it as a representation of everything that is bad with humanity. It was an unwelcome thing, and many were adamant about it not happening in their backyard. Court injunctions and stringent health and safety regulations provided ammunition for local authorities to strike back at these monster events. Often the promoter just gave in, yet Rupp did not. Even though locals tried to stop this event, it still happened. Impressed with Rupp's perseverance, Janis Joplin even agreed to an unscheduled performance.

Palm Beach Pop has been called "Woodstock South," but this festival had everything you didn't want at a love fest besides a fabulously familiar lineup. Like others, the event had its own share of typical problems like cold, wet weather, lack of food, overdoses, arrests, one death, and a shifty promoter. According to Sam Cutler: "This was a straight-ahead rock and roll festival for us. It was opposed by the local community and at the time the government was afraid of pop festivals. There was big paranoia in right-wing conservative American society about where the economic fruits of large festivals were going. The Justice Department thought they were fronts for drug dealers and radical left-wing groups."[31]

Hanley crew member Billy Pratt recalls the festival not being as impressive as others: "This was a smaller festival. It was kind of a gentle affair, and was nicely done. The stage wasn't as big. It was sort of a step down for us. My mother came with me to that festival. She was very proud of the work I had been doing. She stayed at the hotel and was reading all of the articles in the newspaper regarding all of the hippies coming into town."[32]

Nick Burns feels that Hanley never really recovered financially from this event: "The promoter was just a car dealer and worse yet had skipped town.

Some of the rumors that were floating around were that the crew was not getting paid for the gig. I do not think Hanley ever got paid . . . it was bad."[33] Hanley recalls: "David Rupp had his hands full with the local government so there was no real leadership at the event. I never got paid for the festival and if I did it was not much."[34]

After Palm Beach Pop, Burns remembers flying off to work a Moody Blues performance. At the airport he spotted Wavy Gravy in a torn orange jumpsuit with some of his "parts" hanging out. According to Burns they were not going to let Wavy Gravy on the plane: "If I had not had a roll of duct tape in my carry on toolbox they would not have let him on the plane. His jump suit was rather perforated and certain parts of his anatomy were very visible to the clientele at West Palm Beach and they were not happy about it."[35]

Burns adds that Palm Beach Pop was a disaster, yet the camaraderie of people kept it together. The cops caused all the trouble: "There were sheriffs running around busting little chiquitas for marijuana. Whomever was announcing onstage was very specific in telling people to watch who you were talking to. 'If someone offers you something do not take it. They are setting you up.' It was very scary there for a few moments, but the music was very good."[36]

On Monday December 1, the crew packed up the White Dove and Yellow Bird at the cold and wet racetrack. Exhausted, Butkovich had a long drive back to Medford in the semi. "After it was all over I fell asleep for a day and a half or so in my hotel room, jumped back into the truck and headed back to Boston."[37]

In the Yellow Bird, Hair headed out to California for a December 6 concert at the Altamont Speedway. He recalls waiting for his Altamont deployment phone call when he arrived on the West Coast:

> I stayed on with Hanley Sound from Woodstock until after Altamont. We were supposed to do Altamont. I was in the truck in Los Angeles waiting for the word to go from Judi. We were hovering. Then we didn't get the gig for some reason. I really don't know why? But I am glad we didn't! By New Year's Eve of 1969, I was at the Long Beach Arena with Hoyt Axton, Country Joe and the Fish, and Three Dog Night.[38]

Chapter 50

WINTERS END

FROM BETHEL TO BITHLO

The list of canceled and defunct festivals after Woodstock grew even longer by the early 1970s. Efforts to recreate a festival with the same vibe as Woodstock fell flat. In March 1970 Parker Dunham of the *Boston Globe* wrote: "A combination of mismanagement, naiveté, and hostility from local officials has torpedoed several rock festivals. Those still in the works are shrouded in uncertainty."[1]

Festivals promoting a weekend's worth of peace and love accompanied by the industry's top performers were popping up across the country. Without hesitation promoters advertised these ventures without even securing a contract or a sufficient site location. Events that were poorly planned fell apart at the last minute while promoters slipped out the back door just as things got bad. This resulted in advertising outlets refusing to promote upcoming festivals within their publications. In fact *Rolling Stone* magazine would not publish any information on future festivals until promoters were willing to "Verify their promises, including some proof of who's under contract to perform" and address "What kind of arrangements were being made for the thousands of people who are coming to see them."[2]

The Winters End Pop Festival was scheduled to be held on March 27–29 (Easter weekend) in sunny, warm Miami. According to its brochure, the festival promoters attempted to replicate what happened in Bethel: "All the tribes of Woodstock Nation will gather together March 27th, 28th, and 29th, in Miami, Florida, to celebrate Winters End and perform the Rites of Spring. 600 acres of quiet countryside will be the host. Winters End IS the Hog Farm, and communal kitchens, Indian tribal gatherings, trading posts and a bazaar of exotic shops."[3]

For twenty dollars, Winters End promised the average festivalgoer an entire weekend of Woodstock-quality talent like Canned Heat, Joe Cocker, Country

360

Joe and The Fish, Grand Funk Railroad, the Grateful Dead, B.B. King, Richie Havens, The Hog Farm; Sweetwater, Iron Butterfly, Mountain, the Kinks, Little Richard, John Mayall, Steve Miller, Johnny Winter, Sly and the Family Stone, Ten Years After, Ike and Tina Turner. All were slated to play. Yet only a few did.

For even more Woodstock continuity, promoters attempted to bring in Woodstock production alumni to assist with development and production. Hanley Sound was among them. According to underground journalist and one of the festival's promoters, Thomas King Forcade claims in the *San Diego Free Door*, Winters End was intended to be Woodstock all over again: "Mike Lang and Artie Kornfeld two of the Woodstock promoters were paid $5,000 on the hope they could get The Band, Donovan, or Hendrix to play. They didn't. John Morris, formerly of the Fillmore East, handled talent again. Hanley did sound again. Stan Goldstein and Chris Langhart and others were hired to do the stages again."[4]

Because of their outstanding work in Bethel this list of experienced people also included members of the hippy commune the Hog Farm. They were hired to host the free kitchen and help with bad trips. The Hog Farm was now a staple at many of the festivals of the day.

With the Winters End site getting switched a few times, Hanley could smell a problem brewing. Beginning in Miami and rejected there, Winters End moved to Hollywood, Florida, then eventually settled fifteen miles east of Orlando, on US 50, at the 110-acre Econ "Dude" Ranch—a tourist attraction in Bithlo. Hanley recalls: "The entire situation was bizarre. There were a few promoters involved. They could not find a suitable site until days before the event. It was a logistical nightmare."[5]

After hearing of the problems with Winters End, the Hog Farm requested that all "tribes of Woodstock Nation" language be removed from the festival's advertising material. Even so, Wavy Gravy attended. In his book *The Hog Farm and Friends*, he recalls the locals were not very welcoming: "Arrived in Orlando that night to discover the site had been switched several times by as many injunctions. Orange County was up in arms and rock n roll was running scared."[6] A recent law had been passed in Orange County banning rock festivals, stalling the facilitation of the event. However, it did not stop its promoters from still selling tickets.

On Sunday March 22, Hanley Sound employees learned in the *Boston Globe* about the issues the Florida festival was having. The *Globe* reported, "The three Boston ticket outlets for Winters End were told to halt ticket sales, 'until promoters found a suitable place to hold the festival.'"[7] Florida Governor Claude Kirk wanted to pull the plug on the festival and so did Boroda, who was extremely frustrated about the whole situation. "There were so many people down there who didn't believe in the peace movement and festival scene. They were rednecks, cracker jacks. It was the Bible Belt. As soon as they saw long hair they judged you. I could feel I wasn't welcome."[8]

Arriving the week of March 23, Boroda waited in his hotel room. He reported back to Medford via a long-distance flat-rate plan called a WATS Line (Wide Area Telephone Service). To avoid costly charges employees had to ring the office using the name "Dan Michaels." Whoever answered would be signaled to return the call at that number. Boroda remembers: "I was on the phone with Judi a lot as I waited days and days at hotels because of these injunctions. As soon as I received the green light I went and set up."[9] Hanley Sound's office number was 617-395-8151 never to be erased from the minds of those who worked there.

Boroda finally got word late Thursday evening March 26 that the festival was going to be held in Bithlo. He arrived at the ranch the following day with no time to spare. Forcade recalls the bare-bones situation when he arrived at the site between Thursday and Friday: "Gravel which had been ordered the night before to beef up the roads never arrived and all other suppliers were mysteriously 'out.' Scaffolding ordered as an alternative to proper staging mysteriously did not appear until Friday night, and required a $5,000 deposit for some $400 worth of scaffolding."[10] With thousands invading the ranch hourly, it was evident that people were expecting to hear a weekend's worth of music.

On Friday March 27, Boroda was worried that the ranch lacked the production infrastructure to hold a proper festival. As the production staff were trying to get things going, local police stationed around the perimeter of the dude ranch were turning vendors away. Even so, Boroda worked frantically throughout the day and night setting up sound equipment next to a makeshift stage located on the ranch's cow pasture.

As a result, the tall Hanley Sound speaker towers, commonly situated on stage left and right, were now just a fraction of that height. Journalist Bob Fiallo who witnessed the sound company's work at the Palm Beach Pop Festival and Atlanta International Pop Festival saw the difference in Hanley's sound application for Bithlo. Fiallo recalls:

Early on there were some issues because they didn't have a crane to elevate the speakers like they had at West Palm or at Atlanta. The sound towers on either side of the stage were just a couple of stories high, and there were no light towers. Hanley did not have the big speaker towers up like I usually saw at his festivals. There was no fencing or security walls like you would see at other festivals. The stage was just one level off of the ground with scaffolding and plywood. There was plywood on the ground, which it sat upon so it wouldn't sink into the mud. The sound engineer sat about one level above the crowd on scaffolding.[11]

Setting up production for a three-day festival on such short notice was a tense situation. Wavy Gravy recalls the craziness in putting together small cities (festivals) in a fraction of the time: "Too close for comfort. This time there was no beehive of longhairs in labor, it was an instant city in action. Just a few shabby structures, but mostly the mud and the rain took Woodstock to Winters End and made common ground. With Woodstock for mamma, Winters End was a miscarriage."[12]

By late Friday evening more people were coming, and thankfully some provisions made their way in. Forcade remembers, "Many of the technicians went without sleep, worked for nothing, and even spent what they brought down with them to keep the show going."[13] In the scorching Florida heat, a ragtag assemblage of volunteers and crews worked exhaustingly to get the stage and sound up and running. Forcade adds, "volunteers from the audience built the stage, finishing up on Saturday morning, after working all night."[14]

From where Wavy Gravy was stationed, he could see the sound company working away into the Saturday morning dawn: "Still they kept at it. Grew towers like flowers that blossomed in speakers still speechless and solemn awaiting completion. The weavin' of wires and stack up of amps, mountin' of microphones then contact and turn on. This instant arena was ready to rock."[15] Hanley adds: "From what I recall we didn't have a crane to lift the speakers. So we built step scaffolding so we could get them up onto the towers."[16]

The dude ranch had the potential to be a sufficiently bucolic setting for a festival. As the stage was erected in the middle of its cow pasture, trees flowering all around, the strong scent of blooming orange blossoms wafted into the moist swampy air. Situated in Everglades country, everything became wet with dew. But as the cool humid early morning transitioned into an intense afternoon heat, the dew quickly evaporated.

The sound system was up and ready by early Saturday morning. Boroda learned that the local tax collector was seizing all of the accruing festival funds, and imagined he might not get paid. According to Wavy Gravy, "Hanley's men had orders not to turn on the sound until they got their money."[17] The Hog Farm leader recalls a guy named "Butch" of the Florida biker group The Outlaws, hired for security, helping out. Butch drove his Harley over to the sound console with $1,200 for Boroda before the Florida tax commissioner could snatch it from the cash box. "Winters End was on the air and Butch was in the bushes,"[18] adds Wavy Gravy.

Bob Fiallo writing for the *Tampa Tribune* at the time, states in his article "Rock Show Vibrates, and Then Dissipates," that the show almost got canceled over the money issues: "All performances were in question Saturday afternoon when Hanley Sound, the engineering group which set up the massive sound system and lighting equipment, wanted the $1,200 payment before the show could go on. The payment materialized."[19]

When the tax collector finally got to the cash box, thirty-four dollars remained. Boroda recalls: "It was a nightmare when I was on the road. I knew if I didn't bring the money back we wouldn't make payroll. So I refused to turn the sound system on without payment. I became a real animal to the point where at this festival the promoter almost punched me in the face."[20]

Rumors were circulating that most of the bands had canceled. People still came. In true Woodstock fashion, thousands overran the ranch for free by climbing fences. With no means to collect tickets, Winters End ended up being another free event. Wavy Gravy remembers: "The switching of sites made it too late for a fence, so bikers were hired instead of barbed wire, but a lot of people got in for free."[21] Winters End only lasted around twenty-four hours.

As the weekend wore on, promoters were in a hotel room battling the many injunctions against them. Those who were going to financially support the event quickly backed out, while performers who heard of the festival's troubles were dropping like flies. The Hog Farm leader tried to get to the media in time so he could plead to hopeful travelers not to come. When he had the opportunity, Wavy Gravy's message turned into a plea for food and water for the gathering masses. Provisions were slim.

By Saturday, thousands were waiting to hear music. The only performers to actually play Winters End, aside from a number of local acts, were Johnny Winter, Mountain, and the Allman Brothers. When the bands heard that Hanley got paid they also asked for payment, but performed regardless. According to Fiallo: "Then the bands began to ask for cash, and the word was out that the festival headliner, Johnny Winter, wasn't going to play. His band wanted cash but rather than disappoint his fans, Winter jammed with the Allman Brothers, who didn't care about pay, stating they wanted to serve the people."[22]

Even though local police turned away thousands, a crowd between 40,000 to 50,000 managed to gather. With an order to leave the premises, many attendees left by noon Sunday and the numbers eventually dwindled down to a muddied few thousand. According to the St. Petersburg, Florida, newspaper the *Evening Independent*: "The Winters End Rock Music Festival came to an end today without having a beginning. Gates . . . were closed to newcomers at 9:50 pm Sunday. Music makers playing without a permit were told to silence their beat by sunrise."[23] By Monday morning around forty police were sent into the ranch to bust people, pushing the remaining attendees out of the area.

In the end, five of the six promoters, excluding Forcade, were arrested at their hotel for violating the county's ban on festivals without a permit, and interfering with the county's zoning ordinances.[24] According to the *Chicago Tribune*, "Deputies converged on a downtown Orlando motel room that served as the festival headquarters and arrested Mike Foreman, Bert Cohen, Stephon Mashury, Enoch Shachar, Fred Wasserman, and ranch owner James Brown."[25] Winters End was an uncomfortable Easter weekend for most. Yet, for those who came, they slept, sat, and enjoyed a little taste of music.

This event was yet another financial disaster for promoters—and for Hanley, too, who was caught in the wave of anti-festival sentiment moving its way across the country. Just over the horizon, the sound engineer missed out on even more potential revenue as two upcoming Florida festivals called the Seminole Rock Revival and the Great Easter Rock Festival were canceled.[26]

Chapter 51

OH, CANADA

TORONTO ROCK AND ROLL REVIVAL

In the early 1970s Canadians and US citizens shared a similar social and political response to the growing festival movement. Gate-crashing and a free-festival attitude was spreading at various Canadian events. These happenings required wide-open spaces, and many Canadian communities could accommodate that. But such communities were also subject to media scare tactics informing them that disaster areas occur when big festivals happen. Running out of food, water, and other essentials was something they wanted to avoid. Zoning boards and community activists were in full action and, one by one, big events were being court ordered to shut down. Canadian promoters felt the hit.

Johnny Brower, a twenty-three-year-old Canadian promoter, hired Hanley Sound for his Toronto Pop Festival, known as Pop Festival 69. The event was held on June 21–22, 1969, at Varsity Stadium. It attracted over 30,000 festivalgoers a day with acts like the Band and Sly and the Family Stone. Production crew member Dennis "D.D." Hill first met Hanley at the stadium and was impressed:

> Pop Festival 69 in Toronto was one of the largest events in Canada at that point. Later on we put on the Rock and Roll Revival show and Hanley did the sound for that as well. I was blown away by the size and caliber of the Hanley Sound speaker systems. The size of his gear and quality of sound was amazing! When he came with all this gear we were all like wow! Why do you need boxes that big? Of course when you heard the system, then you understood. I went on to use some of the same techniques Bill was using in my own work.[1]

After Hanley's involvement at the Toronto Pop Festival (June) and Woodstock (August), promoters Johnny Brower and Ken Walker (Brower-Walker

Enterprises Ltd) invited Hanley back to Varsity Stadium. They planned a one-day, twelve-hour festival on September 13 called the Toronto Rock and Roll Revival. By reputation alone, Hanley Sound was the obvious choice. Even later, Walker hired Hanley for an event of his own called the Festival Express.

According to Brower, Hanley was the best in the business at the time: "I was at Woodstock and the sound was the greatest. It was the best sound and sound equipment anybody had seen or heard. We had the Toronto Rock and Roll Revival scheduled for September 13th and of course we had to have Bill as our soundman, and like at Woodstock Bill Hanley rose to the occasion."[2] The Toronto Rock and Roll Revival affair featured John Lennon and the Plastic Ono Band among other notables that included the Doors, Alice Cooper, Chuck Berry, Jerry Lee Lewis, Little Richard, Bo Diddley, and others.

Hanley Sound crew members Burns and Butkovich drove the White Dove tractor-trailer to Varsity Stadium and met Hanley with the sound equipment. However, traveling into another country wasn't easy when drugs were involved. Burns explains:

Going across the bridge into Canada was nerve-wracking. Before we left I remember asking David Butkovich if he had anything on him and if so to get rid of it because customs was a bitch. Come to find out he had some shit stuffed in some of the speakers, and of course the customs officials wanted to see what was in the black cabinets. So we had to pull one off of the back loading deck lift gate and spin it around for them. They told me to take the back panel off so I explained to them I couldn't do it and that they would have to. I said that there are probably ninety to one hundred three-inch #8 marine wood screws that have to be taken out. I said we will sit here while you take the back off and that all that's in there is fiberglass insulation and the support structure. So they shined their lights in there and shook them a bit and said okay you guys are clear. When we got inside if I had a gun with me I could have shot Butkovich, I was pissed! Come to find out he had a couple of hundred pounds of grass packed in those cabinets.[3]

Once the two individuals arrived at the stadium, Burns had some difficulty pulling the rig into the venue. Burns recalls, "Trying to get the White Dove into that stadium took me a good forty-five minutes because it had an offset gate so it was a constant back and pull, back and pull."[4] When Burns entered the stadium he could sense the tension in the audience: "When we finally arrived all kinds of hassles were happening because the Vagabond Motorcycle Club decided they wanted to be an escort to the Doors. For a one-day festival, the crowd was not particularly friendly."[5] Later eighty members of the club escorted John and Yoko's limousine from the Toronto airport to the stadium.

This was the first festival that David Marks was left to mix on his own. Packed with over 50,000 people, Hanley decided to leave the console and go to lunch. "Bill walked away from the desk . . . I said 'Hey Bill where're you going?' 'I'm going to find some fried chicken in Toronto,' he said. 'Who's going to mix?—I cried.' Above the crowd noise Bill screamed—'you are!'"[6] Marks recalls from that moment on, it was just him and every rock band he grew up listening to in South Africa: "I wouldn't budge or let anybody near the desk for that entire afternoon. Bill graciously let me handle it. Even the famous 'feedback' incident with Yoko Ono did not deter me from hogging the entire mix."[7] Marks claims that this legendary performance by John Lennon and Yoko Ono left a lasting impression: "Yoko got under a sheet and I didn't know she had a mic! But I heard this turkey warble and when the feedback started I couldn't identify the source. Then suddenly someone shouted at me! 'It's under the sheet!' . . . 'It's under the sheet!'"[8]

There was also shock rocker Alice Cooper who threw a live chicken into the audience. The response was a barrage of dismembered chicken parts flying back onto the stage, a visual Burns will never forget:

> I was about four feet from Alice Cooper on the stage when he threw that chicken into the audience. I saw them wring the neck off of that chicken. After I saw that I retreated backstage. I also came within a whisker of taking my side cutters and cutting the cable to Yoko Ono who was sitting on a burlap bag on stage wailing into the microphone. That was her contribution to the John Lennon performance.[9]

Cooper's manager, Shep Gordon, who was present, recalls Hanley's exemplary work during this event: "Anytime there was a show with Hanley's name on it you knew you were getting good sound."[10]

The following year, in July 1970 Johnny Brower, John Lennon, and Yoko Ono planned to host the Toronto Peace Festival at the Mosport Park Raceway. But due to local fears, their permits were denied. According to Brower, festivals back then not only incited chaos but also terrorized people and communities:

> They thought that these kids were going to go mad and start ravaging cities. The insanity of those days we can laugh at now, but it wasn't funny back then, having to deal with people who looked at these audiences as zombies from the *Night of the Living Dead*. Like at any moment they could snap and start raiding peoples' homes and destroying villages. The stuff we used to hear at some of these town meetings was right out of Mississippi Burning except the hippies were the blacks and the municipalities were the rednecks. It was a scary era from the standpoint of putting on a rock concert.[11]

HANLEY SYSTÈME AUDIO

In May 1970, while Hanley was preparing for the one-day Montreal Pop Festival, he conversed with the *Montreal Gazette* reporter Dave Bist via telephone. The topic of conversation was about a shift occurring within the festival market at the time. From his office in Medford, Hanley explained to the reporter that the "panic reaction" from small-town citizens has cast the future of festivals into doubt: "There are a lot of promoters who haven't handled things politically. The key to these events is keeping the towns calm and telling them just what is going to happen on the site. Policing should be handled by non-regular forces like the California Hog Farm group that worked at Woodstock or the security crew from the Fillmore East in New York."[12]

After Woodstock, as the pop festival arena broke wide open, Hanley proved immune to the mud, rain, excruciating heat, and governmental problems. Festivals, as journalist Herbert Aronoff put it, brought Hanley "Face to face with one of the most frightening spectacles in North America: police versus pop fan." Hanley was known by his colleagues to be very calm during the most tumultuous moments. At a show in Toronto, Hanley actually spent two hours negotiating with a police inspector, begging him not to engage his riot squad on the festivalgoers. Hanley felt strongly about keeping the police out of events like this, as he explained in an interview with Aronoff:

> Police have the wrong idea about festivals. My father is a policeman, so I know. They see them as sports events, with half the crowd cheering one side and the other half opposing them. There is no confrontation at a pop festival. It is not the crowd against the performers. Everybody's there for the same reason. A lot of energy is generated at a pop festival but unless there's an opposing force it stays peaceful energy. That's why police aren't needed at these things. There's no opposing team and no uniforms until they arrive. That's when confrontation begins. At Woodstock it was a realization of what could exist. Only about one percent of what's happening is real. Only about one percent of the musicians are real. But, when it works it helps to prove that people can be peaceful together.[13]

It's clear that Hanley's responsibility as a sound engineer went beyond the mixing console. During some of these events, social and political obligations added additional layers to his role. However, the obligation of providing quality sound was as significant as his technical innovations, creating a legacy that lasted decade after decade, one event after another. If clarity and intelligibility in sound had not been successfully executed at these festivals and demonstrations, the individuals listening would have missed the message, and more importantly the meaning.

Up until then, most audiences probably didn't think much about sound quality, often taking it for granted. Yet, the voice of the counterculture needed a way to be heard, especially at a 500,000-person gathering. With these ideas in mind, some very important questions arise: What if the sound at Woodstock had failed? And would the 100,000-plus demonstrators in Washington have stayed peaceful if they hadn't heard the messages of nonviolence? In a 1970 *Montreal Gazette* article, "A Sound Question: What Binds Nixon and Hoffman?" some possible answers to these questions are analyzed. Herbert Aronoff successfully touches on the essence of Bill Hanley's influence:

> Was there some single strand that binds President Nixon to the revolutionary Abbie Hoffman, to pop stars Sly and The Family Stone? Do Woodstock, the Columbia University riots, and the Washington peace marches have anything in common? Imagine for a minute any or all of the events going wrong. Pull the plug on Nixon and who knows where the country would head? Prevent the messages of peace from being heard in Washington and 60,000 people may have easily been led into violence. Could Woodstock have worked as a mime show? In North America where so little can make so much go wrong, reproducing sound so that everyone can hear and understand what's going down seems to be almost as vitally important. Is it too ridiculously paranoid to say that the hand that holds the plug rules the world?[14]

If not Hanley, then some other company might have pulled off so many important events. However if we look at the sound reinforcement field at the time, it was Hanley who created most of its blueprint and design. Albeit a design, that was mostly worked out through trial and error. Even so it was borrowed, streamlined, and made better by many others. According to Alun Vontillius: "Had it not been for Bill progressing sound out of the 1930s and 40s technologies, there wouldn't be an industry. He applied what was there, adapted it, pulled it all together, and realized the need to do it for large rock and roll audiences."[15]

When in the early 1970s festival locations became restricted, they eventually resurfaced in arenas, ballparks, and stadiums. This created a change in how sound reinforcement was to be applied in years to come. With new developments and advancements in this industry, it meant more competition for the Medford sound company.

Very few were willing to schlep their gear through the mud and the rain like Hanley and his dedicated crew. It was a sacrifice the intrepid engineer made for the next generation of sound engineers, and when the festival market withered away, so did Hanley's business. Hanley explains: "After Woodstock the festival market dried up. I lost a quarter of a million dollars worth of business in three weeks, and I was already three or four hundred thousand dollars in debt. Competition was starting to heat up; other people like the Clair Brothers and Showco were getting into the market. I believed my own bullshit, and I wasn't counting in the competition."[16]

Rusty Brutsché of the influential second-generation Dallas sound company Showco, makes these observations:

I was backstage at Texas Pop and was able to really look at everything Hanley was doing and take it all in. There is no question in my mind that Hanley was the first guy to put it all together doing big gigs with all the equipment and big amplifiers. He was starting to wrestle with the problem of large systems beyond the clubs. It made you realize and helped us realize the ideas of what was possible. What Bill did in the sixties with the big festivals was sort of a sixties phenomenon which quickly turned into touring and arenas. Hockey arenas, and basketball arenas, the business switched into an industry pretty quick. There were festivals here and there after Woodstock. Later on the big stadium tour became fashionable. The thing that really blew up was when arena touring started. This is what created the money. I did a couple of festivals but Showco moved to the big stadiums pretty quickly. The infrastructure at a stadium is so much better than a cornfield or to start from scratch.[17]

Judi Bernstein remembers: "After Woodstock the phone calls died a bit because of what the public saw in Bethel, New York. The promoters were canceling contracts for outdoor events. We lost a lot of business in the weeks after Woodstock."[18] Most who worked for Hanley and felt the successful ride the company was on during these golden years could see and feel the downturn in his business. Bernstein witnessed this firsthand as she was the one answering the phones and directing crews to jobs:

Things were changing so rapidly and the mud from Woodstock was settling. It changed the concert market. Cities and towns were unwilling to have crowds. Woodstock ruined the major concert market for a time and in return ruined the festival sound market. As it became better we now were faced with an influx of new sound companies who were gearing up in the meantime. It was the Clair Brothers, Showco, and English companies as well. Then lighting companies were popping up who also decided they could also do sound. It became extremely competitive in a relatively short time. We saw more riders between groups and sound companies who saw the value in that sort of thing. So there it was, the apple tree was growing and the apples were all being picked.[19]

In 1970 the big question for Hanley Sound was: What does the future hold for the festival? Even though the writing may have been on the wall, the sound engineer remained optimistic. But others like rock critic Mike Jahn pessimistically claimed, "The mass youth gatherings that were once so refreshing like an unexpected sea cruise, are going down like the Andrea Doria."[20]

MONTREAL POP

On Saturday, May 9, 1970, Hanley provided sound for the Montreal Pop Festival. A twelve-hour event at the Forum in Montreal, Quebec, the festival had a modest 10,000 in attendance. The eclectic lineup included the Byrds, Mashmakhan, Robert Charlebois, Frijid Pink, the Amboy Dukes (with Ted Nugent), Frost, the Collectors, and the extremely loud Grand Funk Railroad. In a *Montreal Star* article a journalist notes the power of Hanley's equipment, claiming the music was "very heavy, extremely heavy," with "fortresses of amps and speakers" and "hordes of equipment," adding that it was played at an "unrelenting volume" for "hours and hours."[21]

Aronoff was impressed by the company's ability to balance rock and roll's heavily distorted sound with intelligibility at this event. He claimed that the audience remained mostly peaceful, sprawling about the arena in "obvious comfort" while grooving to the sounds on what was "probably the best sound system (by Hanley of Boston) the city had ever heard."[22]

In French, Montreal Pop's handbill read, "Système Sonar 'Hanley' Sound System" which many claim was a "big improvement" as Hanley's system "reproduced the sound" with "merciless accuracy." Praising the balance of loudness and quality sound through Hanley's system, the *Montreal Star* journalist claimed: "Rock should be loud and dirty but it shouldn't be boring. What blew through those mighty speakers was a fairly broad sampling of what's happening in mainstream pop music these days."[23]

Grand Funk Railroad was known to "knock everyone back with the loudest, most together hard rock ever heard."[24] In doing so they challenged Hanley's equipment. By now the engineer and his crew had realized the potential that new rock music offered, but Grand Funk Railroad was another step up. If Hanley's company were to remain relevant in a new decade, it would need to adapt to a new and rapidly evolving music scene.

THE TRANSCONTINENTAL POP FESTIVAL

In the spring of 1970, Canadian promoter Ken Walker of Eaton-Walker Associates (with Thor Eaton and George Eaton) approached the sound engineer once again. He requested Hanley's expertise for a unique concert experience, a traveling festival called the Transcontinental Pop Festival, better known as the Festival Express (FE). The FE was a series of concerts traveled to by train across Canada, in eleven days from June 24–July 5 1970. These concerts not only provided Canadian festivalgoers a score of mini-pop festivals, it offered performers and crew a ride across the majestic Canadian wilderness. The *San Francisco Chronicle* reported: "The $1-million effort . . . will haul top rock and folk music performers across the Nation nonstop for twelve concerts in four cities."[25]

According to *Billboard*: "The Express—which consists of a 14-car train rented from the Canadian National—kicks off in Montreal on June 24th, moves to Toronto for a two-day festival in the CNE grounds on June 27–28; Winnipeg, July 1st, and Calgary, July 4th. Sound is being managed by Bill Hanley of Boston."[26] Due to some issues with the city of Montreal, the official send-off of the train-fest got pushed to the 27th in Toronto at CNE (Grandstand) Stadium.

Musically the FE was impressive, offering an incredible string of varied acts, including Mountain, the Band, the Grateful Dead, the Flying Burrito Brothers, Buddy Guy, and Janis Joplin, among others. Expecting to sell 50,000 seats for each show, promoters projected around 250,000 people for the entire tour. Although in theory the FE seemed wonderful, it was met with protests. At different locations, attendance was low.

Prior to the train's departure, Hanley Sound crew members Lee Osborne, Billy Pratt, Bruce DeForest, Rick Slattery, Sam Boroda, Nelson Niskanen, David Kingsbury, and Judi Bernstein packed two semi tractor-trailers with the necessary equipment for the event. Once the trailers were filled, they left Medford for the inaugural send-off in Toronto on June 27, 1970. Slattery remembers: "Toronto was slated to set the stage for the first gig. After Toronto we spent six days on the train with all of our equipment."[27]

New to Hanley's crew was New Yorker David Kingsbury, who had dropped out of college in 1969 and was looking for work. Friend Nelson Niskanen referred him to Hanley Sound and by September he was an employee. Assisting with many smaller gigs, Kingsbury's first big show was the FE in 1970:

> Before we left for Toronto, we built a system for the stage, which was like a trailer. One whole side just opened up. Hanley wanted to put it way up in the air so we put these massive pneumatic pistons at each corner. This allowed us

to raise the stage almost ten feet into the air! After doing weeks of prep for this whole thing we all went up to Toronto, put on the festival, then packed it all on the train and traveled to the next stop.[28]

Niskanen recalls that Hanley was so focused on the portable stage, it left no time to work on the tractor-trailer's brakes, claiming that he and others on Hanley's crew made their way to Toronto with a trailer full of heavy equipment that couldn't be stopped: "Hanley was working on the wiring on his portable stage hydraulic system. He kind of ran out of time in terms of getting the brakes working. We piled in and one of the other fellows drove and we headed to Toronto with no brakes. It was kind of scary!"[29]

In attendance was Tom Field, who recalls frantically trying to make the train tour happen from a logistical standpoint:

I was contracted by Hanley to provide the lighting for the train tour. When we got to the railroad yards we saw this special train set up for us. It was amazing! Bill took the Wenger Wagon stage on the tour and we just piggy-backed this along. We were basically bringing a festival on wheels into the stadium a day in advance. This was a major technical feat, and certainly had not been done before. The sound was quite good as well![30]

Although Hanley was not on the tour, he was in Toronto to oversee the crew. The sound engineer wanted to make sure everything left the station successfully. Hanley put DeForest in charge of stage management and Bernstein in charge of production. She adds, "I was the only person on the train representing Hanley Sound."[31] Sam Boroda was the lead sound engineer for the FE.

At least three of the fourteen rail cars were designated for Hanley's production equipment. In total, the FE train was made up of one baggage car, one staff car, one dining car, two lounge cars, two flat cars, two engines, and five sleeper cars. According to Bernstein, providing full stage, sound, and lighting for this event was a monumental task. Slattery recalls: "We also did the recording for Festival Express. I remember we had some electronic feedback at one point. I still have tinnitus!"[32] Lee Osborne was responsible for recording the individual festivals. Boroda explains how Osborne designed the recording setup for the FE:

At each show from the stage there was a split. The microphone feed went to the house mix and the other to the recording. It was like a fly-by-night kind of thing, in those days and we didn't have a splitter. Lee was in the recording trailer with two consoles (side by side) that he and Bill had put together.

It was all custom. These were eight-track Scully tape recorders. All of the pre-amps were Langevin AM16 racks outside of the console connected with an umbilical cord. He was also using a Nagra recorder with Pilotone (60Hz) for film sync. It was essentially the same setup as Woodstock and Lee did it all.[33]

Once the trucks were loaded and reached the specified train yard, the railroad depot loaded the tractor-trailers onto the flatbed. Niskanen remembers: "The truck and load went right onto the flatbed. They tied them down and then we were on to the next stop. When we arrived they pulled them off for us, and we drove them to the stadium site."[34] Slattery adds, "It was all pretty seamless."[35] Boroda explains:

We set up everything for the Festival Express to be as efficient as possible. I am the one who set up all of the racks and mounted the McIntosh amplifiers to them. There were about eighteen of them I think? In the front of the truck was all of the recording equipment. In the back there were about four or five freestanding vertical racks bolted together and to the truck. The racks held the amplifiers. So when you came in to the site you didn't need to do anything. There was nothing to break down. You just opened the back door of the truck and all that was left to do was to bring in the master cable and connect everything.[36]

Bernstein claims that as soon as each event began production, additional manpower was needed: "We had crews of nonunion guys and grabbed volunteer hippies to help us build the sound towers with scaffolding. They helped us load it on a tow motor. The stage was puny back then. It would fit on a flat car when it was folded up."[37]

When crews arrived at each site, the promoter had prearranged for dozens of people to help, with Hanley's crew directing them. According to Kingsbury: "Each of us had an area that we were responsible for. My main job was miking the stage and helping put the stage up in the air. If we got overheated we could go in the tractor-trailer and cool off since Hanley had them refrigerated to keep the amps cool."[38]

What seems to be most memorable for those on this rock and roll rail ride was the overindulgence of drugs and alcohol. According to Tom Field and others, rampant partying was a constant on the train: "We all lived on this train for the entire tour. I remember our lighting designer Jerry Nathanson had his birthday on the train. Janis Joplin appeared at noon on his birthday and placed this bottle of Southern Comfort on his breakfast table. This is how it was."[39]

Bernstein recalls that, "Joplin drank and drank" followed by even more "drinking, jamming and tripping on LSD." She remembers:

There was so much liquor, pot smoke and music. On one of the days of the tour it was the birthday for Ian Tyson of Ian and Sylvia and I was supposed to bring the cake to the stage. Sam Cutler who was managing the Grateful Dead opens a small vile and douses liquid LSD on top of the cake. A lot of people got sick. I should have tripped and fell with the cake, I was stupid.[40]

Niskanen recalls: "One time the train stopped and the bands sent out for a bunch of liquor. They had a party on the bar car accompanied by a jam session with Jerry Garcia, Phil Lesh, Leslie West, and Buddy Guy!"[41] Kingsbury adds: "I remember going to one car and Buddy Guy was doing his thing. Then another car it was the Grateful Dead. But my biggest recollection was how everyone bowed down to Jerry Garcia. He was like the king. Everyone wanted to play with him, listen to him, hang out with him."[42]

Although the production crew and performers did occasionally intermingle, there was some degree of separation. Rick Slattery, twenty-three years old at the time, observes: "The train was a little segregated because we were in the back end and the musicians were in the front end. The crew played cards and hung out, and all the while we could hear them jamming. We didn't care! It was great!"[43] Kingsbury remembers:

We slept in separate cars, but during the day you could go from car to car. In those days the Canadian National Railroad was super posh. We are talking red velvet, conductors that were dressed to the nines. That was the funny thing. These conductors had never seen anything like us! We were these freaks on this train! We were smoking dope and drinking like crazy and they were trying to maintain their dignified stance. I remember this one conductor, he wanted to try pot. But he felt like we needed to try something of his, which was snuff. We got him stoned. It was a pisser![44]

Not everyone indulged in drugs and alcohol with these darlings of rock. Boroda was uninterested. He felt it was important to be focused, especially when the train pulled into its stop at 5:00 a.m.: "I needed to make sure I was there so the production was up and running before concert time."[45]

During the calmer moments, various members of the crew took advantage of two nearly empty baggage cars. While in motion they opened up the doors. In between shows the Hanley Sound crew and others sat and relaxed, watching as the world went by. According to Field: "The experience of sitting on the train going across the Canadian countryside was amazing! The jamming that happened at night on the train was spectacular. This was a once in a lifetime experience!"[46]

Niskanen recalls that the train ride was relaxing. Having never been in a sleeper car before, he claimed the motion was so comforting it rocked you to sleep: "The whole entourage was in these sleeper cars. There were long stretches in between shows and the scenery was spotted with lakes. It was so nice."[47] Boroda claims the experience felt like a traveling circus: "The Festival Express was the same concept as Barnum & Bailey Circus. You put the talent, production people, and gear on separate cars and you just travel from city to city. And when you arrive you bring the stuff into the field and set it up, break it down, load it back on the train, and move on to the next leg."[48]

For Boroda, the FE was a dream come true. Early one morning he was woken up by the halting screech of the train. "At 3:00 or 4:00 a.m. the train just stopped. I got up out of my bunk to investigate what was going on. I looked through the window and Garth Hudson of the Band was playing his saxophone in the middle of the Canadian Rockies. The sound was echoed all over the canyon as the sun arose. It was amazing."[49]

Nevertheless, not all went well on this tour. The situation outside some of the shows was chaotic. Canadian festivalgoers, who wanted to get into the concerts for free, plagued the events as the train made its scheduled stops along the way. Bernstein remembers, "Everyone wanted to get into the events for free!"[50] Slattery recalls: "Toronto went well, but what messed it up were people trying to make it a free experience. The promoters put out a lot of money. People were trying to climb over the fences and the walls."[51] The two-day stop in Toronto included the Grateful Dead (with New Riders of the Purple Sage joining in), the Band, Ian & Sylvia, Buddy Guy, and Janis Joplin. Also performing were Traffic and Ten Years After. These groups did not move on with the train tour.

In Toronto, protesters felt that the promoters were cheating them by setting ticket prices too high. Eventually they crashed gates and fought with local police. According to the *Toronto Daily Star*, on Saturday June 27 at CNE Stadium more than "2,000 youths protesting admission prices tried to storm the gates of the two-day rock music festival."[52] According to the promoters, if the rioting had continued, or worsened, they were prepared to call it off. To quell the angry crowds, Jerry Garcia and the rest of the Grateful Dead agreed to play a free concert the next day at neighboring Coronation Park.

Cash Box reported that Hanley's sound system did not get off to a good start: "Meanwhile back inside the [CNE] grandstand stockade, the many groups were attempting to be heard through an inadequate sound system (. . . the same as was used at Woodstock). Some of the belters, like Montreal's Charlebois could be heard but Eric Anderson was just lost . . ."[53]

After witnessing the happenings on this tour, journalist Joe Fernbacher wrote in the *Spectrum* that at the festivals, gate-crashing, production, and infrastructure were becoming predictable:

Assembly-line rock festivals have a number of highly similar characteristics. Like some poorly written epic drama, each festival contains (sort of like an army survival kit): two or three promoters who are Capitalistic pigs (to quote an oft used phrase), a group of kids fucked up on drugs or trying to get fucked up on drugs, security problems like you were inside a prison camp trying to see the commandant, two light towers that are placed almost exactly like this at Woodstock, a stage that looks slightly the same, and sound work by Hanley . . . Now if you put all this together, hype it up through advertising, rumor, whatever, you will have what we had up in Toronto.[54]

A week later in Calgary, on July 4, Mayor Rod Sykes suggested that promoter Ken Walker allow angry protesters into McMahon Stadium for free. Some say Walker responded by punching the mayor in the face. Kingsbury recalls: "At some point during the tour the promoter gave up trying to make a profit and sort of let people in. We were all so busy working that we weren't really aware about what was going on. I can't say I witnessed it, but I did hear about it."[55] According to Niskanen, crews were somewhat insulated because they were inside stadiums: "They did have some ability to control crowds as these shows were in stadiums. Unlike open-air festivals like Woodstock where there was little or no control, at these events you could. There were fences and security guards. We were mostly taking care of business inside so it didn't really affect us. We didn't really notice it."[56]

In a 1970 *Billboard* article, "Festival Express Marred by Protests, Poor Attendance," a journalist wrote: "The promoters of the recent Festival Express train which organized stadium type pop festivals in Toronto, Winnipeg, and Calgary, lost $350,000. A publicist for the Festival Express said only about 60,000 had attended. The Express was marred from the start by youth demonstrations."[57]

Those on the production crew recall the FE as a "fly by the seat your pants experience." Facilitating a train loaded with equipment required the backbreaking work of unloading, setting up, striking down, and reloading equipment in the middle of volatile environments. Stagehand Dave Bluestein was only eighteen years old on the FE. Referring to himself as the lowest "slug of a roadie" that one could be at the time, he assisted Hanley Sound, claiming: "It was a giant clusterfuck in Toronto! We had two days there. The fact that there were protesters and we were setting it all up on the fly made it a genuine disaster! Once we got it all on the train and off to Winnipeg we knew what we were doing. By the time we got to Calgary the setting up was streamlined."[58]

Bluestein now recalls that the facilitation and systemization of something like the FE had never been done before: "The most important thing to remember is that it was like cavemen compared to how things are done today. It was normal for us then, but today it would have been the most arduous thing anyone could ever do."[59] FE production manager D. D. Hill remembers the tour crew (of one hundred and fifty) working extremely hard to make these events come off successfully: "It was basically our train. We rented it. This had never been done before. It was one of a kind. Anybody that was on the train raved about how incredible it was. When we arrived at the venues people would ask what's it going to sound like, and I reassured them you're not gonna believe how good it's going to sound. For this era it was amazing!"[60]

In typical Woodstock fashion, the FE organizers invited a film crew along for the ride. However, the footage never saw the light of day until the documentary *Festival Express* was finally released to critical acclaim in 2003. Bernstein can be seen briefly in the film. According to Kingsbury: "We kept hearing that the promoters wanted to make a movie. There were people shooting footage everywhere. You rolled into the station and there would be a film crew shooting the trains as they pulled in."[61] Boroda recalls that there were cameramen filming all around him: "It was too bad because they shot so much of the production part of the concert but that stuff always ends up on the cutting-room floor."[62]

On July 5, after the FE tour ended in the city of Calgary, Bernstein reported back to Hanley about the event. He told her to relate to the crew "good job" and gave them each a $300 bonus. Pleased, the sound engineer informed his team that they didn't have to report back to Boston for at least a few days. Some of Hanley's crew remained and explored the Canadian province of Alberta.

Kingsbury remembers: "A few of the crew scored some psychedelics and got really tripped out; they were later picked up by Canadian Mounties. Rick Slattery and I had to go get them. We didn't know much about Mounties, but they turned out to be the nicest guys! They just handed over the tripped-out crew and told us to keep them safe. We were shocked!"[63] Slattery reminisces: "It was heaven looking out at the Canadian wilderness on a train with a bunch of kids and musicians—we had such a good time."[64] The FE made such an impression that Jerry Garcia, lead singer of the Grateful Dead, penned (with Robert Hunter) the song "Might As Well" about the unique experience:

> Great North Special, were you on board?
> You can't find a ride like that no more . . .
> One long party from front to end
> Tune to the whistle going 'round the bend.[65]

Chapter 52

HOT 'LANTA AND LOVE VALLEY

Alan Sisson was a local kid from Braintree, Massachusetts, majoring in engineering at Northeastern University. In the fall of 1969, while he was looking for a job in a college work-study program, his friend Nick Burns gave him a call. The two had met while working at a small psychedelic lighting company in the area. Burns suggested to Sisson that he check out Hanley's bustling business. By the end of the year Sisson made his way out to see the sound engineer. Sisson recalls, "Nick contacted me and having been in a number of rock bands in high school where I always became the tech person, this looked like a great job!"[1]

He was hired immediately, joining Hanley Sound at a time when the festival market was still setting the business aflame. Sisson stayed on for less than a year. In that short time he witnessed some of Hanley's largest and both successful and ill-fated events. According to Sisson:

> I worked with Hanley at the Second Annual Atlanta Pop, New York Pop, and Powder Ridge Festivals at which time I parted ways from his business. When I arrived at Hanley Sound the company was riding the success of Woodstock, and business was booming. The office, under Judi's reign, was full of high energy, yet often chaotic. Harold Cohen was the top tech. However, Bill seemed to me to be far more interested in the tech stuff than he was in the business end of things.[2]

In early 1970 the confident and pony-tailed thirty-three-year-old Hanley was unforgettable, as Sisson recalls: "I remember Hanley as a good natured 'flake.' I have fond memories of him spinning on one foot when he got excited about something, usually a tech thing of some kind."[3]

By now Hanley and his crew were able to deploy production with ease. From pop festivals, to political campaigns, to peace demonstrations, his crew schlepped over $500,000 worth of amplifiers, microphones, and loudspeakers across the country. When Sisson arrived, Hanley was hauling most of his equipment in large tractor-trailers. He saw firsthand how his new boss prepared equipment for a large festival. He remembers:

> Hanley owned a pair of Peterbilt tractors to haul the trailers, which were always full of BIG gear, used for festivals. I was told they previously had a Kenworth that had been stolen, and the insurance money was used to buy two Peterbilt trucks. For the major festivals, I saw Hanley use the JBL D series (D120 12-inch and D140 15-inch) low-end speakers in those big refrigerator size cabinets. We re-coned those speakers at the shop and occasionally on-site. Hanley bought cases of the re-cone kits. We usually used three pieces of theater lighting gel that was scrap from one of the lighting company techs, so we could shim the voice coil when doing a re-cone.[4]

Hanley did not discriminate about jobs by size, headline, politics, or culture. To him all jobs were equally as important as the next. Writer Herbert Aronoff states, "Almost everywhere there is a crowd, almost everywhere there is a head-line, there's Hanley Sound, getting the message across to audiences as big as Woodstock (450,000) or as small as a Knights of Columbus communion break-fast in Medford, Mass."[5]

THE SECOND ATLANTA INTERNATIONAL POP FESTIVAL

Festivals were becoming big business in North America in 1970, but potential host communities became adamantly against them especially in the southern states. On July 3–5, 1970, Hanley Sound supported the Second Annual Atlanta Pop Festival (Atlanta Pop II) in small-town Byron, Georgia.

Quality sound was on promoter Alex Cooley's mind, and Hanley was an obvious choice since the two had already worked together. Itemized in the promoter's Atlanta Pop II Festival production notes, he writes, "Bill Hanley Sound & Stage & Lighting (Atlanta, Newport Jazz & Folk, Miami)."[6] Atlanta Pop II saw over 200,000 in attendance and by now the public backlash of putting on events of this size was getting personal and confrontational for the promoter. Cooley explains:

> The Governor of Georgia hated my guts. I was in a restaurant one time; he came in and turned around and left. We were the whipping boys, no doubt about that. Every now and then though you ran across a policeman who was a kindred spirit. There was only one restaurant in Atlanta that treated me nice; the others would not let me in. I had hair down to my waist then and I was not welcome in places. Being on the road so much we often got turned

away; they poured coffee on the counter and stuff like that. We represented
something bad to them. They didn't understand. There was a great deal of
pushback. You could not go into a municipal building with a rock and roll
band. It was considered music of the devil or something.[7]

By 1970 there was national awareness of the festival movement. The Woodstock documentary was playing in theaters, while newspapers and magazines
hyped these mass-gathering extravaganzas. In locust-like fashion, festivalgoers invaded small-town America and, for a time, it seemed that more were
attending these big events than ever before. Alun Vontillius explains: "News
hype and magazines informed the public about what a couple of hundred
thousand hippies were like for good and bad. The people who were influenced
by Woodstock and other festivals were showing up by now."[8]

Despite resistance, Cooley with great difficulty was trying desperately to find
a location for his upcoming festival. Finally he settled on the Middle Georgia
Raceway in Byron. This was a more pastoral environment than that of his first
festival. The new site was situated close to a pecan field. Vontillius remembers
the early development of Atlanta Pop II: "The Second Annual Atlanta Pop
Festival was moving from location to location until they found someone who
was willing to host it on a pecan farm."[9]

The Woodstock-savvy locals of Byron were apprehensive when they realized
that a horde of hippies might be coming their way. But most accepted the idea
simply because they had no choice. Vontillius recalls: "Byron was a hick little
town, but we experienced very little or negative hostility from the locals. Festival
people were there with good and bad intentions. Some took advantage of the
sexuality and drugs and others were there for the music."[10]

In Woodstock fashion, traffic jams clogged roads, while festivalgoers broke
through surrounding fences and Atlanta Pop II became free. In his article
"Impressions of Atlanta," *Tampa Tribune* journalist Bob Fiallo wrote, "Every
festival was to be another Woodstock, and Atlanta Pop II was no different."[11]

Atlanta Pop II shared many similarities to its predecessor, Atlanta Pop I.
Both were located on raceways, and both were held on July 4 weekends one
year apart. Crowds at both dealt with sweltering 100° heat, and each offered a
lineup of performers that was extraordinary. Fiallo adds: "Woodstock Nation
gathered again for peace, music, love and laughter. And once again, the music
was magic."[12]

Nick Burns noticed a change from the previous year's Atlanta Pop Festival:
"The slippery slope was happening after Woodstock and you could see it at
Atlanta Pop II. The promoters wanted to make as much as they could. It was
all bottom-line and very much a business now. This in my opinion changed
the whole vibe of the festival scene."[13]

When Hanley Sound covered the southern states, the appointed crew traveled in one or two of the company's semis. Another was a twenty-foot International straight truck, referred to as the Jolly Green Giant. Burns recalls: "Usually we would build whatever we had to for the show. Configure what we had to configure. The truck would be loaded up and we headed down three to four days before."[14] Engineers Hanley and Boroda often flew to the event to meet the equipment and team. About a year or two later, after establishing himself in the southern states, Hanley had a truck stocked with equipment waiting in Atlanta. Strategic placement like this allowed Hanley Sound crews to easily facilitate an event at any given time.

Atlanta Pop II production crews had been working at the site about a month or so before the event. When Hanley and his team arrived, they had been putting the finishing touches on the main stage and light towers. Burns recalls when he arrived at a festival site like Atlanta Pop II, the first line of duty was to seek out sufficient power. Hanley remembers: "The power people were fantastic and gave us no hassles. Without them the festival would never have gone off."[15] Burns explains:

The first thing we did was to find out how much electricity we had to work with, which was usually none. Then we needed to find out where to hook up our power boxes. I remember Atlanta Pop II having no power; it was like Woodstock, there was nothing. Atlanta Gas and Light were called and started putting up poles followed by transformers, and that's how we got power.[16]

Vontillius and others who worked with Hanley at the first Atlanta Pop Festival recall a more structured event at Atlanta Pop II: "There was much better communication at this festival. We were able to communicate about the bands more efficiently. Hanley made distinct recommendations on how to mike things. There was more organization on having crew meals and having them together."[17]

The triangulated light towers consisted of three sixty-foot-long telephone poles with two platforms between them. The top platform held the follow spot operator, with Hanley's mixing location on the lower level. Burns, who was perched up so high, remembers it being hot and terrifying: "There were two spotlights on each tower. It wasn't much fun up there. There was something more secure about scaffolding compared to climbing up these extremely tall telephone poles. There were no safety harness belts in those days. You were up there for the duration of the show. It sucked!"[18]

According to Sisson: "This design was an attempt to try and prevent people from climbing up on the towers. From this location I was fortunate to have been able to mix the band 'Mountain'—one of my favorites at the time!"[19] Sisson recalls the challenges of trying to keep things on stage moving along with efficiency by way of a rolling platform system:

> The stage was designed with two rolling platforms on tracks, to allow pre-staging. While one band's gear (drum kit, keyboard, amps) was being set up, another band was playing. One platform would be off to one side of the stage (or the other) while the other platform was center stage. I think Atlanta was the first use of the heavy bass cabinets as well (8 JBL D series speakers per box). I also recall that there was an 8 or 10 ft. plywood fence in front of most of the audience area. The light show was situated behind a screen in the back of the stage.[20]

According to Boroda the "system" idea came from Chip Monck, whom Cooley had hired as the festival's stage director: "It was a whole new design. We used it at West Palm Beach Pop as well. Chip was very good at designing these sorts of things."[21]

Bob Fiallo recalls drugs were as extreme as the high temperatures, which soared to 115°: "The overdose tent and medical staff kept busy day and night treating the foolish, the incautious, the injured and the heat prostrated as make-shift ambulances slid through the crowd evacuating the most serious medical problems. Heat prostration was the REAL problem."[22] Burns remembers: "There was no shade. I used to wear my old army shirts just to keep the sun off. And if you were up on one of the towers running one of the Super Trouper spotlights you had to wear gloves, the metal was too hot to touch. If you were up on the stage, which acted like a big flat reflector, you really got fried."[23] Such soaring temperatures were a serious problem for Hanley's equipment. The sound engineer had trouble keeping the amplifiers cool enough to function. As usual, the engineer learned his lesson through trial and error. Behind the Atlanta Pop II stage Hanley had one of his 54 ft. semis parked, outfitted with a Thermo King refrigeration unit at the front of the trailer. Kept at a distance so the compressor could not be heard, Hanley Sound crews often climbed into the rig to cool off from the oppressive heat. According to Sisson: "Hanley used a trailer with a 'reefer' in it, known as a refrigeration unit. This kept the big Mac 350 tube amplifiers cool. I mean it was 105 degrees in the shade!"[24]

According to crew member Brian Rosen: "Atlanta was hot and we had lot of problems with amps overheating and popping circuit breakers because of it. Bill had the brilliant idea of putting the reefer unit in the trailer! We built a rack at the front of its nose for all of the amps so the backs of amps faced the cooler side. We entered through a plastic curtain. This was classic Bill."[25] Vontillius recalls Hanley's innovation: "You literally had to put on a fur parka to go in. These amps would get very warm and there were thousands of feet of speaker lines (#10 stranded S electrical cable) on rolled spools."[26]

A year earlier, just two weeks after Woodstock, on August 30 through September 1 at the Texas International Pop Festival in Lewisville, Texas, Hanley built a room under the stage out of scaffolding and polyethylene sheeting. The room housed over fifteen 350-watt McIntosh amplifiers, and was equipped with windows and air conditioners to cool it. The air conditioners were stuck through the plastic material. Burns recalls that Texas was hotter than Atlanta, and without air conditioners everything would have fried: "At Texas Pop, Hanley sent us out to a local hardware appliance store and we asked to buy all of their air conditioners. The owner looked at us like we were nuts! We put over $6,000 worth of air conditioners on Hanley's Diners Club and went back and built that room."[27]

Two emerging engineers, Rusty Brutsché and Jack Calmes, were in attendance at Texas Pop and studied what Hanley and his crew were doing. Brutsché acknowledges Hanley's creative ideas with the use of scaffolding and methods of cooling amplifiers. He considers this one of the most innovative things he had seen up to that time. He remembers:

> To get this stuff up fifty feet in the air had never even occurred to me. I was struck with how well that worked being able to get your sound out there. He used massive McIntosh amplifiers. They were 350-watt Mac amps and he had like fifteen of them. He had also built a room out of Visqueen and had scaffolding under the stage. It had a window with air conditioners stuck in the Visqueen to cool the room and amplifiers. The idea of using 4500-watts was unbelievable. Most people in those days thought using 10-watts was all you would ever need for a sound system. The thought of using thousands of watts was a big deal! If he hadn't done what he had done there would not have been all of those festivals.[28]

In 1970 Calmes and Brutsché established one of the most innovative second-generation concert sound and lighting service companies in the country: Showco. Calmes, who was responsible for booking Texas Pop, recalls Hanley's work at the event:

> When I saw how much money Bill Hanley got for doing the sound at the Texas Pop Festival and at Woodstock, it set me thinking. Hanley didn't get burned at either Festival, although the rest of us did. And Hanley was just about the only guy in the country with access to enough equipment needed for the big shows. What's more, although Hanley's sound was superb for the day, it would still go off three or four times a concert. He'd be sitting

there soldering the lines together half an hour before the concert. It was a
joke. He was completely disorganized. And it wasn't that he couldn't teach
somebody to help him run the system; he just wouldn't! So I knew it could
be done better. And if it could be done better, we'd have all the business we
could possibly want.[29]

Hanley was older than the rest of the Atlanta southern crews and far more
experienced. Because of this he was able to take control of situations like
gate-crashing as if it were second nature. According to the *New York Times*: "The
promoters of the event had been charging fourteen dollars a person admission,
but this was discontinued when 14,000 youths, many of whom said they did
not have 'bread,' gathered at the gates and began chanting, 'In, in, in!' Fearing
possible trouble, officials opened the gates to all."[30] Alex Cooley's assistant Kathy
Masterson recalls some of the chaos that ensued after the festival fences came
down: "We never expected the festival to be free. I remember Bill went in like
clockwork. I was busy making sure the acts got paid and got on and off the
stage and Hanley knew what he was doing. He was very professional and knew
what to do when we were all trying to figure this stuff out."[31]

This was the very beginning of the festival business and many promoters
did not have a handle on elements of effective festival production and facilitation like security for crowd control, proper fencing, money collection, ticket
sales, and general infrastructure. Atlanta Pop II was losing money fast. Cooley
hurriedly sent Masterson out to the box offices in an effort to gather up all the
collected money. Masterson relates:

> I remember Alex sending me out to clear out the box offices because the
> fences were coming down. The Galloping Gooses were the security and they
> escorted me while I stuffed all the money in pillowcases. They then threw
> me over the six-foot-high fence so I could run back with the money. That's
> how loose this thing was. It wasn't like Ticketmaster. My ex-husband and I
> used to put the tickets out at record stores and then go back that night and
> pick up the money. We went back to my apartment to count it and it never
> occurred to us that we might be robbed.[32]

Atlanta Pop II was not a total wash. With over thirty acts performing, like
Ten Years After, Grand Funk Railroad, the Allman Brothers, Jimi Hendrix, and
Richie Havens, music dominated the holiday weekend. According to *Billboard*,
"The acoustics were superb and the acts were even better."[33]

At early dawn Monday, July 6' just after things began to wind down on that
blazing hot weekend, Masterson was sitting next to Hanley alongside his tractor-trailer. In the distance they heard Richie Havens open with his rendition of

"Here Comes the Sun" as the sun indeed rose over the festival field. As Hanley prepared to strike his equipment, loading up for his next event, the humid air was saturated with the sweet smell of pecans. According to Masterson: "We were able to pull it off. No one died, and no one got hurt. This festival brought thousands of people together and I will never forget seeing Jimi Hendrix on the Fourth of July with all of those fireworks going off behind him. It is my favorite music memory. I sat in Hanley's sound truck next to the stage and watched. I had an out of body experience."[34]

By 1970 Hanley had been working in sound for almost half of his life. The sound engineer was known for his innovative ideas, expertise, diligent work ethic, and top-of-the-line equipment. Yet the music getting louder became a challenge for him. Hanley continuously tried to keep up with the acoustical impact of loud rock music without compromising clarity and intelligibility. According to Alex Cooley, sound reinforcement was in its infancy and rock and roll was very difficult to manage: "Before Bill groups used to hang little speakers up on little tripods on the side of the stage. The old sound people didn't really understand, but Bill was one of the first who understood what a rock and roll artist was trying to do. They wanted to blast you with their sound and that was part of rock and roll and Bill got it."[35]

After Atlanta Pop II Hanley continued to provide sound reinforcement for events in the southern states. Hanley observes: "The Atlanta Festivals were something special and so very hot. I liked doing the southern festivals because the crowds were great and the sound was always good."[36]

THE LOVE VALLEY ROCK FESTIVAL

On the weekend of July 17–19, 1970, two weeks after Atlanta Pop II, a lesser-known event occurred in North Carolina called the Love Valley Rock Festival. Mayor and founder Andy Barker had created the blueprint for the small cowboy-themed community he named Love Valley. The area was situated in a secluded valley of the Brushy Mountain Range in Northwestern North Carolina. It was yet another southern affair that put Hanley's Yellow Bird refrigerated trailer to good use.

According to Barker, after he learned his children had attended the Second Annual Atlanta Pop Festival, he became curious and asked: "'What in the devil is a rock festival?' They tried to explain it to me, and I said, 'Well, hell, I'll just have one here.' They said, 'Daddy, you don't know what you're talkin' about.' I said, 'Hell, I don't care. We'll do it.' So I decided to do it."[37] When beginning to plan his event, Barker called on Hanley Sound since he had heard of the company's extensive work for festivals in the South.

About fifteen miles north of Statesville, the town of only one hundred residents grew to city like proportions as over 75,000 festivalgoers swarmed in. Barker charged five dollars for admittance to the festival. Still, some tickets were counterfeited, and others figured out to avoid the fee completely by accessing the area via the woods. Those who made it in packed a side of the mountain that surrounded a circular arena designated for rodeos. Campgrounds were full of festivalgoers, as well as the mountain area behind the arena.

Hanley left his crew in charge, as he was in New York that weekend preparing for a festival on Randall's Island. According to the sound engineer, he remembers Judi Bernstein receiving a call from crew member Kelly Sullivan. Hanley recalls:

I guess when Kelly and the crew got to the site and proceeded to lock the wood panels together they noticed that the C clamps were missing, an essential component when locking these panels together. Someone forgot to put them on the truck, so Kelly called Medford and Judi agreed to have them sent down by plane. When Kelly arrived at the airport to pick up the two crates of clamps they were not to be found. They tore the place apart when finally they called the guy in charge. It seems as though he put them in a refrigerator because he thought they were sea clams![38]

Around forty-three bands played the three-day event. Most were regional acts, unlike the impressive lineups at Cooley's Atlanta festivals. Nevertheless, it was an opportunity for emerging performers who did not mind getting paid a pittance to get exposure on a big festival stage. Most notable were the Allman Brothers, who headlined the event. At the time the group had only one album to their credit. The band performed a couple of solid sets over the weekend.

Again drugs were rampant and it was blazing hot. During the festival, audiences were hosed down to keep cool. Like most small southern towns, residents of Love Valley's surrounding areas were unhappy with so many festivalgoers invading their communities. Making headlines more for its size than its music, the Love Valley Rock Festival was considered a moderate success for those who attended.

Chapter 53

A PORTRAIT OF AN INTERRUPTED FESTIVAL AND GRAND FUNK RAILROAD

NEW YORK POP

On the hot and humid weekend of July 17–19, 1970, Hanley and his crew were preparing for a three-day event called the Randall's Island Rock Festival, known as New York Pop. It was held on the outskirts of Spanish Harlem across the Harlem River at the 22,000-seat Downing Stadium. The island was not the most bucolic location for a festival, especially for anyone who decided to sleep there for the weekend. A notice went out to attendees:

> Randall's Island is in the middle of New York's dirtiest river. Large areas of the island are made up of garbage fill. Rats inhabit these areas and come out to play after things quiet down. Rats are bad playmates. Further, Randall's Island is East Harlem Turf. It belongs to the Puerto Rican community. It is their green park. They do not appreciate strangers being on their land, parking their cars on baseball diamond and meadows. They certainly do not appreciate overnight guests.[1]

As at other festivals, New York Pop's fences were taken down by demonstrators and this resulted in financial disaster for its promoters. During the festival a coalition of over twenty radical groups picketed, including the Weather Underground, the Young Lords, and the Black Panthers. By the end of the event, the Young Lords "proved most helpful in maintaining crowd control."[2]

Radical groups demanded that festival promoters pay them and let them in for free. A flyer handed out by the groups claims:

> This concert belongs to the people. For the first time the community whose culture is being packaged will also share in the profits of the concert. The

389

Black Panthers, The Young Lords and a collective of white street groups are now partners to the money this concert will make. The money will be used to enhance and support the community service projects of these groups as well as to provide new bread for various bail bonds.[3]

A *Billboard* article titled "Randall's Fest Hit Problems" reports, "Promoters suffered drastically on the financial level after 'The Young Lords,' a militant Puerto Rican group and various other local action groups demanded and received full control of the event."[4]

New York Pop is an example of a festival trying to find its footing in a shifting cultural scene. In a 1970 *Village Voice* article, Carman Moore wrote about festivals like New York Pop: "Somebody has to prove to me that the festival idea is over and can ever become an institution in the rock world."[5] He added: "Producers acting like they have more money than they really have, rock bands asking for more money than they're really worth, and radical organizations trying to rip off money without studying how you really get that money, all add up to some kind of death knell for the Woodstock promise."[6]

As advertised on the official festival poster, Hanley provided sound, stage, and "closed circuit TV projections on giant screens." Hanley explains that this arrangement was unique for the time: "Randall's Island was one of the first outdoor concerts I used rear stage video projection."[7] Of the long list of performers who were scheduled to play, many refused because of nonpayment, like Ravi Shankar. Sly and the Family Stone didn't even show (they had a reputation in the business for this). As a result, local acts were asked to fill in for the missing performers. According to the *Village Voice*: "Of the feature bands that showed, not much happened but noise. And the noise was no fault of soundman Bill Hanley, amplification was crystal clear almost all the way."[8]

Some notable bands did rise to the occasion, like Mountain and a few others. According to *Billboard*, "The opening bill, July 17th, saw Jimi Hendrix, Grand Funk Railroad, John Sebastian, Steppenwolf and Jethro Tull perform."[9] On July 18 only two groups performed. By July 19, the final night, just three acts played. "By Sunday, New York Pop was officially declared free."[10] Alan Sisson recalls how Hendrix and his entourage arrived at Randall's Island:

The Hendrix crew used a boat (a small cabin cruiser) to get over to the island from the Bronx. They came over during high tide. When it came time to leave, the tide had gone out and the boat was grounded. I was involved in getting them back into Manhattan in an old station wagon that Hanley had backstage. I bought Jimi a pack of cigarettes; he shook my hand as he thanked me.[11]

Working with Hendrix was longtime McCune sound engineer Mike Neal: "I worked with Hanley at the Randall's Island New York Pop Festival. I was traveling with Jimi Hendrix when he played the event. Hanley designed the sound system and as I recall it was a really good system."[12]

While on the island, Hanley's crews were on guard, including Sisson, who remembers the crowd as very hostile: "They were throwing bottles (and other things) at the stage and equipment. I had to climb up on the speaker stacks to clean up the glass and check for damaged speaker cones."[13] Nick Burns, who was on one of the light towers working a Super Trouper follow spot, recalls the event as dangerous. Just below, angry crowds began to shake the scaffolding structure that his heavy spotlights were on. He was worried the tower would topple over. He remembers almost being arrested for projecting a light show off of the nearby Bellevue Psychiatric Hospital:

> At night we were having some fun pulsing the colored spotlights for a light show and this, what looked like an abandoned building, was the perfect backdrop. Come to find out it was a psychiatric hospital! The police came and told us we were agitating the patients and if we didn't stop they were going to pull the plug on the whole festival. We had no idea it was a hospital or we would not have done that.[14]

Since Boroda often sent lighting cues to follow spot operators, he claims that he may be to blame for the creative lighting cues resulting in this mishap. Those working the follow spots like Burns frequently tested out lighting at dusk before a performance. Boroda explains: "We communicated through headphones then and I used to listen to the lighting cues. I had this idea to create a multicolored sunrise on the backdrop screen on the stage. I wanted different colors fading in and out. So I asked Tom and Nick to try it out and they must have been playing around with this on the Super Troupers!"[15]

An event fraught with demonstrations, downed fences, overrun with free-loading gate-crashers, and a coalition of twenty-one radical groups making unreasonable demands made for a really bad scene. But according to *Billboard*, New York Pop could have been worse: "What started out as a political platform for the so-called youth movement in the New York area, turned into an experiment in collecticism [sic] with stagehands local and performers working without fees for the benefit of brotherhood and equality."[16] New York Pop saw around 20,000 in attendance. However, it's suspected that an additional 30,000 got in for free once the gates came down.

GRAND FUNK RAILROAD

One of the bands that performed at New York Pop was the deafeningly loud Grand Funk Railroad (GFR). The ear-piercing Flint, Michigan, group had already put Hanley's expertise to the test. They first gained acclaim for their raucous performances in 1969 and 1970 at the Atlanta Pop Festivals where Hanley mixed. Although the sound engineer often had to manage the creatively distorted and loud fuzz of an artist's expression, nothing quite matched the sound pressure levels of GFR.

In a 1970 *Boston Globe* article, "When Hard Rock Meets an Eardrum," writer Richard Knox asks, "Scientists are beginning to ask what, if any, damage is being done to the ears of rock music fans that immerse themselves repeatedly in the never-never land of noise?"[17] This is a good question to ask a sound engineer, because if any hearing loss occurred in Hanley's lifetime, GFR would have been suspect.

By the early 1970s many new and emerging sound companies were trying their hand in sound reinforcement. Tychobrahe of Hermosa Beach, California, run by founder Bob Bogdanovich and senior engineer Jim Gamble, worked with GFR in those days. Among the West Coast sound company's crew members was New England native David Pelletier. Pelletier recalls Hanley as being the premier sound company, not just in Boston but also across the United States. During the mid-1960s while managing local bands in Boston, Pelletier was catching wind of Hanley's work: "In 1966 I saw the Beatles at Suffolk Downs, and I knew he did the sound for that. Hanley was totally out of my league at the time. I was still a little kid wanting to get into the business."[18]

According to Pelletier, GFR was one band he could never forget because of their booming sound: "When we did the Forum with them, they were really loud. They were the loudest band I had heard up until that time. GFR were going for it and we helped them get there."[19] In a *Los Angeles Times* article, "Bogdanovich Runs a Sound Business," writer Kathy Orloff claims that GFR offered a sound so loud it shook the foundation of wherever they performed:

> The sound is so loud it shakes the floors and rattles ceiling and walls, operating so close to the threshold of absolute pain that it almost disappears into itself, blaring out of endless columns of speakers, pouring out in blazing, careening screams, hurtling forward, wound around through miles of wire and tubes and bits of heavy metal, lost in the electric pitch of sonic amplitude, vibrating and surging, enveloping everything even remotely close, heartbeats absorbing thundering bass to brain waves synched in with tonic horn treble sounds of total consonance.[20]

Grand Funk Railroad's escalating volume pushed concert sound to new technological heights. According to some, at the zenith of GFR's popularity one of their concerts could require at least two days of production accommodations. When the band played an arena, or even an outdoor setting, the sound engineer was often not prepared for it. This disturbed Hanley, who claims the bands dB levels were "off the charts!"

According to Hanley, GFR manager Terry Knight (who often acted as sound engineer) was a real pain in the ass to work with. Hanley recalls that at New York Pop the imposing Knight put together a separate sound system, placing it right next to his, using "Electro-Voice full-range speaker systems assembled into boxes, augmented by bass speakers and a pile of DC300 Crowns."[21]

After GFR's system was set up next to Hanley's, Knight forced his way behind his console, took over the controls, and turned everything up. The group's manager knew exactly how loud he wanted GFR to be, and that was extremely loud. Hanley and his crew had no time to test the group's equipment and were taken aback by the onslaught of distortion. Hanley explains: "We didn't 'A-B' them, Terry Knight just set up, and when the band came out, he turned it on! Christ, was it loud—distorted—but loud! Here I was, proud of my fidelity, but I stood in the audience and listened to it and realized it was a valid experience—distortion!"[22]

In the back of Hanley's mind he was worried. What did this mean for the future of his business? "I was trying to avoid blowing out my loud speakers! It was very loud, like when Dylan went electric at Newport, and I now realized even more that distortion was forever becoming an extension of the artist."[23] Boroda, who was on hand, witnessed the event unfold:

> When Grand Funk said they wanted to use their own PA it scared Hanley. It was the first time this ever happened; they didn't want to use Hanley's equipment. He was afraid they would make us look like idiots. Their system had a lot of SPL but no projection. It was all in your face. You would back up away from the system and hear nothing. They didn't use horns; they were infinite baffle sealed enclosures. The system was garbage. So they played and they tore down their system and we went on with the rest the concert. Bill had to face these sorts of challenges.[24]

According to lead GFR singer/guitarist Mark Farner, Terry Knight was notorious for telling sound guys what to do in the early days: "As far as the distortion we would just go up and run all the knobs wide open and we would melt down a few tubes along the way! Because of the distortion, guys like Hanley were trying to avoid it, and all of the sudden it's part of the sound. I can see how this would be confusing for a technician."[25]

Grand Funk's skyrocket to fame in such a short time did not impress *Village Voice* rock critic Carman Moore, who wrote, "Grand Funk was creative about their noise for a tune or two, but they just aren't into very much musically."[26] Hanley was aware that Grand Funk was looking for a certain sound, but he felt there were better ways to achieve loudness without sacrificing fidelity. He observed in a 1989 article in *Recording Engineer/Producer (RE/P)* magazine:

> It was valid for what the band wanted for their audience. Up to that time I had been successful in promoting fidelity, but who was I to superimpose my values on their music? I really had to scratch my head and wonder how much money to spend on fidelity. Here was this kid with his system full of $28.50 EV horns, and piles of them, two or three hundred of them. He got heavy bass out of his bass cabinets and had incredibly efficient output, and it was a valid experience.[27]

Over time many newer bands didn't hire Hanley as much. The sound engineer was unable to adapt quickly enough to the heavy rock music dominating the evolving musical landscape. Hanley recalls: "The bands weren't hiring me; they were booting me out of jobs because it wasn't what they wanted. They were getting into this distortion thing and everybody loved it. Grand Funk was the top group in the country at the time. If they wanted me, I had to alter my thinking."[28]

By 1970 Hanley was revered as a leader in the sound reinforcement industry. He was even highlighted in *Esquire*'s "Top 100 in Rock" for his "extraordinary" work in sound reinforcement, stating "The Ubiquitous Hanley Sound Inc. (Woodstock was just another weekend's work) makes live performances sound almost as good as records. Even gets top billing in ads for concerts. Did sound for Nixon campaign. Wasn't it important, said Hanley, 'that people heard and understood every word he spoke?'"[29]

After New York Pop, another festival called the Powder Ridge Rock Festival was developing in Connecticut, and as one writer warned: "Powder Ridge looms ahead. The producer says—don't come without a ticket."[30] This was yet another festival fiasco that Hanley Sound was preparing for.

Chapter 54

THE FESTIVAL THAT NEVER WAS

THE UNFESTIVAL

By late summer of 1970 it was business as usual for the Hanley Sound crew. Preparations were under way for the Powder Ridge Rock Festival, yet another mass gathering. It's often referred to as the "festival that never was" because only one artist on its scheduled lineup actually performed. It was to be held on the three-hundred-acre Powder Ridge Ski Area in Middlefield, Connecticut, from July 31 to August 2, 1970.

Little has been written about Powder Ridge. Yet what happened in Middlefield on that weekend is a good snapshot of where the country was. Knowing that their town was on the cusp of being taken over by hippies, the small conservative community pushed back hard. In Robert Santelli's book *Aquarius Rising: The Rock Festival Years*, Middlefield is described as a community "Proud of its three Vietnam War veterans and its Girl Scout troop and its reputation as a town that exhibited a decent moral tradition and a healthy American atmosphere in which young people could grow."[1]

It was almost one year after Woodstock when Middlefield went to battle with the festival's promoters. Public outcry, coupled with political disdain toward this or any festival, made it easy to issue legal injunctions. It didn't help that word was spreading about its promoters. They were known as "Middleton Arts International Inc." and some of the individuals had suspected ties to organized crime. Despite looming injunctions, tickets were being sold and quickly. With so much against it, there was no way a festival of this magnitude could ever get off the ground smoothly.

With opening day closing in, the war between the promoters and town government continued to heat up. Ultimately Judge Aaron J. Palmer denied all requests to have a festival in Middlefield, enacting a court injunction just four days before the event was to take place, on Monday July 27. Remaining hopeful, the sleazy promoters continued to encourage speculative ticket holders to come and simply "wait and see" what happens.

Powder Ridge was dismembered before it grew legs to stand on. Word about the festival's problems had traveled rapidly and performers instinctively stayed away. Regardless, optimistic crowds still came looking for an experience in free love, drugs, and great music. Hopeful attendees arrived and saw that typical festival necessities were lacking, most noticeably food, toilets, and music. Drugs were rampant. Powder Ridge was to end in disappointment.

For twenty dollars attendees could have experienced a weekend full of incredible music, a lineup fit for the history books. Santelli relates, "The promoters promised at least eighteen top-notch acts."[2] In an effort to re-create what happened at Woodstock, Middleton Arts International intentionally booked acts that played in Bethel, like Sly and the Family Stone, Janis Joplin, Mountain, Joe Cocker, Richie Havens, John Sebastian, Ten Years After, folk singer Melanie, among others.

Despite impending legal obstacles, the land had been leased, and Hanley called in. Out of all the performers booked, Melanie Safka was the only one who fearlessly drove out to the ski resort to perform, even though she was told not to. According to Safka, "Powder Ridge represented an opportunity for those in the Northeast who had missed Woodstock to experience the excitement and ebullience of what the media called a countercultural extravaganza of music, drugs, and out and out nudity."[3]

When Hanley agreed to provide sound reinforcement and production for Powder Ridge, he had already been in the business for more than a decade. The seasoned sound engineer seemed unfazed by Powder Ridge's less than favorable forecast. Hanley, used to working with public and civil entities, decided to move forward without a second thought.

By Wednesday July 29, over 8,000 people had arrived at the ski area. On Friday, there were 20,000 or more camped out on the slopes waiting for something to happen. The ill-fated attendees remained hopeful for the slightest chance that Powder Ridge was on. Authorities simply weren't able to shut down the festival. Over 30,000 left the ski resort littered with trash as it closed out on Sunday. One attendee remembers: "We had a good time anyway though. You have heard of the un-cola . . . well this was the unfestival."[4]

THE KIDS OF POWDER RIDGE

On Wednesday July 29 Judge Palmer handed over the power of enforcing the injunction to the Middlesex County State Attorney, Vincent J. Scamporino. This is when things took a turn for the worse. The injunction specified that if any performer played a single note at the festival, it would result in immediate prosecution. According to *Hartford Advocate* writer Robert Piasecki, the warning stated, "Any musicians who violated the injunction and performed at Powder Ridge would be charged with contempt of court and subject to possible fines, imprisonment or both."[5]

That Wednesday, Connecticut state troopers began placing signs throughout the winding narrow roads leading to the ski area. They read "POWDER RIDGE FEST PROHIBITED COURT INJUNCTION," and were designed to discourage and inform the incoming crowds that the festival was canceled. The highway department did the same on the main highways. On Thursday July 30 town roads were closed except for Middlefield residents. Nevertheless, determined festivalgoers ignored the warnings. Rock radio stations informed the public to stay away, while press releases were also sent out. Hanley recalls the media doing a "good job" of keeping everyone up to date, but that they were also responsible for "feeding the public fear."

Hanley and his crew had arrived at the resort on Monday July 27 to facilitate full production. The position of the stage and sound system was planned for the base of the ski resort with the audience perched on the rolling slope looking down. Because of the injunction, Hanley's point source line array sound system that was to be suspended above the stage by crane was never completely set up. According to the sound engineer, this special configuration had never been done at any concert or with any sound system before:

> A crane that we rented was set up for a big long line array. The rig was to be set above and in front of the stage on a 100 ft. boom. I had eight A10s hung above from a fifty-foot jib, and eight HSI [Hanley Sound Inc.] 410s hung down below from a lattice boom held together with a steel frame. It had a four-piece 3/8-inch zinc-coated steel cable with turnbuckles so we could precisely aim the speakers.[6]

Boroda explains: "This idea was the beginning of line array system configuration. So we got the crane and hung these speakers up. However back then not a single person was certified for rigging. You cannot get away with that today without a permit."[7] Sisson recalls the uniqueness of this design as the first use of a "point source" sound system at a festival: "A massive array of speakers hung from a crane above the stage. It was sound coming from a single point, rather than from two stacks of speakers on either side of the stage. Bill wanted to reduce or eliminate phase cancellation allowing for consistent sound across the entire sound field."[8] Scott Holden remembers working on this elaborate structure back at Hanley Sound:

If I recall correctly, what Bill was going to do was hang all of the speakers on a crane and that structure was being built somewhere. But he was also going to make the stage elevated up on scaffolding. I was back at the shop doing all the welding of the enforcing pads that would go underneath the stage. Everyone smoked back then. I remember welding so much for this that when I unzipped my coveralls and reached for my cigarettes and matches, the matches had completely ignited from a spark that hit them. Powder Ridge almost happened, then it was canceled. I remember the politics around this event; it was huge back then. This was the death knell of outdoor concerts in the country.[9]

Harold Cohen was working away at the Medford sound company, busy in the wood shop located next to the main storefront of the building. Hanley had called Cohen and ordered additional 4x10 speaker cabinets to be built. Hanley wanted them sent down to the festival site. Cohen was apprehensive. The engineer was receiving negative reports about Powder Ridge and was concerned about sending any additional equipment.

Even though the speaker cabinets were finished, they never made it to Middlefield. Cohen explains that this was a good thing, because they were too large to fit through the door anyway: "We didn't have any set plans to build these so instead of having side panels set in—we mistakenly put them on the outside. That made the cabinets 1–1/2" (3/4" plywood times 2), which meant they were too wide to fit through the door. Eventually, they were removed from the storefront window next door well after Powder Ridge."[10]

Nelson Niskanen drove from Long Island, New York, to meet Hanley and the rest of the crew at the ski slope. "When I arrived the trucks were there and I helped set up the sound system and stage. We set up the Wenger Wagon at the foot of the ski slope and the kids were piling in while we were doing so. When we finally had all the gear in place and almost ready to go, we were told to take it all down."[11]

With a twenty-foot Ryder truck packed, while on the road Nick Burns got final word not to go to Middlefield. He was supposed to arrive with any additional equipment that Hanley might need. "I was driving with some special light show stuff. So I stopped to make a phone call to Judi on a pay phone. She told me to turn around and head back, the festival was canceled! I never made it to Powder Ridge."[12]

As equipment was being laid out, Powder Ridge was evolving into a tricky situation for Hanley and his team. At Woodstock they had problems, but solutions were made and the show went on. There was no solution in sight for Powder Ridge. Rick Slattery saw things going awry quickly: "Bill and I are setting up the stage and sound system but things started to fall apart. People were telling us to not set it up yet etc. . . . Then things started to go haywire. It became clear at some point that things were getting bad. We were living at the ski lodge waiting to get some work done but we never got to set our complete system up. Most of it was unpacked."[13]

David Kingsbury claims they were ready to go just as the injunction stopped production: "Thousands and thousands of kids showed up. After the injunction it turned really ugly. All these freaks were there with lots of dope. Rows of them set up booths and were out and out selling drugs."[14] After the partial setup, most on staff stayed at the lodge with Hanley. Waiting for the green light, the sound engineer thought there still might be a chance for the event to go on. Kingsbury remembers: "Hanley sent me to guard the trucks because somebody had just slashed the tires on our straight truck. They probably thought we were going to leave after the injunction stopped everything. They wanted to keep us there it seems."[15]

By Wednesday the festival promoters were mysteriously missing, and all of the responsibility was left to the resort owners, brothers Herman and Louis Zemel. Right before dusk, with bullhorn in hand, Louis climbed onto Hanley's stage and attempted to convince the loitering crowd of around 15,000 to leave. Santelli relates, "Amid shouts, catcalls, and boos he described the implications of the court order that prohibited the occurrence of the event."[16] The *New York Times* reported, "'The festival can't go on,' the fifty eight year old Louis Zemel said in a quavering voice, 'The big acts are not going to happen.'"[17]

The Zemels were ordered by Scamporino to cut off all utilities to the ski resort by 8:00 a.m. Thursday July 30, killing any chance of Hanley's PA ever being used. The *Times* adds, "Mr. Zemel spoke last night in compliance with a directive issued earlier by State Attorney Vincent J. Scamporino who has been designated to enforce the injunction."[18]

Even though police presence raised the potential of being arrested, the kids at Powder Ridge denounced all authoritarian rule, and were prepared to deal with whatever was to come. This spirit caught on as Dr. William Abruzzi, the "Woodstock Doctor," took to the stage. Abruzzi, who had seen his share of medical cases at Woodstock, assured attendees that a medical facility would be in place if needed. The crowds liked that.

But with no music and no proper PA to communicate with the audience, Hanley knew there would be trouble. That Thursday the sound engineer was trying to get some of the acts to play, including Melanie, with whom he worked at Woodstock. Hanley recalls the phones were tapped at the resort as he was attempting to get Safka into the festival safely: "Ultimately they caught me trying to bring Melanie up to give the kids some music so we could calm the situation. I didn't want to have a confrontation because I heard the kids had set fire to a jeep up on top of the mountain behind a ski slope. We wanted to give them something to do so things would calm down."[19]

By Friday July 30 most of the roads to the festival site were closed. As Safka listened to the radio she heard that the festival was canceled, but decided to proceed anyway. "I thought that's impossible, so I said I am going! There was a press conference at the lodge and it was decided that it was canceled and that any performer that played would be arrested—do not attempt to go there. So I went. I heard that these news guys were going in and I just went with them in their car. My guitar was in the trunk."[20] According to Kingsbury, Hanley did everything possible to make sure the show went on: "Hanley was a rebel and didn't want to follow authority."[21] Learning that Melanie was going to come regardless of the situation, Hanley began to prepare sound under cover of night.

Having been through Woodstock a year earlier, the defiant and optimistic folk singer was sure that the festival was going to take place. The *New York Times* reported, "The singer who is known professionally only as Melanie, had been among those advertised to perform at the festival and Mr. Scamporino said they risked arrest if they appeared."[22] On Friday evening, noticing a nearby Mr. Softee ice cream truck, Hanley pulled together a makeshift sound system powered by its generators. "The vendors helped us out. We were waiting for Melanie to show up and we needed to get her a sound system ASAP."[23]

Safka recalls: "I arrived at that ski lodge, and there were a lot of people there. I felt like Santa Claus. I was the only performer who played. Hanley hooked me up to a Mr. Softee generator and I sang. It was really cool, and by the time I got out, nobody could catch me. I was in the car and out of there!"[24]

According to Santelli: "Melanie played, to the delight of those gathered at Powder Ridge, and the crowd applauded with heightened enthusiasm. In between songs she told of how she was proud to play her music under such pressing conditions."[25] Kingsbury recalls the special performance: "I remember when Melanie played. It was a spot of light in an otherwise ugly scene."[26] This memorable set by Melanie included some of her most popular songs, including "Lay Down," and "Beautiful People." Melanie was the only signed performer to show up at Powder Ridge.

Safka was never paid or held legally accountable for ignoring the injunction, yet Hanley was. Santelli writes about the risk Melanie and Hanley took to perform at the event: "After Bill Hanley of Hanley Sound instructed his people to ready the stage for her performance, he was arrested and charged for contempt of court."[27]

Ordering Hanley not to turn on his sound system is like telling Edison not to screw in a light bulb; it's just never going to happen. Even though this was a dangerous situation, Hanley ended up in handcuffs for powering on his system. Boroda recalls: "Powder Ridge was a fiasco. They told Hanley not to turn the sound system on and he did, so he got arrested. Hanley told them we needed to have communication with the people but they didn't care. Hanley was anti-system and this was his view of politics. He knew what they would do to him if he turned it on."[28]

As the weekend went on, the atmosphere at the ski resort became increasingly tense. 50,000 kids were waiting to hear music. Instead of authorities removing festivalgoers from the sloping green hill, Scamporino ordered Zemel to shut off the electricity and water. With no food, bathrooms, or music, he hoped this would send a clear message to leave, rather than one of love. But Scamporino eventually gave in only slightly, allowing running water and sanitation—but still no sound system. Given all of the illegally parked cars, traffic, and excessive drug use, he remained resistant.

With the spirit of Woodstock still fresh in people's minds, the festival-attendee hopefuls made do and created their own experience. *New York Times* journalist Joseph Treaster reported: "At one point, 200 or so young men and women fell into step together chanting and pounding out a beat with sticks and cans, shoes, bottles, tambourines and bongo drums. A young man with a harmonica and a couple of flutes joined in."[29]

By late Saturday August 1, any notion of live music ever occurring was nonexistent. The only sounds heard were from car stereos and random acts of acoustic guitar playing. What was evident though, was a scene of excessive drug activity and nudity. For the conservative people of Middlefield this was not taken lightly. According to Sisson, at one point during the event the Hanley Sound crew had to vacate their rooms in the lodge so the police could handle increasing drug issues: "There was no music, so it became a drug festival."[30]

With so much down time during the excruciatingly hot weekend, some of the crew cooled off by taking advantage of a covered pool situated behind a fence in the back of the stage. Kingsbury remembers: "It was so hot. I was part of the crew so we could access the area from the backstage. A few of us snuck in under the plywood cover and swam."[31] The crew member recalls volunteering at the trip-out tent: "I heard about rapes, and bad trips so I wanted to help. I ended up volunteering in the bad trip tent and saw guys trying to take advantage of women that were incapacitated."[32]

According to a writer for the *Hartford Advocate*: "A cloud of marijuana hung over the central portion of the resort last night, and drug dealers sold their wares openly. Some moved through the crowd crying, 'Acid, mescaline, acid, mescaline'... On a busy road near the main office, hawkers intoned like hot dog salesmen in Shea Stadium, 'Get your White Lightning, Electric Kool-Aid here.'"[33]

The drug situation reached crisis level aided by all the down time, something Hanley wanted to avoid. The sound engineer felt he could have rectified the situation by turning on the sound. Hanley claims there was more drug use at this event than at Woodstock and with no proper public address system in place, how could one get the word out effectively to those who faced so many bad trips? But without electricity or a functioning sound system, it wasn't going work: "At Woodstock we had Chip Monck and John Morris on stage acting as emcees. This was significant as it kept the audience informed and calm. At Powder Ridge there was no coordinated system of communication. If my sound system had been set up we could have helped control or at least reduce the amount of drugs being used."[34]

Rick Slattery claims that Powder Ridge was a total mess, with Hanley left holding the bag: "At some point it became obvious that there was something wrong with the promoters and the fact that the groups weren't showing up. I never really understood what actually went down, I know there were money issues. Very quickly the audience was overdosing and going on bad trips. It turned into a refugee camp, i.e., no food etc."[35]

Alan Magary of the *Middletown Press* reported that Scamporino heard that Hanley provided sound for Melanie's performance on late Friday evening: "Scamporino's main concern yesterday, he said, was in shutting off power to the festival audio system."[36] Scamporino grew impatient, and ordered the arrest of the Zemels and Hanley for violating the injunction.

The owners were ordered by phone to check in with the Middlefield firehouse for arrest, while Hanley eventually turned himself in, and was detained onsite. By 2:00 p.m. Monday afternoon, Hanley and the Zemel brothers made their State Superior Court appearance in front of Judge Palmer. Hanley's hometown paper the *Medford Mercury* reported, "Judge Aaron J. Palmer suspended judgment against William F. Hanley, owner of the Hanley Sound Company, 430 Salem St., Medford, in a hearing which followed the attendance of some 30,000 young people at a rock festival that wasn't held."[37]

When Hanley arrived at the police station he recalls seeing phone-tapping equipment. He claimed they were using the devices to monitor communications at the ski resort: "When they brought me to jail I remember seeing six tape recorders running at 15/16 of an inch per second, just sitting there, they were all running and recording. They had the phones tapped and this is how they heard me trying to bring Melanie in. I was brought into a room and interrogated. They let me go the next morning."[38]

According to Sisson, business started to shift in a negative direction for Hanley after Powder Ridge: "I picked Bill up at the police station. As I remember it, Hanley Sound began to go downhill after Powder Ridge. More and more big outdoor events (the kind Hanley had geared up for) failed due to injunctions. Many of us got laid off as the money problems piled up."[39]

Hanley recalls that it was hard to keep up with so many festivals getting squashed. Thirty out of close to fifty large-scale events slated for 1970 were either canceled or shut down:

The government was attacking festivals and I went to jail over Powder Ridge. I got arrested during all of this and it cost me a couple grand to get out of it. I had to get a lawyer and go to court on Monday morning. I had people there to protect my equipment. They put it back in the truck. We had a crane there and all the rigging to put on a large-scale festival, and I didn't get paid for the gig either.[40]

Toward the end of the weekend, the Woodstock Doctor treated at least fifty bad trips every hour. "In one instance the medical tent had over 150 kids suffering from bad trips. The freak out problem would not have been so severe if so much of the acid hadn't been laced with strychnine or mescaline."[41] By Sunday August 2 people were leaving the festival site, and over 300 hundred were arrested on drug charges. "Dr. Abruzzi later told reporters that the lack of music at Powder Ridge was the primary reason why so many young people resorted to such intense drug related activity."[42]

The promoters did not issue refunds to the festivalgoers who had already bought tickets (via mail) for that weekend. They also failed on a promise to host a makeup festival at another venue, replacing the disaster that had just occurred. "The chief promoter of last year's Powder Ridge Rock Festival, held in Middlefield, Conn., is under indictment over the sale of more than 20,000 tickets at twenty dollars each for a concert that never took place. The money has never been refunded."[43] In 1972 promoter Raymond Filiberti was eventually sentenced to four years in jail. According to the *New York Times*, Filiberti was going to serve time for "Lying to the state Attorney General's office about his role in the proceedings."[44]

Although this was Nelson Niskanen's final show of his career with Hanley Sound, he has fond memories about his time with the company: "Hanley was a genius and I will never forget it. At Powder Ridge I ended up meeting a girl who later became my wife. Some friends of mine were at the festival and introduced me to her. Since the concert never happened we just sat around and got to know each other and the rest is history."[45]

The Powder Ridge Rock Festival was dead before it had life. It was the kids of Powder Ridge that were the victims. They were taken advantage of by acts of greed, put on show and corralled onto a ski slope for three days while the world watched. *Life* magazine appointed Powder Ridge as the "festival that never was" but it still left the town of Middlefield scarred. The fear that people had for festivals, left the future of them in doubt.

Hanley's colleague, rock writer Alfred Aronowitz, eloquently penned his opinion about what happened at Powder Ridge in a 1970 *Rolling Stone* article:

And so Powder Ridge stays with us like an open wound. Of course it was shameful for grown men to be so desperate in their greed that they had to rob children. No matter who you are, you can't put on a show without these people and to put on a festival is so much more complicated than merely putting on a show. To hold a festival is to convene an army and anyone who participates in that convention had better be pretty damn sure of the responsibility of its leaders before lending any effort that's going to bring hundreds of thousands of kids into a circumstance that could end with the kind of confrontations that our professional radicals have been trying to provoke. Or, as a matter of fact, into the circumstance which ended with the utter psychic desolation of Powder Ridge. This was bad enough. And I'm not just talking about the acts and their managers and their agents. I'm also talking about all of these soundmen and lighting technicians and production experts and stagehands who are always there, like mercenaries waiting for just any kind of war, all too anxious to fall behind any lame brain who sounds like he's slick enough to get something started.[46]

By 1972 Hanley had seen his share of injunctions and government problems, it was clear that the rock festival he once knew was changing. The sound engineer had lost thousands of dollars because of post-Woodstock festival fear, yet remained determined to honor the moniker bestowed upon him—the Father of Festival Sound. This is a title he rightly earned, yet it had come with much risk. Calamities like the Powder Ridge Rock Festival that he went out on a limb for were exhausting his resources. Festivals were what he was gearing up for all along. His fierce determination and a deep commitment to quality sound kept him moving forward, and he did. He knew he had built the foundation that large-scale festival sound rested on and always hoped for a better future.

Slattery explains: "I think at some point it was at Powder Ridge where things started to go wrong for Hanley. After Woodstock he was on his way up, but about five years after he was way down."[47] Harold Cohen also saw the writing on the wall during this phase of Hanley Sound. Despite Hanley's request, Cohen knew that sending additional equipment to the ski resort would put the company at a great financial risk, something he felt the business could not withstand. He observes:

I heard what was going on at Power Ridge. I told Bill I did not want to send any additional equipment to the site. The primary reason was that virtually every piece of equipment the company owned would be there. I felt the risk was too high should something happen. I also felt it endangered putting the company out of business, for a period of time or forever. Hanley Sound was not a cash-rich business. Bill was not happy with my decision but he wasn't there to override me. In retrospect I made the right decision.[48]

Writer Alfred Aronowitz always had a special affinity for Hanley's work. He claimed: "Nobody loves a pop festival better than Bill Hanley, even if only because they give him a chance to do what he knows how to do best."[49] Aronowitz had been viewing and writing about Hanley's innovations for some time now, and by 1972 he was making some real connections regarding the significance of his expertise in sound reinforcement. The journalist realized that Hanley was not entirely driven by money but also by providing a service for the betterment of humanity. Aronowitz characterized the sound engineer's struggling business in a 1972 *New York Daily News* article:

He's lost a fortune working for promoters who consistently go into the tank, but he keeps on working for them just to keep on doing what he loves to do. I remember him sleeping under his truck at Woodstock. I remember him hanging out at Powder Ridge for days after it had become an obvious lost cause. He has been gassed and arrested doing the sound at youth demonstrations. He's been all out forced into bankruptcy by the unions.[50]

Chapter 55

A FESTIVAL UNDER THE SUN

THE FIRST INTERNATIONAL PUERTO RICO POP FESTIVAL

The Mar y Sol Pop Festival, also known as the First International Puerto Rico Pop Festival, was held in Manatí, Puerto Rico, on April 1–3, 1972. The site for the event encompassed over 400 acres of countryside near Los Tubos Beach in Vega Baja, about thirty miles west of San Juan on the north central shore of the island. Mar y Sol means "Sea and Sun" in English. Such an exotic location near beautiful sandy beaches provided a wonderfully picturesque setting for a pop festival.

Southern promoter Alex Cooley felt that hosting such an event in Puerto Rico seemed like a good idea. He already had three festivals under his belt. With so many ordinances against festivals on the US mainland, Mar y Sol seemed a no-brainer. According to Cooley, he initially became interested in throwing a festival in the Caribbean through an acquaintance of his that had relations in San Juan. For the event, Cooley partnered with other promoters.

Things began to fall apart for Cooley before they even began. When planning for the event in early 1971, he found that acquiring the appropriate permits was complicated. His contact in San Juan had referred him to a source in the Puerto Rican army for proper permits. The individual, who held the rank of colonel, made promises he did not keep, recalls Cooley: "I guess that if there was any mistake, that was the mistake, that I believed him, because he couldn't produce any of those things. All that he was gonna do, all his job was that he was the liaison between me—the promoter/producer of the festival—and the Puerto Rican government."[1]

Cooley flew to New York to begin work on Mar y Sol and placed ads to market the event on the mainland. Once the news hit Puerto Rico that a festival of this magnitude was coming, newspapers negatively hyped it. Cooley recalls: "Once the government turned against us, everybody seemed to get against us. I mean, before, when I was down there everybody was welcoming me with open arms. Radio stations, the advertising people. Everybody was pro-festival. But once those articles started appearing and once government ministers started saying bad things, all that changed."[2]

A week before the festival, the Puerto Rican government requested that Cooley cease moving ahead with Mar y Sol. Eventually they executed a restraining order against the promoter. Cooley hired a lawyer. The promoter fought all the way to an appeals court and won the case.

Regardless of this positive outcome, Cooley was only getting minimal help from the local government and citizens. "Everybody just turned against us and wouldn't help us. I was in a position where I had all these groups coming and it obligated a signed contract for 'em. But I had a situation down there where I couldn't get a piano or an organ. Just nobody would do business with us."[3] Mar y Sol communications director Bill Dial, recalls Cooley moving ahead with the festival that for all intents and purposes was cursed:

> There was an injunction the week before the festival, trying to stop it. Those loyalists among us fought it like crazy, got a hot lawyer, went to court and won our case so the show could go on. I found out years later that Alex was praying for us to fail. He already knew the financial disaster this thing was headed for, and an injunction to prevent it would have been exactly what he needed. I think he was torn. Which is typical of Alex. He knew he was going to lose, that the injunction would help him, but he was also devoted to the cause; what we were doing was right, if financially unwise. He went ahead anyway, and with courage that inspired all of us to do what, in retrospect, must have been very foolish.[4]

Cooley had leased a 429-acre former dairy farm for the event, so canceling the festival was not an option. He and his and partners now had other things to worry about, like moving the infrastructure of a festival to Puerto Rico. The logistics of getting production materials to the remote site proved to be more of a logistical hassle than he ever could have imagined. Cooley recalls that Hanley had "one heck of a time" getting the sound equipment there: "I think he had to rent big freight containers that would be shipped on ocean barges."[5] It was a costly and complicated execution for Hanley and his crew, who flew in from New York while four semi tractor-trailers sailed off by cargo ship.

Hanley remembers: "The sound and lights came by boat as well. The officials went through all of our equipment and made us empty out all of our stuff, because they thought we had dope."[6] Tom Field was subcontracted by Hanley Sound to provide lighting, and recalls the vagueness in planning for Mar y Sol. Field observes: "We weren't sure if it was even getting off of the ground. A couple of things regarding this festival were right out of a Hollywood movie.

So this thing is going to happen and then it's not going to happen and then it's on again. Bill was getting frustrated with the freight forwarding company and Judi was frantic."[7]

Hanley hired Cambridge native, folk enthusiast, and electronics wizard Walter Lenk to help with the event. Hanley first met him in 1965 at the Philadelphia Folk Festival. By the end of 1969, Lenk began assisting with design and repair projects in Medford. He explains:

> My reputation spread and people would come to me to fix their guitar amp, or design this for them, or fix this or build this or answer questions or whatever the heck it was. And that included eventually sound system stuff. . . . I was working for a couple local sound companies, and Bill was one of them. I was building little things for him . . . so I would ride my motorcycle up and do stuff up there or take it back and do it at my workshop in Cambridge. . . . in 1972 he started asking me to do gigs for him.[8]

By 1972 Lenk was on his way to Puerto Rico representing the sound company. He recalls the difficulty of getting to the remote festival site:

> Bill called me up and asked if I could do a gig in Puerto Rico. I said sure! . . . it was winter at the time in Boston. I was supposed to go to Puerto Rico with Bill but he couldn't fly with me last minute so I flew alone. I was a young kid and I didn't care. Bill loaded sound, stage, and lights on four tractor-trailers and had it brought down by sea on a container ship to Puerto Rico. The biggest part of day one was waiting around San Juan at a fried chicken place, waiting for the trailer to clear customs![9]

The First International Puerto Rico Pop Festival was marketed to college-age spring breakers as more of a vacation experience. This unique event could be considered the first "destination" festival of its time. Attendees bought a package that included tickets and travel to the remote beach location. Ticket prices were steep. Special rates had been set at $149.00 for a round-trip ticket from New York to Puerto Rico. This price point included airfare, transportation, and also camping facilities. This was costly for the average gate-crasher who was used to getting into a festival for free.

In a *New York Times* article, "Puerto Rico Fete Draws Thousands," a festival-goer waiting for his luggage at the airport in Puerto Rico explained the difficulty of affording the event: "My friend over there has literally starved himself to save his unemployment checks so he could come here."[10] Kathy Masterson, who worked for Alex Cooley, recalls some of the madness:

Part of the problem was that Alex was trying to bring people from all over the country through a travel agency to Puerto Rico. He tried to make it like a destination festival. Just like they have destination weddings now, it was meant to be a package. I was supposed to meet the plane and get the people off the plane at the airport, then onto buses and out to the festival site. However, I didn't get out to the festival until the day before. It was all pretty much a logistical nightmare, I think.[11]

The lineup for Mar y Sol rivaled other festivals. With big-name headliners—like Black Sabbath, Alice Cooper, the Allman Brothers Band, the Faces, Emerson Lake and Palmer, the J. Geils Band, Dr. John, Fleetwood Mac, Dave Brubeck, B.B. King, and Bloodrock—Mar y Sol looked great. The roster included singer Billy Joel's first major performance. Although some acts didn't show, most performed. Masterson remembers escorting artists from the Normandie Hotel to the festival site via helicopter: "We were all staying at a very plush hotel with all of these golfers and rock stars and it was really funny. Dr. John and Rod Stewart were walking through the lobby and all of these golfers were staring at them, looking like, who are these people?"[12]

It was Holy Week (Easter) in this part of the world and not a good time to throw a rock festival. Such a thing was viewed as sacrilegious to the locals. Puerto Rican citizens were well aware of the nudity, drugs, and other unholy acts festivals like these brought.

Among festivalgoers and crew there was still a hope that Woodstock could be repeated, and Hanley was a direct connection to that event. According to journalist Aronowitz: "Bill Hanley was probably the last remaining person I know to be a walking monument to the spirit of Woodstock. But the fantasy of recreating that spirit at Mar y Sol was doomed as soon as it was born in promoter Alex Cooley's head."[13]

Promoters were expecting well over 50,000, but only 30,000 showed up. Mar y Sol's attendance had been affected greatly when word hit the mainland that the local Puerto Rican government had enforced an injunction on the event. The *New York Times* reported, "The Mar y Sol or Sea and Sun Festival received legal clearance yesterday when a San Juan Superior Court judge reversed this earlier ruling that the gathering could not be held because of alleged illegal sale of drugs at the site."[14] Now with the injunction being overturned, more people began to arrive.

For Hanley, who was used to working under these circumstances, it was just another festival with a court injunction: "They just don't want this many kids together in the same place. It's Washington, the federal government, that's behind all these injunctions."[15]

Attendees who had flown in on Friday March 31 waited for shuttle buses. But the buses didn't arrive because the bus drivers thought the festival was off. This delayed things, forcing Cooley to switch gears and negotiate with local cab companies to accommodate the arrival of thousands. Any possible Friday element of the show was now a wash. It took over three hours to get from the airport to the festival site.

Despite a slow start, people were eager to hear great music and jammed up the small access roads. "The roads leading to the site were filled with cars and hikers asserting that they were only the vanguard of more than 100,000 rock music fans who would be present for what they have long hoped will be Woodstock II."[16] As they arrived, makeshift shelters began to pop up at the farm near Los Tubos Beach.

The stage area was located in a clearing made by local workers who had removed sugar cane fields and palm tree groves. When festivalgoers came into the area, a commune called "The Family" that Cooley and his team had hired was still constructing the site. The promoter brought in the group to help prepare the land, build the stage, and erect sound and lighting towers.[17]

In order to get electricity to the remote festival site, Hanley had to bring in generators. Cooley recalls some of the sound engineer's equipment being used at Woodstock: "Hanley had a LOT of equipment, so I don't know if every same exact piece was what was at Woodstock. But a lot was . . ."[18]

Mar y Sol was held in a very poor area of the Caribbean island. To many it was an overpowering display of American superfluity visited upon intense poverty, immediately creating a sense of divisiveness among locals. A festival stage located within an enclosed wall of corrugated sheet metal didn't help. Walter Lenk refers to the Mar y Sol stage site as a "walled town" with a "gate in the back," comparable to "medieval England."

On both sides of the stage there was a fairly large array of speakers stacked up on scaffolding. Hanley and his crew mixed from a tower out in front of the audience. At one point Lenk was there mixing for the first time: "I learned how to do it by just doing it. Things were very loose back then. A bunch of fairly good acts were booked. I remember we had a special drummer stage monitor that was supposed to blow their hair off."[19]

A last-minute performer, singer songwriter Joseph Lee Hooker, got his chance at stardom when one of the bands failed to show. Hooker witnessed Hanley's team as they switched out the multi-piece jazz-rock fusion band Mahavishnu Orchestra and prepared him for his short set. Hooker recalls:

I showed up at Mar y Sol with no money and only my guitar. It was a miracle I was able to perform. It was Easter Sunday and I went on after the Mahavishnu Orchestra with John McLaughlin. Hanley's crew went from this high-energy band and switched everything out so efficiently. They came up after to fix my microphone and there was no sound check. I remember everything was done so well and it sounded incredible. Hanley's sound was so good you didn't even know they were there.[20]

Within the festival enclosure, Hanley could be found inside his semi tractor-trailer, equipped with a refrigeration unit to cool his tube amplifiers and most likely himself. Working away, the sound engineer spent his time soldering the detailed wiring on his equipment. Physically exhausted, he had not slept since he left the mainland that week. Aronowitz recalls Hanley falling asleep while working: "He was soldering, his droopy eyelids betraying the fact that, as usual he had been awake and working for a couple of days straight. With his fatigue spaced eyes and wispy beard, he looked like some mad genius inventor, patching together his system even while one of the bands were playing. That's what he is, I guess."[21]

From years of working with Hanley, Cooley noticed the sound engineer's "pronounced" work ethic. "I remember one particular time in Puerto Rico, it was hot as blue blazes. We were having the festival, which was inland a little ways from the water and it was murderously hot. But Bill was working, making sure the sound was good. I remember him at two or three in the morning twiddling with things and making things work better."[22]

Getting water to the site was also a problem because the local government cut off the water supply. By using his generators Hanley figured out how to pump water and set up makeshift showers. This provided a much needed source of drinking water for the overheated audience. According to Hanley: "They ran out of water at the event, so I got the generators going to pump water for the festival people. I recall being chased by some local authorities while I was attempting to do this. It was a scary situation."[23]

After Hanley surreptitiously installed the water pump, many recall seeing the sound engineer running through a field with authorities in pursuit. But water began to flow. According to Masterson, the sound engineer was a problem solver:

I remember Bill got us water when they cut off the supply. There were thousands with no water and he managed to get it going! This was typical Bill. He knew how to build it. He rolled with it and was a solution-oriented guy. Bill worked with us during some really disastrous things and always kept his cool. He found a way through every problem. I remember talking to him afterwards saying "oh my god! we survived this" and he responded saying "I didn't think we were going to!" It was that dicey! The sound was gorgeous, however.[24]

It was increasingly evident that the residents of the area did not like "gringos" invading their homeland. Fueled on alcohol, the machete-wielding locals were tearing down American flags. With limited security, fights began to break out, Santelli relates: "The relationship between locals and festivalgoers deteriorated rapidly."[25]

Nevertheless, the Puerto Rican community of Manatí was not shy about making a few dollars. During the festival some of the local residents set up booths selling food, water, and even marijuana at outrageously inflated prices. Fresh water was being sold at a dollar a glassful and wells were drying up fairly quickly.

Toward the end of Mar y Sol, word was getting out that transportation back to the airport was no longer available. This news made festivalgoers worried and angry. By Monday April 3 hordes of tired, frustrated, and sunburned festival mainlanders broke camp and trekked back to the airport walking. Some naïvely thought they could thumb a ride.

A feeling of not being wanted sunk in. With many vacating the area, production people back at the festival site were left scrambling to leave. In the intense heat Hanley and his team struck the sound equipment and hightailed it out of Manatí. According to Lenk: "At the end of Mar y Sol when we were tearing down, the locals came in and salvaged everything they possibly could and took it away."[26] Reflecting on the event, communications director Bill Dial considers Mar y Sol as a "life changing" experience:

> I can look back at it with some amusement, though at the time we were all in danger of rotting in a Puerto Rican jail. On the third day of the festival I was standing in front of the stage, behind the corrugated steel barricade that separated us from the 30,000 badly sun-burned and sun-stroked kids who had come down from New York and New Jersey to listen to music in this Vega Baja pasture. It was 104 degrees F., it was raining, and there was some terrible local salsa band filling time on stage. Hanley, the expert sound designer, was standing next to me looking at the crowd. He said, "You know, if anybody out there is having a good time, he's just not paying attention."[27]

At one point Hanley noticed that the local police force was looking for Cooley. Near the festival's office located on the side facing the beach, a sign spray-painted with the words ALEX COOLEY GO HOME made it clear that the promoter was public enemy number one. Back stage, things were also getting scary for the crew. Masterson recalls: "The authorities really wanted it shut down, and the word was that the FBI was looking for Alex so we were all kind hiding out. We ended up closing the back stage area because things were getting dicey. I remember we actually threw Annie Leibowitz out and she wasn't too happy about that!"[28]

When Cooley caught wind that there was a warrant out for his arrest, he left the festival site immediately. In order to escape without anyone knowing, the promoter was smuggled out of the area on the night of Monday April 3. Well hidden in the back of a Volkswagen bus, he covered himself in trash. The sympathetic and courageous driver sped onto the San Juan Airport runway toward Cooley's departing plane. The promoter nervously jumped on board, crouched down in his seat, and made it out of the country safely avoiding arrest. "We made it through the checkpoints and I literally was let out of the bus and on onto the tarmac and into the plane. I heard they were looking for me back at the festival site. I was out of there. It was a crazy, crazy adventure!"[29]

By Wednesday April 5, Cooley learned that the Puerto Rican government obtained an arrest order against him for failing to meet with the Treasury Department. He owed over $40,000 to the government in taxes. "Also looking for Cooley and his representatives were about one hundred persons who are trying to collect for services rendered at the festival."[30] Hanley also hit a snag while trying to vacate Puerto Rico. The sound engineer had his tractor-trailers impounded when authorities held them for inspection. They were eventually shipped back to Boston.

At the time of Mar y Sol there had been a strong music fan base on the island. Rock and roll was big everywhere, not just on the mainland, and those who loved the music of the day actually wanted Mar y Sol to happen. Internationally known Puerto Rican photographer Glenn Abbott attended the festival. He was hired by local magazine *Buenos* to cover it. A native, Abbott was at the event eight days, arriving before and leaving well after.

Abbott was able to capture some of the only existing photographs of Mar y Sol. An avid and passionate music lover, he had been a longtime employee of the only record store in Puerto Rico. When he had heard the news about Mar y Sol he was overjoyed that a festival was happening in his native land. However, not everyone was overjoyed. Abbott explains:

> For us this was the Puerto Rican version of Woodstock with palm trees and sun. A lot of the bands that were there were not at Woodstock. The problem was that we are big, big Catholics. Drugs, nudity, and rock and roll meant that the locals were freaking out. The only thing that they had to go on was what happened at Woodstock. Free love did not go over well during holy week. We had nationalists putting up resistance against anything that had to do with "gringo" at that time. This all came from the government, they were afraid of the unknown.

From the religious perspective they thought they were going to come and take our women away. It was like blasphemy and everyone was like, send Alex Cooley home! Up until then we never had anything close to that number of people here and never will. In Puerto Rico we were not prepared to cope with that type of crowd at that time. We got over it quickly and it was peace and love. There was always something that was going to go wrong, we had Altamont and things like that. It was an awesome feeling to have all the bands I came to know all here in Puerto Rico. It was unbelievable to see it happening all before my eyes; and as far as I can remember the sound was awesome! The quality of Bill Hanley's sound system still holds up on all of the bootlegs![31]

The story of Mar y Sol is a good cross-cultural overview on how other people in the world felt about festivals at the time. The not-in-my-backyard attitude infecting many Puerto Ricans was similar to how many people on the mainland felt.

Festivals like this were upsetting communities. Mass gatherings that allowed the counterculture to let its freak flag fly grew smaller and smaller. The target of this contempt was often directed toward promoters like Cooley. Although Cooley wanted to make money, he genuinely loved the music. In the end Mar y Sol lost around $200,000.

Hanley also experienced a major financial loss from Mar y Sol. According to him, it was not only hot, but also "very expensive." In a 1972 article called "The Last Rock Festival," journalist Charles Giulliano suggests that the sound engineer became pissed off for not getting paid, demanding his "fee" from the promoter for his sound service. Giulliano writes that Cooley jokingly respond-ed by claiming he would give Hanley "the deed to his house" in order to cover what he owed him.

Although Giulliano's words are speculative, festivals with "days of no toilets or water" and "exhausted, disillusioned broke, angry kids," fueled rock journal-ists with good commentary: "Basically it is repulsive to think that any promoter would contemplate gathering thousands of young people without any form of security to say nothing for transportation, sanitation and canteen facilities."[32] Regardless if this writer's source is credible or not, it's clear that Cooley and Hanley lost a lot of money.

For the time, it was bands that gave youth their music, promoters who pro-vided forums, and Hanley who plugged in the sound—not always a winning combination, yet the formula continued. Cooley and Hanley had no definitive guidebook to refer to when it came to the construct of a festival. No one, not the producer, festivalgoer, sound engineer, or musician, knew exactly what sort of ride he or she was on when it came to events like these. It was uncharted territory. It was a gamble. If a festival failed financially, so would Hanley.

Aronowitz predicted the fate of the festival circuit after Mar y Sol with the sound engineer in mind: "Hanley's also a very soft touch. Like mountain climbers who want to climb bigger mountains, like airplane designers who want to build bigger airplanes, like astronauts who want to travel greater distances, Hanley wants bigger crowds to challenge him. How could I tell him that festivals were over—through—finished—at least for a few years?"[33]

Chapter 56

HANLEY'S SOUTHERN TRUCK, NOLA, AND THE GURU MAHARAJ JI

THE HANLEY SOUND SOUTHERN TRUCK

By early 1971 Hanley Sound crew member Rick Slattery was burning out. He joined the company in early 1969 and had worked nonstop since then. Worried that they might lose a quality worker, Bernstein and Hanley gave him the keys to a twenty-four-foot white straight truck and sent him to Atlanta. Now Slattery had his own vehicle with his own equipment. He moved to the southern state where he then represented the company. Living close to Kathy Masterson and her husband Ron Worsley, Slattery reported on his duties through them: "Basically I did anything that needed to be done down there. From Florida to St. Louis I did them all."[1] Slattery held this position for about a year, after which he took on a job with the country rock band Goose Creek Symphony.

Hanley's connections with Alex Cooley led him to the famed promoter's longtime booker and production manager Ron Worsley. Worsley had connections in Massachusetts, so he and Hanley developed a friendly working relationship. During the early 1970s they often spent downtime vacationing on Cape Cod.

Worsley was instrumental in assisting and managing Hanley's southern crew and storing the truck at various warehouses. According to southern crew member "Crazy" Jim Wiggins who came from a lighting background and a stint at Capricorn Records. "Ron asked me if I wanted to take over Hanley's truck for one hundred dollars a week. I said yes I would do it! I was up for anything that had to do with music. Inside the truck were Hanley's huge speakers, which were a trademark of his at the time."[2]

When Hanley Sound booked a gig that was not in the range of their Northeast office, they needed a logistically feasible and economic solution. At any given time an individual like Jim Wiggins drove to gigs representing the company

416

all over the Southeast: "I drove for Hanley all through Gainesville, Florida, Miami, Florida, North Carolina, and South Carolina."[3]

For a southern show, Hanley or Boroda flew in to engineer a performance. Wiggins recalls the first time he met Hanley at a gig in Gainesville: "Bill flew in and I remember one of the speakers had blown. He took the cover off of the back of the four huge JBL speakers and he showed me how to check the continuity on them so I could find which was blown. He proceeded to show me how to install a new one and check the polarity so they were in phase with the other speakers."[4]

Most often the southern truck was pre-stocked with the all the mikes, speakers, McIntosh amps, etc., needed for a show. The vehicle was parked and secured outside of Wiggins's residence in Decatur, Georgia. If a call came in he could hit the road with ease. Wiggins recalls: "Ron gave me cash per diem for fuel to get to the gig. I didn't have a credit card so if I had any problems I called Ron and he communicated with Judi Bernstein at the Medford office. They would have it all set for me and I simply followed their instructions."

Wiggins handled many shows and tours while he worked for Hanley. The crew member recounts schlepping road cases, setting up speakers, and wiring everything together under Hanley or Boroda's supervision. He recalls:

> I was so into it and glad to have a job! When I was with Hanley I was scared to death. I had long hair and here comes this guy who is older than the rest of us. He didn't really pay much attention to me but pointed out this and that. I listened and did whatever he told me to do. It was a learning experience and the only boring thing was the driving, even though I usually had someone with me. It was always hotel food—no veggies and all meat and potatoes. I didn't mind; I am a southern boy![5]

During the early to mid-1970s, Hanley's twenty-four-foot straight truck based in Atlanta was often sent to the New Orleans Jazz and Heritage Festival. Sometime in 1972 crew member Phil Tripp took over the wheel. "I typically drove the truck for various events. Each year I left from Atlanta for New Orleans and unloaded our gear at the fair grounds for the New Orleans Jazz and Heritage Festival."[6]

THE NEW ORLEANS JAZZ AND HERITAGE FESTIVAL

The inaugural New Orleans Jazz and Heritage Festival (NOJHF) was held on April 22–26, 1970, in Beauregard (Congo) Square, now known as Louis Armstrong Park. Impresario George Wein (Festival Productions, Inc.) of Newport

Jazz and Folk fame was brought in to produce and design the event. The festival initially showcased a gospel tent, four open stages, and modest sound reinforcement.

Attendance grew beyond anyone's expectations, and by 1972 the festival location was moved to the 145-acre New Orleans Fair Grounds and Racetrack. Built in 1872, it was the third oldest in the country. According to Wein: "The Fair Grounds infield was a better site than Congo Square in terms of the area. In 1972 we had erected five separate stages: Soul, Gospel, Jazz, Country/Cajun, and Blues.... The evening concerts, meanwhile, had assumed a more national profile."[7] The larger space proved more suitable for the size of the crowds.

The following year Wein called upon Hanley's expertise, knowing he would require a better sound system for the expansive area. However, when Hanley and his crew arrived in 1973, the infield of the fairgrounds was unused, uneven, and swampy. Hanley recalls: "It was a horror show, no question about it. It was a racetrack ... the mud, and everything."[8]

Among the many stages Hanley and his crew provided sound for, he also supported evening concerts held at the nearby Municipal Auditorium. Crew member Walter Lenk recalls: "At one point Bill was flying a system in a ... semi-circular raked auditorium ... he was way behind the eight ball and he was just levitating stuff as the audience started coming in ... I was left to mix at the beginning, and his one commandment to me was 'When George [Wein] nears the microphone, make sure it's on.'"[9]

Bob Jones, technical director for the Newport Festival, claims Hanley was an obvious choice for NOJHF: "He could mix acoustic instruments properly and was held in high regard because of his work at Newport. So it made sense that we brought Bill to New Orleans for our Jazz Festival as well as several other jazz festival productions."[10] Wein could have contracted other southern sound companies at the time. However, he knew what Hanley was capable of.

The NOJHF offered audiences a variety of musical and cultural presentations, with Hanley Sound servicing several stages. Production person Laura Loughlin recalls riding a bike to get around the festival site more efficiently: "There was a bunch of small stages and the sound was very important. No one was used to hearing stuff that loud, bands didn't have that type of gear back then either. I remember Hanley used to put trucks in different places to block the sound and change wind direction."[11] Lenk adds, "I remember that Kelly Sullivan had found a little miniature motorcycle, and he was riding around the fairgrounds kind of checking everything out and bringing stuff back and running over beer cans, there were a lot of beer cans on the ground."[12]

Hanley and his team worked arduously in preparation for the annual event. With two or three systems at various stages and locations throughout the festival, Hanley Sound crew member Phil Tripp (and others) recall having their hands full:

> For us it was two weeks of hard work. We had to maintain and run sound systems for two big weekend events with additional shows during the week, which included Municipal Auditorium, the riverboat *President*, and Gospel Tent. The outdoor shows were on the weekend, with the indoor events held during the week. Often we would break down a daytime stage to do a night show during the week. To prep for the weekends, we drove the truck into the fairgrounds and set it up at one of the outer stages which were the main stages #1 and #5. We were popular when we set up in the Gospel Tent because we knew how to do them and keep them rolling through. We would set up the system, tweak it for audio, then pull the schedule together and get the performers in and out. Hanley made everyone work the gospel stage first as he felt if an individual could get a large amount of people on off the stage quickly (and on time) then they could do anything.[13]

Performances also occurred on the riverboat *President* that was located several miles away. The vintage 1924 cruise boat was a floating venue for some of the best jazz acts. It boasted three decks offering two shows a night with the first beginning around 6:30 or 7:00 p.m., after the festival was over. The riverboat cruise provided a unique experience for audiences. It was the essence of New Orleans, floating down the Mississippi River at night, winding around the Big Easy and listening to live jazz through Hanley's sound system. According to Hanley's wife, then girlfriend Rhoda Rosenberg, the riverboat performances were something special:

> One year when Kelly and Bill did the riverboat cruise, instead of just playing up on it these musicians started rolling, marching all over the river and the boat ... These musicians are just going, and the boat's packed, I'm surprised it didn't sink. Kelly was so smart ... all of a sudden he was scrambling around. These musicians are going all over the place, and they're marching all around. So Kelly starts pulling wires left and right, and getting everybody on the sound crew to start following them with mics![14]

Tripp eventually worked his way up from the Gospel Tent to the riverboat. For Hanley and his crew, installing a sound system on a boat was a logistical nightmare. According to Tripp, "It was a horror story for power and wiring."[15] Hanley recalls the difficulty of moving heavy equipment on and off the boat from the narrow pier:

We were doing the steamboat, out on the river ... we were backed up, I think we were having a concert, either, the stuff came off the boat, or something happened? Luckily we had the truck loaded, and then all of a sudden we noticed that there were three wheels on the dock, and one wheel over the river. A front wheel over the river! [The driver] couldn't quite judge his right-hand front wheel from the driver's seat. I almost had a heart attack.[16]

According to Hanley, there would be typical equipment production mishaps: "I was just keeping everyone together and making sure all the equipment got to where it had to be, and was all running right, and that they weren't cross firing into one another."[17] Lenk was in charge of fixing broken equipment. In a 2014 oral history conducted by the New Orleans Jazz and Heritage Foundation, he recalls the order in which jobs and equipment were assigned:

Well, the jobs for the kind of festival, for this kind of thing is, you take the stuff out of the truck, and if there's not a list of what goes where, you then assemble systems, take them to the stage, set them up, get them working, figure out how you're gonna deal with rain protection, figure out what you're gonna do overnight to do that kind of thing, and then when the crowd comes, somebody has to work on the stage to deal with the performers, to set up microphones; somebody else has to operate the system, which means mixing it, turning it up and down; then, and that could be one person. Sometimes it was. And then there needs to be somebody floating so that when something happens—somebody has to go away to go to the hospital—somebody has to, you know who has to take over, or you get somebody to take over, or you figure out what's going on. Or if there's an emergency and the system stops running, then somebody has to go and fix it.

... I was really the only person that had technical capability of fixing stuff at that point, so when we got to the point where we had no spares, and it was very clear that something had to be fixed, I was the person that floated into doing that particular job. I don't pretend that there was any set strategy of who was gonna do what, it was more or less who was capable of doing what and what needed to be done at the particular time ... I spent a lot of time that weekend fixing equipment, in a tote board. I set up a little workshop bench and a tote board, and people would bring equipment into the tote board and I would work on it and fix it. And we're talking about stuff that's honky.[18]

Kathy Masterson, who had worked with Hanley at the Atlanta festivals, recalls moving up the NOJHF production ladder. She claims Hanley's courage to include a woman on a production crew was unique for the time. Masterson reflects:

Bill let me manage the stage for him at the Jazz and Heritage Festival. I was the only woman who had done this up until that point. The roadies were very mean to me. They would say 'No chicks on my stage!' . . . there was a lot of sexism. It was a macho thing back then and I was a little girl weighing around 110 pounds. I got pushed around, shoved off of the stage etc. Bill stood up for me one day and yelled, 'You leave her alone, she is just trying to make a living like you are!' They all got sheep faced, because it was THE Bill Hanley yelling at them! After that they all just left me alone. This was before women ever did stuff like this. Back then guys thought women can't do this job. Then I worked stage #2 for a number of years.[19]

Despite the good work Hanley had been known for, by now it was becoming obvious to some that he might have been spreading himself and his business thin. Broken, uncatalogued, and mismanaged equipment was becoming a problem, and on the minds of his crew. In the spring of 1973 Lenk flew down to Louisiana to meet the staff and assist with the festival. When he arrived he noticed Hanley's equipment in fairly poor shape:

Hanley asked me if I would go and work on the New Orleans festival, I said "Sure," and he said, "I'll meet you at the airport and we'll fly down," and I said, "Fine," and then I got a message from Judi that says, "Bill's not coming, go anyway." So I get off to the airport and flew down and Quint [Davis] picked me up at the airport. I went off and booked myself . . . a room in a motel in the French Quarter and then rented a car. . . . What I remember about working at the festival that year was that the equipment was in pretty ratty shape, and it was a matter of essentially designing each system as we went to the stage, and then my function, as I remember it, turned out to be repairing stuff. 'Cause it would break in the middle of the show, and I'd have to have a spare, and so at some points, as soon as I was fixing something, something else would break, and I was running neck-in-neck with what was going on.

. . . the back of the equipment truck . . . was wildly chaotic. It was not as though it had been packed with any thought, and Bill had this habit of having a couple trailers out back of his lot on Salem Street, and when something would come back from a gig it would stay in the trailer, and then things would get cross loaded between the trailers. Very often in a legit operation, or a normal operation, I'm used to stuff being taken off the vehicles and inventoried and then looked over and checked to make sure it works before going back out on the road. And that clearly wasn't happening here.[20]

Nevertheless, the quality of Hanley's sound almost always prevailed. And although the sound engineer might have been overwhelmed, his conviction for good sound remained. Alan Kaufman, NOJHF stage manager, recalls Hanley's focus during the festivals: "Whenever Hanley came in to do a show, it was a BIG sigh of relief. Everything was like clockwork; the trucks got unloaded, everything set on the stage, sound checks etc. I know the bands really loved working with Bill as well. It was some of the clearest and best sound I had ever heard."[21]

Masterson recalls: "Hanley was there from the beginning and helped make that festival what it was. It was his ability to organize so many stages that made the festival grow, it started out small and now it's one of the biggest festivals in the world."[22] Hanley provided sound reinforcement for the event from 1973–76. After Woodstock his business had dwindled, and by then it was a difficult task for the sound company to accommodate a festival so far from Medford. Hanley observes: "I wanted to get more control of the speakers, which we were working on. But then it was, you know, getting to the festival early enough, and trying to get enough money to make ends meet, because it was very expensive for me to go there and carry all that stuff."[23]

Currently under the leadership of festival director Quint Davis, NOJHF continues to see enormous success. Hanley recalls: "We just weren't invited back. Quint [Davis] said that they wanted to use more local companies, and that was the excuse. I learned a lot about the wonderful music coming out of there."[24]

KELLY SULLIVAN

Hanley Sound crew member Kelly Sullivan was known as a high-energy, driven, bleach blond surfer (looking) dude. Sullivan was considered among the elite of the team and was known to be very close to Hanley. Masterson recalls: "Kelly and Bill brought out the best in each other and could almost complete each other's sentences and thoughts. He was the right person for Bill at that time and filled that spot. They were magnificent together."[25]

Over the years Hanley Sound had seen a fair share of unique characters like Sullivan pass through the shop doors. It was a transient type of business. It employed many individuals with varying degrees of technological expertise throughout its existence. According to Boroda, this was not always ideal: "Anyone that Hanley met on the road that he thought could work, he brought in. I said, 'Who the heck is this guy?' No suitcase, no clothes . . . gypsy! I said to myself, 'What is Hanley doing?' Some even stole from him."[26]

Hanley met Sullivan at the NOJHF, where the young production person was working for Quint Davis. Sullivan's previous job was for Ringling Bros. Barnum & Bailey Circus, taking care of the exotic cats, a position referred to as a pussy pusher. Hanley claims that Sullivan would do anything to get the job done, even if meant pulling a truck into a gig by his teeth. Hanley recalls: "What made him special was his drive. He learned whatever he could from me to make the shows happen. He got it done, and he was good at his work. He was very smart and could do most logistical things."[27]

Sullivan became known for his great personality and problem-solving skills. According to Hanley Sound crew member Fritz Postlethwaite, Kelly had a "fake it until you make it" philosophy:

> Kelly wasn't a technical guy at all. He knew a little bit about everything but not a lot about anything. He was a very outgoing, gregarious backslapper, and he was great at it. He kept that place running through sheer will of a smile. There were times when we didn't have money to get the truck out of hock. We would arrive at a gig without speaker cables because we forgot to load [them]. So we had to figure something out and he would just talk somebody into doing something that they had no business doing, and he was great at it. Kelly really kept that place running.[28]

A memorable event was when Sullivan was pulled over while driving the company's tractor-trailer somewhere in the Bible Belt. Unable to produce a license—because he didn't have one—he told the officer that he was on tour with Dolly Parton, so they let him go. Hanley recalls: "If Kelly was at a show in the North and happened to get pulled over, he would say Bob Hope and shift the story."[29] However, Lenk wasn't as convinced as others were regarding Sullivan's capabilities:

> Kelly was a wild self-promoter who could do anything, according to him. I regarded him as a bag of hot air. And sometimes, he caught something and made it work. But he was really good at talking up a game and sometimes he pulled through on things. And sometimes he didn't. He was a fairly wild guy, and just had bunches of crazy appetites for things. Kelly had little technical understanding of things. Electronically speaking, sound system wise. I think Kelly was pretty good at a lot of mechanical stuff. And Bill was too. You know, Bill was really a very good mechanic. And so I think they connected very much on that level. Yeah . . . he was very vibrant . . . he had this headlong rush through life and would partake of things even when they were wildly contraindicated. But I can only judge him on ways that I interacted with him on, and that was, when it came to technical stuff, I just did not have a lot of credence in his ability to deal with things.[30]

Sullivan left Hanley Sound during the mid-1980s. Hanley heard later on that he passed away from HIV. The sound engineer was devastated, as were those closest to him. Hanley reflects: "I always suspected that Kelly lived another life. But it did not bother me. I didn't care at all. He passed away too early, and he is missed."[31] Memory of Sullivan's personality is still strong among those who worked with him.

FRITZ POSTLETHWAITE AND THE GURU MAHARAJ JI

Frederick Russell "Fritz" Postlethwaite Jr. joined Hanley and his team in 1972 and remained until about 1974. As a youth he was the typical gearhead, taking apart radios and other devices. An early lover of music, he played in various high school bands. At some point while networking backstage at the many concerts he frequented, Postlethwaite was introduced to Bruce DeForest, a colleague and occasional crew member of Hanley Sound. DeForest worked a couple of festivals with Postlethwaite in the Midwest and reported back to Hanley regarding his hard work.

In no time, Postlethwaite received a call from Hanley Sound crew member Kelly Sullivan requesting his assistance with some shows, bluntly asking, "Would you like to come and work for Hanley Sound?" Fresh out of college and wanting to be in the entertainment business, Postlethwaite emphatically obliged. Given specific instructions to be at a certain corner at a certain time, he anxiously waited with suitcase in hand, "So I said, 'Sure, of course!' Pretty soon after that I was doing a show with Bob Hope in Terre Haute, Indiana! I never looked back. I was gone for about a decade after that."[32]

Sullivan handed the keys to the tractor-trailer over to the new kid, expecting him to drive, literally throwing him into the business. Postlethwaite remembers: "After that show I did for Bill, I didn't really know what was going to happen. Kelly said, 'Okay, we are heading back to the shop now, you drive!' I had never driven a truck before! I learned how to drive that rig real quick and off I went back to Medford."[33]

A year later, in early November 1973, Postlethwaite and Hanley were at the Houston Astrodome preparing for a spiritual spectacle called "Millennium 73." The three-day festival was slated for November 8–10 and was produced by the Divine Light Mission (DLM). The DLM organization was founded in 1960 by Guru Shri Hans Ji Maharaj in Northern India. Out of all the events in Hanley's history, nothing matched this short tour with the fifteen-year-old guru named Maharaj Ji.

The DLM saw immense growth in the United States during the early 1970s under the leadership of Ji Maharaj's youngest son Guru Maharaj Ji, known as Prem Rawat. This new religious movement in the West was often viewed as a cult; however, for many it was an alternative religion. The charismatic DLM leadership called the movement a church instead of a religion and people flocked.

At the time, Millennium 73 was billed as the "most significant event in human history." According to the *New York Times*, the Maharaj Ji inspired one of the fastest-growing religious movements ever in this country: "The Guru's loyal followers referred to him as the 'Lord of the Universe and King of Kings.'"[34] Eleanor Blau of the *New York Times* wrote that the young "Guru Maharaj Ji" was known to wear a white Nehru suit, while his hair lay "slicked down." Donning the hint of a prepubescent moustache, the Guru often sat perched on top a gold velvet sofa.[35]

Hanley and his crew had seen their fair share of bizarre performances, but this was just another show. According to Hanley, "The Guru was fifteen years old, but did not look it. The whole thing was odd. People were in awe of him and there was this whole mysticism around him. We were on tour with him for a few dates."[36]

Hanley and his team, including Postlethwaite, were at the Astrodome a week prior to the event, installing a large sound system. The stage, at least five levels, stretched sideline-to-sideline based on the DLM's specifications. "It was around fifty yards deep and it was huge," recalls Postlethwaite, who was in in awe of the enormity of the venue. On the lower level, the stage housed a symphony orchestra. The sofa placed at the top was where the Guru sat.

Postlethwaite explains: "We built sound towers that were gigantic. They were 4x8 scaffolding that went at least five levels high. We were sitting on Astroturf, which had been covered by plywood. The Astroturf was real spongy and made getting those 2,000-pound bass bins up on those things pretty scary."[37]

Hanley claims John Tedesco provided the lighting for the event: "John was from the Fillmore East and was one of Chris Langhart's students."[38] In a 1973 journal called *The Divine Times*, Hanley's sound system is noted as an essential component to this entire spiritual event:

> Crucial to communication is the sound system that enables the festival participants to hear Guru Maharaj Ji speak and the airwaves to carry his verbal message, said Larry Bernstein, designer of the Millennium stage, "You've got sound problems in the Astrodome that are astronomical. There's really not a sound company in the United States that's anxious to do that job." However, a sound company was found—Hanley Sound Company. Bill Hanley, who developed the sound system for Woodstock in 1969, attended by 500,000 people, brought in three 4,000 ton clusters of speakers to enable all people to hear Guru Maharaj Ji and all other parts of the program. He is also working with the Blue Aquarius soundman to achieve the right "mix" of sound for the band of Shri Bhole Ji.[39]

Hanley recalls the acoustical challenges within an arena so large: "They have it wrong. It wasn't quite 4,000 tons! It was more like a couple if I remember correctly. This was my first time in the Astrodome and it was difficult directing the sound because it bounced all over. The building was enormous and there was no digital delay system technology like there is now, it was all analog."[40] According to Postlethwaite, the Houston Astrodome was incredible:

This was quite a show. Talk about being thrown into the fire, I had never been into a facility this size. In fact there probably wasn't one in the world as big as the Astrodome. I remember meeting Annie Leibovitz under the stage. I remember Bill saying to me "go ahead take these men and go do something," he gave me like one hundred stagehands et cetera! It was a real interesting experience and one that tested my abilities![41]

Looking back, Hanley remembers it all being very strange: "The Guru really had some faithful followers! They were all over the place! It was an infectious thing; these followers were all living together! There was a lot going on, and all these people were into mysticism."[42] While electronic fireworks burst on an electronic scoreboard, the 20,000 or so followers chanted in praise of the young Guru. Their voices echoed throughout the giant Houston Astrodome.

Amid all of the visual chaos, a musical group called Blue Aquarius led by the brother of the Guru (clad in a glitter suit) played with big band fury. The *New York Times* reported, "Maharaj Ji, the young Indian guru whose followers say is god, talked about inner peace and accepted a golden swan . . . last night in the finale."[43]

In 1975 the Guru requested Hanley's services again for a festival called Hans Jayanti. The rider contract outlined these specifications: "A reliable sound system with excellent quality is needed for this Festival."[44] According to Hanley, "I guess the Guru was really pleased with my work!"[45]

Back at the Medford firm, Postlethwaite witnessed a business in disarray. He would make multiple attempts to sort out the disorganized warehouse and office. Between some of the local Boston shows that Hanley did a good job of bringing in, Postlethwaite and others worked tirelessly at keeping the weathered equipment together:

It was a mess physically, logically, financially, and managerially. For years it looked like Hanley's transient employees had come and gone and took gear out, beat it to death, bring it back, and throw it in the warehouse. It was a mess! Half of the cabinets had blown drivers in them and the wiring was a mess. Eventually we sorted out all the cabinets and amps. We re-sorted the solid-state equipment and tried to make it a complete reliable touring system.

I understand that years and years of festivals was a chaotic time and things deteriorated. It sort of came to an end by the time I came on board.[46]

With broken equipment, trucks in need of repair, and less business coming in, Postlethwaite saw the writing on the wall. After attempting to clean up years of neglect, the young engineer had some new ideas. In order to try change the direction of Hanley's struggling business, Postlethwaite presented the sound engineer with some solutions. With so many universities in the area, he suggested bringing in some local college MBA student interns for advice. He thought they could give the sound engineer a free analysis of his struggling business. Postlethwaite explains, "Bill didn't want to have anything to do with the idea." None of it resonated with Hanley: "No matter how much Judi or I tried to keep the business going, Bill thwarted whatever we tried to do. It was frustrating."[47]

Realizing the direction the business was headed, Postlethwaite decided to leave. "This was the final reason why I left, because of the disorganization. I banged my head against the wall; as soon as you took one step forward, Bill would find a way to take two steps back. Bill could have sold his business and made his name a brand."[48] Because of Hanley's mentorship, Postlethwaite learned the ropes of the sound business. His year and a half with the company eventually launched him into the music industry.

Only months after leaving the Medford firm, Postlethwaite became a freelance audio engineer at the Bottom Line in New York. After a short stint at the club, he went on an extensive US and Canadian tour with the Bee Gees, while working for Jack Weisberg of Weisberg Sound in New York. Following that, he accompanied the 1975 Rolling Thunder Revue Tour with Bob Dylan and friends, working for Bob See of See Factor Industry Inc. Toward the end of the Bee Gees tour, Postlethwaite was introduced to the rock band KISS. Seeing their potential, he eventually became the band's audio engineer and tour manager during the group's early years.

Postlethwaite is credited on a number of chart-topping albums. He worked with some of the biggest names in the production world including Biff Dawes, Wally Heider, Dave Hewitt (Record Plant) and Eddie Kramer (Electric Lady). He also worked with artists like Bob Seger, Sammy Hagar, Cheap Trick, AC/DC, John Mellencamp, .38 Special, the J. Geils Band, Judas Priest, Rush, Styx, Ted Nugent, and Uriah Heep.

Chapter 57

HANLEY'S INVENTION

THE MAGIC STAGE

Dozens of companies borrowed Hanley's techniques and innovations in sound by the end of 1960s. His standards became commonplace in this newly established industry, although it took years before his influence was fully comprehended. When he saw the work of his competitors, he decided to shift gears slightly, focusing on a more complete production package to accompany his sound reinforcement service.

Hanley had an idea to create a staging system that could be deployed immediately and with minimal effort, decreasing the laborious setup time for an event. And thus Magic Stage Inc. was born. According to Hanley, "It was a unique company with unique services."[1] A separate entity from Hanley Sound Inc., it was an idea that took many years of trial and error to develop.

The origin of the Magic Stage idea began in 1964 when Hanley purchased the portable Showmobile Mobile Stage Canopy, better known as the Wenger Wagon. Its intended use was for parks and mid-sized festivals. Hanley observes, "I felt then that I needed to sell a more complete concert production package for my clients."[2] Although the Wenger Wagon could be modified into different configurations, it had to be set up manually. The sound engineer was frustrated by its limitations and saw a need for a more expedient way to deploy a stage. Hanley recalls, "At the time it was a constant challenge to set up current systems with efficiency."[3]

According to Jerry Wenger of the Wenger Corporation, the Showmobile unit never got as elaborate as Hanley's Magic Stage: "The first Showmobile designs did not have any hydraulics in them. They only had a winch and cables that allowed you to crank it up. It was years later that we added a battery and affixed them with various hydraulic cylinders to lift various parts. It kept changing over the years."[4]

Finding that the Wenger Wagon was too low, in 1967 the sound engineer developed and designed a hydraulic system that could raise the stage an additional four feet. "The Wenger Wagon was efficient but it only got three feet off of the ground,"[5] recalls Hanley. From here the Magic Stage concept began to develop.

After his tenure at Newport, in 1968 Hanley became inspired. He began to design a more detailed and complete version of this staging system idea, one that could efficiently cover full production for large events. At the time he saw a need for other services that included lighting, staging, roof structures, and video. Hanley noticed existing production services were primitive, costly, and time consuming. "By the time I came up with the idea for the Magic Stage I had years of experience with the design and installation of large-scale production systems. I saw a real need for the systemization, simplification, and automation for all of these types of production services. In my opinion the Magic Stage was the first system addressing this need."[6]

When Hanley needed to give his Wenger Wagon additional height for George Wein's Midwest Jazz Festivals, his New York colleague Bruce DeForest was there to assist. According to Hanley: "DeForest was an assistant to rock writer Al Aronowitz and was also the manager of the band the Myddle Class. He was a soft-spoken guy and came to work for us in Medford. Bruce was out on the road with us as well."[7] Hanley remembers DeForest as an important figure within the sound company's history.

According to Postlethwaite, DeForest was a gearhead like Hanley, so it made sense the two had a relationship: "DeForest worked with hydraulics once, and wanted to figure out how they functioned, so then he became an expert on it. He tinkered with stages until he knew how they worked. Bill picked up on this because he was kind of a jack-of-all-trades. I suspect that Bill hired Bruce because he was one of those tinkerers, too."[8]

DeForest helped Hanley raise the Wenger Wagon another eight feet into the air by putting a hydraulic system into it. Hanley recalls: "This really led to the idea of the 'Magic Stage.' Bruce helped significantly with this. When the marketplace changed, it forced me into full production. Up until this time I considered myself a person who would bring high-quality audio to the marketplace, but it needed more."[9]

By 1969 the sound engineer had adapted a suspension system for cross-country towing by tractor-trailer and eventually improved the hydraulic system, allowing him to raise the stage to twelve feet. According to the *Medford Daily Mercury*: "The setting up of a stage for major public events, whether outdoors or in large arenas, takes several days. To cope with these problems Hanley has invented a portable stage apparatus which may be transported over the road and erected immediately at its site of use with the assembly of much fewer parts than normally."[10]

In its day Hanley's Magic Stage was often referred to as a "traveling theater" that could be "opened by one man in an hour."[11] When in full use the sound engineer intended that one individual could load it, haul it, and set it up in less than twenty minutes. Hanley's wife, Rhoda, recalls the beginning stages of development on the unit: "Bill brought his plans to a mechanical and structural engineer to get the specs finalized. The engineer was amazed at the accuracy with which Bill designed this machine, having no formal training!"[12]

Parts of the Magic Stage were constructed out of a modified truck body, with the height of its foldable side panels at about ten feet. The side panels could be extended to around thirty to forty additional feet to match the width of the roof. Hanley notes, "All of the parts of the Magic Stage were built into it. It unfolded mechanically using an intricate system of hydraulics."[13] When opened, the extended part of the stage was supported by the wheels of the truck as well as legs that were mounted on pivots. To engage the unit, the truck was affixed with hydraulic power devices and controls to manipulate the various moving parts, around 120 hydraulic cylinders and 5,000 feet of hose.

Once extended, the stage floor was made up of large plywood boards. The unit was equipped with a roof structure that was in line with the current highway and travel restrictions of forty to fifty feet in length. Customizable for full production, the electrical sound and light systems and other devices could be operated from built-in controls. A writer for the *Mercury* highlights its significance to the music industry:

> Hanley's apparatus may travel over the road to a distant place, such as an outdoor festival, and may be erected as a massive stage within hours or minutes after arrival, instead of days. All important parts and accessories are tied together so that the possibility of loss or damage is reduced considerably. Labor and supervisory personnel are at a minimum; and after the stage has served its purpose, it may be restored very quickly to its original condition and driven away as a truck. Furthermore, a number of similarly constructed trucks may be used together to extend the length of the stage.[14]

The evolution of the Magic Stage design was ongoing and eventually there were two that Hanley designed, customized, and modified. He was constantly adapting the units to accommodate various concert situations, resulting in thirteen patents of his designs.

In 1971 he designed and built a special roof extension to expand the roof size. In 1972 he designed and built a portable fifty-foot stage five feet off the ground at the Cleveland Stadium. According to Hanley in a Magic Stage manual: "This moved on tracks to permit the stage to slide off the field, a distance of 380 feet, in fifteen minutes to allow the start of a Cleveland Browns football game."[15]

Hanley continued to modify his creations, and by 1974 he designed and built what he referred to as the "Electric Super Stage," claiming it was an "all-weather stage" and could be distinguished by a mechanical system that allowed the roof to be assembled on the ground and then raised to "any needed height." This design boasted a 120x45-foot deck with "minimum sight interference" and included a roof area of 60x40 feet: "This stage and roof were capable of being assembled in twelve hours and is capable of being transported on the highway by tractor trailer."[16]

In 1975 the sound engineer commissioned a model design for a new version of the Magic Stage called the "Power Stage." This design was for a humungous device that could be set up in even less time than previous stages. The intended complete size of the unit accommodated a whopping 88x44-foot deck with an 80x50-foot roof. "A special feature of the stage is that it could be set at any level between eight feet and twelve feet above the ground," Hanley explained. The deck could be stored in two forty-four-foot trailers and the roof stored in a third trailer.[17]

In 1976 former crew member Dana Puopolo dropped by the sound company to see what his old boss was up to. Puopolo had left Hanley Sound a few years earlier.

> That night, Hanley showed me a wooden prototype of the magic stage. As you walked in, the model was located in a big open room, the "workroom" as we called it. I thought it was a good idea but I wasn't sure there was a market for it anymore. He sunk a lot of money into building one. I heard it hung out at a hanger at Logan airport for years. I am sure that took a lot of money out of the business.[18]

By the mid- to late 1970s, newer sports arena complexes were being constructed. Arena rock was on the rise. Acts were drawing larger crowds, allowing promoters and artists to make more money than ever before. Hanley observes: "The development of the new sport complexes, where a lot of these events were scheduled, forced me to respond to the houses' demand to create theatrical staging quickly."[19]

Hanley realized that the days of the early stadium setups, which required fifty to sixty men working several days, were no longer effective. "A new concept of construction was needed to allow the sporting arenas the availability of nearly instantaneous transformation, from stadium to theater, and my Magic Stage could solve that problem."[20]

The Magic Stage was fully developed and functioning by 1978–79. Hanley recalls, "The first use of the full-size Magic Stage was at the Lowell University Spring Festival."[21] The Magic Stage saw many shows on the Eastern Seaboard.

During its short time in full operation, many well-known acts performed on its large deck including Aerosmith, Natalie Cole, Willie Nelson, the Grateful Dead, and the Beach Boys.

Those who witnessed the device precariously unfolding claim it was amazing. In 1979 it had been used as a camera platform for Pope John Paul II's arrival in the United States. In November 1980, Hanley was awarded one of the final thirteen patents issued for his invention. All thirteen of his patent designs were granted with no changes. Throughout the later years of his career, Hanley's focus was on the development of his Magic Stage.

JOHN SCHER AND GIANTS STADIUM

Promoters across the country were noticing Hanley's Magic Stage by the late 1970s. New Jersey concert promoter and producer John Scher was one of them. Scher had hired Hanley for sound services in the early 1970s and was impressed by his work. According to Scher it was around this time that the music business was evolving into an industry and becoming a "professional" thing: "Yes, there were shows in the 1950s and 1960s but it wasn't until the 1970s that it started to become a real industry."[22] Scher recalls Hanley Sound as being one of the first professional sound companies he ever worked with:

> I met Bill right around this time. He was a really good guy and a little bit of an insane genius. I suspect he spent a lot of time in the laboratory devising things and looking to improve them. Bill was the first who took the sound business to a professional level. Bill provided something that the acts liked and from the very beginning had the vision of what live concerts could become. He did this in the sound business and later in the outdoor staging area. I used Hanley for one of the shows that I put on in the early part of my career. I hired him for venues that didn't have their own sound system, which were most of them. I used him for college gymnasiums, outdoor summer shows, parks, raceways, etc.[23]

Scher saw a need for the Magic Stage and felt it was a good fit for his anticipated large-scale concerts at the newly constructed Giants Stadium in the Meadowlands Sports Complex in East Rutherford, New Jersey. "Putting a stage up for outdoor shows quickly when sometimes you to had squeeze in between baseball games or other events was difficult and costly. Bill invented this idea to roll in a stage on a truck and be able to set up in an hour or two instead of a day or two."[24]

For the promoter, Giants Stadium was a perfect solution for bands drawing bigger audiences. According to the *New York Times*, Scher was growing tired of older venues where rock concerts were held: "John Scher, the principal promoter of Sunday's concert, had a summer series for five years in the old Roosevelt Stadium in Jersey City, but now he describes that place with accuracy as a horrible, antiquated facility."[25] Scher also held concerts using Hanley at Raceway Park in Englishtown, New Jersey, and at Madison Square Garden. At the time he felt that there were "no regular outdoor facilities for bands that were doing large-scale summer tours."[26]

On a beautiful Sunday morning, June 25, 1978, Hanley and his crew drove his impressive rolling theater into the 76,891-seat arena. The Beach Boys, the Steve Miller Band, Pablo Cruise, and Stanky Brown performed on the deck of the Magic Stage that day. Nearly at capacity, this was the first rock concert ever held at the venue. Drummer for the Beach Boys, Dennis Wilson, was so impressed by Hanley's creation that he wrote the sound engineer a letter the following year,

January 22, 1979

Dear Bill,

Just a note to tell you how impressed I was by your portable stage at the Beach Boys Meadowlands concert last year. I look forward to hearing from you soon to discuss our 1979 tour schedule.

Until then!

Dennis Wilson[27]

On August 6, 1978, Hanley and his crew were back at Giants Stadium for an Aerosmith, Ted Nugent, Journey, and Mahogany Rush concert. Aerosmith manager David Krebs (Leber, Krebs Inc.) claims: "The band found no fault with the Magic Stage and the roadies said it was a sturdy and solid platform." He observed that the Magic Stage filled a "real need in the industry."[28]

On Labor Day weekend, Saturday, September 2, 1978, the Grateful Dead performed at Giants Stadium with the New Riders of the Purple Sage. John Scher recalls that at the time Hanley's Magic Stage was truly "revolutionary." In a note to the sound engineer the promoter wrote: "Everyone within my organization was amazed when the truck rolled in and produced a stage in such a short time. I believe the shows ran superbly because of the Magic Stage. Today it is possible to say that with your automated system you have opened the doors and dates of the ballparks and arenas to a realistic 'day of show' production."[29]

Hanley remembers: "From what I recall, all three of these shows in 1978 were nearly sold out, with a total gross exceeding two million dollars!"[30] Hanley and his Magic Stage were invited back the following year. Scher adds that because of Hanley's design he was able to secure dates for shows in 1979 that had been previously denied because of "tight production schedules."[31]

On July 2, 1978, at the Mardi Gras Festival in Richmond, Virginia, singer Natalie Cole performed on the Magic Stage in front of 40,000 fans. Renowned black promoter Teddy Powell (TP Productions) describes Hanley's stage: "Let me be the first to say the Magic Stage is a winner. I was thrilled to use the machine on the 4th of July weekend in Virginia. It certainly proved to be an important asset at that outdoor concert. The Magic Stage saved me time and money, two critical problems with show business productions."[32]

It was Hanley's vision that he could deploy several Magic Stage units across the country at any one time. The engineer also saw its potential outside of the entertainment market. The promotional manual for the Magic Stage suggests: "The Magic Stage had been marketed within the entertainment industry, since that was the reason for its creation. However, the Magic Stage could serve a variety of functions. As examples, it could be used as a portable emergency shelter, an instantaneous hospital site and as a folding heliport on limited deck space of seagoing vessels. Its full potential market has not yet been exploited."[33]

Because of the intricacy of its design, the Magic Stage often needed refinement. As a result the unit sometimes didn't function properly and was not fully roadworthy until the mid- to late 1970s. According to Phil Tripp, who was with the company till about 1975, the Magic Stage was a brilliant idea but had a lot of problems: "It unfolded but it was not as reliable as it should have been. There were problems with the lighting grids and it came at a fairly high price. Traveling with this ate up a lot of gas as well. With Bill he could not be competitive because you could rent scaffolding locally and build the stage much more cheaply."[34]

Although an innovative concept at the time, the stage eventually became antiquated for large arena settings. Hanley recalls: "The industry became Hollywood. The arenas and stages kept getting bigger and I could only design something so big, I was limited by the tractor-trailer size."[35] Scher explains that Hanley solved a problem by inventing something, but it may have come too late: "It never took off and it seemed like he was tinkering with it forever. Requirements for stages, especially for places like Giants Stadium, were much bigger than his stage could provide. It can only be so big on that truck. I think it lost its initiative so to speak. The industry grew and everything became bigger."[36]

Although some on Hanley's crew saw its potential, they claim that he wasted way too much time and money on the project and should have focused on what he was known for—sound. For the few who stuck it out until those final days of the company's existence, they say the Magic Stage drained Hanley Sound of its revenue. According to Rick Slattery, "Bill is my friend, but lots of people told him that stage was a bad idea when it started."[37]

Woodstock colleague Stan Goldstein saw how focused Hanley was on the unit. This was at a time when he claims the music industry was "moving at lightning speed." Although Goldstein felt the concept was a very advanced idea, he saw his friend spending a large amount of money, time, and energy on the project. "Guess what, there was no longer a need for it, and no desire for it, mainly because the festival business as it was, evaporated. So with it, also went Bill Hanley."[38]

Hanley was still refining his design in 1980. In a *Boston Globe* article, the hopeful sound engineer reflected: "Right now our focus is on industry automation, primarily for staging and audio. The industry is tired of wasting time and hours required to set up these things."[39] But by then, need for a device like the Magic Stage had all but disappeared.

The units now lay exposed in the sound engineer's yard, rusting away. An engineer's inventions are like children, and for Hanley the Magic Stage was one of them: "Nobody has ever made a machine like this before and no one has since."[40]

Chapter 58

SOUND OF SILENCE

THE COMPETITION IS CLOSING IN

The 1973 Watkins Glen Summer Jam Festival in New York State was one the largest to occur after Woodstock. Despite the Mass Gathering Act enacted a couple of years earlier, over 600,000 showed up on July 28 at the Grand Prix Raceway outside Watkins Glen, New York, to hear the Grateful Dead, the Band, and the Allman Brothers. Hanley recalls: "Governor Rockefeller had been behind this law against big festival gatherings. There were health department regulations after Woodstock. Because of his Mass Gathering Act promoters had to deal with their state's health department."[1]

Watkins Glen reputedly had the largest festival audience ever, but Hanley did not provide sound reinforcement for the event. It was the sound engineer's old Fillmore East colleague Bill Graham behind the complete production. Graham's FM Productions provided the lighting, stage, and the more than 50,000-watt sound system for the memorable gathering.

An event of this size had a real impact on Hanley's business since only few years earlier he had been on a short list of choices for sound reinforcement. In his book, *Aquarius Rising: The Rock Festival Years*, Robert Santelli comments on Graham's FM productions and its state-of-the-art sound system:

> Graham's FM Productions had been contracted to employ the Digital Audio Delay Line system, a computerized sound system designed so that people sitting up front and near the towers of speakers would not be blasted into the universe. It also enabled people sitting way in the back to hear the music just as clearly as those closer to the stage. With such a system, sets of speakers are set up a hundred yards apart. The first set of speakers receives the sound from the stage and relays it back to the second set. This set rebounds the sound to the third set. All this occurs with split-second precision. It is not discernible

to the human ear that there is a micro lapse in the sound. For both the sound check and the actual concert, the system worked like a charm.[2]

It was clear to those at Hanley Sound that by now sound systems were becoming more accessible and easier to deploy. Hanley's hard work had opened the eyes of many concert promoters and producers, who now felt they could put on a complete production. Watkins Glen also demonstrated a level of professionalism that did not exist at previous festivals. According to the *New York Times*, promoters Jim Koplik and Shelly Finkel were proud of their accomplishment: "The experienced promoters also take pride in believing they have planned a festival that will incorporate all the best features of the large rock gatherings in the past and eliminate the faults."[3]

Hanley Sound remained somewhat relevant until about 1975. By then many other firms had surfaced. A fresh, new, and better live sound industry market of companies sprouted all over the country. Some of these second-generation outfits included the Clair Brothers, Tycobrahe, Showco, Sound Image, Dawson Sound, Maryland Sound, Community Audio, and Meyer Sound. Most, if not all, developed new ideas in packaging, touring, and product development.

There was no longer a need to own or take tractor-trailers to every show, especially for smaller, local events. These were costly assets that could have been sold or redistributed within the company for other resources. The Clair Brothers, one of the largest and most lucrative sound companies to emerge from this period of sound reinforcement, platformed off of Hanley's successes and failures. President Roy Clair reflects on how he and his brother Gene learned from Hanley's mistakes:

> I think Hanley got headstrong. I think he thought he was bigger than the groups he was working for. Bill doing the festivals, and the peace marches where he wasn't getting paid because they fell apart, didn't help his business I am sure. I saw Hanley not getting paid. We did a festival once and didn't get paid. After that, I always demanded to be paid up front for our work. You need to pay your men, pay for the truck, and everything else, and the next thing you know you start to work out of a hole. Bill Hanley had a big speaker system and a big truck. Clair had a small speaker system that made a lot of noise. Our cabinets put out the same amount of sound but you could put them in a van or a smaller truck. Bill had to use a semi and had to charge accordingly. We made our systems smaller by virtue of putting the same amount of speakers in a smaller cabinet. So if you wanted fast cheap and loud, some people hired us. When I saw what was going on in the business at that time, I thought to myself there is plenty of room.[4]

Because of increased competition, promoters and bands now had cheaper and varied options. The application of live sound reinforcement had come a long way from the days of Bill's childhood basement in Medford. This new industry that Hanley helped create was now growing into a multimillion-dollar empire. Showco founder Rusty Brutsché recalls the tipping point:

Bill couldn't get beyond the inventor stage. There is not any systemization or packaging. Everything is individual pieces of gear all stacked up and wired together. In order to make a business out of it you had to make it where you did a lot of shows and have a lot of systems and a lot of people. You can only mix one gig at a time. You have to be able to train people on more of a production line. You have to think about how to make it all work and hold it together. It took us years to figure out how to package this stuff so it wouldn't just fall apart. I went through horror story after horror story of just opening up the truck and seeing pieces inside the truck broken apart by vibrations from bumps in the road.[5]

Hanley's crew members and colleagues reflect on why the company faded like it did. Some say that with the right decision making, Hanley Sound could have transitioned more smoothly after Woodstock. In most instances these individuals agree that Hanley was best at being a sound engineer and not a manager.

SAM BORODA: Bill was very welcome to new ideas. But the big failure was that he was taking on more than he could chew and the equipment was in bad shape. So when other companies like McCune Sound and the Clair Brothers and Stan Miller with Stanal Sound came along with no technical problems, promoters liked that. The equipment was at the limit and was not well maintained. This scared the crap out of promoters because he was fixing equipment right up until show time. It was like flying a plane and fixing it while it's in the air. The equipment was unreliable and I was sent on the road with a lot of these problems. However you have to give Bill credit. He was the one who came up with the ideas when no one else did.[6]

RICK SLATTERY: It was heartbreaking to see Bill meteor up after Woodstock and then succumb to poor managerial skills, and incompetent workers, which killed him. People would steal from him. One guy was busted in Buffalo for drugs with a sound system I needed for a gig in Indiana. The wheel bearing on a truck seized up and made us an hour late for a show and on and on it went. Not being able to do jobs because of broken equipment, people getting busted etc. It doesn't take long for clients to figure out that there were other alternatives and by this time there were competitors.[7]

STAN GOLDSTEIN: Bill was a gearhead. He was not the clever businessman. Bill's great strength was his vision and his dedication in fulfilling that vision— not of a business empire, but of providing this ephemeral high-quality sound. This ideal was not meant to be accompanied by a company that wanted to make a profit. There was also the fact that Bill prepared himself for an industry that collapsed around him. The festival business died. So Bill had made an enormous investment in preparing to supply the equipment to an industry that didn't need what he had prepared to do. Bill had purchased tractors and trailers ready to roll with all of these giant systems that were no longer being hired. Bill had also conceived of systems to make festivals themselves more practical. He conceived of a portable staging system for which he would be the vendor. He poured in a tremendous amount of money, time, and energy into the Magic Stage, which was a very advanced idea for an instant stage concept.[8]

CHIP MONCK: Bill always bought for the future. What he did was "AH! Here is an inexpensive trailer! I'll buy it!" Bill was thinking everything was going to continue to get better and always going to improve, that anybody and everybody is going to want his services. The point is, when you have all these trailers, there is a considerable cost involved in that. Beyond that, what happens when the work starts to slack off? What happens when the McCunes get a little better or the Clair Brothers get a little better? When their marketing is a little more astute? When it isn't as simple as just showing up with a soldering iron? Bill got caught planning for the future, building for the future, and not having the invitation or the workload he expected.[9]

CHRIS LANGHART: Bill was not a personnel manager. In the beginning it didn't seem that there was anyone else but Hanley because it was called Hanley Sound. There was no money to do anything with to grow and the ability to keep track of it. Bill was very close to the chest regarding money and the number of bodies. It did occur that someone drove off of his lot in Boston with a Kenworth trailer filled with his equipment and he never saw it again. It was definitely harum-scarum in the money department. There was no one to keep track of all the money because it wasn't important to him, it was sound that was important to him.[10]

DANA PUOPOLO

In 1974, having just graduated from the University of Massachusetts with a degree in engineering, Dana Puopolo was looking for a job. After a friend found a Hanley Sound ad in the local white pages, Puopolo became interested, walked into Hanley's shop, and got hired. "There was no sign on the door so I walked in and there was Judi at the front desk, so I asked for a job."[11]

Hanley assessed prospective employees' skills by giving them different tasks in electronics. The sound engineer asked Puopolo to build a mike cable as an employment test. Puopolo recalls: "I soldered it and it was acceptable. So he said okay you can work full time in the summer and we will pay you one hundred dollars a week. That sounded good to me! This is how I started working for Bill. Right off the street, just my W2 forms, no application or anything."[12]

When Puopolo arrived, he saw a fairly disorganized space. As he walked through the front door of the sound company he saw Hanley's office off to the left. The windows looked onto Salem Street; the curtains were drawn. While surveying the area, he was impressed by the sound engineer's messy desk, which was piled high with various trade magazines and journals. Hanley's equipment was either stored in a warehouse across town, or in one of the large broken-down trucks out back. Puopolo explains: "One of the eighteen wheelers had two wheels that didn't work. So they elevated them off the ground and secured them with chains so the back of the truck had four wheels on one side and two wheels on the other side. He also had an overloaded Buick station wagon that they used to haul stuff in. I heard the wheel bearings burned out because of the weight from hauling winches."[13]

On June 28, 1974, Puopolo and another crew member were ordered to bring the Jolly Green Giant, a twenty-foot International straight truck with no parking brake, to Carnegie Hall for a Newport Jazz event: "I used this truck but the parking brake didn't work. I had to use chucks underneath the wheels in case it rolled. When I brought it to Carnegie Hall I had to jam the truck against a light pole so it wouldn't roll. Bill used to say "the equipment didn't have to be perfect, just close enough for rock and roll." This is what we had to live with then."[14]

Most often Puopolo repaired blown amplifiers and other equipment for Hanley. He also helped load one of the semi tractor-trailers driven by Kelly Sullivan, who handled most of larger national events. According to Puopolo: "Kelly hardly worked at the office. He would come in and ask me to fix this and fix that. We loaded him up and he would be gone for two weeks."[15]

As Puopolo worked he began to catch wind of the company's legacy. "Everybody knew that Bill had done Woodstock but no one really talked about it. We were busy trying to keep the business from going under. We were concerned with having enough gas to get to the next gig."[16]

For those who held on, they saw the once bustling sound company getting smaller. More local projects than larger national ones were coming in. However, an occasional larger job emerged sometimes from George Wein, who would throw Hanley work. According to Puopolo, Hanley Sound was really shaping up to be a hand-to-mouth business then. He recalls that Hanley's mother Mary Hanley, the company's longtime bookkeeper, was in charge of writing checks: "The days of the huge festival had fallen, when Hanley got a big check, paid off all of the employees, paid off all the bills, and by then it was back to the way it was. When we would do a bunch of small stuff and get paid, it was time to pay all the bills again. There wasn't a lot of reserve. You could tell his mother was worried."[17]

By 1974 Hanley could no longer live off his Woodstock reputation. Creditors didn't care. They only wanted their bills paid in full and on time. Most who worked for Hanley Sound were making only a fraction of their worth and they all knew it. Puopolo remembers Hanley could not keep up with the competition:

Bill didn't have the capitalization and didn't have modern equipment compared other emerging sound companies. He was still using Crown DC300s and old McIntosh tube amplifiers, at the time it was not state-of-the-art, the stuff was old. The competitors were using more efficient equipment that played louder and cleaner. They saw what Hanley was doing and did it better. They were more organized and ran it like a business. Bill didn't have a business plan.[18]

In April 1975 Puopolo felt it was time to move on, ultimately finding his way into commercial radio and broadcasting. According to him, those who held on to their positions at Hanley Sound didn't do so for the money. They did it because Hanley inspired loyalty in people. "People would work for him just for the sake of working there and helping out. Everyone liked him and wanted to help out and keep him successful. Bill lived for that company. I think if he didn't have it he would have withered up and died. Bill inspired us all."[19]

Puopolo stayed local after leaving Hanley Sound. One evening in early 1977 he was driving by and paid his old boss a friendly visit. When he knocked on the familiar door, Hanley happily greeted him and invited him in. After a closer look, Puopolo realized that the sound engineer had a strange wire running from his shop over to the insurance company directly across the street. "By now he was in dire straits and they shut off his phone because he hadn't paid the bill. They wouldn't install a line in his office until he paid up, so his insurance broker installed a line in their office for him. He ran a wire across the street from the roof of his insurance agent, then over to the roof of his place, then down into his office. You could see the wire clearly, but he was pretty proud of it. I said, Bill, you can't run wires over a public street!"[20]

MONEY SOON COME

It was not only the festival market drying up and changing direction that suppressed Hanley's once bustling business; it was his lack of business sense. Too many times of not getting paid and fronting huge sums of money to ship equipment for canceled festivals was finally taking its toll. In October 1975 Hanley was hired for a concert in the Caribbean island nation of Jamaica. He was asked to provide sound reinforcement for a large event called the Dream Concert at Kingston's National Stadium, featuring Bob Marley and Stevie Wonder. A Jamaican newspaper claimed: "A symbolic meeting of America's soul and Jamaican reggae is the description of . . . the meeting of America's Stevie Wonder and Jamaica's Bob Marley on the October 4th Dream Concert."[21]

The event evolved into a political tug-of-war between the promoter and the Jamaican government. It became costly for all involved, including Hanley, who recalls: "Those who sponsored the event ended up in somewhat of a financial disaster."[22] Bernstein was adamantly against sending the equipment that far. "I begged Bill not to send the equipment until he had a substantial deposit. He wouldn't listen to me."[23] Hanley wanted to broaden the expanse of his business, so he made the executive decision to travel the distance along with four tractor-trailers full of equipment, stored freight style, on cargo ships.

This event was another full Hanley production. The local Jamaican papers touted the stage to be "seven feet high, 50 feet wide and 60 feet long. Only thirty five percent of the bleachers at the stadium will be used according to the concert facilitators. The sound is from the United States."[24] According to Hanley: "I freighted all of my equipment down there. My equipment stayed for three days and no one showed up! There was a lot of in-fighting which stalled production. I did get paid something but I don't know what."[25] Back in Medford, Bernstein called Hanley every day to see if he received payment from the promoters:

> Finally I got him on the phone and asked if he got any money, and Bill says, "soon come." That's what they told me—"soon come." He said, down here, if you are not talking about ganja, the answer to everything is "soon come"! If you order food and you call two hours later because it hasn't arrived they tell you—"soon come." I don't remember if the concert happened or if the money "soon came."[26]

THE FINAL YEARS OF HANLEY SOUND

Frustrated and disappointed, between 1975 and 1977 Sam Boroda, Judi Bernstein, and Harold Cohen left the company. According to Cohen, Hanley Sound could have been saved, recalling during the early to mid-1970s Hanley had refused an attractive business investment from Tom Driscoll. Driscoll had his hand in concert promotion in the late 1960s including the Second Miami Pop Festival, where he saw the potential of what Hanley was doing. Impressed, on June 13, 1969, Driscoll hired Hanley for a series of weekend summer rock concerts called the Magic Circus at the Hollywood Palladium in Los Angeles. The promoter was heir to Driscoll's, one of the leading suppliers of fresh berries in the world.

As Hanley's lack of business sense suffocated his company, the berry tycoon stepped in and made the sound engineer an offer of around $10,000 for a 51 percent stake in the company. According to Cohen, many within the company were disappointed with Hanley for not jumping at the opportunity:

> If Bill had gone along with what Tom wanted to do I wouldn't be working right now. There was an agreement written up and even though he would have 51 percent control of the company he was not interested in running Hanley Sound. He was interested in making it financially stable and expanded. There was talk about a hotel, and a manufacturing facility. There was also talk about having multiple locations similar to music stores. We already had a relationship with all of the equipment and product lines so it seemed like a logical step to go. The agreement said for the first three years Bill would be on a salary and after that if he didn't like what was going on he could leave and go into business himself again in direct competition. Bill always professed that he wanted to put an umbrella over everyone's head and this was the opportunity. When push came to shove he didn't do it, he was greedy. Bill didn't have any background in business. There was no one around that really knew business, including Bill's family.[27]

Bernstein and Cohen's plea with Hanley to take the offer and sign the agreement fell on deaf ears. According to Bernstein: "I couldn't convince him. None of us could. Bill wanted to be the boss and wanted to own the whole thing all the time. He didn't want to give up control of the company. If Bill agreed, it would have put a roof over everyone's head and given us the money we needed to survive. It would have given him the freedom to do what he did best."[28] According to Cohen, five years later Hanley confessed that he regretted the decision to not partner with the businessman.

For those who remained, the impending outcome of this once successful sound company was evident. For the many who dedicated themselves to Hanley's vision including Judi Bernstein, they all felt that Hanley Sound should have been more successful than it was. Bernstein recalls, "Everybody who was working for Bill felt like we were promised the world."[29]

By the end of the decade Hanley's company was in a slow and steady down-turn. Even though some jobs trickled in, the company's golden years had passed. When Bernstein finally left, Hanley took it hard. For almost twelve years he had relied on his female compatriot, friend, and loyal colleague:

> We were so tied to Bill and his family it made it difficult to leave, but I was angry. Bill couldn't afford to pay us any longer and I felt like I was left high and dry. Suddenly there was no business and all of these other companies came into the picture. It was a real fight for business. By this time Bill was off in his head with the Magic Stage and this wasn't conducive to business. I didn't have any choice but to leave.[30]

In 1977 Kelly Sullivan and his girlfriend Laura Loughlin stepped in after Bernstein departed. Hanley had met them at the New Orleans Jazz and Heritage Festival a few years prior. They were instrumental in keeping the business alive until its near demise. Loughlin moved to New England to be closer to Sullivan, who was already residing in the area and acquired a job with the struggling business.

Many recall Sullivan and Loughlin as the reason Hanley Sound kept afloat during its last years of existence. Hanley remembers: "Kelly and Laura helped us keep it together after Woodstock and after the festival marketplace fell apart. They were unbelievable."[31]

Loughlin was very organized and that is what Hanley Sound needed. According to her, when she arrived in her packed Mustang from Charlottesville, Virginia, the office and business were in complete disarray: "My role at Hanley Sound was receptionist, answering the phones, keeping a calendar for all the gigs, maintaining receipts and general office stuff. When I got there the place was a shit show! It was dirty, dark, depressing, and had no electricity. There was no leadership, no marketing, and no branding."[32]

Although the company was still in business by the late 1970s, Hanley's crew was smaller, matched by an even smaller cash flow. One of the last of the sound company's memorable gigs was on October 1, 1979, when Pope John Paul II arrived in Boston. Loughlin remembers: "This was a high profile event for us. I remember being so close to the Pope; it was an incredible experience."[33] According to Hanley, coming from an Irish Catholic family, this event made his mother and father very proud. Loughlin and Sullivan remained at Hanley Sound until the early eighties. She observes: "Kelly was a genius and loved Bill, but Bill was no front man, so Kelly became the front and back office. However, Hanley Sound could not keep up with the times. A lot of the gear was not working and the trucks were broken down. Hanley would have been a household name if he had got patents on some of his ideas. He was a genius of a pioneer."[34]

By 1984 Hanley finally realized that he needed to close the doors of Hanley Sound. He explains his thoughts on his once bustling sound company in a *Medford Citizen* article: "There are a lot more people doing the type of work I do these days—there is much more competition. Couple this with the fact that there are not as many large scale concerts being produced these days, and you can see where the market has tightened considerably."[35]

Looking out his office window, across the busy street, Hanley could see the two-bay garage where his business began in 1957. Thoughts of the cold winter nights he spent working with Terry, old friend Phil Evans, brothers Tom and Dick Hawko, and Phil and Ed Robinson ran through his mind. It was 1987. Times had changed. Hanley pulled the shade, and turned off the lights for the last time at 430 Salem Street. He sold the property shortly thereafter.

A coffeeshop now operates in the sound company's former location. The space and overall structure are still in existence today with most of the exterior remaining the same. With help from individual donations and the City of Medford, I was able to have an historical marker erected in 2013 celebrating the sound company's impact. To this day Hanley wishes he had never sold the building.

Most people who work in the contemporary sound reinforcement or concert production business acknowledge the innovations of this avant-garde company and its ingenious leader. Many claim that Hanley cared more for the people his sound system served than pursuing monetary rewards. If this is the case, the outcome of this story makes perfect sense. Sound engineer Stan Miller says that what happened to Hanley can happen very easily:

> Bill early on had this inquisitive nature. I just don't think he allowed it to go beyond where he was at that point. He didn't turn the page and go on to the next thing. Bill also did not latch on to a specific client that he could have been involved with for a long time. I had a lot of them for many years. This forces you to turn the page and move on. What happened to Bill happens to a lot of people who are entrepreneurs.[36]

Ken Lopez, chair of the Music Industry Program at The Thornton School of Music at the University of Southern California, feels that if Hanley had been focused on money, he possibly might not have accomplished all that he did:

> Hanley was a dreamer and idealist. So much of his energy was diverted into the politics of the time. I think this is part of the reason that he didn't become a business empire. I think you have to give Bill credit for putting his money where his mouth was. While he was aiding social causes, other people were building their empires. I think it took a combination of entrepreneurial

risk-taking and a salesmanship type of person to pull it off back then. Bill isn't built like that, that's not Bill.[37]

Bill Hanley's legacy is complex and significant. He dedicated his life to audiences so they could enjoy clarity and audibility. Because of this, to this day his impact continues to be felt. Yet he was more than just a sound engineer. He had a deep, selfless, and earnest concern for each audience member sitting in the way back, seats we have all occupied at one time or another. But no matter where you were during one of Hanley's performances, the last seat in the house was likely the best seat in the house.

Hanley and his wife Rhoda enjoy a busy life on the north shore of Boston. His son, William J. Hanley, is a second-generation concert production engineer, carrying the family name forward. At eighty-two years of age, Bill Hanley's convictions about sound have not changed. Every Tuesday evening you can catch the sound engineer adjusting levels for world-class jazz musicians at a cabaret-style jazz show called Jockos in Methuen, Massachusetts. Every Friday night he attends swing dancing in Cambridge at Boston Swing Central. At both locations word eventually caught on that they are in the presence of someone truly special.

EPILOGUE

Although no longer in the sound business, the Hanley Sound brand will forever be remembered in the multi-million-dollar sound reinforcement industry. His innovations are as relevant now as they ever were during the company's existence. For this reason Hanley's story is one of success. This is impressive considering that Hanley Sound closed its doors over thirty years ago. His position as the "Father of Festival Sound" within the history of sound reinforcement commands no price tag.

During the time he was in business, Hanley accomplished much and influenced many. For decades, he led those around him into new and uncharted territory based on his innovations and ideas. It is my observation that he would not have been able carry out these amazing feats without the tireless efforts of his crew and family. With this in mind, it is important in closing to review the technological achievements that the Hanley Sound Company produced in its twenty-year life span.

Hanley's innovations that shaped the sound industry have now become standards of best practice. As we have learned, he created systems of sound based on existing recording, cinema, and high-fidelity technologies. By doing so he disrupted the commonplace use of public address systems for projecting music and introduced new ways of hearing. These innovative developments were implemented in the following manner beginning in the late 1950s into the early 1970s:

- The introduction of large Voice of the Theater cinema speakers that were often raised into the air.
- The use of construction-grade scaffolding for speaker tower deployment.
- The introduction and use of ganged McIntosh MC 3500 vacuum tube and transistorized solid state Crown DC 300 power amplifiers in concert settings.
- Front of house (FOH) sound engineer and sound console positioning.
- The construction, use, and placement of footlight stage monitor systems.

- Facilitation of multimiking individual instruments (Shure Unidyne), most evident in concert and orchestral settings.
- The introduction of the Shure SM57 microphone to the Lyndon B. Johnson inauguration in 1965 (still in use for inaugurations today).
- The company's modifications of the Shure SM57 microphone resulting in the microphone company's development of the SM58 most commonly used in rock and roll now.
- The Hanley brothers custom built Belden "snakes" developed out of a need to organize, simplify, and systemize complicated miking from a long distance.
- The use of the CM winch to hoist large speakers for line array positioning. These are now referred to as "concert hoists." The current "CM Lodestar" brand is a derivative of Hanley's 1969 modification and has been widely used in the concert industry ever since.
- The creation of the "Hanley Hula Hoop," a large circular speaker array used at Madison Square Garden.
- The deployment of large sound systems with bi-amplification, and passive crossover.
- Early elements of the systemization and packaging of touring sound systems.
- Early touring deployment of air ride, refrigerated tractor-trailers with lift gate mechanisms.
- The design and construction of security wall systems for large festivals.
- The use of closed-circuit television and large-screen video displays at the Fillmore East, Newport Jazz (NY), and Philadelphia Folk Festivals.
- The automated hydraulic stage design called the "Magic Stage."

Hanley's *modus operandi* was more about delivering quality sound to the masses and less about the monetary rewards his innovations probably deserved. But money did not drive him, and he led by innovation, solely for the passion of doing it. In the course of my research, Hanley has expressed that he should have become a nonprofit 501(c). As far as I can tell, at the height of the company's success, it did make a substantial profit, although the exact figures are not available. However, some claim that as quickly as Hanley made money, he spent it on new equipment.

Often, innovators are not adept at balancing books or even networking. What if Hanley had partnered with someone who was more knowledgeable at running a successful business? A relationship like that might have allowed him to focus on what he did best. We will never know where Hanley Sound could have gone if this had been the case.

In August 2019 Hanley will be called to Woodstock once again. Only this time his speakers and other artifacts will grace the halls of the Bethel Woods Center for the Arts in a special exhibition celebrating the festival's fiftieth anniversary. If only Richie Havens and Bill Hanley could meet one last time, I am sure the folk singer would raise his cup and salute the man that gave him a voice on that remarkable opening day.

As we celebrate this important milestone, Hanley should be acknowledged as one of the great technical innovators of sound reinforcement in the twentieth century. His influence on the craft has formed how we listen to live music. Because of this, Bill Hanley's impact will continue to shape popular music culture in the future.

APPENDIX: SYMPOWOWSIUM ATTENDEES

According to Lisa Law, those who were in attendance are the following: Ken Elton Kesey, Peter Yarrow, Michael Butler, William F. Hanley Jr., Bobby Steinbrecher, Paul A. Rothchild, Lewis J. Weinstock, Ken Babbs, Michael Vosse, John B. Sebastian, Rock Scully, Star High-Roamer, Tom Law, Steve Samuels, Cyrus Renais Faryar, Lisa Law, Carl A. Gottlieb, Paul Krassner, Michael Lang, Bonnie Jean (Froggie) Romney, Tom Watson, Bill Nordhoff, Carlos Lanigam, Edward (Butch) Arthur Sweeney, Patrick Sullivan, Ruffin Cooper Jr., Mac Pate, Cahan Laughlin (Travis T. Hipp), Tom Barry, Anne (Spunky) Liggett, Bobbie (Flash) Miller, Jere (Wacco) Brian, Jill Stensland, Michael Alan Carl, John King, Mona Vakil, Danu Smith, Christopher Cowing, Mary Esposito, Elizabeth Ryan, Bruce M. Rappapott, Bruce Gilbert, Stan Dewey, Kenneth Lee Karpe, Eddie Heath, Richard Holmes, David Ehrlich, Don Barshay, Victor Maimudes, Maggie Denver and kids, Ralph (Boots) Henry Santini, Ed Phelan, Jerold Schultz, Jerry Hopkins (*Rolling Stone*), Ralph Burris, Milan Melvin and wife, Joyce Mitchell, Rari Reim, Jean Nichols, and Eloy Hernandez and family.

NOTES

PROLOGUE

1. Tom Fox, "The First 15 Minutes." *Philadelphia Daily News*, September 19, 1970.
2. "Hijacker Terrorized Boston Stewardess." *Boston Globe*, September 1970.
3. Bill Hanley, interviews by author, Merrimac, Massachusetts, 2011–16.
4. Hanley, interview.
5. "Hijacker Terrorized Boston Stewardess."
6. Bill Hanley, interview.
7. Bill Hanley, interview.
8. Fox, "The First 15 Minutes."
9. Fox, "The First 15 Minutes."
10. Fox, "The First 15 Minutes."
11. Fox, "The First 15 Minutes."
12. Fox, "The First 15 Minutes."
13. Adam Zoll, "Peace, Love and Amplification." *Medford Citizen*, August 11, 1994.

CHAPTER 1: FIRST GENERATION, SECOND GENERATION

1. "Lansing Heritage." Audioheritage.com. http://www.audioheritage.org/html/history/lansing/altec.htm (accessed August 9, 2013).
2. Steven E. Schoenherr, "Motion Picture Sound: Part 1." Aes.org. http://www.aes.org/aeshc/docs/recording.technology.history/motionpicture1.html (accessed August 9, 2013).
3. John Aldred, "100 Years of Cinema Loudspeakers." Filmsound.org. http://filmsound.org/articles/amps/loudspeakers.htm (accessed August 9, 2013).
4. John Eargle and Mark Gander, "Historical Perspectives and Technology Overview of Loudspeakers for Sound Reinforcement." *Audio Engineering Society*, vol. 52, no. 4 (2004).

CHAPTER 2: THE NOT SO GLAMOROUS LIFE OF A SOUND ENGINEER

1. Donna Halper, email correspondence with author, December 30, 2013.
2. David Marks, "A Sound African Safari: From Live Peace in Toronto to the Thin End of Wedgies in Soweto." 3rdearmusic.com. http://www.3rdearmusic.com/hyarchive/hiddenyearsstory/pasafari.html (accessed March 27, 2013).
3. Bill Hanley, interview.
4. Jackson Browne (co-written by Bryan Garofalo), "The Load-Out." *Running on Empty*. Asylum Records, 1977.
5. Rusty Brutsché, phone interview by author, July 24, 2012.

CHAPTER 3: A BOY WHO LOVED SOUND

1. Dee Morris, *Medford: A Brief History*. Charleston, NC: History Press, 2009, 10.

2. Dee Morris, 46.

3. Dee Morris, 44.

4. Dee Morris, 44.

5. Bill Hanley, interview.

6. "Accident Victim Buried in Malden." *Boston Globe*, April 2, 1924.

7. Bill Hanley, interview.

8. Susan Hanley Campbell, interview by author, Burlington, MA, July 21, 2011.

9. Ted Ashby, "Wired for Politics: Audio Enthusiast Lives for Sounds." *Boston Globe*, October 2, 1964.

10. Patricia Hanley Hughes, interview by author, Burlington, MA, July 21, 2011.

11. Bill Hanley, interview.

12. Patricia Hanley Hughes, interview.

13. Bill Hanley, interview.

14. Terry Hanley, interview by author, Woburn, MA, July 20, 2011.

15. Charles Boeckman, *And the Beat Goes On: A Survey of Pop Music in America*. Washington, DC: Luce, 1972, 81.

16. Bill Hanley, interview.

17. Bill Hanley, interview.

18. Terry Hanley, interview.

19. "A Day from My Diary," January 16, 1955. Patricia Hanley Hughes archive.

20. Bill Hanley, interview.

21. Susan Hanley Campbell, interview.

22. Terry Hanley, interview.

23. Susan Hanley Campbell, interview.

24. Bill Hanley, interview.

25. Terry Hanley, interview.

26. Jack Moore, interview by author, Methuen, MA, July 1, 2013.

27. Patricia Hanley Hughes, interview.

28. Bill Hanley, interview.

29. "WW2 Draft Registration Card," November 19, 2012. Richard Howe collection.

30. Nathan Cobb, "The Sounds of Rock for the Multitudes." *Boston Globe*, August 1, 1970.

31. Terry Hanley, interview.

CHAPTER 4: POOR SOUND ALL AROUND

1. Patricia Hanley Hughes, interview.

2. Terry Hanley, interview.

3. Ken Lopez, phone interview by author, November 11, 2011.

4. Richard R. Smith, William Koon, and Tom Wheeler, *Fender: The Sound Heard 'Round the World*. Fullerton, CA: Garfish, 1995, 7.

5. Bill Hanley, interview.

6. Bill Hanley, interview.

7. "History of the Hampton Beach Casino." Hampton.lib.nh.us. http://www.hampton.lib.nh.us/hampton/history/casino/index.htm (accessed June 18, 2012).

8. "History of the Hampton Beach Casino."

9. Bill Hanley, interview.

10. Ken Lopez, interview.

11. Harold Cohen, interviews by author, 2011–16.

12. Herbert H. Wise, *Professional Rock and Roll*. New York: Amsco, 1967.

CHAPTER 5: ORGAN MUSIC AND TRAINING

1. George Pyche, interview by author, Haverhill, MA, August 11, 2011.

2. George Pyche, interview.

3. Bal-A-Roue Newsletter, "The Bal-A-Roue? For Health's Sake." 1952. National Roller Skating Museum collection.

4. Bal-A-Roue Newsletter, "The Bal-A-Roue?"

5. Bal-A-Roue Newsletter, "The Bal-A-Roue?"

6. Bill Hanley, interview.

7. Bill Hanley, interview.

8. "Parnelli Innovator Honoree, Father of Festival Sound." Fohonline.com. http://www.fohonline.com/index.php?option=com_content&task=view&id=579&Itemid=1 (accessed June 29, 2015).

9. Terry Hanley, interview.

10. Barbara Hanley, interview by author, Burlington, MA, July 21, 2011.

11. "America's Outstanding Rink, The Bal-A-Roue Welcomes Benny Aucoin." Bal-A-Roue Newsletter no. 2, p. 1 (1952). National Roller Skating Museum collection.

12. Bill Hanley, interview.

13. George Pyche, interview.

14. Bill Hanley, interview.

15. Barbara Hanley, interview.

16. Bill Hanley, interview.

17. Bill Hanley, interview.

18. Bill Hanley, interview.

19. Bill Appleyard, interview by author, Medford, MA, 2012.

20. Bill Hanley, interview.

21. Harold Cohen, interview.

22. "Boston's Radio Shack a Pioneer in Hi-Fi." *Daily Boston Globe*, September 30, 1956.

23. Edward McGrath, "How Joe DeMambro Ran $350 into Millions." *Boston Globe*, May 1, 1960.

24. FOH Online, "Parnelli Innovator Honoree."

25. Bill Hanley, interview.

26. Bill Hanley, interview.

27. Leo Beranek, interview by author, Weston, MA, July 17, 2012.

CHAPTER 6: 430 SALEM STREET

1. Cobb, "The Sounds of Rock for the Multitudes."

2. Bill Hanley, interview.

3. Ed Robinson, interview by author, Amesbury, MA, 2016.

4. Ed Robinson, interview.

5. Phil Robinson, interview by author, Kingston, MA, January 2016.

6. Bill Hanley, interview.

7. Ed Robinson, interview.

8. Patricia Hanley Hughes, interview.

9. Phil Robinson, interview.

10. Robert Glynn, "Young Men, Big Job." *Boston Evening Globe*, December 11, 1964.

11. Phil Robinson, interview.

12. Bill Hanley, interview.

13. Glynn, "Young Men, Big Job."

14. David Roberts, phone interview by author, June 4, 2012.

15. Cobb, "The Sounds of Rock for the Multitudes."

16. Bill Hanley, interview.

17. Bill Hanley, interview.

18. Terry Hanley, interview.

19. Bill Hanley, interview.

20. Tom Convery, interview by author, Medford, MA, July 17, 2012.

21. Ray Fournier, interview by author, Sanford, ME, July 15, 2011.

22. Terry Hanley, interview.

23. Bill Hanley, interview.

24. Bill Hanley, interview.

CHAPTER 7: JAZZ!

1. John H. Fenton, "Boston's Art Festival: Four Day Show." *New York Times*, May 13, 1956.

2. Bill Hanley, interview.

3. Bill Hanley, interview

4. David Klepper, email message to author, February 1, 2016.

5. Bill Hanley, interview.

6. David Klepper, email message.

7. Bill Hanley, interview.

8. Polly Webster, "Mobile Radio Studio Serves Fairs, Shows." *Boston Globe*, January 16, 1955.

9. Bill Hanley, interview.

10. Joe Boyd, *White Bicycles: Making Music in the 1960s*. London: Serpent's Tail, 2006, 46.

11. George Wein and Nate Chinen, *Myself Among Others*. Cambridge, MA: Da Capo, 2003, 129.

12. Wein and Chinen, *Myself Among Others*, 130.

13. Wein and Chinen, *Myself Among Others*, 134.

14. Wein and Chinen, *Myself Among Others*, 134.

15. John Sdoucos, interview by author, Newport, RI, August 7, 2011.

16. Wein and Chinen, *Myself Among Others*, 135.

17. Wein and Chinen, *Myself Among Others*, 138.

18. Wein and Chinen, *Myself Among Others*, 137.

19. William Keogh, "Jazz Fans Fill Newport Casino as Festival Ends." *Providence Journal*, July 19, 1954.

20. Howard Taubman, "Newport Rocked by Jazz Festival." *New York Times*, July 19, 1954, 21.

21. Howard Taubman, "Newport Festival." *New York Times*, July 25, 1954.

22. Taubman, "Newport Festival."

23. Wein and Chinen, *Myself Among Others*, 145.

24. Harold Schonberg, "Music: Jazz Festival." *New York Times*, July 16, 1955, 13.

25. Wein and Chinen, *Myself Among Others*, 146.

26. Wein and Chinen, *Myself Among Others*, 146.

27. Wein and Chinen, *Myself Among Others*, 85.

28. Bill Hanley, interview.

29. Bill Hanley, interview.

30. Bill Hanley, interview.

31. Bill Hanley, interview.

32. Bill Hanley, interview.

33. Bill Hanley, interview.

34. Wein and Chinen, *Myself Among Others*, 83.

35. Bill Hanley, interview.

36. Bill Hanley, interview.

37. "Riot in Newport Puts End to Jazz Festival." *Chicago Tribune*, July 4, 1960.

38. "Riot in Newport."

39. Bill Hanley, interview.

40. Bob Jones, interview by author, Newtown, CT, July 18, 2013.

41. Bill Hanley, interview.

42. Bill Hanley, interview.

43. Terry Hanley, interview.

44. Bill Hanley, interview.

45. "Theater on the Charles." *Boston Globe*, 1959. Hanley archive.

46. Cyrus Durgin, "Boston Arts Center's Home Rises in Brighton." *Boston Globe*, June 17, 1959.

47. "Three Shakespearian Plays to Open Boston Arts Center." *Daily Boston Globe*, April 19, 1959.

48. Bill Hanley, interview.

49. Phil Robinson, interview.

50. Ed Robinson, interview.

51. Phil Robinson, interview.

52. Bill Hanley, interview.

53. Ed Robinson, interview.

54. Cyrus Durgin, "Remainder of McBAC Theater Attractions Are Announced." *Boston Globe*, June 16, 1960.

55. "Tent Theaters Fate to Be Set on Thursday." *Boston Globe*, December 19, 1961.

56. Cameron Dewar, "Wein Unveils Big Plans For Castle Hill Concert." *Billboard*, March 6, 1961.

57. Phil Robinson, interview.

58. Terry Hanley, interview.

CHAPTER 8: NEWPORT NEEDS A HOME

1. Bill Hanley, interview.

2. Bill Hanley, interview.

3. Chip Monck, interviews by author, 2011–17.

4. Robert Kiernan, interview by author, Salisbury, VT, May 22, 2013.

5. Bill Hanley, interview.

6. Robert Kiernan, interview.

7. Bill Hanley, interview.

8. Chip Monck, interview.

9. Chip Monck, interview.

10. Bill Hanley, interview.

11. Robert Kiernan, interview.

12. Jerry Wenger, phone interview by author, December 18, 2018.

13. Jerry Wenger, phone interview.

14. Jerry Wenger, phone interview.

15. Robert Kiernan, interview.

16. Chip Monck, interview.

17. Chip Monck, interview.

18. Bill Hanley, interview.

19. Wein and Chinen, *Myself Among Others*, 262.

20. Chip Monck, interview.

21. Wein and Chinen, *Myself Among Others*, 232.

22. D. Neale Adams, "Master of the 1½ Ton Whisper Catcher." August 1, 1966. Hanley archive.

23. Wein and Chinen, *Myself Among Others*, 329.

24. J. F. Morton, *Backstory in Blue: Ellington at Newport '56*. New Brunswick, NJ: Rutgers University Press, 2008, 261.

25. Bill Hanley, interview.

26. Wein and Chinen, *Myself Among Others*, 262.

27. Wein and Chinen, *Myself Among Others*, 329.

28. "Festival Efforts Start in Newport." *New York Times*, December 17, 1964.

29. Simon George, "Wein: Dynamic Impresario." *Billboard*, July 17, 1965, 42.

30. Bill Hanley, interview.

31. Michael Steinberg, "Met Opera Joins Newport Festival." *Boston Globe*, May 25, 1966.

32. "Hanley Financial Report." February 1, 1967. Hanley archive.

33. Robert Glynn, "Young Men, Big Job." *Boston Evening Globe*, December 11, 1964.

34. George Wein, interview by author, Newport, RI, August 7, 2011.

35. Albert Murray, "Gumbo." *Jazz: A Film by Ken Burns*. WGBH. January 18, 2001.

CHAPTER 9: ORCHESTRA SOUND

1. Wein and Chinen, *Myself Among Others*, 267.

2. Wein and Chinen, *Myself Among Others*, 267. Operas do not have movements, so Wein's terminology is incorrect.

3. Barry McKinnon, "RE/P Files: An Interview with Bill Hanley." Prosoundweb.com. https://www.prosoundweb.com/channels/livesound/re_p_files_an_interview_with_bill_hanley/ (accessed September 6, 2017).

4. McKinnon, "RE/P Files."

5. McKinnon, "RE/P Files."

6. Rachel E. Lyons, "Bill Hanley Interview." *New Orleans Jazz & Heritage Foundation*, October 30, 2014.

7. Harold Cohen, interview.

8. McKinnon, "RE/P Files."

9. Bill Hanley, interview.

10. Bill Hanley, interview.

11. Bill Hanley, interview.

12. Bill Hanley, interview.

13. Lenny Litman, "Al Hirt Show Sounds Good." *Pittsburgh Press*, August 25, 1970.

14. Litman, "Al Hirt Show Sounds Good."

15. Sam Boroda, interviews by author, 2017–18.

16. "City and Suburban Life." *Pittsburgh Press*, June 10, 1971. Hanley archive.

17. "City and Suburban Life."

18. "Columbia's Festival of Stars." *Cash Box*, 1972. Hanley archive.

19. "Columbia's Festival of Stars."

20. "Columbia's Festival of Stars."

21. Sam Boroda, interview.

22. "Columbia's Festival of Stars."

23. Dan Coughlin, "Orchestra Warms Up." *Cleveland Plain Dealer*, May 4, 1972.

24. Coughlin, "Orchestra Warms Up."

25. "Cincinnati Symphony Orchestra Correspondence." 1972. Hanley archive.

26. Jim Joseph, "Skyrockets Play 'Spectacular' Finale." *Cincinnati Post*, September 11, 1972.

27. Gail Stockholm, "Al Hirt's Trumpet Talks at Stadium." *Cleveland Plain Dealer*, May 4, 1972.

28. Stockholm, "Al Hirt's Trumpet."

CHAPTER 10: FOLK AND DISTORTION

1. Wein and Chinen, *Myself Among Others*, 313.

2. Wein and Chinen, *Myself Among Others*, 314.

3. "Folkniks Tab Newport Folk Bash Huge Hype for Clan." *Billboard*, July 27, 1959.

4. Stephen Calt, "National Folk Festival." *Sing Out!*, 1970.

5. "When the Philadelphia Folk Festival Was Held in Paoli." Tredyffrin Easttown Historical Society, History Quarterly Digital Archives, vol. 18, no. 3 (1980): 83–88.

6. "Carnival Ordinance Bans Philadelphia Folk Festival." *Billboard*, February 13, 1965, 8.

7. Bill Hanley, interview.

8. Bill Hanley, interview.

9. Steve Twomey, "Folk Festival Has Grown Quickly." *Philadelphia Inquirer*, August 27, 1974.

10. Peter Yarrow, interview by author, Worcester, MA, April 1, 2012.

11. Peter Yarrow, interview.

12. D. Neale Adams, "Master of the 1½ Ton Whisper Catcher." August 1, 1966. Hanley archive.

13. Adams, "Master of the 1½ Ton."

14. Adams, "Master of the 1½ Ton."

15. Peter Yarrow, interview.

16. Peter Yarrow, interview.

17. Loyal Jones and John M. Forbes, *Minstrel of the Appalachians: The Story of Bascom Lamar Lunsford*. Boone, NC: Appalachian Consortium, 1984, 64.

18. Jones and Forbes, 51.

19. Jones and Forbes, 51.

20. Bill Hanley, interview.

21. Elijah Wald, *Dylan Goes Electric!: Newport, Seeger, Dylan, and the Night That Split the Sixties*. New York: Dey Street, 2016, 3.

22. Leo Beranek, interview.

23. Paul Williams, *The Crawdaddy! Book: Words (and Images) from the Magazine of Rock*. Milwaukee, WI: Hal Leonard, 2002, 38.

24. Wald, *Dylan Goes Electric!*, 3

25. Wein and Chinen, *Myself Among Others*, 333.

26. Bill Hanley, interview.

27. Bill Hanley, interview.

28. Bob Jones, interview.

29. Bob Jones, interview.

30. Robert J. Lurtsema, "On the Scene." *Broadside* 4, no. 11 (1965).

31. Lurtsema, "On the Scene."

32. Bob Jones, interview.

33. Elijah Wald, interview by author, Medford, MA, August 2015.

34. David Dann, "Michael Bloomfield at Newport." Mikebloomfieldamericanmusic.com. http://www.mikebloomfieldamericanmusic.com/newportinset.htm (accessed April 27, 2017).

35. Bill Hanley, interview.

36. Bill Hanley, interview.

37. Wald, *Dylan Goes Electric!*, 259.

38. Wald, 259.

39. Elijah Wald, interview.

40. Boyd, *White Bicycles*, 105.

41. BBC Radio 4. "What Happened When Bob Dylan Plugged In?" Bbc.co.uk. http://www.bbc.co.uk/programmes/po2zonfn (accessed April 27, 2017).

42. Boyd, *White Bicycles*, 105.

43. BBC Radio 4, "What Happened When Bob Dylan Plugged In?"

44. Bill Hanley, interview.

45. Elijah Wald, interview.

46. *Pete Seeger,* No Direction Home. *2005. Paramount Pictures, DVD.*

47. Elijah Wald, interview.

48. Peter Yarrow, interview.

49. Wein and Chinen, *Myself Among Others*, 334.

50. Bob Jones, interview.

51. "Dylan Mixes Bag of Tricks at Music Fest." *Billboard*, September 11, 1965. 16.

52. Nora Ephron and Susan Edmiston. "Bob Dylan Interview." New York, 1965. In *Bob Dylan: The Essential Interviews*, ed. Jonathan Cott. New York: Simon and Shuster, 2006, 47.

53. Richard Alderson, phone interview by author, March 22, 2017.

54. Richard Alderson, phone interview.

55. Richard Alderson, phone interview.

56. Richard Alderson, phone interview.

57. Richard Alderson, phone interview.

58. Richard Alderson, phone interview.

59. Richard Alderson, phone interview.

60. Richard Alderson, phone interview.

61. Richard Alderson, phone interview.

62. Richard Alderson, phone interview.

63. Richard Alderson, phone interview.

64. Richard Alderson, phone interview.

65. Wein and Chinen, *Myself Among Others*, 334.

66. Harold Cohen, interview.

67. Adams, "Master of the 1½ Ton."

68. Adams, "Master of the 1½ Ton."

69. Adams, "Master of the 1½ Ton."

70. Adams, "Master of the 1½ Ton."

71. Adams, "Master of the 1½ Ton."

72. Adams, "Master of the 1½ Ton."

73. Adams, "Master of the 1½ Ton."

74. Karl Atkeson, phone interview by author, December 3, 2011.

75. Ray Fournier, interview.

76. Harold Cohen, interview.

77. Bill Hanley, interview.

CHAPTER 11: THE INAUGURATION OF PRESIDENT LYNDON B. JOHNSON

1. Bill Hanley, interview.

2. Ashby, "Wired for Politics."

3. Bill Hanley, interview.

4. Bill Hanley, interview.

5. Glynn, "Young Men, Big Job."

6. "Medford Firm Wins Contract for Jan. 20 Audio Services: LBJ's Inaugural Voice to Have Local 'Sound.'" *Medford Mercury*, 1964. Hanley archive.

7. "Medford Firm Wins."

8. Phil Robinson, interview.

9. Bill Hanley, interview.

10. Phil Robinson, interview.

11. Glynn, "Young Men, Big Job."

12. Glynn, "Young Men, Big Job."

13. David Roberts, interview.

14. David Roberts, interview.

15. Bill Hanley, interview.

16. Bill Hanley, interview.

17. Lewis S. Goodfriend, phone interview by author, August 2, 2014.

18. Lewis S. Goodfriend, phone interview.

19. Terry Hanley, interview.

20. "Shure Records a Page in History." *Sound Scope* 2, no. 1 (March 1965).

21. Bill Hanley, interview.

22. Terry Hanley, interview.

23. Bill Hanley, interview.

24. David Roberts, interview.

25. Bill Hanley, interview.

26. Michael Pettersen, interview by author, AES, New York, October 22, 2011.

27. David Roberts, interview.

28. Bill Hanley, interview.

29. Phil Robinson, interview.

30. "1965 Inauguration letter." Hanley Archive.

31. David Roberts, interview.

32. "Shure Records a Page in History."

33. Shure Americas. "History: The Evolution of an Audio Revolution." Shure.com. http://www
.shure.com/americas/about-shure/history (accessed May 26, 2013).

34. Michael Pettersen, interview.

35. Bill Hanley, interview.

36. Dinky Dawson, interview by author, Plymouth, MA, September 16, 2011.

37. Michael Pettersen, interview.

38. Terry Hanley, interview.

39. Terry Hanley, interview.

40. Harold Cohen, interview.

41. Michael Pettersen, interview.

42. Glynn, "Young Men, Big Job."

CHAPTER 12: HANLEY SOUND NEEDS A MANAGER

1. Judi Bernstein-Cohen, interview by author, Medford, MA, July 27, 2011.

2. "Youth Strong on Talent Agency Scene, Too." *Cash Box*, 1968.

3. "Youth Strong on Talent Agency Scene, Too."

4. Judi Bernstein-Cohen, interview.

5. John Sdoucos, interview.

6. John Sdoucos, interview.

7. Judi Bernstein-Cohen, interview.

8. Linda J. Greenhouse, "R 'n' R-For Love or Money." *Harvard Crimson*, October 27, 1966.

9. Judi Bernstein-Cohen, interview.

10. Judi Bernstein-Cohen, interview.

11. Judi Bernstein-Cohen, interview.

12. Judi Bernstein-Cohen, interview.

13. David Roberts, interview.

14. Judi Bernstein-Cohen, interview.

15. Judi Bernstein-Cohen, interview.

16. Bill Hanley, interview.

17. Judi Bernstein-Cohen, interview.

18. Judi Bernstein-Cohen, interview.

19. Judi Bernstein-Cohen, interview.

20. Judi Bernstein-Cohen, interview.

21. Judi Bernstein-Cohen, interview.

CHAPTER 13: GO-GO DANCERS, BATMAN, AND THE BEACH BOYS

1. Richard Goldstein, "Pop Eye: Soundblast '66." *Village Voice*, June 6, 1966.

2. Goldstein, "Pop Eye."

3. Bill Hanley, interview.

4. Randal, Jonathan. "Stadium Bases Are Loaded with Go-Go Girls." *New York Times*, June 11, 1966.

5. Randal, "Stadium Bases Are Loaded."

6. Goldstein, "Pop Eye."

7. Goldstein, "Pop Eye."

8. Goldstein, "Pop Eye."

9. Goldstein, "Pop Eye."

10. "Tear Gas Quells Manning Bowl Rioters." *Daily Evening Item*, June 25, 1966.

11. Ray Fournier, interview.

12. Bill Hanley, interview.

13. "Manning Field." Ci.lynn.ma.us. http://www.ci.lynn.ma.us/attractions_manningfield.shtml (accessed May 27, 2013).

14. Steve Krause, "You can't always get what you want." Itemlive.com. https://www.itemlive.com/2016/06/24/you-cant-always-get-what-you-want/ (accessed August 30, 2017).

15. "Batman Concert Zonks." Unknown source, June 1966. Hanley archive.

16. Robert Sherman, "Batman to the Rescue in Shea Caper." *New York Times*, June 26, 1966.

17. Sherman, "Batman to the Rescue."

18. Sherman, "Batman to the Rescue."

19. "Batman Concert Zonks."

20. Terry Hanley, interview.

21. Gilkenson, Steve. "Beach Boys Give Concert at Rhode Island Auditorium." Unknown source, 1966. Hanley archive.

22. Bill Hanley, interview.

23. Gilkenson, Steve. "Beach Boys Give Concert."

24. Bill Hanley, interview.

25. Bill Yerkes, phone interview by author, 2012.

26. Bill Yerkes, phone interview.

27. Bill Yerkes, phone interview.

28. David Roberts, interview.

29. David Roberts, interview.

30. David Roberts, interview.

31. Bill Hanley, interview.

CHAPTER 14: HANLEY AND THE FAMILY BAND

1. Paul Cowsill, interview by author, Providence, RI, August 10, 2011.

2. Judy Klemesrud, "The Singing 'Mini-Mom' with a Big Family." *New York Times*, August 5, 1968.

3. Alfred G. Aronowitz, "The Cowsills in Concert Widen the Generation Gap." *New York Times*, December 29, 1967.

4. Paul Cowsill, interview.

5. Paul Cowsill, interview.

6. Paul Cowsill, interview.

7. Bob Cowsill, interview by author, Providence, RI, August 10, 2011.

8. Paul Cowsill, interview.

9. Bob Cowsill, interview.

10. "Cowsills to Record First 'Live' Album." *Cash Box*, December 21, 1968.

11. Harold Cohen, interview.

12. Harold Cohen, interview.

13. Harold Cohen, interview.

14. Harold Cohen, interview.

CHAPTER 15: HANLEY AND THE REMAINS

1. Brett Milano, *The Sound of Our Town: A History of Boston Rock and Roll*. Beverly, MA: Commonwealth Editions, 2007, 21.

2. "Matthews Arena." Nuhuskies.com. https://nuhuskies.com/sports/2010/1/28/matthewsarena .aspx (accessed August 14, 2013).

3. Fred Goodman, *The Mansion on the Hill: Dylan, Young, Geffen, Springsteen, and the Head-On Collision of Rock and Commerce*. New York: Times Books, 1997, 10.

4. Milano, *The Sound of Our Town*, 21.

5. John Sdoucos, interview.

6. Milano, *The Sound of Our Town*, 25.

7. Brian Fitzgerald, "Rock Reincarnation: Boston Band the Remains Has its New Day." Bu.edu. https://www.bu.edu/bridge/archive/2002/11–01/remains.htm (accessed September 1, 2017).

8. "A Session with the Remains." Reocities.com. http://www.reocities.com/nemsbook/r/remains .htm (accessed September 2, 2017).

9. Bill Hanley, interview.

10. John Sdoucos, interview.

11. Terry Hanley, interview.

12. Bill Briggs, interview by author, Amesbury, MA, September 19, 2011.

13. Bill Briggs, interview.

14. "A Session with the Remains."

15. Judi Bernstein-Cohen, interview.

16. Goodman, *The Mansion on the Hill*, 14.

17. Vern Miller, phone interview by author, July 15, 2013.

18. Vern Miller, phone interview.

19. John Sdoucos, interview.

20. Bill Briggs, interview.

21. Goodman, *The Mansion on the Hill*, 14.

22. Jon Landau, "Too Loud for One Crowd, But Not for Their BU Fans." *Crawdaddy!* 1967.

23. Landau, "Too Loud."

24. Milano, *The Sound of Our Town*, 25.

25. Vern Miller, interview.

26. Barry Tashian, phone interview by author, September 2, 2011.

27. Barry Tashian, phone interview.

28. Barry Tashian, phone interview.

29. Barry Tashian, phone interview.

30. Barry Tashian, *Ticket to Ride: The Extraordinary Diary of The Beatles' Last Tour*. Nashville: Dowling Press, 1997, 1.

31. Vern Miller, interview.

32. Bill Kopp, "Day of the Remains, Part One." Blog.musoscribe.com. http://blog.musoscribe.com/index.php/2010/11/01/day-of-the-remains-part-one/ (accessed August 30, 2017).

33. Vern Miller, interview.

34. Barry Tashian, *Ticket to Ride*.

35. Bill Hanley, interview.

36. "What Happened to the Coffee?" *The Heights Review* vol. XLVIII, no. 8 (17 November 1967).

37. James R. Beniger, "Chuck Berry: Old-Time Music Grows Old." *Harvard Crimson*, November 14, 1967.

38. Bill Hanley, interview.

39. Lawrence Azrin, "The '60's Boston Scene." Punkblowfish.com. http://punkblowfish.com/BostonClubsSixties.html (accessed August 31, 2017).

40. "The Concert Promoters Association." *Pollstar.* https://vimeopro.com/user23430836/promoters-hall-of-fame-interviews (accessed October 21, 2017).

CHAPTER 16: THE BEATLES FACE A FEVERISH PITCH

1. The Beatles, *The Beatles Anthology*. San Francisco: Chronicle Books, 2000, 186.

2. Tony Bramwell and Rosemary Kingsland. *Magical Mystery Tours: My Life with the Beatles.* New York: Thomas Dunne Books, 2005, 197.

3. Tashian, *Ticket to Ride*.

4. Vern Miller, interview.

5. Tashian, *Ticket to Ride*, 83.

6. Erik 4-A, "Interview with Glenn D. White." *Tape Op: The Creative Music Recording Magazine*, January 14, 2004.

7. 4-A. "Interview with Glenn D. White."

8. Bob Spitz, *The Beatles: The Biography*. New York: Little, Brown, 2005, 480.

9. Spitz, *The Beatles*, 480.

10. Spitz, *The Beatles*, 578.

11. Spitz, *The Beatles*, 578.

12. Ashby, "Wired for Politics."

13. Duke Mewborn, interview by author, November 23, 2015.

14. Duke Mewborn, interview.

15. Duke Mewborn, interview.

16. Chuck Gunderson, *Some Fun Tonight!: The Backstage Story of How the Beatles Rocked America: The Historic Tours of 1964–1966*. Milwaukee, WI: Backbeat Books, 2016, 60.

17. Gunderson, *Some Fun Tonight!*

18. Duke Mewborn, interview.

19. Duke Mewborn, interview.

CHAPTER 17: BEATLES BY CHANCE

1. Tony Bramwell and Rosemary Kingsland. *Magical Mystery Tours: My Life with the Beatles.* New York: Thomas Dunne Books, 2005, 166.

2. Tashian, *Ticket to Ride.*

3. Bill Briggs, interview.

4. Vern Miller, interview.

5. Bill Briggs, interview.

6. Tashian, *Ticket to Ride*, 7.

7. Barry Tashian, interview.

8. Vern Miller, interview.

9. Barry Tashian, interview.

10. Judith Sims, "On Tour with The Beatles." *Teen Set*, December 1966.

11. Barry Tashian, interview.

12. Bill Hanley, interview.

13. Bill Briggs, interview.

14. Bill Hanley, interview.

15. Vern Miller, interview.

16. Ernie Santosuosso, "Top Honors at Beatle Concert Go to Protective Fence." *Boston Globe*, August 19, 1966.

17. Vern Miller, interview.

18. Jonathan Gould, *Can't Buy Me Love: The Beatles, Britain, and America.* New York: Harmony Books, 2007, 340–41.

CHAPTER 18: SHEA STADIUM SOUND

1. The Beatles, *The Beatles Anthology.* San Francisco: Chronicle Books, 2000, 186.

2. Roy Clair, phone interview by author, July 30, 2013.

3. Bill Hanley, interview.

4. Murray Schumach, "Shrieks of 55,000 Accompany Beatles: 55,000 Fill Shea to Hear Beatles." *New York Times*, August 16, 1965.

5. Matt Hurwitz, "The Beatles Made Music History at Shea Stadium." Variety.com. http://variety.com/2016/music/spotlight/beatles-music-history-shea-stadium-1201861551/ (accessed April 8, 2016).

6. Hurwitz, "The Beatles Made Music History."

7. Spitz, *The Beatles*, 577.

8. "1965 Beatles at Shea Stadium Production Sheet." Laurie Jacobson collection.

9. "Audio Engineering Society List of Members." Audio Engineering Society, New York, 1951.

10. Hurwitz, "The Beatles Made Music History."

11. Donald J. Plunkett, "In Memoriam." *Journal of the Audio Engineering Society* 31, no. 1/2 (1983).

12. Schumach, "Shrieks of 55,000 Accompany Beatles."

13. Hurwitz, "The Beatles Made Music History."

14. Barry Tashian, interview.

15. Harold Cohen, interview.

16. Harold Cohen, interview.

17. Harold Cohen, interview.

18. Bill Hanley, interview.

19. Robert K. Sanford, "The Beatles Sing in the Rain for Wet, Enthusiastic Audience—23,000 Pay to Hear Them—First Aid Stations Busy." *St. Louis Dispatch*, August 22, 1966.

20. Sanford, "The Beatles Sing."

21. Barry Tashian, interview.

22. Bill Briggs, interview.

23. Glenna Syse, "Beatles: A Sound Analysis." *Chicago Sun-Times*, 1966.

24. Tashian, *Ticket to Ride*.

25. Barry Tashian, interview.

26. Barry McKinnon, "An Interview with Bill Hanley." http://www.prosoundweb.com/article/re_p_files_an_interview_with_bill_hanley.

27. Bill Hanley, interview.

28. McKinnon, "An Interview with Bill Hanley."

29. "Parnelli Innovator Honoree, Father of Festival Sound."

30. Barry Tashian, interview.

31. Barry Tashian, interview.

32. Barry McKinnon, "Woodstock: An Audio Milepost." *Pro Sound News* (November 1988): 71–72.

33. Bill Hanley, interview.

34. McKinnon, "An Interview with Bill Hanley."

35. McKinnon, "Woodstock: An Audio Milepost."

36. Bill Hanley, interview.

37. Sid Bernstein and Aaron Arthur. *It's Sid Bernstein Calling: The Amazing Story of the Promoter Who Made Entertainment History*. Middle Village, NY: Jonathan David, 2002, 200.

38. Jonathan Gould, *Can't Buy Me Love: The Beatles, Britain, and America*. New York: Harmony Books, 2007, 347.

39. Harold Cohen, interview.

40. Adam Zoll, "Peace, Love and Amplification." *Medford Citizen*, August 11, 1994.

41. Paul L. Montgomery, "The Beatles Bring Shea to a Wild Pitch of Hysteria." *New York Times*, August 24, 1966.

42. Vern Miller, interview.

43. Bill Hanley, interview.

44. Barry Tashian, interview.

45. Gould, *Can't Buy Me Love*, 247.

46. Bill Hanley, interview.

CHAPTER 19: THE COMPETITION

1. Maureen Droney, "Necessity Mothers Invention." *Mix Magazine*, November 1, 2004.

2. Abe Jacob, interview by author, New York, June 13, 2013.

3. Abe Jacob, interview.

4. Abe Jacob, interview.

5. "About, Harry McCune Sound Service." Mccune.com. http://mccune.com/about-harry-mccune-san-francisco (accessed September 04, 2017).

6. Mike Neal, Skype interview by author, June 28, 2012.

7. Mike Neal, Skype interview.

8. Mike Neal, Skype interview.

9. Mike Neal, Skype interview.

10. Mike Neal, Skype interview.

11. Mike Neal, Skype interview.

12. Dan Healy, phone interview by author, April 14, 2013.

13. Dan Healy, phone interview.

14. John Meyer, phone interview by author, July 2, 2013.

15. "History, Swanson Sound Service." Swansonsound.com. http://www.swansonsound.com/history.htm (accessed August 9, 2013).

16. Droney, "Necessity Mothers Invention."

17. Dan Healy, interview.

18. John Meyer, interview.

19. Roy Clair, interview.

20. Robert Meagher, interview by author, Rochester, VT, May 30, 2012.

21. Jim Gamble, phone interview by author, August 22, 2013.

22. Ken Lopez, interview.

23. Dan Healy, interview.

24. Jim Gamble, interview.

25. Jim Gamble, interview.

26. Robert Meagher, interview.

27. Robert Meagher, interview.

28. Karl Meagher, interview by author, Rochester, VT, May 30, 2012.

29. R. J. Tinkham, "Sound Recording and Reinforcing at the Monterey Jazz Festival." *Audio Magazine*, December 1, 1958.

30. Robert Meagher, interview.

31. Roxon, Lillian, "Rock Fete without Mud, Dope, Tickets." *Chicago Tribune*, August 22, 1971.

32. Jim Gamble, interview.

33. Ken Lopez, interview.

34. Karl Meagher, interview.

35. Robert Meagher, interview.

36. Karl Meagher, interview.

37. Bill Hanley, interview.

38. Stan Miller, interviews by author, 2012–17.

39. Stan Miller, interview.

40. Stan Miller, interview.

41. Stan Miller, interview.

42. Stan Miller, interview.

43. Stan Miller, interview.

44. Stan Miller, interview.

45. "In Features: Stan Miller, Digital Sound Pioneer." Fohonline.com. http://www.fohonline.com/home/20-features/2947-stan-miller-digital-sound-pioneer.html (accessed March 11, 2018).

46. Kevin Young, "Live Sound: In Profile: Stan Miller—Continuing to Explore the Possibilities." Pro Sound Web. June 6, 2011. www.prosoundweb.com/topics/production/in_profile_stan_miller_continuing_to_explore_the_possibilities/.

47. Kevin Mitchell, "Stanley Miller: A Life of Sound." Fohonline.com. http://www.fohonline.com/cat-blog/49-bio/12787-stanley-miller-a-life-of-sound.html (accessed March 11, 2018).

48. Stan Miller, interview.

49. Bill Hanley, interview.

50. Shannon Slaton, Mixing a Musical: Broadway Theatrical Sound Mixing Techniques. New York: Routledge, 2011.

51. Robert Kiernan, interview.

52. Robert Kiernan, interview.

53. Robert Kiernan, interview.

54. Chip Monck, email correspondence with author, September 18, 2013.

55. Robert Kiernan, interview.

56. Robert Kiernan, interview.

57. Robert Kiernan, interview.

58. Robert Kiernan, interview.

59. Roy Clair, interview.

60. Bill Hanley, interview.

61. Bill Hanley, interview.

62. Kevin Young, "In Profile: Dinky Dawson, Concert Sound Pioneer." Fohonline.com. http://www.prosoundweb.com/topics/audio/in_profile_dinky_dawson_concert_sound_pioneer (accessed April 13, 2017).

63. Charlie Watkins, "History, Watkins Electric Music." Wemwatkins.co.uk. http://www.wemwatkins.co.uk/history.htm.

64. Watkins, "History, Watkins Electric Music."

65. "The Seventh National Jazz and Blues Festival." Ukrockfestivals.com. http://www.ukrockfestivals.com/1967-windsor-festival.html (accessed September 6, 2017).

66. David Pelletier, interview by author, Salem, MA, August 14, 2011.

67. Bill Hanley, interview.

CHAPTER 20: BIG APPLE! BIG SOUND!

1. Bill Hanley, interview.

2. Kevin M. Mitchell, "Grandfather of Rock 'n' Roll Productions." ChipMonck.com. http://www.chipmonck.com/making_it_happen.htm.

3. Chip Monck, interview.

4. Mike Jahn, "Chambers Brothers, Soul Group, Are Fillmore East's Headliners." *New York Times*, April 7, 1969.

5. George Chambers, phone interview by author, September 15, 2013.

6. Joe Chambers, phone interview by author, September 15, 2013.

7. Lester Chambers, Skype interview by author, March 1, 2013.

8. Lester Chambers, Skype interview.

9. Willie Chambers, Skype interview by author, July 11, 2013.

10. Julius Chambers, phone interview by author, October 8, 2013.

11. Julius Chambers, phone interview.

12. Willie Chambers, interview.

13. Lester Chambers, interview.

14. Willie Chambers, interview.

15. Joe Chambers, interview.

16. Roy Clair, interview.

17. Julius Chambers, interview.

18. Joe Chambers, interview.

19. Terry Hanley, interview.

20. David Roberts, interview.

21. Bill Robar, phone interview by author, February 1, 2016.

22. Bill Hanley, interview.

23. Billy Pratt, phone interview by author, March 1, 2017.

24. Terry Hanley, interview.

25. Terry Hanley, interview.

26. Terry Hanley, interview.

27. "Steve Paul's 'Scene' Goes TV: 2-Hour Metromedia Telecaster." *Cash Box*, September 2, 1967.

28. "Generation, a Psychedelic Club, to Bow in 'Village.'" *Billboard*, March 16, 1968.

29. Bill Hanley, interview.

CHAPTER 21: DOWN IN THE VILLAGE

1. Morris Dickstein, *Gates of Eden: American Culture in the Sixties*, New York: Basic Books, 1977. 189.

2. Melanie Safka, interview by author, Londonderry, NH, March 10, 2012.

3. Danny Kalb, interview by author, Londonderry, NH, November 16, 2012.

4. Suze Rotolo, *A Freewheelin' Time: A Memoir of Greenwich Village in the Sixties,* New York: Broadway Books, 2008, 363.

5. Tom Rush, interview by author, Marblehead, MA, April 28, 2012.

6. Rotolo, *A Freewheelin' Time*, 363.

7. Rotolo, *A Freewheelin' Time*, 363.

8. Rotolo, *A Freewheelin' Time*, 361.

9. Steve Katz, interview by author, Londonderry, NH, November 16, 2012.

10. Karl Atkeson, interview.

11. Bill Hanley, interview.

12. Jesse McKinley, "Howard Solomon, 75, Owner of Famed Village Nightclub." *New York Times,* June 16, 2004.

13. Roy Blumenfeld, interview by author, Londonderry, NH, November 16, 2012.

14. McKinley, "Howard Solomon," 75.

15. Bill Hanley, interview.

16. Bill Hanley, interview.

17. Bill Hanley, interview.

18. Bill Hanley, interview.

19. Bill Hanley, interview.

20. Robert Sontag, email message to author, June 7, 2013.

21. Robert Sontag, email message.

22. Dan Sullivan, "Mothers of Invention at the Garrick." *New York Times*, May 25, 1967.

23. McKinley, "Howard Solomon," 75.

24. Toni Ruiz, "Fred Neil Chronology." *Zonnet.nl.* http://www.home.zonnet.nl/jim2873/fredneil/chronology.html (accessed May 28, 2013).

25. Howard Solomon, "Everybody's Talkin' Message Board." *Love.torbenskott.dk.* http://love.torbenskott.dk/forum/topic.asp?TOPIC_ID=4793 (accessed July 9, 2015).

26. Bill Hanley, interview.

27. Solomon, "Everybody's Talkin.'"

28. Bill Hanley, interview.

29. Jim Reeves, "Autobiography." *Reevesaudio.com.* http://www.reevesaudio.com/me.html (accessed 2012).

30. Reeves, "Autobiography."

31. Rich Weidman, The Doors FAQ: All That's Left to Know about the Kings of Acid Rock. Milwaukee, WI: Backbeat Books, 2011.

32. Bill Hanley, interview.

33. Jim Reeves, phone interview by author, May 5, 2012.

34. Reeves, "Autobiography."

35. Bill Hanley, interview.

36. "Doors' Ondine Act Should Open Chart Doors." *Billboard*, March 25, 1967: 22.

37. Jesse Colin Young, interview by author, Scituate, MA, May 13, 2012.

38. Lowell Levinger, Skype interview by author, March 16, 2014.

39. Lowell Levinger, Skype interview.

40. Jesse Colin Young, interview.

41. "Do the Acts or Instruments Capture Fancy of the Teens?" *Billboard*, November 26, 1966: 79.

42. "Do the Acts or Instruments?," 79.

43. Tom P. Steinbleker, "A Top Banana of Rock." *New York Times*, December 3, 1967.

44. Jesse Colin Young, interview.

45. Jesse Colin Young, interview.

46. Jesse Colin Young, interview.

47. Paul Williams, *The Crawdaddy! Book: Words (and Images) from the Magazine of Rock*. Milwaukee, WI: Hal Leonard, 2002, 42.

48. Steve Katz, interview.

49. Herb Gart, phone interview by author, September 15, 2012.

50. Herb Gart, phone interview.

51. Herb Gart, phone interview.

52. Bill Hanley, interview.

CHAPTER 22: MORE MONITOR, PLEASE!

1. Bill Hanley, interview.

2. Jack Casady, interview by author, Plymouth, NH, July 16, 2012.

3. Bill Hanley, interview.

4. Herbert H. Wise, *Professional Rock and Roll*. New York: Amsco, 1967.

5. Bill Hanley, interview.

6. Ellen Sander, "The Sweet Country Sounds of Buffalo Springfield." *New York Times*, October 6, 1968.

7. Bill Hanley, interview.

8. Sarah Benzuly, "Local Crew: McCune Sound." Mixonline.com. http://www.mixonline.com/news/live-gear/local-crew-mccune-sound/368480 (accessed September 24, 2017).

9. Sander, "The Sweet Country Sounds."

CHAPTER 23: A FRENCHMAN AND THE SECOND AVENUE SHUFFLE

1. Judi Bernstein-Cohen, interview.

2. Sam Boroda, interview.

3. Sam Boroda, interview.

4. Sam Boroda, interview.

5. John B. Cooke, *On the Road with Janis Joplin*. New York: Berkley Books, 2015, 127.

6. Sam Boroda, interview.

7. Scott Holden, interview.

8. Sam Boroda, interview.

9. Sam Boroda, interview.

10. Bill Hanley, interview.

11. Bill Hanley, interview.

12. Al Alvarez, "Anderson Theatre." Cinematreasures.org. http://cinematreasures.org/theaters/31581 (accessed May 5, 2017).

13. Claire Sedore, "Broadway and Off Broadway Theatres." Worldtheatres.com. http://world-theatres.com/broadway-theatres (accessed May 28, 2013).

14. John Glatt, *Rage & Roll: Bill Graham and the Selling of Rock*. Secaucus, NJ: Carol, 1993, 88.

15. John Morris, interview by author, New York, September 24, 2011.

16. Chris Langhart, interview by author, New York, September 24, 2011.

17. Bill Hanley, interview.

18. Glatt, *Rage & Roll*, 88.

19. Glatt, *Rage & Roll*, 88.

20. Howard Smith, "Scenes." *Village Voice*, February 22, 1968.

21. Sam Andrew, interview by author, Londonderry, NH, August 8, 2012.

22. Harold Cohen, interview.

23. Bill Hanley, interview.

24. "Life: Chronology." Janisjoplin.net. http://www.janisjoplin.net/life/chronology/#year_1968 (accessed October 1, 2017).

25. Henry Ellis, "The Unbelievable But True Story of Janis Joplin's Appearance at Winter Carnival '69." Franklinpierce.edu. http://www.franklinpierce.edu/alumni/Article_Janice_Joplin _Appearance_at_Winter_Carnival.htm (accessed October 1, 2017).

26. Jerry Pompili, phone interview by author, January 29, 2011.

27. Jerry Pompili, phone interview.

28. John Morris, interview.

29. Joshua White, interview by author, New York, February 7, 2012.

30. Bill Hanley, interview.

CHAPTER 24: THE VICTORY SPECIAL

1. Russell Freeburg, "Nixon Hails Apollo in Ohio." *Chicago Tribune*, October 23, 1968.

2. "Nearly 2,000 Hear Nixon at Deshler: GOP Candidate Hits Demo Administration HHH," *Bryan Times*, October 23, 1968.

3. "Nearly 2,000 Hear Nixon."

4. Bob Goddard, interview by author, New York, September 24, 2011.

5. Sam Boroda, interview.

6. Freeburg, "Nixon Hails Apollo in Ohio."

CHAPTER 25: THE WHITE MAN'S APOLLO

1. John Glatt, *Live at the Fillmore East and West: Getting Backstage and Personal with Rock's Greatest Legends*. Guilford, CT: Lyons Press, 2014, 128.

2. Glatt, *Live at the Fillmore East and West*, 125.

3. John Morris, interview.

4. Bill Graham and Robert Greenfield, *Bill Graham Presents: My Life inside Rock and Out*. New York: Doubleday, 1992, 231.

5. Chris Langhart, interview.

6. Marty Bennett, "N.Y. Rock n' Roll Scene Makes Great Leap Upward With Fillmore East Bow." *Variety*, March 14, 1968.

7. John Morris, interview.

8. Graham and Greenfield, 231.

9. Jerry Pompili, interview.

10. Joe McDonald, interview by author, Somerville, MA, May 9, 2012.

11. Amalie R. Rothschild and Ruth Ellen Gruber, *Live at the Fillmore East: A Photographic Memoir*. New York: Thunder's Mouth Press, 1999, 9.

12. John Morris, interview.

13. Bill Robar, interview.

14. Terry Hanley, interview.

15. Bill Robar, interview.

16. "May–September 1968 Jefferson Airplane Performance List." *Rock Prosopography 101*. Rockprosopography.com. http://rockprosopography101.blogspot.com/2009/07/may-september -1968-jefferson-airplane.html (accessed May 9, 2017).

17. Dick Butler, phone interview by author, January 24, 2016.

18. Bill Thompson, Jeffersonairplane.com. http://www.jeffersonairplane.com/ja-air/viewtopic .php?f=4&t=1714 (accessed May 9, 2017).

19. Jonathan Cott, "American Rock n' Roll Bands Descend on England." *Rolling Stone*, October 26, 1968.

20. Bill Hanley, interview.

21. Bill Hanley, interview.

CHAPTER 26: IT'S LOUD!

1. Morris Dickstein, *Gates of Eden: American Culture in the Sixties*. New York: Basic Books, 1977, 190.

2. Mark Farner, interview by author, Waltham, MA, July 28, 2012.

3. Lester Chambers, interview.

4. Judi Bernstein-Cohen, interview.

5. Bill Hanley, interview.

6. Bennett, "N.Y. Rock 'n' Roll Scene Makes Great Leap Upward."

7. Harold Cohen, interview.

8. Harold Cohen, interview.

9. Graham and Greenfield, *Bill Graham Presents*, 231.

10. Richard Goldstein, *Village Voice*, April 18, 1968.

11. Glatt, *Live at the Fillmore East and West*, 129.

12. Bill Hanley, interview.

13. Graham and Greenfield, 231.

14. John Chester, interview by author, High Bridge, NJ, September 24, 2011.

15. John Chester, interview.

16. Glatt, *Live at the Fillmore East and West*, 129.

17. John Chester, interview.

18. John Chester, interview.

19. John Chester, interview.

20. John Chester, interview.

21. John Chester, interview.

22. John Chester, interview.

23. Pridden, Bob, interview by author, Boston, MA, November 17, 2012.

24. Dinky Dawson, interview.

25. Kevin Young, "In Profile: Dinky Dawson, Concert Sound Pioneer." Fohonline.com. http:// www.prosoundweb.com/topics/audio/in_profile_dinky_dawson_concert_sound_pioneer (accessed April 13, 2017).

26. Bill Hanley, interview.

27. Glatt, *Live at the Fillmore East and West*, 145.

28. Bill Hanley, interview.

29. Joshua White, interview.

30. Billy Pratt, interview.

31. Billy Pratt, interview.

32. Bill Hanley, interview.

33. Clayton Riley, "The Golden Gospel of Reverend Ike." *New York Times*, March 9, 1975.

34. Bill Hanley, interview.

35. Rick Slattery, interviews by author, 2011–16.

36. Bill Hanley, interview.

37. Bob Goddard, interview.

38. Judi Bernstein-Cohen, interview.

39. Chris Langhart, interview.

CHAPTER 27: SIGHT AND SOUND AT THE FILLMORE

1. Joshua White, interview.

2. Joshua White, interview.

3. Joshua White, interview.

4. Joshua White, interview.

5. Greg Zinman, "Joshua Light Show 1967–68." Joshualightshow.com. http://www.joshualight show.com/about-classic/joshua-light-show-1967–68 (accessed 2012).

6. Barbara Bell, "You Don't Have to be High." *New York Times*, December 28, 1969.

7. Alfred Aronowitz, "Powder Ridge: What Is to Be Done?" *Rolling Stone*, December 10, 1970.

8. Bill Hanley, interview.

9. Joshua White, interview.

10. Jerry Pompili, interview.

11. Bill Hanley, interview.

12. Sam Andrew, interview.

13. Howard Smith, "Janis Joplin's Face, Seven Feet High—Groovy!" *Village Voice*, February 20, 1969.

14. Bill Hanley, interview.

15. Jerry Pompili, interview.

16. Karl Atkeson, interview.

17. Annie Fisher, "Riffs." *Village Voice*, October 1968.

18. Mike Jahn, "Psychedelic Music Still Keeps the Jefferson Airplane Flying." *New York Times*, December 2, 1968.

19. Joshua White, interview.

CHAPTER 28: GETTING PAID

1. "Hanley Financial Report February 1 1967." Hanley archive.

2. Judi Berstein-Cohen, interview.

3. Bill Hanley, interview.

4. "Generation, a Psychedelic Club, to Bow in 'Village.'"

5. Harold Cohen, interview.

6. Ray Fournier, interview.

7. Sam Boroda, interview.

8. Mike Jahn, "Last Rites Are Merry Ones at the Fillmore East." *New York Times*, June 29, 1971.

9. Graham and Greenfield, *Bill Graham Presents*, 337.

10. Annie Fisher, "Riffs." *Village Voice*, January 30, 1969.

CHAPTER 29: THE FESTIVAL EXPERIENCE

1. Bill Mankin, "We Can All Join In: How Rock Festivals Helped Change America." *Like the Dew*, March 4, 2012. https://likethedew.com/2012/03/04/we-can-all-join-in-how-rock-festivals-helped-change-america/.

2. Harvey Cox, "In Praise of Festivity." *Saturday Review*, October 25, 1969: 26.

3. Harvey Cox, "In Praise of Festivity," 25.

4. Robert Santelli, *Aquarius Rising: The Rock Festival Years*. New York: Dell, 1980, 2.

5. John Petkovic, "Record Rendezvous: Cleveland Cradle of Rock n' Roll Sits Empty, Awaits New Life." *Cleveland Plain Dealer*, July 9, 2017.

6. Jude Sheerin, "How the World's First Rock Concert Ended in Chaos." Bbc.com. http://www.bbc.com/news/magazine-17440514?SThisFB (accessed 2017).

7. Sheerin, "How the World's First Rock Concert Ended in Chaos."

8. Sheerin, "How the World's First Rock Concert Ended in Chaos."

9. Glynn, "Young Men, Big Job."

10. Bob Jones, interview.

11. Dickstein, *Gates of Eden*, 186.

12. Howard Smith, "Scenes." *Village Voice*, February 22, 1968.

13. Santelli, *Aquarius Rising*, 228.

14. Michael Lang, interview by author, Woodstock, NY, January 16, 2012.

15. Chris Langhart, interview.

16. David Crosby, "Music Is Love" (written by David Crosby/Neil Young/Graham Eric Nash). *If I Could Only Remember My Name* (Atlantic Records, 1971).

17. *Debra Thal, "Goose Lake vs Blues Festival." Michigan Daily, August 4, 1970.*

18. Mike Jahn, "Rock Festivals: Sometimes You Can Even Hear Music." *New York Times*, July 11, 1971.

CHAPTER 30: THE FESTIVAL PHENOMENON BEGINS

1. Michael Getz and John Dwork, *The Deadhead's Taping Compendium, Volume 1: An In-Depth Guide to the Music of the Grateful Dead on Tape, 1959–1974*. Vol. 1. New York: Holt Paperbacks, 1998, 9.

2. Tom Wolfe, *The Electric Kool-Aid Acid Test*. 1968; reprint, New York: Bantam Books, 1999, 392.

3. Oli Warwick, "The Genius of Don Buchla, by the Synth Disciples He Inspired." *FACT Magazine: Music News, New Music*. http://www.factmag.com/2016/09/26/the-genius-of-don-buchla-suzanne-ciani-morton-subotnick-surgeon-alessandro-cortini/ (accessed March 16, 2018).

4. Wolfe, *The Electric Kool-Aid Acid Test*, 258–63.

5. Mel Lawrence, phone interview by author, March 15, 2012.

6. Mel Lawrence, phone interview.

7. Mel Lawrence, phone interview.

8. Santelli, *Aquarius Rising*, 16.

9. Harvey Kubernik, *A Perfect Haze: The Illustrated History of the Monterey International Pop Festival*. Solana Beach, CA: Santa Monica Press, 2012.

10. Mel Lawrence, interview.

11. Kevin D. Greene, "The Greatest Music Festival in History." Nytimes.com. https://www.nytimes.com/2017/06/13/opinion/the-greatest-music-festival-in-history.html?_r=0 (accessed June 17, 2017).

12. Ben Sisario, "Monterey Pop, the Rock Festival That Sparked It All, Returns." *New York Times*, April 14, 2017. https://www.nytimes.com/2017/04/14/arts/music/monterey-pop-festival-50th-anniversary.html (accessed June 30, 2017).

13. Stephen K. Peeples, "Monterey International Pop Festival: The Book." Stephenkpeeples.com. http://stephenkpeeples.com/news-and-reviews/monterey-international-pop-festival-book/#8 (accessed June 17, 2017).

14. Lou Adler, quoted in Kubernik, *A Perfect Haze*, 46.

15. Lou Adler, quoted in Kubernik, *A Perfect Haze*, 46.

16. Beth Harris, "Backstage Beefs, Onstage Magic: Monterey Pop 50 Years Later." Chicagotribune.com. http://www.chicagotribune.com/lifestyles/travel/ct-monterey-pop-50-years-later-20170608-story.html (accessed June 17, 2017).

17. Mel Lawrence, interview.

18. Abe Jacob, interview.

19. Abe Jacob, interview.

20. Ken Lopez, interview.

21. Steve Katz, interview.

22. Roy Blumenfeld, interview.

23. Ken Lopez, interview.

24. Williams, *The Crawdaddy! Book*, 194.

25. Robb Baker, "The Sound." *Chicago Tribune*, August 17, 1969. Retrieved June 10, 2017.

26. Kubernik, *A Perfect Haze*, 159.

CHAPTER 31: ALL ROADS LEAD TO WOODSTOCK

1. Mel Lawrence, interview.

2. "Miami Pop Festival Production Notes Excerpts, December 1968." Tom and Barbara Rounds collection.

3. Stan Goldstein, phone interview by author, December 2, 2011.

4. Mel Lawrence, interview.

5. Ellen Sander, "The Miami Festival: An Inspired Bag of Pop." *New York Times*, January 12, 1969.

6. Harold Cohen, interview.

7. Harold Cohen, interview.

8. Harold Cohen, interview.

9. Sam Boroda, interview.

10. Sander, "The Miami Festival: An Inspired Bag of Pop."

11. Harold Cohen, interview.

12. Jean Young and Michael Lang, *Woodstock Festival Remembered*. New York: Ballantine Books, 1979.

13. Young and Lang, *Woodstock Festival Remembered*.

14. Stan Goldstein, interview.

15. Mel Lawrence, interview.

16. Young and Lang, *Woodstock Festival Remembered*, 11.

17. Stan Goldstein, interview.

18. Stan Goldstein, interview.

19. Bill Hanley, interview.

20. Bob Spitz, *Barefoot in Babylon: The Creation of the Woodstock Music Festival, 1969*. New York: Viking, 1979, 57.

21. Lee Mackler-Blumer, interview by author, New York, September 24, 2011.

22. David Marks, Skype interview by author, November 11, 2011.

23. Michael Lang with Holly George-Warren, *The Road to Woodstock*. New York: Ecco, 2009, 79–80.

24. Young and Lang, *Woodstock Festival Remembered*, 12.

25. Michael Lang, interview.

26. Mel Lawrence, interview.

27. Artie Kornfeld, "Kornfeld-Lang Adventures." Artiekornfeld-Woodstock.com. http://www.artiekornfeld-woodstock.com/Kornfeld-Lang%20Adventures.htm (accessed August 25, 2013).

28. Joel Rosenman, phone interview by author, March 2013.

29. Ticia Bernuth-Agri, phone interview by author, September 13, 2013.

30. Judi Bernstein-Cohen, interview.

31. Mel Lawrence, interview.

32. Bill Hanley, interview.

33. Jayne Wilson, phone interview by author, March 3, 2018.

34. Jayne Wilson, phone interview.

35. Jayne Wilson, phone interview.

36. Bill Hanley, interview.

CHAPTER 32: THE GOLDEN AGE OF THE ROCK FESTIVAL

1. Mankin, "We Can All Join In."

2. Cobb, "The Sounds of Rock for the Multitudes."

3. Santelli, *Aquarius Rising*, 266–68.

4. Barry Fey and Steve Alexander, *Backstage Past*. Phoenix, AZ: Lone Wolfe Press, 2011, 38.

5. Chip Monck, interview.

6. Fey and Alexander, *Backstage Past*, 42.

7. Bill Hanley, interview.

8. Bill Hanley, interview.

9. Santelli, *Aquarius Rising*, 97.

10. Santelli, *Aquarius Rising*, 97.

11. Bill Hanley, interview.

12. Santelli, *Aquarius Rising*, 98–99.

13. Fey and Alexander, *Backstage Past*, 44.

14. Santelli, *Aquarius Rising*, 99.

15. Bill Hanley, interview.

16. Fey and Alexander, *Backstage Past*, 44.

17. Bill Hanley, interview.

18. Wein and Chinen, *Myself Among Others*, 340.

19. Wein and Chinen, *Myself Among Others*, 341.

20. Wein and Chinen, *Myself Among Others*, 103.

21. John S. Wilson, "Unruly Rock Fans Upset Newport Jazz Festival." *New York Times*, July 7, 1969.

22. David Marks, interview.

23. Sander, "The Miami Festival: An Inspired Bag of Pop."

24. Maria Lameiras, "Legendary Music Promoter Ushered Rock & Roll into Atlanta." Eyecenter.emory.edu. http://www.eyecenter.emory.edu/alumni/15-giving-alex-cooley.htm (accessed November 28, 2015).

25. Alex Cooley, interview by Darlene Kelley, Atlanta, March 25, 2013.

26. Bill Hanley, interview.

27. "Alex Cooley Presents: The Live Music Experience, The Atlanta International Pop Festival." http://www.alexcooley.com/fest-atlpop1.html (accessed June 17, 2013).

28. Kathy Masterson, phone interview by author, October 2015.

29. Paul Beeman, "Music Fans Stay Orderly Despite Heat, Wine, Drugs." *Atlanta Journal-Constitution*, July 5, 1969.

30. Alex Cooley, interview.

31. Lameiras, "Legendary Music Promoter Ushered Rock & Roll into Atlanta."

32. Alun Vontillius, phone interview by author, October 2015.

33. Alun Vontillius, phone interview.

34. Alun Vontillius, phone interview.

35. Alun Vontillius, phone interview.

36. Beeman, "Music Fans Stay Orderly."

37. Paul Beeman, "Pop's the Thing Despite Heat at Hampton." *Atlanta Journal-Constitution*, July 7, 1969.

38. Nick Burns, interviews by author, 2013–15.

39. Beeman, "Pop's the Thing."

40. Beeman, "Pop's the Thing."

41. Beeman, "Music Fans Stay Orderly."

42. Mark Farner, interview.

43. Alex Cooley, interview.

CHAPTER 33: WOODSTOCK

1. Lang and George-Warren, *The Road to Woodstock*, 79–80.

2. Lang and George-Warren, *The Road to Woodstock*, 36–37.

3. "The Maverick Festival." Newpaltz.edu. http://www.newpaltz.edu/museum/exhibitions/maverick/maverick_festival.htm (accessed August 28, 2013).

4. "The Maverick Festival."

5. "The Maverick Festival."

6. Benjamin Genocchio, "100 Years Ago, When the Arts Found Woodstock." Nytimes.com. http://www.nytimes.com/2003/08/14/arts/100-years-ago-when-the-arts-found-woodstock.html (accessed October 29, 2017).

7. Alf Evers, *Woodstock: History of an American Town*. Woodstock, NY: Overlook Press, 1987, 654.

8. Evers, 665.

9. Jim Abbott, *Jackson C. Frank: The Clear, Hard Light of Genius: A Memoir*. Brooklyn, NY: Ba Da Bing Records, 2014, 112.

10. "Hippy Festival Upstate Is Cool Amid the Bonfires." *New York Times*, September 4, 1969.

11. Michael Lang, "1969: The (Other) Woodstock Festival, Woodstock Arts: It Happened in Woodstock." Rootsofwoodstock.com. http://rootsofwoodstock.com/tag/michael-lang (accessed 2015).

12. Cyril Caster, email message to author, December 22, 2013.

13. Lang and George-Warren, *Road to Woodstock*, 39.

14. Judi Bernstein-Cohen, interview.

15. Jan Hodenfield, "After Woodstock: Money and Smiles." *Rolling Stone*, October 4, 1969.

16. Hodenfield, "After Woodstock."

17. Rob Cox, "Woodstock: Peace, Love and Red Ink." Money.cnn.com. http://money.cnn.com/2009/08/14/news/companies/woodstock_no_profit.breakingviews/ (accessed June 21, 2017).

18. Patrick Lydon, "A Joyful Confirmation that Good Things Can Happen Here." *New York Times*, August 24, 1969.

19. Lydon, "A Joyful Confirmation."

20. Bill Hanley, interview.

21. Hodenfield, "After Woodstock."

CHAPTER 34: BY THE TIME HANLEY GETS TO WOODSTOCK

1. Ticia Bernuth-Agri, interview.

2. Bill Hanley, interview.

3. Bill Hanley, interview.

4. Ticia Bernuth-Agri, interview.

5. Ticia Bernuth-Agri, interview.

6. Lang and George-Warren, *The Road to Woodstock*, 116.

7. Mel Lawrence, interview.

8. Lang and George-Warren, *The Road to Woodstock*, 117.

9. Ticia Bernuth-Agri, interview.

10. Lang and George-Warren, *The Road to Woodstock*, 118.

11. Ticia Bernuth-Agri, interview.

12. Mel Lawrence, interview.

13. Mel Lawrence, interview.

14. Ticia Bernuth-Agri, interview.

15. Mel Lawrence, interview.

16. Lang and George-Warren, *The Road to Woodstock*, 119.

17. Lang and George-Warren, *The Road to Woodstock*, 120.

18. Lang and George-Warren, *The Road to Woodstock*, 120.

19. Joel Rosenman, John Roberts, and Robert Pilpel, *Young Men with Unlimited Capital: The Story of Woodstock*. Houston: Scrivenery Press, 1999, 99.

20. Young and Lang, *Woodstock Festival Remembered*, 18.

21. Lang and George-Warren, *The Road to Woodstock*, 122.

22. Bill Hanley, interview.

23. Bill Hanley, interview.

24. Lang and George-Warren, *The Road to Woodstock*, 128.

25. Brian Rosen, phone interview by author, February 2015.

26. Bill Robar, interview.

27. Dick Butler, interview.

28. Sam Boroda, interview.

CHAPTER 35: THE WOODSTOCK CREW

1. Sam Boroda, interview.

2. Harold Cohen, interview.

3. Harold Cohen, interview.

4. Sam Boroda, interview.

5. Sam Boroda, interview.

6. David Marks, interview.

7. David Marks, interview.

8. David Marks, interview.

9. David Marks, interview.

10. Billy Pratt, interview.

11. Nelson Niskanen, phone interview by author, March 1, 2017.

12. Terry Hanley, interview.

13. Terry Hanley, interview.

14. Harold Cohen, interview.

15. Rick Slattery, interview.

16. Chris Langhart, interview.

CHAPTER 36: GOING UP THE COUNTRY

1. Sam Boroda, interview.

2. Judi Bernstein-Cohen, interview.

3. Alfred Aronowitz, "Woodstock's Aquarian Playing Fields: An Alfalfa Slopes Date with Destiny." *New York Post*, August 12, 1969.

4. Aronowitz, "Woodstock's Aquarian Playing Fields."

5. Young and Lang, *Woodstock Festival Remembered*, 22.

6. Nick Burns, interview.

7. Nick Burns, interview.

8. Joel Makower, *Woodstock: The Oral History*. New York: Doubleday, 1989, 174.

9. Scott Holden, interview by author, Melrose, MA, March 6, 2013.

10. Scott Holden, interview.

11. Dick Butler, interview.

12. Nick Burns, interview.

13. Nick Burns, interview.

14. Nick Burns, interview.

15. David Marks, interview.

16. David Marks, interview.

17. David Marks, interview.

18. Brian Rosen, interview.

19. Lang and George-Warren, *The Road to Woodstock*, 128.

20. Brian Rosen, interview.

21. Bill Robar, interview.

22. Bill Robar, interview.

23. Rosenman, Roberts, and Pilpel, *Young Men with Unlimited Capital*, 133.

24. Rosenman, Roberts, and Pilpel, *Young Men with Unlimited Capital*, 133.

25. Sam Boroda, interview.

26. Sam Boroda, interview.

27. Scott Holden, interview.

28. Spitz, *Barefoot in Babylon*, 354.

29. Sam Boroda, interview.

30. Sam Boroda, interview.

31. Bill Hanley, interview.

32. David Crosby and Carl Gottlieb, *Long Time Gone: The Autobiography of David Crosby*. New York: Doubleday, 1988, 161.

33. Mick Richards, "Interview with Bill Belmont." *Virtue Films*, 1994–2013.

34. Billy Pratt, interview.

35. Tom Field, interview by author, Salem, MA, March 13, 2013.

36. John Chester, interview.

37. John Chester, interview.

38. Bob Goddard, interview.

39. Judi Bernstein-Cohen, interview.

40. Judi Bernstein-Cohen, interview.

41. Judi Bernstein-Cohen, interview.

CHAPTER 37: NO CONSOLE, NO FOOD

1. Bill Hanley, interview.

2. Tom Field, interview.

3. Harold Cohen, interview.

4. Harold Cohen, interview.

5. Harold Cohen, interview.

6. Scott Holden, interview.

7. Nelson Niskanen, interview.

8. Scott Holden, interview.

9. Bill Hanley, interview.

10. Harold Cohen, interview.

11. Brian Rosen, interview.

12. Harold Cohen, interview.

13. Harold Cohen, interview.

14. Lang and George-Warren, *The Road to Woodstock*, 79.

15. Sam Boroda, interview.

16. Judi Bernstein-Cohen, interview.

17. Brian Rosen, interview.

18. Billy Pratt, interview.

19. Nelson Niskanen, interview.

20. Bill Hanley, interview.

21. Brian Rosen, interview.

22. Scott Holden, interview.

23. Dick Butler, interview.

24. Nelson Niskanen, interview.

25. Nick Burns, interview.

CHAPTER 38: STAGE ANNOUNCEMENTS

1. John Morris, interview.

2. John Morris, interview.

3. Judi Bernstein-Cohen, interview.

4. Mitchell, Kevin M. "Grandfather of Rock n' Roll Productions." ChipMonck.com. http://www.chipmonck.com/making_it_happen.htm (accessed 2014).

5. John Morris, interview.

6. Joel Rosenman, interview.

7. "The Whiz Kids of Woodstock." *Louisville Courier Journal*, September 21, 1969.

8. Joshua White, interview.

9. Joe McDonald, interview.

10. Elliot Tiber and Tom Monte, *Taking Woodstock*. Garden City, NY: Square One, 2007, 201.

11. Nelson Niskanen, interview.

12. Judi Bernstein-Cohen, interview.

13. Judi Bernstein-Cohen, interview.

14. Judi Bernstein-Cohen, interview.

15. Judi Bernstein-Cohen, interview.

16. Spitz, *Barefoot in Babylon*, 432.

17. Spitz, *Barefoot in Babylon*, 432.

18. Spitz, *Barefoot in Babylon*, 432.

CHAPTER 39: THE NUTS, BOLTS, AND OCCASIONAL ZAP!

1. Tim Brookes, *Guitar: An American Life*. New York: Grove Press, 2006.

2. Blair Jackson and David Gans, *This Is All a Dream We Dreamed: An Oral History of the Grateful Dead*. New York: Flatiron Books, 2016, 131–32.

3. "Tired Rock Fans Begin Exodus." *New York Times*, August 18, 1969.

4. Bill Hanley, interview.

5. Nick Burns, interview.

6. Jack Casady, interview.

7. Makower, *Woodstock: The Oral History*, 278–79.

8. "The 'Young Men with Capital' Who Started Woodstock." Npr.org. http://www.npr.org/templates/transcript/transcript.php?storyId=111887950 (accessed June 22, 2017).

9. "The 'Young Men with Capital.'"

10. Chip Monck, interview.

11. McKinnon, "Woodstock: An Audio Milepost."

12. Chip Monck, interview.

13. Jackson and Gans, *This Is All a Dream We Dreamed*, 131–32.

14. Joshua White, interview.

15. Bob Pridden, interview.

16. Jack Casady, interview.

CHAPTER 40: WOODSTOCK, HOW SWEET THE SOUND . . . SYSTEM

1. "Bethel Pilgrims Smoke 'Grass' and Some Take LSD to 'Groove.'" *New York Times*, August 18, 1969.

2. Nick Ercoline and Bobbi Ercoline, interview by author, Bethel, NY, August 15, 2015.

3. Ercoline and Ercoline, interview.

4. Makower, *Woodstock: The Oral History*, 147.

5. Ken Kessler, *McIntosh: ". . . For the Love of Music . . ."* Binghamton, NY: McIntosh Laboratory, 2006, 234.

6. Kessler, 234.

7. John Morris, interview.

8. Spitz, *Barefoot in Babylon*, 56–57.

9. Joe McDonald, interview.

10. Makower, *Woodstock: The Oral History*, 147.

11. Bill Hanley, interview.

12. Makower, *Woodstock: The Oral History*, 147.

13. Spitz, *Barefoot in Babylon*, 370.

14. McKinnon, "Woodstock: An Audio Milepost."

15. McKinnon, "Woodstock: An Audio Milepost."

16. David Jacobs, "Woodstock Goes Gold." *EQ Live* 5, 1994. 60–64.

17. Kessler, *McIntosh*, 231.

18. Kessler, *McIntosh*, 231.

19. Kessler, *McIntosh*, 231.

20. Kessler, *McIntosh*, 231.

21. Jacobs, "Woodstock Goes Gold."

22. Jacobs, "Woodstock Goes Gold."

23. Sam Boroda, interview.

24. Chip Monck, interview.

25. Chip Monck, interview.

26. Bill Hanley, interview.

27. Sarah Benzuly, "It Was Only 40 Years Ago." Mixonline.com. http://www.mixonline.com/news/tours/it-was-only-40-years-ago/368168 (accessed May 28, 2017).

28. Jacobs, "Woodstock Goes Gold," 60–64.

29. Sam Boroda, interview.

30. Sam Boroda, interview.

31. Nick Burns, interview.

32. Bill Hanley, interview.

33. Jacobs, "Woodstock Goes Gold," 60–64.

34. Nelson Niskanen, interview.

35. Sam Boroda, interview.

36. Bill Hanley, interview.

37. Sam Boroda, interview.

38. Sam Boroda, interview.

39. Eddie Kramer, "Woodstock Interview by Mike Hobson." *Acoustic Sounds News*, August 18, 2004.

40. Bill Hanley, interview.

41. Bill Hanley, interview.

42. Chris Morton, "Eddie Kramer—From Jimi Hendrix to Woodstock." Floydslips.blogspot.com. http://floydslips.blogspot.com/2000/01/eddie-kramerfrom-jimi-hendrix-to.html (accessed June 10, 2017).

43. Eric Blackstead, producer. *Woodstock*, Cotillion, 1970.

44. "1969 Warner Brothers Letter." Hanley archive.

45. Bill Hanley, interview.

46. McKinnon, "Woodstock: An Audio Milepost."

47. Bill Hanley, interview.

48. Michael Wadleigh, Skype interview by author, August 23, 2013.

49. Nathan Cobb, "Weekend Records." *Boston Globe*, July 17, 1970.

50. Kessler, *McIntosh*, 231.

51. Kessler, *McIntosh*, 231.

52. Kessler, *McIntosh*, 233.

53. Kessler, *McIntosh*, 233.

54. Benzuly, "It Was Only 40 Years Ago."

55. Nelson Niskanen, interview.

56. Bill Hanley, interview.

57. Bill Hanley, interview.

58. Mike Evans and Paul Kingsbury, *Woodstock: Three Days That Rocked the World*. New York: Sterling, 2009, 56–57.

59. Chris Langhart, interview.

60. Santelli, *Aquarius Rising*, 171–72.

61. David Curry, "Deadly Day for the Rolling Stones." *Canberra Times*, December 5, 2009.

62. Santelli, *Aquarius Rising*, 173.

63. Robert Greenfield, *S.T.P.: A Journey through America with the Rolling Stones*. New York: Saturday Review Press, 1974. 70.

64. Bill Hanley, interview.

CHAPTER 41: I'M GOING HOME

1. Harold Cohen, interview.

2. Jocko Marcellino, interview by author, Warwick, RI, February, 18, 2017.

3. Nick Burns, interview.

4. Sam Boroda, interview.

5. Sam Boroda, interview.

6. Harold Cohen, interview.

7. Harold Cohen, interview.

8. Harold Cohen, interview.

9. Sam Boroda, interview.

10. Sam Boroda, interview.

11. Scott Holden, interview.

12. Dick Butler, interview.

13. Nick Burns, interview.

14. Nick Burns, interview.

15. Nick Burns, interview.

16. Scott Holden, interview.

17. Rick Slattery, interview.

18. Dick Cavett, *The Dick Cavett Show*. New York: NBC, August 19, 1969, season 3, episode 37.

19. Cavett, season 3, episode 37.

20. Rick Slattery, interview.

21. Cavett, season 3, episode 37.

22. Bill Hanley, interview.

23. David Marks, interview.

24. David Marks, "The Turtles Last Tour—Part 2." *3rd Ear Music: Hidden Years Archive Project*. 3rdearmusic.com. http://www.3rdearmusic.com/hyarchive/hiddenyearsstory/turtletour2.html (accessed May 30, 2013).

25. Marks, "The Turtles Last Tour."

26. Marks, "The Turtles Last Tour."

27. Heather Dugmore, "Free Peoples Concert." *Wits Review*, October 2017.

28. Dugmore, "Free Peoples Concert."

29. Dugmore, "Free Peoples Concert."

30. Dugmore, "Free Peoples Concert."

31. Marks, "The Turtles Last Tour."

32. Marks, "The Turtles Last Tour."

33. Jim Tarbell, phone interview by author, January 28, 2016.

34. Graham and Greenfield, *Bill Graham Presents*, 336.

35. Bill Hanley, interview.

36. Bernstein and Arthur. *It's Sid Bernstein Calling*, 224.

CHAPTER 42: MADISON SQUARE GARDEN AND THE UNIONS

1. "New Garden Opening in 1967 to Cater to Fan." *New York Times*, January 2, 1966.

2. "New Garden Opening."

3. Harold C. Schonberg, "Music: Echoless Caverns; Felt Forum, Meant for Sport, Lacks Bass." *New York Times*, February 4, 1968.

4. Schonberg, "Music: Echoless Caverns."

5. Gary Keller, phone interview by author, September 30, 2012.

6. Bill Hanley, interview.

7. Agis Salpukas, "Irving M. Felt, 84, Sports Impresario, Is Dead." *New York Times*, September 24, 1994.

8. David Hughes, phone interview by author, October 2012.

9. John Scher, phone interview by author, August 4, 2017.

10. Mike Jahn, "20,000 Hear Doors Give Rock Concert in a Packed Garden." *New York Times*, January 25, 1969.

11. George Knemeyer, "Doors' Sound System Adds to Challenge of Traveling." *Billboard*, February 28, 1970.

12. Knemeyer, "Doors' Sound System."

13. Robert Christgau, "Are the 'Blind' Keeping Faith?" *New York Times*, July 20, 1969.

14. Lucian Truscott IV, "Blind Faith." *Village Voice*, July 17, 1969.

15. Mike Jahn, "Blind Faith Group Sings." *New York Times*, July 14, 1969.

16. Truscott, "Blind Faith."

17. Bill Hanley, interview.

18. Gary Keller, interview.

19. Bill Hanley, interview.

20. David Hughes, interview.

21. "Donovan advertisement." Hanley archive.

22. Mike Jahn, "Garden (Madison Sq.) Is Ideal for the Ever-Florid Donovan." *New York Times*, October, 18, 1969.

23. Gary Keller, interview.

24. Bill Hanley, interview.

25. Gary Keller, interview.

26. Bill Hanley, interview.

27. Donna Halper, email message to author, January 1, 2014.

28. John Scher, interview.

29. David Hughes, interview.

30. Nick Burns, interview.

31. Dick Butler, interview.

32. McKinnon, "Woodstock: An Audio Milepost," 71–72.

33. John Scher, interview.

34. Brian Rosen, interview.

35. Bill Hanley, interview.

36. "An Interview with Bob See." Plsn.com. http://plsn.com/articles/plsn-interview/bob-see/ (accessed October 4, 2015).

37. Richard Cadena, *Electricity for the Entertainment Electrician and Technician*. Waltham, MA: Focal Press, 2014.

38. Nick Burns, interview.

39. Dick Butler, interview.

40. Dick Butler, interview.

41. Sam Boroda, interview.

42. Marks, "The Turtles Last Tour."

43. Marks, "The Turtles Last Tour."

44. John S. Wilson, "Joan Baez Turns a Packed Garden into a Living Room." *New York Times*, August 9, 1969.

45. Marks, "The Turtles Last Tour."

46. Bill Hanley, interview.

47. Wilson, "Joan Baez."

48. Brian Rosen, interview.

49. Brian Rosen, interview.

50. Alfred Aronowitz, "The Band at the Felt Forum." *New York Post*, December 29, 1969.

51. Brian Rosen, interview.

CHAPTER 43: THE GREATEST ROCK AND ROLL BAND IN THE WORLD

1. Robert Christgau, "The Rolling Stones." Entry in *The Rolling Stone History of Rock and Roll*. New York: Random House, 1980, 198–99.

2. "Hanley invoice." Hanley archive.

3. Greenfield, *S. T.P.*, 10.

4. Loraine Alterman and John Burks, *Rolling Stone*, December 27, 1969. Issue 49.

5. "The Rolling Stones Chronicle 1969." Timeisonourside.com. http://www.timeisonourside .com/chron1969.html (accessed October 8, 2015).

6. Alterman and Burkes, *Rolling Stone*.

7. G. Brown, *Colorado Rocks!: A Half-century of Music in Colorado.* Westwinds Press, 2004.

8. Robert Hilburn, "Rolling Stones Date." *Los Angeles Times*, 1969.

9. Hilburn, "Rolling Stones Date."

10. Fred Kirby, "To Promoter Stein, Theaters the Thing for Rock Shows to Spring." *Billboard*, November 15, 1969.

11. Joel Selvin, *Altamont: The Rolling Stones, The Hells Angels, and the Inside Story of Rock's Darkest Day.* New York: Harper Collins, 2016.

12. Francis X. Clines, "16,000 at Madison Square Garden Shout with Joy in Reaction to Sounds of Rolling Stones." *New York Times*, November 28, 1969.

13. Bill Hanley, interview.

14. Bill Hanley, interview.

15. Gary Keller, interview.

16. Gary Keller, interview.

17. Bill Hanley, interview.

18. Gary Keller, interview.

19. Bill Hanley, interview.

20. Chris Langhart, interview.

21. Bob Goddard, interview.

22. Bob Goddard, interview.

23. David Hughes, interview.

24. Gary Keller, interview.

25. Sam Cutler, Skype interview by author, May 17, 2012.

26. Bill Hanley, interview.

27. Rick Slattery, interview.

28. Gary Keller, interview.

29. David Hughes, interview.

30. Bill Hanley, interview.

31. Jack Weisberg, phone interview by author, May 5, 2012.

32. Fritz Postlethwaite, phone interview by author, June 14, 2013.

CHAPTER 44: DEMONSTRATION SOUND

1. Tom Field, interview.

2. Ron Eyerman and Andrew Jamison, *Music and Social Movements: Mobilizing Traditions in the Twentieth Century.* Cambridge: Cambridge University Press, 1998, 107–8.

3. Peter Doggett, *There's a Riot Going On: Revolutionaries, Rock Stars, and the Rise and Fall of the '60s.* Edinburgh: Canongate, 2007, 6.

4. Eyerman and Jamison, *Music and Social Movements*, 106.

5. Jeff Kisseloff, *Generation on Fire: Voices of Protest from the 1960s: An Oral History.* Lexington: University Press of Kentucky, 2007, 204.

6. Bill Hanley, interview.

7. Bradford Lyttle, "A Report and Comments from the Viewpoint of a Practical Organizer: Washington Action November 13–15, 1969." Self-published, 1969, 23.

8. Lyttle, "A Report and Comments," 23.

9. Lyttle, "A Report and Comments," 23.

10. Bradford Lyttle, phone interview by author, September 15, 2011.

11. Bill Hanley, interview.

12. Lyttle, "A Report and Comments," 23.

13. Bradford Lyttle, interview.

14. Lyttle, "A Report and Comments," 23.

15. Lyttle, "A Report and Comments," 23.

16. Bill Hanley, interview.

17. Henry Kissinger and Clare Boothe Luce, *The White House Years*. Boston: Little, Brown, 1979, 288.

18. Kissinger and Luce, *The White House Years*, 288.

19. Lucy G. Barber, *Marching on Washington: The Forging of an American Political Tradition*. Berkeley: University of California Press, 2002, 196.

CHAPTER 45: THE MARCH AGAINST DEATH

1. David N. Hollander, "Boston: 100,000 Rally." *Harvard Crimson*, October 16, 1969.

2. Lyttle, "A Report and Comments," 23.

3. Lyttle, "A Report and Comments," 23.

4. Bill Hanley, interview.

5. Fred Halstead, *Out Now! A Participant's Account of the American Movement against the Vietnam War*. New York: Monad Press, 1978, 515.

6. John Herbers, "250,000 War Protesters Stage Peaceful Rally in Washington." *New York Times*, November 16, 1969.

7. Herbers, "250,000 War Protesters."

8. Peter Yarrow, interview.

9. Herbers, "250,000 War Protesters."

10. "Moratorium Sidelights." *Boston Globe*, November 16, 1969. Hanley archive.

11. "Moratorium Sidelights."

12. "Galaxy of Stars to Make Capital an M (Music & Moratorium) Day." *Billboard*, November 22, 1969.

13. Rob Kirkpatrick, *1969: The Year Everything Changed*. New York: Skyhorse, 2009, 208.

14. "Galaxy of Stars."

15. Bradford Lyttle, interview.

16. Halstead, *Out Now!*, 514.

17. Campbell Hair, phone interview by author, September 15, 2012.

18. Bill Hanley, interview.

19. Alfred Aronowitz, "The DC March: Many Followed, None Led." *New York Post*, November 17, 1969.

20. Rick Wurpel, email message to author, March 24, 2015.

21. Campbell Hair, interview.

22. Aronowitz, "The DC March."

23. David Butkovich, phone interview by author, February 12, 2013.

24. Butkovich, interview.

25. Bill Hanley, interview.

26. Wavy Gravy, interview by author, Bridgeport, CT, July 21, 2012.

27. Bradford Lyttle, interview.

28. Peter Yarrow, interview.

29. Al Braden, phone interview by author, May 15, 2012.

30. Campbell Hair, interview.

31. Bradford Lyttle, interview.

32. Bradford Lyttle, interview.

33. Bill Hanley, interview.

34. Peter Yarrow, interview.

35. Peter Yarrow, interview.

CHAPTER 46: STUDENT STRIKE

1. Halstead, *Out Now!*, 537.

2. Jerry M. Lewis and Thomas R. Hensley, "The May 4 Shootings at Kent State University: The Search for Historical Accuracy." kent.edu. https://www.kent.edu/may-4-historical-accuracy. (accessed July 3, 2013).

3. Bradford Lyttle, "May Ninth." Self-published, 1970, 16.

4. Barber, *Marching on Washington*, 187–88.

5. Barber, *Marching on Washington*, 187.

6. Halstead, *Out Now!*, 549.

7. Bradford Lyttle, interview.

8. Bradford Lyttle, interview.

9. Bill Hanley, interview.

10. Lyttle, "May Ninth," 16.

11. Peter Yarrow, interview.

CHAPTER 47: MAY DAY

1. "Volunteers Press War Protest Plan." *New York Times*, April 11, 1971.

2. "Volunteers Press War Protest Plan."

3. Bradford Lyttle, interview.

4. Barber, *Marching on Washington*, 204.

5. Bob Ruff, "Mayday Tribe Disrupts DC—12,000 Arrested." *The Heights*, vol. 1, no. 29, May 7, 1971.

6. Barber, *Marching on Washington*, 205.

7. Lyttle, "May Ninth," 16.

8. Lyttle, "May Ninth," 16.

9. Ruff, "Mayday Tribe Disrupts DC."

10. "Washington Protests Set." *Toledo Blade*, April 13, 1971.

11. Ruff, "Mayday Tribe Disrupts DC."

12. Barber, *Marching on Washington*, 205.

13. Ruff, "Mayday Tribe Disrupts DC."

14. Barber, *Marching on Washington*, 205.

15. Rhoda Rosenberg, interview by author, Merrimac, MA, July 28, 2011.

16. Rhoda Rosenberg, interview.

17. Bradford Lyttle, interview.

18. "Washington Protests Set."

19. Barber, *Marching on Washington*, 206.

20. Bradford Lyttle, interview.

21. Bradford Lyttle, interview.

22. Bradford Lyttle, *The Autobiography of Bradford Lyttle*. Chicago: Unpublished, 2012.

CHAPTER 48: A SYMPOWOWSIUM AND THE FATE OF THE FESTIVAL

1. Wavy Gravy, interview by author, Bridgeport, CT, July 21, 2012.
2. Wavy Gravy, interview.
3. Marks, "The Turtles Last Tour."
4. Marks, "The Turtles Last Tour."
5. Marks, "The Turtles Last Tour."
6. Marks, "The Turtles Last Tour."
7. Marks, "The Turtles Last Tour."
8. Lisa Law, email message to author, February 28, 2013.
9. Bill Hanley, interview.
10. Reno Kleen, phone interview by author, March 13, 2013.
11. Reno Kleen, interview.
12. Tom Law, interview by author, New York, February 3, 2013.
13. Reno Kleen, interview.
14. Santelli, *Aquarius Rising*, 151–52.
15. Tom Law and Reno Kleen, "Coming to Meet." *Sympowowsium Report* 1, 1969.
16. Reno Kleen, interview.

CHAPTER 49: WOODSTOCK SOUTH

1. Tom Law and Reno Kleen, "Coming to Meet."
2. David Butkovich, interview.
3. David Butkovich, interview.
4. Campbell Hair, interview.
5. Campbell Hair, interview.
6. David Butkovich, interview.
7. Campbell Hair, interview.
8. Ben Franklin, "Rock Festival a Show of Youthful Good Manners." *New York Times*, December 1, 1969.
9. Charles Ralls, "Judge Gives Ok on Rock Festival." *Palm Beach Daily News*, November 21, 1969.
10. "Cold Downpour Fails to Dampen Spirits at Florida Rock Festival." *New York Times*, November 30, 1969.
11. Franklin, "Rock Festival."
12. David Butkovich, interview.
13. Campbell Hair, interview.
14. Hugh Romney, *The Hog Farm and Friends*. New York: Links, 1974, 104.
15. Sam Boroda, interview.
16. Alun Vontillius, interview.
17. Bill Hanley, interview.
18. Bill Hanley, interview.
19. Bill Hanley, interview.
20. Bill Hanley, interview.
21. Bill Hanley, interview.
22. Campbell Hair, interview.
23. Romney, *The Hog Farm*, 105.
24. "Rolling Stones Haggle on Lighting." *Palm Beach Post*, November 26, 1969.
25. Bill Hanley, interview.
26. Alun Vontillius, interview.
27. Nick Burns, interview.

28. Romney, *The Hog Farm*, 106.

29. Sam Cutler, "Rolling Stones: West Palm Beach Pop Festival." Recording, unknown source. December 1, 1969.

30. Romney, *The Hog Farm*, 106.

31. Sam Cutler, interview.

32. Billy Pratt, interview.

33. Nick Burns, interview.

34. Bill Hanley, interview.

35. Nick Burns, interview.

36. Nick Burns, interview.

37. David Butkovich, interview.

38. Campbell Hair, interview.

CHAPTER 50: WINTERS END

1. Parker Donham, "Promoters' Dreams of Another Woodstock Become Nightmare." *Boston Globe*, March 22, 1970.

2. *Rolling Stone* 58, May 14, 1970.

3. "1970 Winters End Festival Brochure." Kane personal collection.

4. Thomas King Forcade, "Winter's End: Free the Rock & Roll Six." *San Diego Free Door*, April 23, 1970: 8–12.

5. Bill Hanley, interview.

6. Romney, *The Hog Farm*, 116.

7. Donham, "Promoters' Dreams."

8. Sam Boroda, interview.

9. Sam Boroda, interview.

10. Forcade, "Winter's End."

11. Bob Fiallo, phone interview, July 17, 2017.

12. Romney, *The Hog Farm*, 116.

13. Forcade, "Winter's End."

14. Forcade, "Winter's End."

15. Romney *The Hog Farm*, 117.

16. Bill Hanley, interview.

17. Romney, *The Hog Farm*, 117.

18. Romney, *The Hog Farm*, 117.

19. Bob Fiallo, "Rock Show Vibrates, and Then Dissipates." *Tampa Tribune*, March 30 1970.

20. Sam Boroda, interview.

21. Romney, *The Hog Farm*, 117.

22. Fiallo, "Rock Show Vibrates."

23. "Dude Ranch Rock Festival Fades Away During Night." *Evening Independent*, March 30, 1970.

24. "Dude Ranch Rock Festival."

25. "Florida Rock Show Promoters Arrested." *Chicago Tribune*, March 30, 1970.

26. Donham, "Promoters' Dreams."

CHAPTER 51: OH, CANADA

1. D.D. Hill, phone interview by author, June 18, 2012.

2. Johnny Brower, phone interview by author, June 27, 2012.

3. Nick Burns, interview.

4. Nick Burns, interview.

5. Nick Burns, interview.

6. David Marks, interview.

7. David Marks, interview.

8. David Marks, interview.

9. Nick Burns, interview.

10. Shep Gordon, Skype interview by author, March 9, 2016.

11. Johnny Brower, interview.

12. Dave Bist, "Could Woodstock Happen Again? Well . . ." *Montreal Gazette*, May 9, 1970.

13. Herbert Aronoff, "A Sound Question: What Binds Nixon and Hoffman." *Montreal Gazette*, May 30, 1970.

14. Aronoff, "A Sound Question."

15. Alun Vontillius, interview.

16. McKinnon, "Woodstock: An Audio Milepost."

17. Rusty Brutsché, interview.

18. Judi Bernstein-Cohen, interview.

19. Judi Bernstein-Cohen, interview.

20. Jahn, "Rock Festivals: Sometimes You Can Even Hear Music."

21. "Newspaper clipping." *Montreal Star*, May 11, 1970. Hanley archive.

22. Herbert Aronoff, "Pop Festival: Twelve Happy Hours." *Montreal Gazette*, May 11, 1970.

23. "Newspaper clipping," *Montreal Star*.

24. "Newspaper clipping," *Montreal Star*.

25. "The Rolling Rock Festival." *San Francisco Chronicle*, May 2, 1970.

26. "Festival Express Rolls June 24." *Billboard*, May 16, 1970.

27. Rick Slattery, interview.

28. David Kingsbury, phone interview by author, March 2017.

29. Nelson Niskanen, interview.

30. Tom Field, interview.

31. Judi Bernstein-Cohen, interview.

32. Rick Slattery, interview.

33. Sam Boroda, interview.

34. Nelson Niskanen, interview.

35. Rick Slattery, interview.

36. Sam Boroda, interview.

37. Judi Bernstein-Cohen, interview.

38. David Kingsbury, interview.

39. Tom Field, interview.

40. Judi Bernstein-Cohen, interview.

41. Nelson Niskanen, interview.

42. David Kingsbury, interview.

43. Rick Slattery, interview.

44. David Kingsbury, interview.

45. Sam Boroda, interview.

46. Tom Field, interview.

47. Nelson Niskanen, interview.

48. Sam Boroda, interview.

49. Sam Boroda, interview.

50. Judi Bernstein-Cohen, interview.

51. Rick Slattery, interview.

52. "Metro Police Cool Off Rock Festival Gate-Crashers." *Toronto Daily Star*, July 29, 1970.

53. "Toronto's Festival Express Fails to Meet Expectations." *Cash Box*, July 7, 1970.

54. Joe Fernbacher, "Swinging Rock Music Redeems Synthetic Atmosphere." *The Spectrum* (Toronto), July 2, 1970.

55. David Kingsbury, interview.

56. Nelson Niskanen, interview.

57. "Festival Express Marred by Protests, Poor Attendance." *Billboard*, July 25, 1970.

58. Dave Bluestein, phone interview by author, May 10, 2012.

59. Dave Bluestein, phone interview.

60. D.D. Hill, interview.

61. David Kingsbury, interview.

62. Sam Boroda, interview.

63. David Kingsbury, interview.

64. Rick Slattery, interview.

65. "Might As Well" (lyrics by Robert Hunter, music by Jerry Garcia). *The Complete Annotated Grateful Dead Lyrics*. Annotated by David Dodd. New York: Free Press, 2005, 258.

CHAPTER 52: HOT 'LANTA AND LOVE VALLEY

1. Alan Sisson, email message to author, December 2, 2013.

2. Alan Sisson, email message.

3. Alan Sisson, email message.

4. Alan Sisson, email message.

5. Aronoff, "A Sound Question: What Binds Nixon and Hoffman."

6. "Cooley Papers." Georgia State University, special collections.

7. Alex Cooley, phone interview by author, March 25, 2013.

8. Alun Vontillius, interview.

9. Alun Vontillius, interview.

10. Alun Vontillius, interview.

11. Bob Fiallo, "Impressions of Atlanta." *Tampa Tribune*, July 11, 1970.

12. Fiallo, "Impressions of Atlanta."

13. Nick Burns, interview.

14. Nick Burns, interview.

15. Bill Hanley, interview.

16. Nick Burns, interview.

17. Alun Vontillius, interview.

18. Nick Burns, interview.

19. Alan Sisson, email message.

20. Alan Sisson, email message.

21. Sam Boroda, interview.

22. Fiallo, "Impressions of Atlanta."

23. Nick Burns, interview.

24. Alan Sisson, email message.

25. Brian Rosen, interview.

26. Alun Vontillius, interview.

27. Nick Burns, interview.

28. Rusty Brutsché, interview.

29. David Oedel, "Sound Management." *D Magazine*, January 1, 1980.

30. "Drug Cases Swamp Doctors in Georgia at Rock Festival." *New York Times*, July 5, 1970.

31. Kathy Masterson, interview.

32. Kathy Masterson, interview.

33. Tom Williams, "Atlanta Pop Fest a Gate Crusher." *Billboard*, July 18, 1970.

34. Kathy Masterson, interview.

35. Alex Cooley, interview.

36. Bill Hanley, interview.

37. Parke Puterbaugh, "Remembering N.C.'s Woodstock." *Greensboro News & Record*, July 10, 2010.

38. Bill Hanley, interview.

CHAPTER 53: A PORTRAIT OF AN INTERRUPTED FESTIVAL AND GRAND FUNK RAILROAD

1. "Randall's Island Letter." Kane personal collection.

2. "Randall's Fest Hit Problems." *Billboard*, August 8, 1970.

3. "Randall's Island Letter."

4. "Randall's Fest Hit Problems."

5. Carman Moore, "A Study in Mistrust." *Village Voice*, 1970.

6. Moore, "A Study in Mistrust."

7. Bill Hanley, interview.

8. Moore, "A Study in Mistrust."

9. "Randall's Fest Hit Problems."

10. "Randall's Fest Hit Problems."

11. Alan Sisson, email message to author, December 2, 2013.

12. Mike Neal, interview.

13. Alan Sisson, email message.

14. Nick Burns, interview.

15. Sam Boroda, interview.

16. "Randall's Fest Hit Problems."

17. Richard Knox, "When Hard Rock Meets and Eardrum." *Boston Globe*, 1970.

18. David Pelletier, interview.

19. David Pelletier, interview.

20. Kathy Orloff, "Bogdanovich Runs a Sound Business." *Los Angeles Times*, October 11, 1970.

21. Barry McKinnon, "An Interview with Bill Hanley." http://www.prosoundweb.com/article/re_p_files_an_interview_with_bill_hanley.

22. McKinnon, "An Interview with Bill Hanley."

23. Bill Hanley, interview.

24. Sam Boroda, interview.

25. Mark Farner, interview.

26. Moore, "A Study in Mistrust."

27. McKinnon, "An Interview with Bill Hanley."

28. McKinnon, "An Interview with Bill Hanley."

29. "Top 100 in Rock." *Esquire*, July 1, 1970. Hanley archive.

30. Moore, "A Study in Mistrust."

CHAPTER 54: THE FESTIVAL THAT NEVER WAS

1. Santelli, *Aquarius Rising*, 196.

2. Santelli, *Aquarius Rising*, 196.

3. Santelli, *Aquarius Rising*, 196.

4. "30,000 Inch Away from Powder Ridge." United Press International. Hanley archive.

5. Robert Piasecki, "Plenty of Sex and Drugs, But No Rock 'n' Roll: What Happens When the Courts Block a Concert and 15,000 Members of the Drug-Taking Free-Love Crowd Show Up to Party?" *Hartford Advocate*, August 25, 2005.

6. Bill Hanley, interview.

7. Sam Boroda, interview.

8. Alan Sisson, email message to author, December 2, 2013.

9. Scott Holden, interview.

10. Harold Cohen, interview.

11. Nelson Niskanen, interview.

12. Nick Burns, interview.

13. Rick Slattery, interview.

14. David Kingsbury, interview.

15. David Kingsbury, interview.

16. Santelli, *Aquarius Rising*, 200.

17. "Resort Owner Tells 15,000 Fans Rock Festival Will Not Be Held." *New York Times*, July 30, 1970.

18. "Resort Owner Tells 15,000 Fans."

19. Bill Hanley, interview.

20. Melanie Safka, interview.

21. David Kingsbury, interview.

22. "Music Is Heard at Rock Festival." *New York Times*, August 1, 1970.

23. Bill Hanley, interview.

24. Melanie Safka, interview.

25. Santelli, *Aquarius Rising*, 200.

26. David Kingsbury, interview.

27. Santelli, *Aquarius Rising*, 200.

28. Sam Boroda, interview.

29. Joseph Treaster, "30,000 Swarm to Ski Area despite Ban on Rock Fest." *New York Times*, July 31, 1970.

30. Alan Sisson, email message.

31. David Kingsbury, interview.

32. David Kingsbury, interview.

33. Piasecki, "Plenty of Sex and Drugs."

34. Bill Hanley, interview.

35. Rick Slattery, interview.

36. Alan Magary, "Happenings on the Hill." *Middletown Press*, July 31, 1970.

37. *Medford Mercury*, August 4, 1970.

38. Bill Hanley, interview.

39. Alan Sisson, email message.

40. Bill Hanley, interview.

41. Santelli, *Aquarius Rising*, 203.

42. Santelli, 201.

43. Jahn, "Rock Festivals: Sometimes You Can Even Hear Music."

44. "4-Year Term Given to Rock Promoter." *New York Times*, January 7, 1972.

45. Nelson Niskanen, interview.

46. Alfred Aronowitz, "Powder Ridge: What Is to Be Done?" *Rolling Stone*, December 10, 1970.

47. Rick Slattery, interview.

48. Harold Cohen, interview.

49. Alfred Aronowitz, "Pop News." *New York Daily News*, April 1, 1972.

50. Aronowitz, "Pop News."

CHAPTER 55: A FESTIVAL UNDER THE SUN

1. Reniet Rivera Ramirez, "Interview with Alex Cooley." Marysolpopfestival.com. http://www.marysolpopfestival.com/interviews/alex-cooley-interview (accessed 2015).

2. Ramirez, "Interview with Alex Cooley."

3. Ramirez, "Interview with Alex Cooley."

4. "Alex Cooley Presents: The Live Music Experience, Mar y Sol Puerto Rico International Pop Festival." http://www.alexcooley.com/fest-atlpop1.html (accessed June 17, 2013).

5. Alex Cooley, interview.

6. Bill Hanley, interview.

7. Tom Field, interview.

8. Rachel E. Lyons, "Walter Lenk Interview." *New Orleans Jazz & Heritage Foundation*, November 11, 2014.

9. Walter Lenk, interview by author, Cambridge, MA, December 10, 2012.

10. Les Ledbetter, "Puerto Rico Fete Draws Thousands." *New York Times*, April, 1 1972.

11. Kathy Masterson, interview.

12. Kathy Masterson, interview.

13. Alfred Aronowitz, "Pop News." *New York Daily News*, April 1, 1972.

14. Ledbetter, "Puerto Rico Fete."

15. Santelli, *Aquarius Rising*, 230.

16. Ledbetter, "Puerto Rico Fete."

17. Santelli, *Aquarius Rising*, 230.

18. Ramirez, "Interview with Alex Cooley."

19. Walter Lenk, interview.

20. Joseph Lee Hooker, phone interview by author, October 2017.

21. Aronowitz, "Pop News."

22. Alex Cooley, interview.

23. Bill Hanley, interview.

24. Kathy Masterson, interview.

25. Santelli, *Aquarius Rising*, 232.

26. Walter Lenk, interview.

27. "Alex Cooley Presents: Mar y Sol Puerto Rico International Pop Festival."

28. Kathy Masterson interview.

29. Alex Cooley, interview.

30. Manny Suarez, "Police Seek Pop Festival Organizer." *The Star*, April 6, 1972.

31. Glenn Abbott, phone interview by author, February 2012.

32. Charles Giulliano, "Last Rock Festival."

33. Aronowitz, "Pop News."

CHAPTER 56: HANLEY'S SOUTHERN TRUCK, NOLA, AND THE GURU MAHARAJ JI

1. Rick Slattery, interview.

2. Jim Wiggins, phone interview by author, April 2015.

3. Jim Wiggins, phone interview.

4. Jim Wiggins, phone interview.

5. Jim Wiggins, phone interview.

6. Phil Tripp, phone interview by author, November 2015.

7. Wein and Chinen, *Myself Among Others*, 366.

8. Rachel E. Lyons, "Bill Hanley Interview." *New Orleans Jazz & Heritage Foundation*, October 30, 2014.

9. Rachel E. Lyons, "Walter Lenk Interview." *New Orleans Jazz & Heritage Foundation*, November 11, 2014.

10. Bob Jones, interview.

11. Laura Loughlin, email message to author, December 2015.

12. Lyons, "Walter Lenk Interview."

13. Phil Tripp, interview.

14. Rachel E. Lyons, "Rhoda Rosenberg Interview." *New Orleans Jazz & Heritage Foundation*, October 30, 2014.

15. Phil Tripp, interview.

16. Lyons, "Bill Hanley Interview."

17. Lyons, "Bill Hanley Interview."

18. Lyons, "Walter Lenk Interview."

19. Kathy Masterson, interview.

20. Lyons, "Walter Lenk Interview."

21. Alan Kaufman, phone interview by author, December 2015.

22. Kathy Masterson, interview.

23. Lyons, "Bill Hanley Interview."

24. Lyons, "Bill Hanley Interview."

25. Kathy Masterson, interview.

26. Sam Boroda, interview.

27. Bill Hanley, interview.

28. Fritz Postlethwaite, interview.

29. Bill Hanley, interview.

30. Lyons, "Walter Lenk Interview."

31. Bill Hanley, interview.

32. Fritz Postlethwaite, interview.

33. Fritz Postlethwaite, interview.

34. Eleanor Blau, "Revered Guru, 15, Brings His Mission Here." *New York Times*, July 28, 1973.

35. Blau, "Revered Guru."

36. Bill Hanley, interview.

37. Fritz Postlethwaite, interview.

38. Bill Hanley, interview.

39. "The Perfect Master in the Age of McLuhan." *Divine Times*, Millennium '73 ed., 1973.

40. Bill Hanley, interview.

41. Fritz Postlethwaite, interview.

42. Bill Hanley, interview.

43. Blau, "Revered Guru."

44. "1975 Sound Specifications Rider, Hans Jayanti Festival." Hanley archive.

45. Bill Hanley, interview.

46. Fritz Postlethwaite, interview.

47. Fritz Postlethwaite, interview.

48. Fritz Postlethwaite, interview.

CHAPTER 57: HANLEY'S INVENTION

1. "Magic Stage Manual." Hanley archive.

2. "Magic Stage Manual."

3. Bill Hanley, interview.

4. Jerry Wenger, interview.

5. Bill Hanley, interview.

6. Bill Hanley, interview.

7. Bill Hanley, interview.

8. Fritz Postlethwaite, interview.

9. Bill Hanley, interview.

10. *Medford Daily Mercury*, January 2, 1981. Hanley archive.

11. Stacy V. Jones, "Patents." *New York Times*, November 15, 1980.

12. Rhoda Rosenberg, interview.

13. Bill Hanley, interview.

14. *Medford Daily Mercury*.

15. "Magic Stage Manual."

16. "Magic Stage Manual."

17. "Magic Stage Manual."

18. Dana Puopolo, phone interview by author, September 30, 2015.

19. "Magic Stage Manual."

20. "Magic Stage Manual."

21. "Magic Stage Manual."

22. John Scher, interview.

23. John Scher, interview.

24. John Scher, interview.

25. John Rockwell, "The Pop Life." *New York Times*, Jun 23, 1978.

26. Rockwell, "The Pop Life."

27. "Dennis Wilson Letter." Hanley archive.

28. "Magic Stage Manual."

29. "John Scher Letter." Hanley archive.

30. "Magic Stage Manual."

31. "John Scher Letter."

32. "Teddy Powell Letter." Hanley archive.

33. "Magic Stage Manual."

34. Phil Tripp, interview.

35. Bill Hanley, interview.

36. John Scher, interview.

37. Rick Slattery, interview.

38. Stan Goldstein, interview.

39. Jim Cowen, "MA Who's Who." *Boston Globe*, November 15, 1980.

40. Bill Hanley, interview.

CHAPTER 58: SOUND OF SILENCE

1. Bill Hanley, interview.

2. Santelli, *Aquarius Rising*, 250.

3. Les Ledbetter, "Rock Promoters Expect 150,000 at Watkins Glen Fete." *New York Times*, July 19, 1973.

4. Roy Clair, interview.

5. Rusty Brutsché, interview.

6. Sam Boroda, interview.

7. Rick Slattery, interview.

8. Stan Goldstein, interview.

9. Chip Monck, interview.

10. Chris Langhart, interview.

11. Dana Puopolo, phone interview.

12. Dana Puopolo, phone interview.

13. Dana Puopolo, phone interview.

14. Dana Puopolo, phone interview.

15. Dana Puopolo, phone interview.

16. Dana Puopolo, phone interview.

17. Dana Puopolo, phone interview.

18. Dana Puopolo, phone interview.

19. Dana Puopolo, phone interview.

20. Dana Puopolo, phone interview.

21. "When American 'Soul,' Reggae Meet." *Daily Gleaner*, September 8, 1975.

22. Bill Hanley, interview.

23. Judi Bernstein-Cohen, interview.

24. "Merry Go Round." *Daily Gleaner*, September 12, 1975.

25. Bill Hanley, interview.

26. Judi Bernstein-Cohen, interview.

27. Harold Cohen, interview.

28. Judi Bernstein-Cohen, interview.

29. Judi Bernstein-Cohen, interview.

30. Judi Bernstein-Cohen, interview.

31. Bill Hanley, interview.

32. Laura Loughlin, interview.

33. Laura Loughlin, interview.

34. Laura Loughlin, interview.

35. Stephen Freker, "Woodstock + 15, Hanley's Still in Tune." *Medford Citizen*, August 21, 1984.

36. Stan Miller, interview.

37. Ken Lopez, interview.

INDEX

ABOUT THE AUTHOR

John Kane is faculty in the design and media department at the New Hampshire Institute of Art. He is the author of *Pilgrims of Woodstock: Never-Before-Seen Photos*. He is currently working on the documentary *Last Seat in the House*, about Bill Hanley. Learn more about his research at www.thelastseatinthehouse.com.

Printed in the USA
CPSIA information can be obtained
at www.ICGtesting.com
LVHW041739280124
769859LV00015B/236

9 781496 826800